The Quantum Theory of Light

of Light

Third Edition

RODNEY LOUDON

Department of Electronic Systems Engineering
University of Essex

OXFORD

UNIVERSITY PRESS

OXFORD

UNIVERSITY PRESS

Great Clarendon Street, Oxford OX2 6DP

Oxford University Press is a department of the University of Oxford.
It furthers the University's objective of excellence in research, scholarship,
and education by publishing worldwide in

Oxford New York

Auckland Cape Town Dar es Salaam Hong Kong Karachi
Kuala Lumpur Madrid Melbourne Mexico City Nairobi
New Delhi Shanghai Taipei Toronto
With offices in
Argentina Austria Brazil Chile Czech Republic France Greece
Guatemala Hungary Italy Japan South Korea Poland Portugal
Singapore Switzerland Thailand Turkey Ukraine Vietnam

Oxford is a registered trade mark of Oxford University Press
in the UK and in certain other countries

Published in the United States
by Oxford University Press Inc., New York

© Oxford University Press 1973, 1983, 2000

The moral rights of the author have been asserted

Database right Oxford University Press (maker)

Reprinted 2010

All rights reserved. No part of this publication may be reproduced,
stored in a retrieval system, or transmitted, in any form or by any means,
without the prior permission in writing of Oxford University Press,
or as expressly permitted by law, or under terms agreed with the appropriate
reprographics rights organization. Enquiries concerning reproduction
outside the scope of the above should be sent to the Rights Department,
Oxford University Press, at the address above

You must not circulate this book in any other binding or cover
And you must impose this same condition on any acquirer

ISBN 978-0-19-850176-3

Printed in the United Kingdom by
Lightning Source UK Ltd., Milton Keynes

Preface

As in previous editions, the purpose of the book remains the provision of a detailed account of the basic theory needed for an understanding of the quantum properties of light. The material is intended to bridge the gap between standard quantum mechanics, electromagnetic theory and statistical mechanics, as taught at undergraduate level, and the theory needed to explain experiments in quantum optics. The development of the quantum theory of light presented here is thus governed by the needs of experimental interpretation, but only a few representative experiments are discussed in any detail and the reader must look elsewhere for more complete accounts of the observations of quantum optical phenomena (for example, Bachor, H.-A., *A Guide to Experiments in Quantum Optics* (Wiley–VCH, Weinheim, 1998)).

The aim throughout is to give the simplest and most direct account of the basic theory. Some of the changes from the second edition result from attempts to improve obscure derivations and to remove mistakes but most are made in response to experimental developments in the subject over the past 18 years or so. For example, the emergence of parametric down-conversion as a key process in the generation of nonclassical light necessitates not only a treatment of the nonlinear process itself but also of a range of topics that includes photon pair states, two-photon interference, homodyne detection and squeezed light. Again, the central role of the beam splitter in both single and two-photon interference experiments requires a careful account of its quantum-optical properties.

Much of the material is based on lecture courses given by the author to final-year undergraduates, to first-year graduate students and to more advanced post-graduates and research workers. Previous editions of the book have been widely used as texts for the different levels of course. This new edition is also designed to serve as a textbook rather than as a monograph, with treatments of the theory that include some shortcuts and omissions of more advanced derivations (see Barnett, S.M. and Radmore, P.M., *Methods in Theoretical Quantum Optics* (Clarendon Press, Oxford, 1997) for the more detailed mathematics). Well over 100 problems are distributed through the text to encourage students to use the theory themselves. There is no serious attempt to cite all the original sources of the various concepts and theories in quantum optics but, instead, references are given to books and papers that the author finds clear and useful as aids to understanding. Quantum optics is very well served by more advanced monographs, reviews and journal articles (for a comprehensive listing at the time of writing, see Slosser, J.J. and Meystre, P., Resource letter: CQO-1: Coherence in quantum optics, *Am. J. Phys.* **65**, 275–86 (1997)).

The scope of the material is restricted in various other ways but, most importantly, in the frequency range of the electromagnetic radiation for which the calculations are valid. The lower end of the range, at about 10^{13} Hz in the far infrared, is fixed by the requirement that no significant thermal excitation of radiation occurs at room temperature. The upper end, at about 10^{18} Hz in the X-ray region, is determined by the requirement that the photon energies should correspond to electron velocities less than about one-tenth the velocity of light. Frequencies within this range are loosely termed 'visible' or 'optical', although only light whose frequency lies in a narrow band around 5×10^{14} Hz is strictly visible.

There are further restrictions in the assumed nature of the matter in interaction with the light. This is generally taken as a random distribution of atoms or molecules in a gaseous or liquid state, except for the discussion of nonlinear quantum optics where some of the more important processes occur only in noncentrosymmetric materials. There is no discussion of the quantum-optical properties of solitons, which form a somewhat separate topic (see Sizmann, A. and Leuchs, G., The optical Kerr effect and quantum optics, *Prog. Opt.* **39**, 369–465 (1999)). There is also no coverage of quantum effects in optical communications and information processing, particularly the burgeoning fields of quantum cryptography and quantum computing, where the current rapid progress of research makes any account obsolete within a couple of years or so. These, and other offshoots of quantum optics in phases of rapid growth, are better treated in topical reviews than in textbooks devoted to basic concepts.

Within these limitations, it is hoped that the book will encourage students to learn about the field of quantum optics, with its unique combination of observable effects that illuminate the most basic properties of quantum-mechanical systems and applications that promise striking advances in information technology and other areas.

East Bergholt, Suffolk, England   R.L.
March 2000

Acknowledgements

The author owes a great debt for helpful advice to many colleagues, including M. Babiker, A.J. Bain, S.M. Barnett, G. Barton, K.J. Blow, M.J. Collett, J.H. Eberly, M.P. van Exter, H. Fearn, J. Fraile, P. Grangier, M. Harris, N. Imoto, E. Jakeman, O. Jedrkiewicz, J. Jeffers, P.L. Knight, V. Lembessis, G. Leuchs, F. De Martini, D.T. Pegg, F. Persico, S.J.D. Phoenix, E.R. Pike, E.A. Power, T.J. Shepherd, A. Sizmann, A. Squire, H. Walther, J.P. Woerdman, and R.G. Woolley. He is especially grateful to S.M. Barnett, who read through the entire manuscript and made a large number of most useful suggestions for the removal of mistakes and clarification of obscurities. The support and encouragement of Mary Loudon in the production of the book are deeply appreciated.

Permission to reproduce or adapt tables and diagrams was generously granted by the following authors and publishers: Fig. 1.16, M. Harris and the American Institute of Physics; Fig. 1.18, K.A.H. van Leeuwen and the American Institute of Physics; Figs. 3.4 and 3.5, A. Squire Ph.D. thesis; Fig. 5.6, M.G. Raymer and the Optical Society of America; Fig. 5.9, H.J. Kimble and the Optical Society of America; Fig. 5.14, J. Mlynek and Nature, copyright (1997) Macmillan Magazines Ltd., with help from G. Breitenbach; Figs. 5.18 and 5.19, P. Grangier and EDP Sciences; Fig. 6.2, W. Martienssen and EDP Sciences; Fig. 6.3, L. Mandel and the American Institute of Physics; Fig. 8.6, S. Ezekiel and the American Institute of Physics; Fig. 8.9, H. Walther and Elsevier Science, copyright (1997), with help from W. Lange; Fig. 8.11, R. Blatt, unpublished data; Fig. 8.15, J.F. Clauser and the American Institute of Physics; Fig. 8.17, M. Gavrila and the American Institute of Physics; Fig. 9.3, H.D. Simaan, unpublished calculation: Figs. 9.5 and 9.6, the Optical Society of America; Figs. 9.12 and 9.13, Institute of Physics Publishing.

Contents

Introduction: The photon   1

1 Planck's radiation law and the Einstein coefficients   3

   1.1   Density of field modes in a cavity   4
   1.2   Quantization of the field energy   7
   1.3   Planck's law   10
   1.4   Fluctuations in photon number   13
   1.5   Einstein's A and B coefficients   16
   1.6   Characteristics of the three Einstein transitions   19
   1.7   Optical excitation of two-level atoms   23
   1.8   Theory of optical attenuation   27
   1.9   Population inversion: optical amplification   31
   1.10   The laser   35
   1.11   Radiation pressure   40
        References   44

2 Quantum mechanics of the atom–radiation interaction   46

   2.1   Time-dependent quantum mechanics   46
   2.2   Form of the interaction Hamiltonian   49
   2.3   Expressions for the Einstein coefficients   52
   2.4   The Dirac delta-function and Fermi's golden rule   57
   2.5   Radiative broadening and linear susceptibility   60
   2.6   Doppler broadening and composite lineshape   65
   2.7   The optical Bloch equations   68
   2.8   Power broadening   72
   2.9   Collision broadening   76
   2.10   Bloch equations and rate equations   79
        References   81

3 Classical theory of optical fluctuations and coherence   82

   3.1   Models of chaotic light sources   83
   3.2   The lossless optical beam-splitter   88
   3.3   The Mach–Zehnder interferometer   91
   3.4   Degree of first-order coherence   94
   3.5   Interference fringes and frequency spectra   100
   3.6   Intensity fluctuations of chaotic light   103

3.7   Degree of second-order coherence   107
3.8   The Brown–Twiss interferometer   114
3.9   Semiclassical theory of optical detection   117
      References   123

4 Quantization of the radiation field   125

4.1   Potential theory for the classical electromagnetic field   126
4.2   The free classical field   130
4.3   The quantum-mechanical harmonic oscillator   133
4.4   Quantization of the electromagnetic field   139
4.5   Canonical commutation relation   144
4.6   Pure states and statistical mixtures   148
4.7   Time development of quantum-optical systems   153
4.8   Interaction of the quantized field with atoms   155
4.9   Second quantization of the atomic Hamiltonian   162
4.10 Photon absorption and emission rates   168
4.11 The photon intensity operator   173
4.12 Quantum degrees of first and second-order coherence   176
      References   178

5 Single-mode quantum optics   180

5.1   Single-mode field operators   181
5.2   Number states   184
5.3   Coherent states   190
5.4   Chaotic light   199
5.5   The squeezed vacuum   201
5.6   Squeezed coherent states   206
5.7   Beam-splitter input–output relations   212
5.8   Single-photon input   216
5.9   Arbitrary single-arm input   221
5.10 Nonclassical light   227
      References   231

6 Multimode and continuous-mode quantum optics   233

6.1   Multimode states   234
6.2   Continuous-mode field operators   237
6.3   Number states   242
6.4   Coherent states   245
6.5   Chaotic light: photon bunching and antibunching   248
6.6   The Mach–Zehnder interferometer   251
6.7   Photon pair states   253
6.8   Two-photon interference   260
6.9   Squeezed light   265

6.10 Quantum theory of direct detection   271
6.11 Homodyne detection   278
6.12 The electromagnetic vacuum   284
      References   286

7 Optical generation, attenuation and amplification   288

7.1   Single-mode photon rate equations   289
7.2   Solutions for fixed atomic populations   292
7.3   Single-mode laser theory   297
7.4   Fluctuations in laser light   304
7.5   Travelling-wave attenuation   310
7.6   Travelling-wave amplification   319
7.7   Dynamics of the atom–radiation system   324
7.8   The source-field expression   328
7.9   Emission by a driven atom   331
      References   337

8 Resonance fluorescence and light scattering   339

8.1   The scattering cross-section   340
8.2   Resonance fluorescence   344
8.3   Weak incident beam   348
8.4   Single-atom resonance fluorescence   352
8.5   Quantum jumps   360
8.6   Two-photon cascade emission   365
8.7   The Kramers–Heisenberg formula   371
8.8   Elastic Rayleigh scattering   374
8.9   Inelastic Raman scattering   378
      References   381

9 Nonlinear quantum optics   383

9.1   The nonlinear susceptibility   383
9.2   Electromagnetic field quantization in media   389
9.3   Second-harmonic generation   393
9.4   Parametric down-conversion   398
9.5   Parametric amplification   404
9.6   Self-phase modulation   411
9.7   Single-beam two-photon absorption   417
9.8   Conclusion   425
      References   426

Index   429

Introduction: The photon

The use of the word 'photon' to describe the quantum of electromagnetic radiation can lead to confusion and misunderstanding. It is often used in the context of interference experiments, for example Young's slits, in such phrases as 'which slit does the photon pass through?' and 'where do the photons hit the screen when one of the slits is covered up'. The impression is given of a fuzzy globule of light that travels this way or that way through pieces of optical equipment or that light beams consists of streams of the globules, like bullets from a machine gun. Lamb has even argued that there is no such thing as a photon [1] and he has proposed that the word should be used only under licence by properly qualified people! It is, however, difficult to disagree with some of his concerns.

Nevertheless, the word is extremely convenient and its avoidance often leads to lengthy circumlocutions. The adoption of the photon by the quantum-optics community is widespread and the present book follows current usage, with sometimes imprecise statements that could amount to misuse of the word. The intention of this Introduction is to limit the damage that might otherwise occur by briefly explaining the concept of the photon as used in the text. It should be mentioned that the word itself was invented by Lewis [2], with a meaning quite different from that adopted in subsequent work. The relevant history is well reviewed by Lamb [1].

The idea of the photon is most easily expressed for an electromagnetic field confined inside a closed optical resonator, or perfectly-reflecting cavity. The field excitations are then limited to an infinite discrete set of spatial modes determined by the boundary conditions at the cavity walls. The allowed standing-wave spatial variations of the electromagnetic field in the cavity are identical in the classical and quantum theories but the time dependences of each mode are governed by classical and quantum harmonic-oscillator equations, respectively. Unlike its classical counterpart, a quantum harmonic oscillator of angular frequency ω can only be excited by integer multiples of $\hbar\omega$, the integers n being eigenvalues of the oscillator number operator. A single spatial mode whose associated harmonic oscillator is in its nth excited state unambiguously contains n photons. Each photon has a more-or-less uniform spatial distribution within the cavity, proportional to the square modulus of the complex field amplitude of the mode function. The single-mode photons are thus delocalized.

These ideas are often extended to open optical systems, with apparatus of finite size but no identifiable cavity. The discrete standing-wave modes of the closed cavity are replaced by discrete travelling-wave modes that propagate from sources to detectors. The excitation of one photon in a single travelling mode is frequently considered in the discussion of interference experiments, for example Young's slits or the Mach–Zehnder interferometer. Each spatial mode in these systems includes input light waves, *both* paths through the interferometer, and output waves appropriate to the geometry of the apparatus. A one-photon excitation in such a mode is distributed over the entire interferometer, including both internal paths. Despite the absence of any localization of the photon, the theory provides a relation between the input and output spatial distributions, equivalent to a determination of the interference fringes. The fringes are the same in the classical and quantum theories, essentially because the spatial modes are identical. The single-mode picture is not strictly valid in the conditions of practical one-photon interference experiments but its use has acquired respectability from some distinguished contributions to the discussion, for example Dirac [3] and Frisch [4].

The frequency separations beween discrete spatial modes are much smaller than any other characteristic frequencies for light beams in most open systems. The discrete modes then effectively condense to a continuum and the systems are more rigorously and conveniently treated by a continuous-mode theory. The typical quantum-optical experiment produces one- or two-photon excitations described by a spatial wavepacket, with some degree of localization. The wave-packet function is expressed as an integral over contributions from waves with a range of frequencies, or wavevectors. It is no longer so straightforward to explain what is meant by a 'photon'; the level of excitation of the system continues to be represented by a number operator with integer eigenvalues, but the mean energy of the one-photon wavepacket is given by $\hbar$ times an average over its frequency components. The one-photon state has, however, the important and distinctive property that it can produce only a single current pulse in the ionization of a photodetector. The concept thus survives as an operational definition in terms of photon detection and it provides a useful qualititative description of the nature of the state.

As a general comment on the notion of interference in quantum mechanics, the effect occurs in general for experiments in which the probability of a given observation is given by the square modulus of a sum of two or more probability amplitudes [5]. Each such probability amplitude represents a contribution to a given output state from the same input state. In the customary photon description of quantum-optical interference experiments, it is never the photons themselves that interfere, one with another, but rather the probability amplitudes that describe their propagation from the input to the output. The two paths of the standard interference experiments provide a simple illustration, but more sophisticated examples occur in higher-order measurements covered in the main text.

References

[1]   Lamb, W.E., Jr., Anti-photon, *Appl. Phys. B* **60**, 77–84 (1995).
[2]   Lewis, G.N., The conservation of photons, *Nature* **118**, 874–5 (1926).
[3]   Dirac, P.A.M., *The Principles of Quantum Mechanics* (Clarendon Press, Oxford, 1930).
[4]   Frisch, O.R., Take a photon, *Contemp. Phys.* **7**, 45–53 (1965).
[5]   Feynman, R.P., Leighton, R.B. and Sands, M., *The Feynman Lectures on Physics* (Addison–Wesley, Reading, MA, 1965).

1 Planck's radiation law and the Einstein coefficients

The quantum theory of light began in 1900 when Planck [1] found that he could account for measurements of the spectral distribution of the electromagnetic energy radiated by a thermal source by postulating that the energy of a harmonic oscillator is quantized. That is, a harmonic oscillator of angular frequency ω can have only energies that are integer multiples of the fundamental quantum $\hbar\omega$, where $\hbar = h/2\pi$ and h is Planck's original constant. In 1905, Einstein [2] showed how the photoelectric effect could be explained by the hypothesis of a corpuscularity of electromagnetic radiation. The quantum of radiation was named a *photon* much later [3], in 1926. The work of Planck and Einstein stimulated much of the early development of quantum mechanics.

Another main stream in the initial formulation of the quantum theory was concerned with the interpretation of atomic spectral lines. The interaction of electromagnetic radiation with atoms was discussed by Einstein [4] in 1917. His theory of the absorption and emission of light by an atom depends upon simple phenomenological considerations but it leads to predictions that are reproduced by the more formal quantum mechanics developed later. The use of quantum theory is, of course, essential for even a gross description of the nature of the states of an atom and the theory is outstandingly successful in its ability to predict the finest details of atomic energy-level structures.

The use of the quantum theory is not, however, essential for the description of many of the properties of visible light. The formal quantization of the electromagnetic field was performed by Dirac [5], who showed in 1927 that the wavelike properties of the field could be preserved in conjunction with the concepts of creation and destruction of photons, but classical theories continued to provide adequate interpretations of the observed properties of light beams. Thus, in 1909, soon after the early contributions of Planck and Einstein, Taylor [6] failed to find any changes from the classical fringes of a Young interferometer when the light source was so feeble that only one photon at a time was present in the apparatus. Higher-order experiments on the interference between optical intensities rather than amplitudes, performed in 1956 by Hanbury Brown and Twiss [7], could also be explained in terms of classical electromagnetic theory. The photoelectric effect itself was shown [8] to be well described by the so-called *semiclassical theory*, in which the atomic part of the experimental system is treated by quantum theory but classical theory is used for the radiation.

The quantum theory of light, or the full quantum theory, in which quantum mechanics is applied both to the radiation and to the atoms with which it interacts, came into its own after the invention of the laser by Maiman [9] in 1960. The operation of the laser itself is largely explained by the semiclassical theory but the high strength and coherence of its emitted beam has enabled a series of experiments that can only be interpreted with use of the full quantum theory. The field of *quantum optics* was essentially born with the development of quantum theories of optical coherence and of the states of the radiation field by Glauber [10] in 1963. There were attempts to retain the semiclassical theory by ingenious adaptations to match some of the new experiments, but the burgeoning of measurements that can only be explained by the full quantum theory has rendered these attempts futile.

The quantum theory of light is an essential tool in the interpretation of measurements and in the understanding of the basic principles of quantum mechanics. It provides the most reliable available descriptions of the whole range of optical experiments, despite the frequent applicability of semiclassical theory. The many quantum-optical experiments of recent years include observations of manifestations of the most fundamental quantum phenomena, for example single-photon interference, basic measurement theory (§5.10), Bose–Einstein statistics (§6.8) and the fluctuation properties of the vacuum radiation field (§§5.5 and 6.12). The applications of the quantum theory of light to these and other important experiments often involve elegantly simple calculations.

The quantization of the radiation field is carried out in Chapter 4. The first three chapters are devoted to topics that form useful components of the quantization procedure or provide classical or semiclassical predictions for comparison with the more general and reliable outcomes of the full quantum theory. The main concerns of the present chapter are the thermal excitation of electromagnetic radiation and the basic kinds of interaction that can occur between light and atomic transitions, beginning with the theories of Planck and Einstein. These phenomena are closely related, as the absorption and emission of light by atoms provide the mechanisms by which the amount of excitation of the electromagnetic radiation can adjust to the conditions of thermal equilibrium. However, the Einstein theory of absorption and emission also applies to systems not in thermal equilibrium, for example a gas illuminated by an external source or an optical amplifier driven by a pump that puts atoms into excited states.

1.1 Density of field modes in a cavity

It is convenient to begin with the theory of electromagnetic radiation confined inside an optical cavity. This system has the advantage that the electromagnetic field can be excited only in discrete spatial modes of the cavity and the resulting calculations are simplified. Some optical experiments do indeed employ a confined region of space, often in the form of a Fabry–Perot cavity, with the laser as prime example, and for these it is appropriate to use the discrete-mode formalism. Most experiments, however, have no identifiable cavity, but rather the optical energy flows from sources through some kind of interaction region to a set of detectors. Even experiments that do use a cavity generally have light passing in and out, and the system as a whole is not contained within the cavity. Nevertheless, the results of discrete-mode theory are usually independent of the size, shape and nature of the assumed cavity, and they can be applied to more general unconfined optical systems. This model is adopted for the earlier part of the book, and the often-hypothetical cavity is not removed until Chapter 6.

Planck's law expresses the spectral distribution of electromagnetic radiation in the interior of a cavity in thermal equilibrium at temperature T. The radiation is called *black-body radiation*; its frequency distribution is the same as that radiated by a perfectly-black body at temperature T. The calculation of the frequency distribution breaks up into two distinct parts. We first consider the spatial dependence of the field in the cavity and derive an expression for the number of its different modes of excitation. In the second part, we consider the time dependence of the field and calculate the energy carried by each mode at temperature T. To make the calculations as simple as possible, we choose a cubic cavity of side L, with axes defined as in Fig. 1.1. The walls of the cavity are assumed perfectly conducting, and the tangential components of the electric field $\mathbf{E}(\mathbf{r},t)$ must accordingly vanish at the boundaries.

The first part of the calculation is entirely classical and we show in Chapter 4 that the spatial dependence of the electromagnetic field is not affected by the quantization. The electric field in empty space must satisfy the wave equation

$$\nabla^2\mathbf{E}(\mathbf{r},t) = \frac{1}{c^2}\frac{\partial^2\mathbf{E}(\mathbf{r},t)}{\partial t^2}, \tag{1.1.1}$$

where c is the velocity of light, together with the Maxwell equation

$$\nabla.\mathbf{E}(\mathbf{r},t) = 0. \tag{1.1.2}$$

The solution that satisfies the boundary conditions has components

$$\begin{aligned}
E_x(\mathbf{r},t) &= E_x(t)\cos(k_x x)\sin(k_y y)\sin(k_z z) \\
E_y(\mathbf{r},t) &= E_y(t)\sin(k_x x)\cos(k_y y)\sin(k_z z) \\
E_z(\mathbf{r},t) &= E_z(t)\sin(k_x x)\sin(k_y y)\cos(k_z z),
\end{aligned} \tag{1.1.3}$$

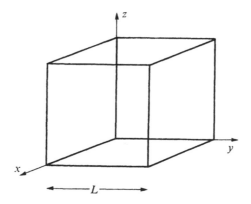

Fig. 1.1. Geometry of the optical cavity.

where $\mathbf{E}(t)$ is independent of position and the wavevector $\mathbf{k}$ has components

$$k_x = \pi v_x/L, \quad k_y = \pi v_y/L, \quad k_z = \pi v_z/L \qquad (1.1.4)$$

with

$$v_x, v_y, v_z = 0, 1, 2, 3, \ldots \qquad (1.1.5)$$

The integers v are further restricted, in that only one of them can be zero, as the field $\mathbf{E}(\mathbf{r},t)$ in the cavity vanishes if two or three are zero. The allowed wavevectors can be plotted as a lattice of points in three dimensions, with lattice constant π/L. Figure 1.2 shows the lattice for values of the integers v up to 4.

It is easily verified that $\mathbf{E}(\mathbf{r},t)$ given by eqn (1.1.3) satisfies the boundary conditions, for example, the x component vanishes at $y = 0$ or L and at $z = 0$ or L. The boundary conditions could be equally-well satisfied with the cosines replaced by sines but the Maxwell equation (1.1.2) could not then be satisfied within the cavity. For the solution (1.1.3), this equation leads to the constraint

$$\mathbf{k}.\mathbf{E}(t) = 0. \qquad (1.1.6)$$

This is just the condition for $\mathbf{E}(t)$ to be at right-angles to $\mathbf{k}$ and there are two independent directions of the field for each allowed wavevector. These two transverse polarization directions are indicated by a label $\lambda = 1, 2$.

The boundary conditions force the wavevector components to take the discrete values specified by the integers in eqn (1.1.5). Each set of the three integers v

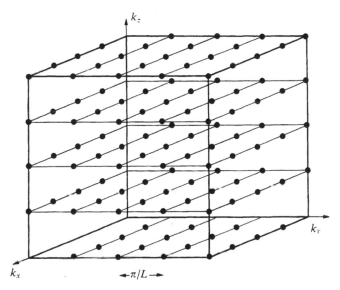

Fig. 1.2. The allowed wavevectors $\mathbf{k}$ for a cubic cavity of edge L. Note the absence of points with two or three zero components.

and the polarization label λ defines a spatial mode of the radiation field. Any excitation of the electromagnetic field in the cavity can be expressed as a linear sum of the contributions of these field modes. We need an expression for the number of field modes that have the magnitude of their wavevector between the values k and $k + \mathrm{d}k$. This is just the number of lattice points in the octant of a spherical shell bounded by the radii k and $k + \mathrm{d}k$. For k very much larger that the lattice constant π/L, the required number is

$$\tfrac{1}{8}\left(4\pi k^2 \mathrm{d}k\right)(\pi/L)^{-3} \times 2, \qquad (1.1.7)$$

where the final factor takes account of the two polarizations.

The density $\rho(k)\mathrm{d}k$ of field modes, defined as the number of modes per unit volume of cavity having their wavevector in the specified range, is obtained from eqn (1.1.7) as

$$\rho(k)\mathrm{d}k = k^2 \, \mathrm{d}k / \pi^2 . \qquad (1.1.8)$$

This result holds generally and it is independent of the nature of the cavity used in its derivation. The angular frequency ω of a mode is related to its wavevector in the usual way by

$$\omega = ck . \qquad (1.1.9)$$

Thus eqn (1.1.8) is converted into an expression for the density $\rho(\omega)\mathrm{d}\omega$ of modes having their frequency between ω and $\omega + \mathrm{d}\omega$,

$$\rho(\omega)\mathrm{d}\omega = \omega^2 \, \mathrm{d}\omega / \pi^2 c^3 . \qquad (1.1.10)$$

Later chapters contain expressions that must be summed over the field modes. With the mode densities of eqns (1.1.8) and (1.1.10), such summations can be converted to integrations over k or ω,

$$\sum_{\mathbf{k}} \sum_{\lambda=1,2} \rightarrow \int \mathrm{d}k \left(Vk^2/\pi^2\right) \rightarrow \int \mathrm{d}\omega \left(V\omega^2/\pi^2 c^3\right), \qquad (1.1.11)$$

where V is the cavity volume. These simple conversions apply when the function to be summed or integrated is independent of the directions of both the polarization and the wavevector. They also assume that the function varies on scales that are much larger that the separations between the wavevectors and frequencies of the discrete modes of the cavity.

1.2 Quantization of the field energy

So much for the spatial dependence of the electromagnetic field. The second stage of the calculation determines the amount of energy stored in each field mode at temperature T. An equation for the time dependence of the electric field of the mode is obtained by substitution of the solution (1.1.3) into the wave equation (1.1.1), with the use of eqn (1.1.9).

$$\partial^2 \mathbf{E}(t)/\partial t^2 = -\omega^2 \mathbf{E}(t). \tag{1.2.1}$$

This is a simple-harmonic-motion equation, whose solution is taken as

$$\mathbf{E}(t) = \mathbf{E}(0)e^{-i\omega t}, \tag{1.2.2}$$

where $\mathbf{E}(0)$ is a constant vector which, in classical electromagnetic theory, can have any arbitrary magnitude and phase.

According to classical theory, the energy contained in the radiation field in the cavity is given by the integral

$$\mathcal{E}_R = \tfrac{1}{2} \int_{\text{cavity}} dV \left(\varepsilon_0 \mathbf{E}(\mathbf{r},t)^2 + \mu_0^{-1} \mathbf{B}(\mathbf{r},t)^2 \right), \tag{1.2.3}$$

where the electric permittivity and magnetic permeability of free space are related to the velocity of light by

$$c = (\varepsilon_0 \mu_0)^{-1/2}. \tag{1.2.4}$$

Here $\mathbf{E}(\mathbf{r},t)$ and $\mathbf{B}(\mathbf{r},t)$ are the real electric and magnetic field vectors, related by the Maxwell equation

$$\nabla \times \mathbf{E}(\mathbf{r},t) = -\partial \mathbf{B}(\mathbf{r},t)/\partial t. \tag{1.2.5}$$

For a specific field mode, the electric field is given by eqn (1.1.3) and the real part of eqn (1.2.2). The corresponding magnetic field components are derived with the use of eqn (1.2.5). The classical theory of the field energy is developed in greater detail in §4.2.

Planck's quantization hypothesis is introduced at this point. In classical theory, the field energy given by eqn (1.2.3) can have any positive value, as $\mathbf{E}(0)$ in eqn (1.2.2) can have any magnitude. However, the field $\mathbf{E}(t)$ satisfies the harmonic oscillator equation (1.2.1) and, if this oscillator is treated quantum-mechanically rather than classically (the details are given in §4.3), then its energy takes only the discrete values

$$E_n = \left(n + \tfrac{1}{2}\right)\hbar\omega, \qquad n = 0, 1, 2, 3, \ldots. \tag{1.2.6}$$

We accordingly assign these allowed values to the electromagnetic radiative energy of the mode as represented by eqn (1.2.3),

$$\mathcal{E}_R = \left(n + \tfrac{1}{2}\right)\hbar\omega. \tag{1.2.7}$$

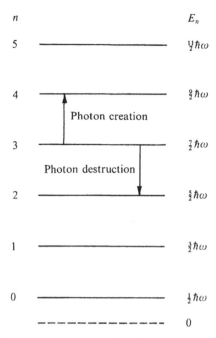

Fig. 1.3. The six lowest energy levels of a quantum harmonic oscillator.

This quantization condition clearly puts restrictions on the possible magnitudes of the amplitude $E(0)$ in eqn (1.2.2). However, it is not necessary to consider these consequences for the present; we treat the fields as classical quantities and impose the quantization only on the field energy. At a deeper level, one must represent the electromagnetic fields by quantum-mechanical operators, as in §4.4, but for many problems the approach outlined above is adequate.

The essence of the quantum theory of the radiation field is thus the association of a quantum harmonic oscillator with each mode of the field. The energy levels of such an oscillator are shown in Fig. 1.3. When the mode energy is given by eqn (1.2.7), the corresponding oscillator is in its nth excited state. For $n = 0$ the oscillator is in its ground state, but a finite amount of energy $\hbar\omega/2$ is still present in the field. This is the *zero-point energy* of the oscillator and its significance is discussed in §6.12. For most experiments, however, the factor that governs the observations is the degree of excitation above the ground-state. The nth excited state has n quanta of energy $\hbar\omega$ in addition to the zero-point energy. The quanta are called *photons*, and one speaks of n photons being excited in the mode of the radiation field. A photon is said to have been created (destroyed) when the electromagnetic energy in the mode is increased (decreased) by a single quantum. Figure 1.4 shows the lowest levels of the harmonic oscillators associated with the first ten field modes in order of increasing frequency. The figure does not show the two independent transversely-polarized modes for each of the oscillators.

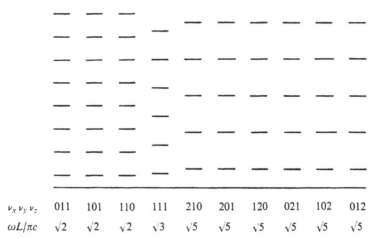

$\nu_x\,\nu_y\,\nu_z$	011	101	110	111	210	201	120	021	102	012
$\omega L/\pi c$	$\sqrt{2}$	$\sqrt{2}$	$\sqrt{2}$	$\sqrt{3}$	$\sqrt{5}$	$\sqrt{5}$	$\sqrt{5}$	$\sqrt{5}$	$\sqrt{5}$	$\sqrt{5}$

Fig. 1.4. Harmonic oscillator levels for the modes that correspond to the ten points closest to the origin in Fig. 1.2.

1.3 Planck's law

In thermal equilibrium at temperature T, the probability $P(n)$ that the mode oscillator is thermally excited to its nth excited state is given by the usual Boltzmann factor

$$P(n) = \frac{\exp(-E_n/k_BT)}{\sum\limits_n \exp(-E_n/k_BT)}. \tag{1.3.1}$$

The zero-point energy cancels when the quantized energy expression (1.2.6) is substituted and, with the shorthand notation

$$U = \exp(-\hbar\omega/k_BT), \tag{1.3.2}$$

the thermal probability becomes

$$P(n) = U^n \Big/ \sum_{n=0}^{\infty} U^n. \tag{1.3.3}$$

The denominator is an easily-summed geometric series,

$$\sum_{n=0}^{\infty} U^n = \frac{1}{1-U}, \tag{1.3.4}$$

and hence

$$P(n) = (1 - U)U^n. \tag{1.3.5}$$

The mean number $\langle n \rangle$ of photons excited in the field mode at temperature T is therefore

$$\langle n \rangle = \sum_n nP(n) = (1 - U)\sum_n nU^n = (1 - U)U\frac{\partial}{\partial U}\sum_n U^n = \frac{U}{1-U}, \tag{1.3.6}$$

and with the use of eqn (1.3.2)

$$\langle n \rangle = \frac{1}{\exp(\hbar\omega/k_B T) - 1}. \tag{1.3.7}$$

This important result is the Planck thermal excitation function. It is plotted in Fig. 1.5 as a function of the oscillator or photon frequency ω.

The results (1.1.10) and (1.3.7) give, respectively, the number of radiation field modes per unit volume whose frequencies lie in the range ω to $\omega + d\omega$, and their mean energy $\langle n \rangle \hbar\omega$ at temperature T in excess of the zero-point energy. These expressions combine to give Planck's law for the mean energy density of

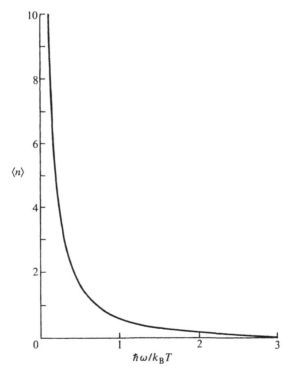

Fig. 1.5. Mean number $\langle n \rangle$ of photons of frequency ω that are thermally excited at temperature T.

the radiation in the modes of frequency ω at temperature T in the form

$$\langle W_T(\omega)\rangle d\omega = \langle n\rangle \hbar\omega\rho(\omega)d\omega = \frac{\hbar\omega^3}{\pi^2 c^3} \frac{d\omega}{\exp(\hbar\omega/k_B T)-1}. \qquad (1.3.8)$$

The radiative energy density $\langle W_T(\omega)\rangle$ is the energy per unit volume per unit angular frequency range in the black-body spectrum and its units are $Jm^{-3}s$. It is plotted as a function of $\hbar\omega/k_B T$ in Fig. 1.6.

Problem 1.1   Prove that the maximum value $\langle W_T(\omega)\rangle_{max}$ of the energy density and the frequency ω_{max} at which it occurs are related by

$$\langle W_T(\omega)\rangle_{max} = \left(\omega_{max}^2/\pi^2 c^3\right)\left(3k_B T - \hbar\omega_{max}\right). \qquad (1.3.9)$$

Show by numerical trial and error or by drawing a rough graph that the value of the frequency is given approximately by

$$\omega_{max} = 2.8 k_B T/\hbar. \qquad (1.3.10)$$

This is known as Wien's displacement law.

Planck's law takes on slightly simpler forms at high and low temperatures. Thus the exponential can be expanded at high temperatures to give

$$\langle W_T(\omega)\rangle \approx \omega^2 k_B T/\pi^2 c^3 \quad (k_B T \gg \hbar\omega). \qquad (1.3.11)$$

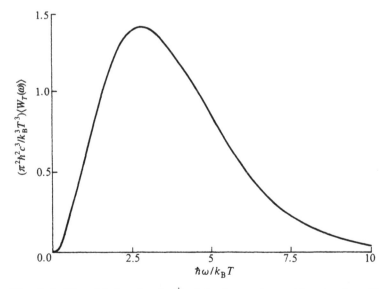

Fig. 1.6. Planck's law for the black-body spectrum at temperature T.

This form of radiation law was derived by Rayleigh [11] in 1900, shortly before Planck's formulation of the more general law. Rayleigh's law is the classical limit obtained when Planck's constant $\hbar$ tends to zero. The exponential in eqn (1.3.8) becomes very large at low temperatures, where the radiative energy density is given by Wien's formula

$$\langle W_T(\omega)\rangle \approx \left(\hbar\omega^3/\pi^2 c^3\right)\exp(-\hbar\omega/k_B T) \quad (k_B T \ll \hbar\omega). \tag{1.3.12}$$

Both approximate forms of Planck's law break down when $\hbar\omega$ is comparable to $k_B T$.

The total energy density of the photons in the cavity is obtained by integration of eqn (1.3.8) as

$$\int_0^\infty d\omega \langle W_T(\omega)\rangle = \frac{k_B^4 T^4}{\pi^2 c^3 \hbar^3}\int_0^\infty \frac{x^3 dx}{e^x - 1} = \frac{\pi^2 k_B^4 T^4}{15 c^3 \hbar^3}, \tag{1.3.13}$$

where use is made of a standard integral [12]. The proportionality of the total energy to the fourth power of the temperature is the Stefan–Boltzmann radiation law, formulated in 1879.

An impressive demonstration of Planck's radiation law is obtained from satellite measurements [13] of the spectrum of thermal radiation present in space, the so-called cosmic background radiation. The measurements fit the theoretical blackbody spectrum for a temperature $T = 2.728 \pm 0.004$ K with uncertainties that are much smaller than the line thickness in Fig. 1.6..

Problem 1.2   Derive an expression similar to eqn (1.3.13) but for the number of photons per unit volume excited in a cavity at temperature T. Show with the help of another standard integral [12] that the cosmic background radiation contains about 5×10^5 photons per litre.

1.4 Fluctuations in photon number

The occurrence of photon absorption and emission processes causes the numbers of photons in each mode of the radiation field in the cavity to fluctuate. The fluctuations take place on a characteristic time scale that is discussed in detail in §3.1. However, some average properties of the fluctuations can be deduced without any knowledge of the time scales involved. We make use of the *ergodic* property of the fluctuations in thermal light [14], considered further in §3.3, which ensures that time averages are equivalent to averages taken over a large number of exactly similar systems, each maintained in a fixed state. The fictitious collection of identical systems is called an *ensemble*; the members of the ensemble are distributed among their various possible states in accordance with the appropriate probability distribution for the system considered.

For the photons in a particular field mode of a cavity, the ensemble consists of

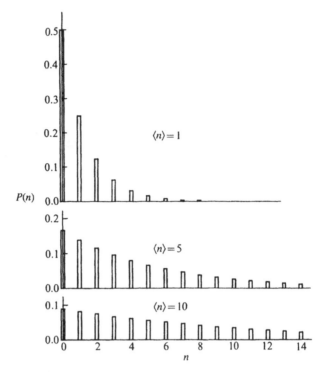

Fig. 1.7. Planck photon-number distributions for three values of the mean number $\langle n \rangle$.

of the same field mode in a large number of identical cavities. Each cavity in the ensemble has some fixed number of photons, where the fraction of cavities that contain n photons is determined by the function $P(n)$ given in eqn (1.3.5). We have already evaluated the mean photon number $\langle n \rangle$ using eqn (1.3.6); the result obtained in eqn (1.3.7) can equally well be regarded as an ensemble average or an average over time of the number of photons in the given mode of the single real cavity. The period over which the time average is taken must clearly be much longer than the characteristic time scale of the fluctuations.

The expression in eqn (1.3.6) for the mean photon number is rearranged as

$$U = \langle n \rangle / (1 + \langle n \rangle),$$ (1.4.1)

and substitution in eqn (1.3.5) gives

$$P(n) = \frac{\langle n \rangle^n}{(1 + \langle n \rangle)^{1+n}}.$$ (1.4.2)

This form of the Planck photon probability distribution as a function of the mean photon number $\langle n \rangle$ is particularly convenient for graphical illustrations and some examples are shown in Fig. 1.7. The probability $P(n)$ is the distribution of readings obtained from a large number of successive measurements of the number of photons in the selected field mode. It is seen that $n = 0$ always has the largest probability of occurrence, and $P(n)$ falls off monotonically with increasing n, a consequence of the decreasing Boltzmann probability of occupation for the higher levels of the mode harmonic oscillator. Note that the probability distributions show no particular feature at $n = \langle n \rangle$.

The Planck probability function $P(n)$ is sometimes called the *thermal* distribution or the *geometric* distribution. It is often useful to characterize a distribution by its factorial moments. The rth factorial moment is defined as

$$\langle n(n-1)(n-2)...(n-r+1) \rangle = \sum_n n(n-1)(n-2)...(n-r+1)P(n), \qquad (1.4.3)$$

where r is any positive integer. The first factorial moment is the same as the mean value of the number of excited photons evaluated in eqn (1.3.6). The same method of differentiation is used to obtain the higher factorial moments.

Problem 1.3   Prove that the rth factorial moment of the Planck probability distribution, eqn (1.3.5) or (1.4.2), is

$$\langle n(n-1)(n-2)...(n-r+1) \rangle = r!\langle n \rangle^r. \qquad (1.4.4)$$

The size of the fluctuation in photon number is characterized by the *variance* of the distribution, defined as

$$(\Delta n)^2 = \sum_n (n - \langle n \rangle)^2 P(n) = \langle n^2 \rangle - \langle n \rangle^2. \qquad (1.4.5)$$

The second factorial moment obtained from eqn (1.4.4) gives

$$\langle n(n-1) \rangle = 2\langle n \rangle^2, \qquad (1.4.6)$$

and hence

$$(\Delta n)^2 = \langle n \rangle^2 + \langle n \rangle. \qquad (1.4.7)$$

Equivalently, the root-mean-square deviation of the distribution is

$$\Delta n = \left(\langle n \rangle^2 + \langle n \rangle \right)^{1/2}. \qquad (1.4.8)$$

The magnitude of the fluctuation in n, given by Δn, is approximately equal to $\langle n \rangle$ for $\langle n \rangle \gg 1$. A similar equality of fluctuation to mean value is found in the classical theory of §3.6. However, the fluctuation given by eqn (1.4.8) always exceeds the mean value to some extent and the additional contribution is associated with the particle-like nature of the photons. The wide spread of the distributions is evident in Fig. 1.7.

The repeated measurements of n envisaged in these derivations are realized in practice by *photocount experiments*, which are described in more detail in §§3.9 and 6.10. As is discussed there, the measured photocount distribution reproduces the statistical results derived above only for idealized experiments with perfectly efficient detectors, where each individual measurement of the number of photons is made effectively instantaneously, that is, in a time much shorter than the characteristic time scale of the fluctuations. Any real measurement has a finite resolving time and the results obtained depend upon its size relative to the fluctuation time scale.

1.5 Einstein's A and B coefficients

The number of photons in a thermal cavity changes with time because of their absorption and emission by the atoms or molecules in the cavity walls. We now consider the basic interaction processes between electromagnetic radiation and atoms. They can be treated by means of a simple phenomenological theory due to Einstein [4]. The theory allows a qualitative understanding of a wide variety of radiative processes, for example, the absorption and scattering of light by atoms and the generation and amplification of light beams by lasers. The Einstein theory is based on some physically reasonable postulates concerning the absorption and emission of photons by atoms. The postulates can all be justified by quantum-mechanical treatments of the interaction processes, as presented in §4.10. However, no formal quantum mechanics is used in the Einstein theory, although the atomic energy levels are assumed discrete and it is convenient, but not essential, to regard the electromagnetic field as quantized in photons.

It is simpler to consider the interaction of radiation, not with the atoms in the cavity walls, but with some atoms or molecules contained in the interior of the cavity. Suppose that a gas of N identical atoms is placed in the cavity, each atom having a pair of bound-state energy levels E_1 and E_2, and let

$$\hbar\omega = E_2 - E_1. \tag{1.5.1}$$

Energy-conserving processes can occur in which photons of frequency ω are emitted or absorbed by atoms that make transitions between the two states. The two atomic levels are allowed to be multiplets with degeneracies g_1 and g_2, and the mean numbers of atoms in the two multiplet states are denoted by the level populations N_1 and N_2. It is assumed that no atoms are in any other states, so that

$$N_1 + N_2 = N. \tag{1.5.2}$$

The levels are illustrated in Fig. 1.8.

A closed cavity that contains atoms and thermally-excited radiation is not a very interesting system from an experimental point of view. A more sophisticated arrangement is needed if the interaction between atoms and radiation is to be measured. For example, we could transmit a beam of radiation across the cavity

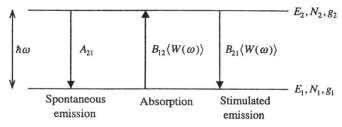

Fig. 1.8. The three Einstein radiative transitions.

and measure the fraction of its initial intensity that survives. Thus, in a typical experiment, the mean energy density of radiation at frequency ω consists of the thermal contribution given by eqn (1.3.8) plus a contribution from external sources of electromagnetic radiation, denoted by a subscript E, and the total energy density is

$$\langle W(\omega)\rangle = \langle W_T(\omega)\rangle + \langle W_E(\omega)\rangle. \tag{1.5.3}$$

This quantity is again the mean radiative energy per unit volume per unit angular frequency range, with units $Jm^{-3}s$. The energy density of the externally produced radiation generally varies with position in the cavity and it does not have the spatially homogeneous and isotropic distribution of the thermal radiation. For example, it may take the form of a parallel beam of light whose strength falls off with distance of propagation, but these spatial characteristics are ignored for the present. It is however an essential requirement for the validity of the Einstein theory that the total energy density is a slowly-varying function of ω in the vicinity of the atomic transition frequency.

The probabilities of photon absorption and emission are defined as follows. Consider a single atom in state 2. There is a finite probability that the atom spontaneously falls into the lower state 1 and emits a photon of energy $\hbar\omega$. The probability per unit time for occurrence of this process is denoted

A_{21} (spontaneous emission rate). $\hspace{3cm}$ (1.5.4)

Now consider an atom in state 1. In the absence of any radiation of frequency ω there is no way in which the atom can pass into state 2, as it is impossible to conserve energy in such a transition. However, in the presence of radiation, the upwards transition $1 \to 2$ can proceed by absorption of a photon $\hbar\omega$. The transition is assumed to occur at a rate proportional to the radiative energy density, and the probability per unit time for this process is denoted

$B_{12}\langle W(\omega)\rangle$ (absorption rate). $\hspace{2cm}$ (1.5.5)

These two processes are intuitively reasonable, but it is not so obvious that the presence of the radiation should also enhance the rate of transition from the upper state to the lower state. However, we shall show that such an enhancement must occur and the radiation induces or stimulates additional transitions $2 \to 1$ at a rate proportional to the radiative energy density. The probability per unit time for this third radiative process is denoted

$$B_{21}\langle W(\omega) \rangle \quad \text{(stimulated emission rate)}. \qquad (1.5.6)$$

The transition rates for all three processes are shown in Fig. 1.8.

The Einstein A coefficient and the two B coefficients defined above are independent of the radiative energy density. They depend only on the properties of the two atomic states, in ways determined by the formal quantum mechanics of §2.3. For the present, they are treated as phenomenological parameters.

Consider the influences of the three transition rates on the atomic level populations. It is assumed that the total number of atoms is sufficiently large for the individual absorptions and emissions to produce smooth time dependences in the populations, whose rates of change are given by

$$dN_1/dt = -dN_2/dt = N_2 A_{21} - N_1 B_{12} \langle W(\omega) \rangle + N_2 B_{21} \langle W(\omega) \rangle. \qquad (1.5.7)$$

This population rate-equation is the main tool in the application of the Einstein theory to radiative processes and several applications are made in the following sections. A simple special case is the steady state, where the population time-derivatives vanish and eqn (1.5.7) reduces to

$$N_2 A_{21} - N_1 B_{12} \langle W(\omega) \rangle + N_2 B_{21} \langle W(\omega) \rangle = 0. \qquad (1.5.8)$$

This is easily solved for the atomic populations, in conjuction with eqn (1.5.2).

Some important relations between the three Einstein coefficients are derived by solution of the steady-state equation (1.5.8) for conditions of thermal equilibrium, where the external contribution to the radiative energy density in eqn (1.5.3) is removed. The solution for the thermal energy density is

$$\langle W_T(\omega) \rangle = \frac{A_{21}}{(N_1/N_2) B_{12} - B_{21}}. \qquad (1.5.9)$$

The mean numbers of atoms in the two levels in thermal equilibrium are related by Boltzmann's law

$$\frac{N_1}{N_2} = \frac{g_1 \exp(-E_1/k_B T)}{g_2 \exp(-E_2/k_B T)} = \frac{g_1}{g_2} \exp\left(\frac{\hbar\omega}{k_B T}\right), \qquad (1.5.10)$$

and substitution in eqn (1.5.9) gives

$$\langle W_T(\omega) \rangle = \frac{A_{21}}{(g_1/g_2)\exp(\hbar\omega/k_B T)B_{12} - B_{21}}. \qquad (1.5.11)$$

This expression for the thermally-excited radiative energy density follows from the definitions of the Einstein coefficients and an application of the Boltzmann distribution to the atomic level populations. However, the result must be consistent with Planck's law for the same physical quantity, given by eqn (1.3.8). The two expressions are identical at all temperatures T only if

$$g_1 B_{12} = g_2 B_{21} \qquad (1.5.12)$$

and

$$\left(\hbar\omega^3/\pi^2 c^3\right)B_{21} = A_{21}. \qquad (1.5.13)$$

The three Einstein coefficients are therefore interrelated, and the transition rates between a pair of levels can all be expressed in terms of a single coefficient. It is evident from eqn (1.5.11) that consistency between the Einstein theory and Planck's law could not be achieved without the presence of the stimulated emission process. It is shown §4.10 that this process arises naturally in the quantum theory of photon emission.

The above relations between the Einstein coefficients are derived by consideration of a cavity in thermal equilibrium, where the radiative energy is homogeneous and isotropic in space. They hold generally for any spatially-isotropic distribution of radiative energy density. However, the external light beams used in experiments do not usually have this property, as in the example of a parallel light beam. Nevertheless, the relations (1.5.12) and (1.5.13) continue to apply for the interaction of light with a gas or fluid of atoms or molecules whose orientations are randomly distributed. The interaction of the radiation with the gas as a whole is then isotropic, even though the interaction with a single atom or molecule may be anisotropic. The orientation-averaged B coefficients are independent of the geometry of the light beam used in an experiment. These remarks do not necessarily cover the interaction of light with matter in the solid state, where the constituent atoms or molecules may be locked in a common orientation to give anisotropic optical properties. We treat here only atoms or molecules in a gas or fluid where optical isotropy prevails.

Finally, we note another way of expressing the thermal photon probability distribution that is easily obtained from eqns (1.3.2), (1.3.5) and (1.5.10) as

$$P(n) = \left(1 - \frac{g_1 N_2}{g_2 N_1}\right)\left(\frac{g_1 N_2}{g_2 N_1}\right)^n. \qquad (1.5.14)$$

The distribution is thus expressed entirely in terms of quantities related to the atomic energy levels.

1.6 Characteristics of the three Einstein transitions

A wide variety of radiative processes can be treated by solution of the rate equation

(1.5.7), using the relations (1.5.12) and (1.5.13) between the Einstein coefficients. However, it is useful first to examine the natures and relative magnitudes of the three basic kinds of radiative transition so that appropriate approximations can be made for specific processes. We begin with a comparison of the two emission processes in a thermal cavity.

The expression (1.5.11) for the thermal energy density is rewritten with the use of eqns (1.3.7) and (1.5.12) as

$$\langle W_T(\omega) \rangle = A_{21} \langle n \rangle / B_{21}. \tag{1.6.1}$$

The sum of the thermally-stimulated and spontaneous emission rates is thus

$$B_{21} \langle W_T(\omega) \rangle + A_{21} = A_{21}(\langle n \rangle + 1), \tag{1.6.2}$$

and the ratio of the two rates is given by the mean thermal photon number $\langle n \rangle$ at the transition frequency ω. It is seen from Fig. 1.5 that

$$\langle n \rangle \approx 1 \quad \text{for} \quad \hbar\omega/k_B T \approx 0.7 \tag{1.6.3}$$

and, for a cavity maintained at room temperature, this corresponds to radiation of angular frequency

$$\omega \approx 3 \times 10^{13} \text{ s}^{-1}, \tag{1.6.4}$$

in the far-infrared region of the spectrum. Hence for radiation of lower frequencies than this, corrresponding to the microwave and radiofrequency regions,

$$\hbar\omega \ll k_B T \quad \text{and} \quad A_{21} \ll B_{21} \langle W_T(\omega) \rangle, \tag{1.6.5}$$

while for radiation in the near-infrared, visible, ultraviolet and X-ray regions,

$$\hbar\omega \gg k_B T \quad \text{and} \quad A_{21} \gg B_{21} \langle W_T(\omega) \rangle. \tag{1.6.6}$$

The thermally-stimulated emission rate is thus much larger than the spontaneous rate for frequencies small compared to eqn (1.6.4), but for frequencies large compared to eqn (1.6.4), the spontaneous rate far exceeds the rates of thermally-stimulated emission and of absorption of thermal radiation.

The main concern of this book is with optical experiments that use electromagnetic radiation in the near-infrared, visible and ultraviolet regions of the spectrum, where the inequality (1.6.6) applies. The thermally-excited energy density at visible frequencies is very small for experiments at room temperature. For frequencies much higher than that in eqn (1.6.4), it is accordingly a very good approximation to ignore the thermal energy density. The total energy density in eqn (1.5.3) then represents just the contribution from external sources of light. The E subscript can be dropped and the external energy density is henceforth denoted simply by $\langle W \rangle$, with the frequency ω understood. For numerical estimates we use an angular frequency

$$\omega \approx 3 \times 10^{15} \text{ s}^{-1} \qquad (1.6.7)$$

with photon energy

$$\hbar\omega \approx 3 \times 10^{-19} \text{ J} \qquad (1.6.8)$$

as typical values in the visible range.

The absorption and stimulated emission associated with light beams from the external sources can, of course, make important contributions to the radiative processes that occur with visible light. It is interesting to consider the relative sizes of the two emission rates that occur for the light sources available in practice. We define a *saturation* radiative energy density, so called for reasons that are clarified in the following section, as

$$W_s = \hbar\omega^3 / \pi^2 c^3. \qquad (1.6.9)$$

It follows from eqn (1.5.13) that

$$W_s B_{21} = A_{21}, \qquad (1.6.10)$$

and the saturation energy density is that which equalizes the rates of stimulated and spontaneous emission. For the visible frequency in eqn (1.6.7), the saturation radiative energy per unit volume in a frequency range $d\omega$ is

$$W_s d\omega \approx 10^{-14} d\omega \text{ Jm}^{-3}. \qquad (1.6.11)$$

The strength of a light source is usually expressed in terms of its intensity, obtained by multiplication of the energy density by the velocity of light. The corresponding saturation intensity density is defined as

$$I_s = cW_s = cA_{21} / B_{21} = \hbar\omega^3 / \pi^2 c^2, \qquad (1.6.12)$$

and the numerical value of the saturation intensity for the frequency given in eqn (1.6.7) is

$$I_s d\omega \approx 3 \times 10^{-6} d\omega \text{ Wm}^{-2}. \qquad (1.6.13)$$

Thus for the example of a conventional light source with an angular frequency bandwidth of $6 \times 10^{10} \text{ s}^{-1}$, the beam intensity is

$$\int d\omega I_s \approx 1.8 \times 10^5 \text{ Wm}^{-2}. \qquad (1.6.14)$$

The strongest available conventional light sources have intensities that are at least an order of magnitude smaller than this saturation intensity. For such light beams, almost all of the photons that are absorbed by the atoms are reradiated by spontaneous emission. Laser light sources can easily have intensities that exceed the

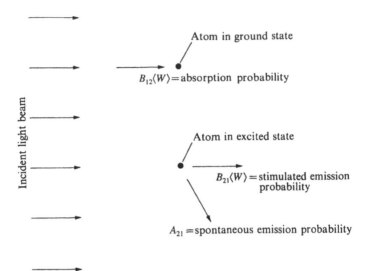

Fig. 1.9. Directional properties of the three Einstein transitions.

saturation value, but their emission spectra are often much narrower than those of the absorbing atoms and the Einstein theory may not apply, as is discussed in detail in §§2.7 and 2.10.

The directional properties of the three Einstein processes are illustrated in Fig. 1.9, and the most important feature is the distinction between the two kinds of emission. The light produced by stimulated emission from excited atoms appears in the same cavity mode as that of the incident light that causes the emission to take place. The emitted light also has the same phase properties as the incident beam. Thus stimulated emission tends to amplify the intensity of the incident beam while maintaining its other properties unchanged. These properties of the stimulated emission are proved in §§4.10 and 7.2. The light produced by spontaneous emission, on the other hand, is completely independent of the incident beam and it can excite any mode of the cavity that satisfies energy conservation. Thus the propagation direction, polarization and phase of the spontaneously-emitted light are arbitrary.

As a light beam traverses the atomic gas, photons are removed by the absorption process. The atoms thus excited eventually return to their lower states and they can replace photons in the beam by stimulated emission. If only these two processes occurred, the beam would pass through the atoms with no change in intensity once a steady state had been achieved. However, as some of the atoms return to the lower state by spontaneous emission, a fraction

$$\frac{A_{21}}{A_{21} + B_{21}\langle W \rangle} = \frac{1}{1 + (\langle W \rangle / W_{\mathrm{s}})} \tag{1.6.15}$$

of the absorbed energy is re-emitted in random directions, only a small part being accidentally in the same mode as the incident beam. Spontaneous emission thus causes a scattering of light and a consequent attenuation of the beam. Such scattering is the microscopic source of the apparent absorption of a light beam during its passage through an atomic gas.

1.7 Optical excitation of two-level atoms

Atomic excited states in the visible frequency range have negligible thermal populations at room temperature. We consider atoms that have a single low-lying state, the atomic ground state, and excited states at visible or higher frequencies. With the lower level in Fig. 1.8 taken to be the ground state, all N atoms are in this state in thermal equibrium. Selective excitation of a particular state can often be achieved by irradiation of the atoms with light from an external source whose frequencies ω satisfy the resonance condition (1.5.1) for only one atomic transition. The system may then behave essentially as a two-level atom with the level populations related by eqn (1.5.2). Less simple processes may involve three or four atomic energy levels, as in the optical amplifier treated in §1.9.

The two-level atom is particularly simple, and it becomes simpler still when both levels are nondegenerate, so that the B coefficients are equal according to eqn (1.5.12). The subscripts on the A and B coefficients can then be omitted, and the rate equation (1.5.7) simplifies to

$$dN_1/dt = -dN_2/dt = N_2 A + (N_2 - N_1)B\langle W \rangle, \tag{1.7.1}$$

where $\langle W \rangle$ is the mean energy density at frequency ω in a light beam supplied by the external sources. It is assumed in what follows that this beam excites a single mode of the cavity with well-defined direction and polarization. The radiative energy density is also assumed in the present section to be a fixed quantity, independent of time and position in the cavity.

We consider first the steady state solution of eqn (1.7.1), where the condition of eqn (1.5.8) reduces to

$$N_2 A + (N_2 - N_1)B\langle W \rangle = 0, \tag{1.7.2}$$

or with the use of eqn (1.6.10)

$$N_2 W_s + (N_2 - N_1)\langle W \rangle = 0. \tag{1.7.3}$$

The mean atomic populations in the steady-state are obtained with the use of eqn (1.5.2) as

$$
\begin{aligned}
N_1 &= \frac{A + B\langle W \rangle}{A + 2B\langle W \rangle} N = \frac{W_s + \langle W \rangle}{W_s + 2\langle W \rangle} N \\
N_2 &= \frac{B\langle W \rangle}{A + 2B\langle W \rangle} N = \frac{\langle W \rangle}{W_s + 2\langle W \rangle} N.
\end{aligned}
\tag{1.7.4}
$$

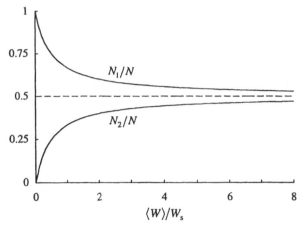

Fig. 1.10. Steady-state atomic populations as functions of the radiative energy density.

Figure 1.10 shows these atomic populations as functions of the radiative energy density. The populations have linear variations with the energy density when this is much smaller than the saturation value, but the dependence becomes nonlinear for $\langle W \rangle \approx W_s$, and both populations tend to $N/2$ for energy densities much larger than the saturation value. The nonlinear behaviour is called *saturation* of the atomic transition, hence the nomenclature of the saturation energy density. Saturation effects can usually be neglected for experiments that use ordinary light beams but they can be important in the theories of experiments that use laser light sources and in the theory of operation of the laser itself.

The mean rates of the three Einstein transitions for the steady-state gas as a whole are obtained with the use of eqn (1.7.4) as

absorption:
$$N_1 B \langle W \rangle = NA \frac{(W_s + \langle W \rangle) \langle W \rangle}{(W_s + 2 \langle W \rangle) W_s} \tag{1.7.5}$$

stimulated emission:
$$N_2 B \langle W \rangle = NA \frac{\langle W \rangle^2}{(W_s + 2 \langle W \rangle) W_s} \tag{1.7.6}$$

spontaneous emission:
$$N_2 A = NA \frac{\langle W \rangle}{W_s + 2 \langle W \rangle} . \tag{1.7.7}$$

Figure 1.11 shows the mean transition rates per atom, normalized by the spontaneous emission rate, as functions of the radiative energy density. It is seen that both the absorption and stimulated emission rates approach a common linear dependence for large radiative energy densities. However, because of the saturation of the upper level population at $N/2$, shown in Fig. 1.10, the spontaneous emission rate approaches the value $A/2$, independent of the radiative energy density.

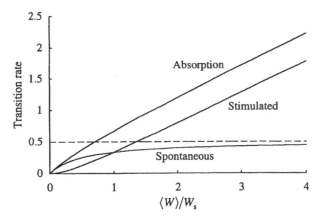

Fig. 1.11. Mean rates of the three Einstein transitions in units of the A coefficient as functions of the radiative energy density.

Now consider the time dependences of the mean atomic populations, which are readily obtained by solution of eqn (1.7.1) with the use of eqn (1.5.2).

Problem 1.4   Prove that the general solution of eqn (1.7.1) is

$$N_1(t) = \left\{ N_1(0) - N \frac{W_s + \langle W \rangle}{W_s + 2\langle W \rangle} \right\} \exp\left[-\left(A + 2B\langle W \rangle\right)t\right]$$

$$+ N \frac{W_s + \langle W \rangle}{W_s + 2\langle W \rangle}. \tag{1.7.8}$$

The solution for $N_2(t)$ follows from eqn (1.5.2).

In conditions where all N atoms are in their ground states at time $t = 0$, when the external light beam is turned on, the mean number of atoms in their excited states at later times is

$$N_2(t) = N \frac{\langle W \rangle}{W_s + 2\langle W \rangle} \left\{ 1 - \exp\left[-\left(A + 2B\langle W \rangle\right)t\right] \right\}. \tag{1.7.9}$$

The dependence of the excited state population on the time is illustrated in Fig. 1.12. The population grows linearly at short times but it approaches the steady-state value N_2 in eqn (1.7.4) asymptotically at long times. Energy is transferred from the light to the atoms when the beam is first turned on and the atomic excited-state population grows. The transfer of energy ceases when the steady state is achieved, and the only effect of the interaction between light and atoms is then a redistribution of the propagation directions of the radiation.

If the external light beam is now turned off again, the excited atoms return to their ground states and their excitation energy is converted to radiation by the spontaneous emission process. Let $t = 0$ be redefined as the instant at which the

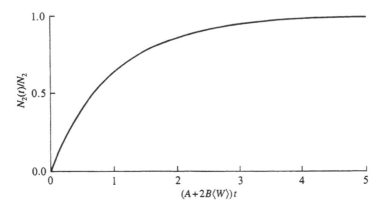

Fig. 1.12. Number of excited atoms as a function of the time t.

incident beam is removed. With the beam energy density $\langle W \rangle$ set equal to zero, the rate equation (1.7.1) reduces to

$$dN_2/dt = -N_2 A,\qquad(1.7.10)$$

with solution

$$N_2(t) = N_2(0)\exp(-At).\qquad(1.7.11)$$

Each decaying atom emits a quantum $\hbar\omega$, and the emitted light intensity falls off with the same exponential time dependence. Observation of this fluorescent emission is an experimental means of measuring the Einstein A coefficient. The reciprocal of A, denoted

$$\tau_R = 1/A,\qquad(1.7.12)$$

is the *fluorescent* or *radiative lifetime* of the atomic transition considered.

Problem 1.5   Derive expressions for the rates of change of the atomic excitation energy as functions of the time for
(i) an atom initially in its ground state, illuminated by a beam of energy density $\langle W \rangle$
(ii) an atom initially in its steady illuminated state, with the beam switched off at time $t = 0$.
Verify by integration over the time that the total energy transfers are equal but opposite in sign.

Problem 1.6   The light beam that illuminates the atoms is now repeatedly turned on and off, with the same duration τ for the on and off periods. The mean number of excited atoms settles into a regular pattern after many on/off cycles have taken place. Sketch the expected form of this regular variation and show that the maximum number of excited atoms is

$$N_2(t)_{max} = N \frac{\langle W \rangle}{W_s + 2\langle W \rangle} \frac{1 - \exp\left[-(A + 2B\langle W \rangle)\tau\right]}{1 - \exp\left[-2(A + B\langle W \rangle)\tau\right]}. \quad (1.7.13)$$

Derive the limiting forms of this expression for $\tau \to \infty$ and $\tau \to 0$ and explain them in physical terms.

The results for the atomic population derived above refer to the mean number of atoms in the excited state. Alternatively, N_2/N can be regarded as the probability of finding a selected atom in its excited state or as the fraction of time spent in the excited state. Of course, a particular atom makes repeated upward and downward transitions as it absorbs and emits photons, these transitions occurring at random time intervals in accordance with the known absorption and emission probabilities. Observations of single-atom absorption and emission are considered in §§8.4 and 8.5. For an N-atom gas, the numbers of atoms in the different levels fluctuate around their mean values even in the steady state but the atomic population fluctuations are not treated here.

Problem 1.7   An atom with transition frequency given by eqn (1.6.7) and radiative lifetime 10^{-7} s is illuminated by a light beam whose energy density equals the saturation value. What fraction of the time does the atom spend on average in its excited state in steady-state conditions? What are the average numbers in each second of (i) absorptions, (ii) spontaneous emissions and (iii) stimulated emissions?

1.8 Theory of optical attenuation

The processes considered in the preceding section assume a radiative energy density $\langle W \rangle$ that is independent of position. The attenuation of a light beam causes its energy density and intensity to fall off with propagation distance. In the present section we derive expressions for the attenuation based on the Einstein theory of atomic transitions. We retain the assumption that only the atomic ground state and a single excited state are significantly populated, but the optical intensity is now a function of the coordinate z parallel to the beam propagation direction. The treatments in this and the following section assume a travelling-wave light beam, in contrast to the cavity-mode standing waves of previous sections.

It is useful to precede the microscopic derivation with a summary of the standard theory of attenuation based on the macroscopic Maxwell equations. The beam intensity is determined by the Poynting vector

$$\mathbf{I}(z,t) = \varepsilon_0 c^2 \mathbf{E}(z,t) \times \mathbf{B}(z,t). \quad (1.8.1)$$

Textbooks of electromagnetic theory or optics [15,16] show that the mean intensity $\langle I \rangle$ of a light beam of frequency ω varies with propagation distance in an attenuating dielectric material according to

$$\langle I(z) \rangle = \langle I(0) \rangle \exp[-K(\omega)z], \tag{1.8.2}$$

where

$$K(\omega) = 2\omega\kappa(\omega)/c. \tag{1.8.3}$$

The quantity $K(\omega)$ is called the *attenuation coefficient* at frequency ω. The *extinction coefficient* $\kappa(\omega)$ is defined in conjunction with the *refractive index* $\eta(\omega)$ via the relation

$$[\eta(\omega) + i\kappa(\omega)]^2 = \varepsilon(\omega) \equiv \varepsilon'(\omega) + i\varepsilon''(\omega). \tag{1.8.4}$$

The *dielectric function* $\varepsilon(\omega)$ embodies the electrical properties of the material and the prime and double primes indicate its real and imaginary parts respectively. For frequencies ω at which $\varepsilon''(\omega)$ is nonzero, the extinction coefficient $\kappa(\omega)$ is also nonzero and attenuation of the light beam occurs in accordance with eqns (1.8.2) and (1.8.3).

The plethora of functions that describe the dielectric material is completed by the *linear susceptibility* $\chi(\omega)$, related to the dielectric function $\varepsilon(\omega)$ by

$$\varepsilon(\omega) = 1 + \chi(\omega). \tag{1.8.5}$$

The susceptibility $\chi(\omega)$ provides the basic description of the optical properties of the dielectric and its form can be calculated for specific models of the material; an example of the calculation is given in some detail in §2.5. The form of the dielectric function then follows trivially from eqn (1.8.5) and the experimentally-measurable refractive index and extinction coefficient are obtained by solution of the real and imaginary parts of eqn (1.8.4).

We now consider the same process of attenuation of a light beam in propagation through an atomic gas from a microscopic viewpoint. Our aim is to relate the attenuation rate, and hence the attenuation coefficient, to the coefficients of the Einstein theory. The gas is assumed to be sufficiently dilute that its refractive index is close to unity and the free-space forms of the atomic absorption and emission rates given in §1.5 need not be corrected for the dielectric properties of the gas. We suppose a steady state to have been achieved and the condition of eqn (1.7.2) applies, so that

$$N_2 A = (N_1 - N_2) B \langle W \rangle. \tag{1.8.6}$$

The left-hand side expresses the rate at which photons are scattered out of the beam by spontaneous emission, while the right-hand side expresses the rate of loss of photons from the beam by absorption minus the rate at which they are returned to the beam by stimulated emission. These two ways of computing the change in beam photon number must be identical in the steady state.

We use the right-hand side of eqn (1.8.6) to calculate the rate of attenuation of the beam, but the descriptions of the atomic transition and of the light beam must first be refined. For the atoms, it is assumed so far that they all have the same sharply-defined transition frequency ω. However, as discussed in §§2.5 and 2.6, there is always some statistical spread in the frequencies of the photons that atoms can absorb and emit, even when the same pair of states is considered for each atom. The distribution of frequencies is described by a *lineshape function* $F(\omega)$, defined so that $F(\omega)d\omega$ is the fraction of transitions for which the photon frequency lies in a small range $d\omega$ around ω. The lineshape function is normalized such that

$$\int_0^\infty d\omega\, F(\omega) = 1. \tag{1.8.7}$$

Some typical functions are illustrated in Fig. 2.4. For the light beam, the energy density $\langle W \rangle$ in the attenuating gas is now a function not only of the frequency ω but also of the propagation coordinate z. It follows from eqn (1.7.4) that the atomic populations are also functions of z, although the thermal motion of the atoms in a gas tends to homogenize them.

With these preliminaries, the net rate of change of the beam energy density at coordinate z is obtained from the right-hand side of eqn (1.8.6) as

$$\frac{\partial \langle W \rangle}{\partial t} = -\frac{N_1 - N_2}{V} F(\omega) B \langle W \rangle \hbar \omega, \tag{1.8.8}$$

where V is the sample volume. The relation is converted by the use of eqn (1.7.4) to

$$\frac{\partial \langle W \rangle}{\partial t} = -\frac{W_s}{W_s + 2\langle W \rangle} \frac{N}{V} F(\omega) B \langle W \rangle \hbar \omega, \tag{1.8.9}$$

where the saturation energy density is defined by eqn (1.6.9).

The attenuation coefficient of eqns (1.8.2) and (1.8.3), however, governs the *spatial* rate of change of the mean intensity density $\langle I \rangle$, whereas the Einstein theory provides eqn (1.8.9) for the *time* dependence of the mean energy density $\langle W \rangle$. The two densities are related by simple scaling with the velocity of light

$$\langle I \rangle = c \langle W \rangle, \tag{1.8.10}$$

as in the relation (1.6.12) between the saturation densities. The derivatives of the densities are related by consideration of the energy loss in a thin slab of the gas perpendicular to the propagation direction. The equivalence of the representations of the loss in terms of the time rate of change of $\langle W \rangle$ and the spatial rate of change of $\langle I \rangle$ gives the relation

$$\partial \langle W \rangle / \partial t = \partial \langle I \rangle / \partial z. \tag{1.8.11}$$

Conversion of eqn (1.8.9) to the intensity density instead of the energy density leads, after some rearrangement, to

$$\frac{1}{\langle I \rangle}\left(1 + \frac{2\langle I \rangle}{I_s}\right)\frac{\partial \langle I \rangle}{\partial z} = -K(\omega), \tag{1.8.12}$$

where

$$K(\omega) = NF(\omega)B\hbar\omega/Vc. \tag{1.8.13}$$

Problem 1.8   Prove that the general solution of eqn (1.8.12) is

$$\ln\left(\frac{\langle I(z)\rangle}{\langle I(0)\rangle}\right) + 2\frac{\langle I(z)\rangle - \langle I(0)\rangle}{I_s} = -K(\omega)z. \tag{1.8.14}$$

Show that

$$\langle I(0)\rangle - \langle I(z)\rangle \approx \frac{\langle I(0)\rangle I_s K(\omega)z}{I_s + 2\langle I(0)\rangle} \tag{1.8.15}$$

for distances z such that the intensity is only slightly changed from its initial value.

The general solution simplifies considerably in two limiting cases. All ordinary light beams have intensities that are much smaller than the saturation intensity, as is discussed in §1.6. The second term on the left-hand side of eqn (1.8.14) can then be neglected and the terms that remain are rearranged in the form

$$\langle I(z)\rangle = \langle I(0)\rangle \exp[-K(\omega)z]. \tag{1.8.16}$$

This is identical to the macroscopic solution in eqn (1.8.2) and it shows that $K(\omega)$ in eqn (1.8.13) is indeed the microscopic expression for the attenuation coefficient in the Einstein theory. It is seen that $K(\omega)$ varies with frequency in a similar way to the distribution of transition frequencies but multiplied by an additional factor of ω. Integration of eqn (1.8.13) with the use of eqn (1.8.7) gives

$$\int_0^\infty d\omega \frac{K(\omega)c}{\hbar\omega} = \frac{NB}{V}, \tag{1.8.17}$$

so that the numerical value of B for a given transition can be determined from measurements of the attenuation coefficient.

In the opposite limit of a very strong light beam, whose intensity is much larger than the saturation value, it is the first term on the left-hand side of eqn (1.8.14) that can be neglected. The solution then simplifies to

$$\langle I(z)\rangle - \langle I(0)\rangle = -\tfrac{1}{2}I_s K(\omega)z = -\left(NF(\omega)A\hbar\omega/2V\right)z, \tag{1.8.18}$$

where eqns (1.6.12) and (1.8.13) have been used. The beam intensity thus falls linearly with propagation distance at a rate determined by the A coefficient, instead of the more rapid exponential fall-off determined by the B coefficient for low intensities.

The decrease in attenuation rate as the atomic transition experiences saturation is caused by the approach of the spontaneous emission rate to the maximum value of $A/2$, shown in Fig. 1.11. Further increases in the beam intensity do not produce proportionate increases in the scattering of energy out of the beam, and the fractional change in $\langle I\rangle$ in traversing the gas decreases. The gas is itself said to be *saturable*; it becomes progressively more transparent with increasing intensity, a phenomenon known as *bleaching*. The macroscopic theory of attenuation based on Maxwell's equations does not apply in saturation conditions. Saturation must obviously be avoided in measurements of the attenuation coefficient if the results are used to determine the B coefficient by eqn (1.8.17).

1.9 Population inversion: optical amplification

If the number N_2 of excited atoms can be made larger than the number N_1 of atoms in their ground states, it follows from eqn (1.8.8) that the beam intensity grows with distance through the atomic gas. The dependence is similar to eqn (1.8.16) except that the attenuation coefficient is now negative. The condition $N_2 > N_1$ is known as *population inversion*, or sometimes as a negative-temperature condition as application of Boltzmann's law in eqn (1.5.10) implies $T < 0$. Of course, population inversion cannot occur in thermal equilibrium and it is shown in §1.7, particularly Fig. 1.10, that it is not achieved in the Einstein theory of resonant absorption of light by a two-level atom. However, population inversion can be achieved for a pair of energy levels in experiments that use additional states of the atom. The simplest process makes use of three states and the theory of the three-level optical amplifier is outlined here.

Figure 1.13 shows the arrangement of transitions for three-level amplification. The ground state is now labelled as 0 and the active levels 1 and 2 are the first and second excited states, all assumed to be nondegenerate. A light beam or other source of energy, known as the *pump*, excites the transition $0 \to 2$ at a mean rate denoted R, so that level 2 is populated to some extent. The fraction N_2/N of the total number of atoms pumped up in this way is usually very small, a typical figure being one in a million, so that the transition $0 \to 2$ is far from the saturation condition. An atom in level 2 emits light by making downward transitions $2 \to 1$ or $2 \to 0$, and an atom in level 1 decays back into the ground state 0. We show that the condition $N_2 > N_1$ can be achieved for appropriate values of the Einstein coefficients. It is then possible to amplify an incident light beam of energy density $\langle W\rangle$ at the frequency ω of the transition between levels 1 and 2.

The equations that govern the transitions in a three-level system are simple generalizations of the rate equations for a two-level atom. The relevant transitions

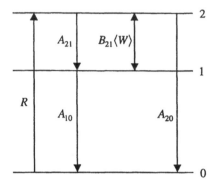

Fig. 1.13. Atomic energy levels and transition rates for a three-level optical amplifier.

are all shown in Fig. 1.13, where the subscripts on the Einstein coefficients denote the pairs of levels to which they refer. We assume that only the three levels shown are populated, so that

$$N_0 + N_1 + N_2 = N. \tag{1.9.1}$$

The refractive index of the gas is assumed to be close to unity, and the three rate equations are

$$dN_2/dt = N_0R - N_2A_{21} - N_2A_{20} - (N_2 - N_1)B_{21}\langle W \rangle, \tag{1.9.2}$$

$$dN_1/dt = N_2A_{21} - N_1A_{10} + (N_2 - N_1)B_{21}\langle W \rangle, \tag{1.9.3}$$

and

$$dN_0/dt = -N_0R + N_1A_{10} + N_2A_{20}. \tag{1.9.4}$$

The sum of the three rates is zero in accordance with eqn (1.9.1).

All of these time derivatives are set equal to zero in steady-state conditions, when eqns (1.9.3) and (1.9.4) give

$$N_2(A_{21} + B_{21}\langle W \rangle) = N_1(A_{10} + B_{21}\langle W \rangle) \tag{1.9.5}$$

and

$$N_1A_{10} + N_2A_{20} = N_0R. \tag{1.9.6}$$

It is immediately evident from eqn (1.9.5) that the condition

$$A_{21} < A_{10} \tag{1.9.7}$$

must be satisfied in order for N_2 to be larger than N_1. In words, the atoms excited into state 2 must decay relatively slowly into state 1, whence they fall rapidly back into the ground state 0. Equations (1.9.5) and (1.9.6) are readily solved for the level populations, and the resulting expression for their difference is

$$N_2 - N_1 = \frac{NR(A_{10} - A_{21})}{A_{10}(A_{20} + A_{21}) + (A_{10} + A_{20})B_{21}\langle W \rangle},$$

(1.9.8)

where N_0 is replaced by N to an excellent approximation, in view of the very small excitation of the upper levels. The structure of the population difference is similar to that for the two-level atom obtained from eqn (1.7.4). Thus for small values of the beam energy density, the difference is independent of $\langle W \rangle$, but saturation of the transition between levels 1 and 2 occurs for large values of $\langle W \rangle$, when the level populations tend to equality.

The time dependence of $\langle W \rangle$ is still given by solution of eqn (1.8.8), with the B coefficient taken as B_{21} and the population difference substituted from eqn (1.9.8). The condition (1.9.7) for population inversion is assumed to hold, so that the right-hand side of eqn (1.8.8) is positive and the light beam is amplified. The mean energy density $\langle W \rangle$ is converted to a mean intensity density $\langle I \rangle$ and the time dependence is converted to a spatial dependence by the steps given in eqns (1.8.10) and (1.8.11). The resulting differential equation can be written in a form very similar to eqn (1.8.12) as

$$\frac{1}{\langle I \rangle}\left(1 + \frac{\langle I \rangle}{I_s'}\right)\frac{\partial \langle I \rangle}{\partial z} = G(\omega),$$

(1.9.9)

where the *gain coefficient* at frequency ω is defined by

$$G(\omega) = \frac{R(A_{10} - A_{21})}{A_{10}(A_{20} + A_{21})} \frac{NF(\omega)B_{21}\hbar\omega}{Vc},$$

(1.9.10)

similar to the attenuation coefficient of eqn (1.8.13). The saturation intensity is a generalization of the quantity defined in eqn (1.6.12), given by

$$I_s' = \frac{cA_{10}(A_{20} + A_{21})}{(A_{10} + A_{20})B_{21}}.$$

(1.9.11)

The Einstein coefficient A_{20} is often much smaller than the others, and this expression for I_s' then reduces to I_s in eqn (1.6.12), but in general its numerical value differs from that in eqn (1.6.13). The factor of 2 difference between eqns (1.8.12) and (1.9.9) is a consequence of the difference in detail between the rate equations for the two and three level systems.

The solution of eqn (1.9.9) is

$$\ln\left(\frac{\langle I(z) \rangle}{\langle I(0) \rangle}\right) + \frac{\langle I(z) \rangle - \langle I(0) \rangle}{I_s'} = G(\omega)z,$$

(1.9.12)

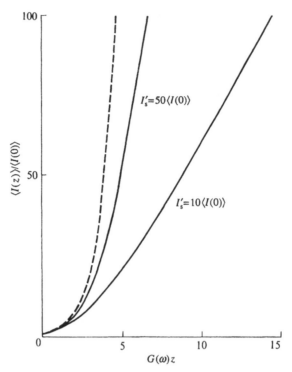

Fig. 1.14. Growth of intensity for a light beam in an amplifying medium for the two values of I_s' indicated. The broken curve is a plot of eqn (1.9.13).

similar to eqn (1.8.14), and this general relation again simplifies in two limiting cases. When the beam intensity is much smaller than the saturation intensity, the second term on the left-hand side of eqn (1.9.12) can be neglected and the growth of the beam is exponential, with

$$\langle I(z)\rangle = \langle I(0)\rangle \exp[G(\omega)z]. \tag{1.9.13}$$

When the beam intensity is much larger than the saturation intensity, the first term on the left-hand side of eqn (1.9.12) can be neglected and the growth is linear, with

$$\langle I(z)\rangle = \langle I(0)\rangle + I_s' G(\omega)z. \tag{1.9.14}$$

Figure 1.14 shows the forms of the general solution for two values of I_s', with the regions of exponential and linear growth clearly visible.

The spontaneous emission that can occur between levels 2 and 1 produces photons whose wavevectors are distributed over all spatial directions and only a small fraction have a wavevector coincident with the amplified beam. However, these photons are produced irrespective of the presence or absence of the incident light and they represent a noise component in the amplified beam. Their effect, which is ignored in the above derivation, is evaluated in §7.6.

Problem 1.9   Consider the three-level atoms shown in Fig. 1.13 with $A_{20} = 0$ and the beam of frequency ω switched off, $\langle W \rangle = 0$. If all of the atoms are initially in their ground states when the pump is switched on at time $t = 0$, derive expressions for the numbers of atoms in levels 1 and 2 at later times t. Show that the population of level 1 at short times t is approximately

$$N_1(t) \approx \tfrac{1}{2} N R A_{21} t^2,$$

(1.9.15)

with N_0 again replaced by N.

1.10 The laser

An amplifying medium is the active ingredient of a laser light source. The gas laser additionally uses an optical cavity to reflect the light beam to and fro through the gas and so build up a high intensity from a gain coefficient $G(\omega)$ that typically has a very small value. Figure 1.15 shows the form of a symmetrical Fabry–Perot cavity with two identical plane-parallel partially-transmitting mirrors separated by a distance L. The properties of the complex amplitude reflection and transmission coefficients, $\mathcal{R}$ and $\mathcal{T}$, of the mirrors are considered in detail for a lossless beam splitter in §3.2, where it is shown that they satisfy

$$|\mathcal{R}|^2 + |\mathcal{T}|^2 = 1.$$

(1.10.1)

The figure also shows the notation for the internal and external fields produced by an incident light beam. The light waves are here taken as travelling, with complex electric fields of the form

$$E(z,t) = E \exp(-i\omega t + ikz),$$

(1.10.2)

in contrast to the standing waves of eqn (1.1.3).

Consider first the transmission properties of an empty cavity. For a linearly-

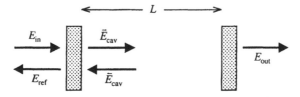

Fig. 1.15. Geometrical arrangement of a Fabry–Perot cavity showing the notation for internal and external electric fields.

polarized light beam of wavevector k that is propagated parallel to the axis, the leftwards and rightwards travelling fields inside the cavity at the left-hand mirror are related by the propagation phase factor over twice the cavity length,

$$\bar{E}_{cav} = \mathcal{R}\vec{E}_{cav} \exp(2ikL). \tag{1.10.3}$$

The input and output fields at the two mirrors are related by the usual tangential boundary conditions. Thus at the left-hand mirror

$$\vec{E}_{cav} = \mathcal{T}E_{in} + \mathcal{R}\bar{E}_{cav} \quad \text{and} \quad E_{ref} = \mathcal{R}E_{in} + \mathcal{T}\bar{E}_{cav}, \tag{1.10.4}$$

while at the right-hand mirror

$$E_{out} = \mathcal{T}\bar{E}_{cav} \exp(ikL). \tag{1.10.5}$$

These four equations are readily solved for the internal and transmitted fields in terms of the incident field, with the results

$$\vec{E}_{cav} = \mathcal{T}E_{in}/\mathcal{D}, \quad \bar{E}_{cav} = \mathcal{R}\mathcal{T}E_{in} \exp(2ikL)/\mathcal{D} \tag{1.10.6}$$

and

$$E_{out} = \mathcal{T}^2 E_{in} \exp(ikL)/\mathcal{D}, \tag{1.10.7}$$

where the denominator is

$$\mathcal{D} = 1 - \mathcal{R}^2 \exp(2ikL). \tag{1.10.8}$$

Standard Fabry–Perot theory is mainly concerned with the relation between output and input intensities. The beam intensity is given by the Poynting vector of eqn (1.8.1), where the electric and magnetic fields are related by eqn (1.2.5). The physical electric field is taken as the real part of the expression in eqn (1.10.2), and the average of the intensity over a cycle of the oscillation at frequency ω is

$$\bar{I} = \tfrac{1}{2}\varepsilon_0 c|E|^2. \tag{1.10.9}$$

The cycle average is here denoted by an overbar and statistical fluctuations in the intensity, denoted by angle brackets, are not explicitly included in the present calculation. The empty cavity input–output relation is obtained from eqn (1.10.7) as

$$\bar{I}_{out} = \frac{|\mathcal{T}|^4 \bar{I}_{in}}{|\mathcal{T}|^4 + 4|\mathcal{R}|^2 \sin^2\left(kL + \phi_{\mathcal{R}}\right)}, \tag{1.10.10}$$

where $\phi_{\mathcal{R}}$ is the phase of the reflection coefficient. The phase angle is very close to π for highly-reflecting mirrors, so that $\mathcal{R}$ is real and $\phi_{\mathcal{R}}$ can be removed from eqn (1.10.10). The maximum transmitted intensity is then

$$\bar{I}_{out} = \bar{I}_{in} \quad \text{for} \quad k = \pi v/L \quad \text{with} \quad v = 1, 2, 3,\ldots, \tag{1.10.11}$$

where these wavevectors define the longitudinal spatial modes of the Fabry–Perot cavity, analogous to the three-dimensional modes defined by the wavevectors in eqn (1.1.4).

The above discussion refers to the empty cavity and we now need to consider the effect of filling it with an amplifying medium. The frequency ω is assumed to resonate with the transition of the inverted atomic population described in §1.9 and the field in the cavity thus experiences gain. The change in phase as the light propagates from one end of the cavity to the other is now accompanied by an increase in its amplitude, represented by the replacement

$$\exp(ikL) \rightarrow \exp\left(ikL + \tfrac{1}{2}gL\right), \tag{1.10.12}$$

where g will be determined by comparison with eqn (1.9.9). The refractive index of the medium is assumed to be little different from unity and the wavevector k is assumed to coincide with a transmission maximum, as defined by eqn (1.10.11). The fractional increase gL in the intensity of the light as the beam propagates once along the length of the cavity, known as the *single-pass gain*, is very much smaller than unity (see the comment after eqn (1.10.22) below). The replacement in eqn (1.10.12) thus becomes

$$\exp(ikL) \rightarrow \pm\left(1 + \tfrac{1}{2}gL\right), \tag{1.10.13}$$

to a very good approximation, where the overall sign depends on whether the integer v in eqn (1.10.11) is even or odd. The input–output relation from eqns (1.10.7) and (1.10.8) is thus approximated by

$$E_{out} = \pm\frac{\mathcal{T}^2\left(1 + \tfrac{1}{2}gL\right)}{1 - \mathcal{R}^2(1 + gL)} E_{in} \approx \pm\frac{\mathcal{T}^2}{|\mathcal{T}|^2 - |\mathcal{R}|^2 gL} E_{in}. \tag{1.10.14}$$

Here $\mathcal{R}$ is again assumed real and its modulus signs are redundant.

The small single-pass gain is determined by solution of eqn (1.9.9), expressed in terms of the cycle-averaged intensity in the cavity, correct to first order in z,

$$\bar{I}_{cav}(z) = \left\{1 + \frac{G(\omega)z}{1 + \left(\bar{I}_{cav}(0)/I_s'\right)}\right\} \bar{I}_{cav}(0), \tag{1.10.15}$$

so that

$$g = \frac{G(\omega)}{1 + \left(\bar{I}_{cav}/I_s'\right)} = \frac{G(\omega)}{1 + \left(|E_{cav}|^2/|E_s|^2\right)}, \tag{1.10.16}$$

where $|E_s|$ is the saturation field amplitude, related to the saturation intensity as in eqn (1.10.9). The field in the cavity includes both the rightwards and leftwards travelling waves and its spatially-averaged square modulus is

$$|E_{cav}|^2 = |\vec{E}_{cav}|^2 + |\bar{E}_{cav}|^2 = \frac{1+|\mathcal{R}|^2}{|\mathcal{T}|^2}|E_{out}|^2, \tag{1.10.17}$$

with use of eqns (1.10.6) and (1.10.7), together with the small value of the single-pass gain. Insertion of the form of g into the inverted input–output relation (1.10.14) gives

$$E_{in} = \pm\left\{1 - \frac{|\mathcal{R}|^2 G(\omega)L}{|\mathcal{T}|^2 + \left(1+|\mathcal{R}|^2\right)\left(|E_{out}|^2 / |E_s|^2\right)}\right\} E_{out}. \tag{1.10.18}$$

All of the main properties of the laser follow from this relation between the input and output fields. The lasing state of the system corresponds to a self-sustaining oscillation of the amplifying cavity in which there is a nonzero output field even in the absence of any input field. The energy for the output in this case is entirely provided by the atomic pump, represented by the rate R in Fig. 1.13. The two solutions of eqn (1.10.18) with E_{in} set equal to zero are

$$|E_{out}| = 0 \quad \text{and} \quad |E_{out}|^2 = \frac{|\mathcal{R}|^2 G(\omega)L - |\mathcal{T}|^2}{1+|\mathcal{R}|^2}|E_s|^2. \tag{1.10.19}$$

The second solution represents the lasing state. It is put into a simpler approximate form by use of the typical values of the mirror reflection and transmission coefficients for a gas laser,

$$|\mathcal{R}|^2 = 0.95 \quad \text{and} \quad |\mathcal{T}|^2 = 0.05. \tag{1.10.20}$$

The reflection coefficient can thus be replaced by unity to a good approximation and the lasing solution from eqn (1.10.19) becomes

$$|E_{out}|^2 = \left(\frac{G(\omega)L}{|\mathcal{T}|^2} - 1\right)\frac{|\mathcal{T}|^2}{2}|E_s|^2. \tag{1.10.21}$$

This solution is clearly valid only for gain coefficients sufficiently high that

$$G(\omega)L \geq |\mathcal{T}|^2, \tag{1.10.22}$$

where the gain defined by the equality in this relation is the *threshold condition* for laser action. Lasing occurs when the inequality is satisfied because the gain in the optical intensity of the light beam in a single traverse of the cavity exceeds the loss of intensity at each encounter with a mirror. With the size of the transmission coefficient given in eqn (1.10.20), it is seen that the single-pass gain is indeed much smaller than unity for gains up to a few times the threshold value, as assumed in the above derivation.

The *cooperation parameter* of the laser is defined as

$$C = G(\omega)L/|T|^2. \tag{1.10.23}$$

As the gain coefficient given by eqn (1.9.10) is proportional to the pumping rate R, the parameter C provides a convenient measure of the pumping with respect to the threshold value. Thus, with the fields in eqn (1.10.21) converted to intensities in accordance with eqn (1.10.9), the laser output intensity from each end of the Fabry–Perot cavity is

$$\bar{I}_{\text{out}} = \tfrac{1}{2}(C-1)|T|^2 I_s', \tag{1.10.24}$$

where an expression for the saturation intensity is given in eqn (1.9.11). The laser is, of course, an important source of light in many practical applications and it is also important in many experimental studies of the basic quantum-optical properties of light. Its influence pervades much of the material in the subsequent chapters of the book.

The device also acts as an optical amplifier, whose properties are described by the relation (1.10.18) with a nonzero input field restored. As the laser above threshold has a nonzero output field even in the absence of an input field, it is convenient to specify the amplification for small values of the input field by the *differential gain*, defined as

$$G_{\text{diff}} = \left(d|E_{\text{out}}|/d|E_{\text{in}}| \big|_{|E_{\text{in}}|=0} \right)^2. \tag{1.10.25}$$

The differential gains below and above threshold are readily evaluated.

Problem 1.10 Show that the differential gains are

$$G_{\text{diff}} = 1/(1-C)^2 \quad \text{for } C < 1 \text{ below threshold} \tag{1.10.26}$$

and

$$G_{\text{diff}} = C^2/4(C-1)^2 \quad \text{for } C > 1 \text{ above threshold}, \tag{1.10.27}$$

with the mirror reflection coefficient again set equal to unity.

The variations of the differential gain with C are shown in Fig. 1.16 together with experimental points in the range $C > 1$ [17]. Large differential gains occur in the region of the laser threshold but the gain falls below unity, corresponding to attenuation, for pumping rates that exceed twice the threshold value.

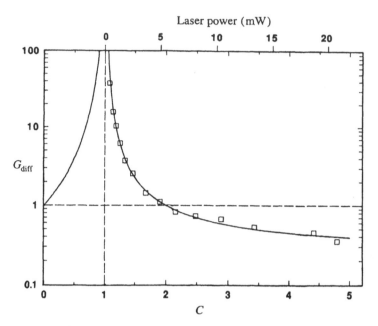

Fig. 1.16. Variation of the differential gain of a laser amplifier with the normalized pumping rate C. The curves show the theory and the points show experimental values for an Ar^+ laser. (After [17])

The spontaneous emission into the lasing mode of the Fabry–Perot cavity is ignored in the above derivations. Although the spontaneous emission rate into a single mode is small compared to the stimulated emission rate for a laser above threshold, the spontaneous emission is important below threshold and it also determines the spectrum of the laser light above threshold. These effects are treated in §§7.3 and 7.4.

1.11 Radiation pressure

The photons that make up a travelling electromagnetic wave of wavevector $\mathbf{k}$ each carry a momentum $\hbar\mathbf{k}$. Their behaviour in this respect is similar to that of other particles in quantum mechanics, and it is an example of the de Broglie relation between particle wavelength and momentum. The total momentum is conserved in any interaction of the photons with other particles, and a striking example is the scattering of photons by electrons, where the conservation of momentum was first demonstrated by Compton in 1927. He used X-ray photons and free electrons, where the effects of photon momentum are particularly important, but the effects are significant even in the interactions of the lower-energy visible photons with relatively heavy atoms. The most accurate measurements of the momentum are made by determining the torsional force on a suspended mirror illuminated off-centre by a laser beam [18], and the expression $\hbar\mathbf{k}$ for the momentum of a photon of wavevector $\mathbf{k}$ has been verified to high accuracy.

The transfer of momentum from light to material particles is equivalent to a force or pressure exerted by the radiation on matter. As a first example of radiation pressure, we return for the moment to the thermal cavity considered in §§1.1 to 1.3. The specular reflection of a photon by the wall at an angle of incidence θ transfers a momentum $2\hbar k \cos\theta$ in the direction normal to the wall. The number of field modes given by eqn (1.1.7) is generalized to include the distribution over different angles θ as

$$\tfrac{1}{8}\left(2\pi k^2 \sin\theta \mathrm{d}k\mathrm{d}\theta\right)\left(\pi/L\right)^{-3}\times 2. \tag{1.11.1}$$

Problem 1.11 Show that the total pressure, or force per unit area, on the walls of a cavity maintained at temperature T is

$$\mathcal{P} = \tfrac{1}{3}\int_0^\infty \mathrm{d}\omega\langle W_T(\omega)\rangle, \tag{1.11.2}$$

where the thermal energy density is given by eqn (1.3.8).

This relation between pressure and total energy density also follows from a consideration of the thermodynamics of the photon gas in the cavity [19].

In contrast to the isotropic radiation in a thermal cavity, our main interest is the effects of the photon momentum in a parallel beam of light in interaction with an atomic gas. Consider the transfers of momentum in the three basic Einstein transitions represented in Fig. 1.9. In the absorption process, an amount of momentum $\hbar k$ is transferred from the light beam to the absorbing atom, which acquires a velocity $\hbar k/M$ parallel to the beam, where M is the atomic mass. If the excited atom subsequently decays by stimulated emission, the emitted photon carries away a momentum $\hbar k$ parallel to the incident beam, and the atom loses the momentum it had previously gained.

If, however, the atom subsequently decays by spontaneous emission, the direction of the momentum carried away by the emitted photon may lie anywhere in the complete 4π solid angle centred on the atom. The atom thus acquires a recoil velocity in some random direction and there is, on the average, no cancellation of the atomic velocity component previously acquired by photon absorption. Thus each photon absorption that is followed by spontaneous emission transfers an average momentum $\hbar k$ to the atom in the direction of the light beam. This is so whether or not the atom–radiation system has reached a steady state, and it contrasts with the transfer of mean excitation energy to the atomic gas as a whole, which ceases in the steady state as emphasized in §1.7.

The rate of change of the mean total momentum $\mathbf{P}$ of the atomic gas in the presence of radiative energy density $\langle W \rangle$ at the frequency of a two-level atomic transition is thus proportional to the difference between the rates of absorption and stimulated emission,

$$\mathrm{d}\mathbf{P}/\mathrm{d}t = \left(N_1 - N_2\right)B\langle W\rangle\hbar\mathbf{k}. \tag{1.11.3}$$

This quantity is always positive for a two-level atom, but negative momentum transfers can occur for transition schemes that use three or more levels, where N_2 can be greater than N_1. We consider here only the two-level system.

In the steady state, where the numbers of atoms in the two states are given by eqn (1.7.4), the momentum transfer rate becomes

$$\frac{d\mathbf{P}}{dt} = \frac{NB\langle W\rangle W_s}{W_s + 2\langle W\rangle}\hbar\mathbf{k}. \tag{1.11.4}$$

The transfer rate tends towards a saturation value

$$d\mathbf{P}/dt = \tfrac{1}{2}NBW_s\hbar\mathbf{k} = \tfrac{1}{2}NA\hbar\mathbf{k} \quad \left(\langle W\rangle \gg W_s = A/B\right) \tag{1.11.5}$$

for very strong light beams. The reliance of the momentum transfer effect on the existence of the spontaneous emission process thus produces an upper limit on the transfer rate for beam strengths such that almost half the atoms are excited. Further increase in the beam strength causes little change in the transfer rate once the saturation condition is achieved.

The steady-state momentum transfer rate of eqn (1.11.4) is equivalent to an average radiation force acting on each atom, of magnitude

$$\mathcal{F} = \frac{B\langle W\rangle W_s}{W_s + 2\langle W\rangle}\hbar k. \tag{1.11.6}$$

An applied light beam drives the atoms in a cavity parallel to the beam direction to give a higher atomic density at the light exit window than at the entrance window. If the beam is propagated parallel to the z axis and $N(z)$ is the atomic density at coordinate z, the spatial distribution of the atoms at temperature T has the form

$$N(z) = N(0)\exp\left(Fz/k_B T\right). \tag{1.11.7}$$

This spatial dependence is observed experimentally [20], with an increase in density of 50% over a propagation distance of 200 mm.

The time dependence of the atomic momentum in an illuminated gas is obtained by solution of eqn (1.11.3).

Problem 1.12 Consider a gas of two-level atoms that are all in their ground states at time $t = 0$, when a parallel light beam of mean energy density $\langle W\rangle$ is switched on. Show that the mean momentum of the gas at time t is

$$\mathbf{P} = \frac{N\langle W\rangle\hbar\mathbf{k}}{W_s + 2\langle W\rangle}\left\{At + \frac{2\langle W\rangle}{W_s + 2\langle W\rangle}\left(1 - \exp\left[-\left(A + 2B\langle W\rangle\right)t\right]\right)\right\}. \tag{1.11.8}$$

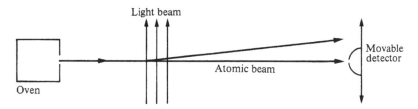

Fig. 1.17. Arrangement for the observation of the deflection of an atomic beam by interaction with a light beam.

Discuss the limiting forms of the result at both short and long times.

The most direct observations of momentum transfer are made when the atoms are also in the form of a beam, which is sent at right angles through a strong light beam. The experimental arrangement is represented in Fig. 1.17. Those atoms that absorb photons are deflected from their original paths. Absorption followed by stimulated emission produces a small lateral shift in the atomic trajectory but no change in its direction. The main effect occurs, as before, for absorption followed by spontaneous emission, where the atom gains on average a momentum $\hbar\mathbf{k}$ perpendicular to its original path. The resulting deflection angle is typically of order 10^{-5} rad per absorption–emission event.

Figure 1.18 shows some measurements [21] of the deflection of a beam of ^{20}Ne atoms by a resonant beam of visible light from a dye laser. The part of the distribution centred on zero deflection results from other isotopes of neon that do not have transitions in resonance with the laser frequency and thus do not experience the momentum transfer. The part of the distribution centred on a deflection of 0.4 mm results from the momentum transfer produced from an average number of about 20 absorption and spontaneous emission events. The energy density $\langle W \rangle$ of the light beam used in the experiments was about $W_s/4$. The breadth of the distribution of deflected atoms is greater than that of the undeflected atoms because different atoms experience different numbers of absorption–spontaneous emission events, although the additional broadening is mitigated to some extent by the rather unusual statistics of resonance fluorescence emission, as discussed in §8.4.

Problem 1.13   Calculate the mean angle of deflection for the experimental results shown in Fig. 1.18, assuming an average atomic velocity of 1000 ms^{-1}.

For much stronger beams of light, where $\langle W \rangle \gg W_s$ and the atomic transition is fully saturated, the rate of absorption–spontaneous emission events is of the order of one half the Einstein A coefficient. The number of such events is thus approximately equal to $A/2$ times the transit time of the atom through the light beam. For the typical values of an atomic velocity of 1000 ms^{-1}, an interaction

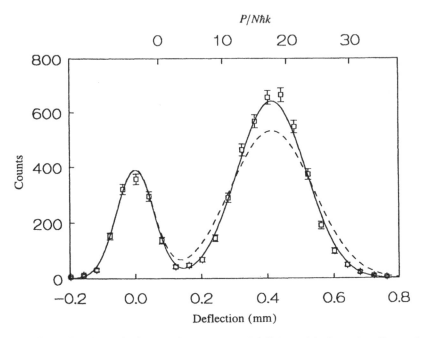

Fig. 1.18. Deflection of a beam of neon atoms by light, with Gaussian fits to the experimental points. The differences between the solid and dashed distributions are discussed in §8.4. (After [21])

path of length 2 mm, and an A coefficient of 10^8 s^{-1}, the average number of momentum quanta $\hbar\mathbf{k}$ transferred to the atom is about 100. Note that the saturation prevents any significant increase in the number of spontaneous emissions for yet stronger light beams.

Several important applications of laser sources employ the radiation pressure forces exerted by the light on atoms and other material particles. Thus light beams are used to trap and cool atoms and to manipulate micron-sized particles, for example biological specimens, for observations on microscope slides [22]. The latter application, known as *optical tweezers*, relies on more sophisticated radiation force effects, associated with inhomogeneous light beams, which are not included in the simple discussion given above.

References

[1]   Planck, M., On the improvement of Wien's equation for the spectrum, *Verh. dtsch. phys. Ges.* Berlin **2**, 202–4, (1900); On the theory of the energy distribution law of the normal spectrum, *ibid* **2**, 237–45 (1900).

[2]   Einstein, A., On a heuristic point of view about the creation and conversion of light, *Ann. Physik* **17**, 132–48 (1905).

[3]   Lewis, G.N., The conservation of photons, *Nature* **118**, 874–5 (1926).

[4] Einstein, A., On the quantum theory of radiation, *Phys. Z.* **18**, 121–8 (1917).

References [1], [2] and [4] are reprinted in English with the above titles in ter Haar, D., *The Old Quantum Theory* (Pergamon, Oxford, 1967).

[5] Dirac, P.A.M., The quantum theory of the emission and absorption of radiation, *Proc. R. Soc. Lond. A* **114**, 243–65 (1927).

[6] Taylor, G.I., Interference fringes with feeble light, *Proc. Camb. Phil. Soc.* **15**, 114–5 (1909).

[7] Brown, R.H. and Twiss, R.Q., Correlation between photons in two coherent beams of light, *Nature* **177**, 27–9 (1956).

[8] Scully, M.O. and Sargent, M., The concept of the photon, *Physics Today* **25**, (3), 38–47 (1972), and earlier references therein.

[9] Maiman, T.H., Stimulated optical radiation in ruby, *Nature* **187**, 493–4 (1960).

[10] Glauber, R.J., The quantum theory of optical coherence, *Phys. Rev.* **130**, 2529–39 (1963); Coherent and incoherent states of the radiation field, *Phys. Rev.* **131**, 2766–88 (1963).

[11] Rayleigh, Lord, The law of complete radiation, *Phil. Mag.* **49**, 539–40 (1900).

[12] Dwight, H.B., *Tables of Integrals and Other Mathematical Data*, 4th edn (Macmillan, New York, 1961).

[13] Fixsen, D.J., Cheng, E.S., Gales, J.M., Mather, J.C., Shafer, R.A. and Wright, E.L., The cosmic microwave background spectrum from the full *COBE* FIRAS data set, *Astrophys. J.* **473**, 576–87 (1996).

[14] Goodman, J.W., *Statistical Optics* (Wiley, New York, 1985).

[15] Jackson, J.D., *Classical Electrodynamics*, 3rd edn (Wiley, New York, 1999).

[16] Born, M. and Wolf, E., *Principles of Optics*, 6th edn (Cambridge University Press, Cambridge, 1997).

[17] Loudon, R., Harris, M., Shepherd, T.J. and Vaughan, J.M., Laser-amplifier gain and noise, *Phys. Rev. A* **48**, 681–701 (1993).

[18] Jones, R.V. and Leslie, B., The measurement of optical radiation pressure in dispersive media, *Proc. R. Soc. Lond. A* **360**, 365–71 (1978).

[19] Landau, L.D. and Lifshitz, E.M., *Statistical Physics*, part 1, 3rd edn (Pergamon Press, Oxford, 1980) §63.

[20] Bjorkholm, J.E., Ashkin, A. and Pearson, D.B., Observation of resonance radiation pressure on an atomic vapor, *Appl. Phys. Lett.* **27**, 534–7 (1975).

[21] Hoogerland, M.D., Wijnands, M.N.J.H., Senhorst, H.A.J., Beijerinck, H.C.W. and van Leeuwen, K.A.H., Photon statistics in resonance fluorescence: results from an atomic-beam deflection experiment, *Phys. Rev. Lett.* **65**, 1559–62 (1990).

[22] Chu, S., Laser manipulation of atoms and particles, *Science* **253**, 861–6 (1991); Laser trapping of neutral particles, *Scientific American* **266**, (2), 48–54 (1992).

2 Quantum mechanics of the atom–radiation interaction

The calculations of the preceding chapter display some of the inadequacies of the Einstein theory of interaction of light with atoms. The theory itself gives no prescription for computing the values of the A and B coefficients appropriate to a given atomic transition. For this one must look to the quantum-mechanical theory of transition probabilities. As light beams are generated by emission from atoms and measurements on light beams ultimately depend on the absorption of light by atoms, it is essential to cover the theory of the atomic interaction processes, even though the main interest of the book is the quantum nature of the light itself. More seriously, the Einstein theory is applicable only for broad-band illumination of the atom, where the energy in the incident light is distributed smoothly across the frequencies within the linewidth of the atomic transition. By contrast, many experiments with laser sources use light beams whose frequency distributions are much narrower than atomic transition linewidths.

The present chapter begins with a quantum-mechanical calculation of the Einstein B coefficient. A semiclassical theory is used, with a classical representation of the electromagnetic field but a quantum-mechanical description of the atomic states. Identical results are obtained when both the light and the atom are treated quantum-mechanically (see §4.10). We assume that there is no thermal excitation of the atomic excited states, and that only the ground state and a single excited state, selected by the frequency of the light, need be considered. The atoms can then be treated as effectively having only two levels, with consequent simplifications in the theory.

The interpretation of experiments with light beams of narrow frequency spread requires a more general theory of the effects of atom–radiation interaction. Such a theory is provided by the optical Bloch equations, derived in §2.7. It is shown that the behaviours of the atomic populations predicted by the optical Bloch equations agree with those of the Einstein theory in the limit of broad-band incident light. However, quite different behaviours occur for narrow-band incident light. It is also shown how various line-broadening mechanisms affect the atom–radiation interaction. These produce changes in the time dependences of the atomic-level populations and they generate characteristic shapes for the spectral lines seen in absorption and emission by atomic transitions.

2.1 Time-dependent quantum mechanics

The Einstein B coefficient describes the rate in time at which absorption and stimulated emission processes occur. The corresponding quantum theory is accordingly based on the time-dependent Schrödinger equation [1,2]

$$\hat{\mathcal{H}}\,\Psi(\mathbf{r},t) = i\hbar\partial\Psi(\mathbf{r},t)/\partial t. \tag{2.1.1}$$

Here, $\hat{\mathcal{H}}$ is the quantum-mechanical Hamiltonian, in general time-dependent, and $\Psi(\mathbf{r},t)$ is the time-dependent wavefunction. A circumflex accent is used throughout to signify quantum-mechanical operators.

For the problem of an isolated atom with no radiation present, the Hamiltonian $\hat{\mathcal{H}}_A$ contains a sum of the kinetic and potential energies of the constituent particles of the atom, and it has no explicit time dependence. It is shown by textbooks on quantum mechanics [1,2] that the wave equation (2.1.1) has solutions of the form

$$\Psi_i(\mathbf{r},t) = \psi_i(\mathbf{r})\exp(-iE_it/\hbar) \tag{2.1.2}$$

for a time-independent Hamiltonian. The time-dependent wavefunction $\Psi(\mathbf{r},t)$ thus splits into a product of a wavefunction $\psi_i(\mathbf{r})$ that is independent of the time and a time-dependent phase factor. Substitution of eqn (2.1.2) for $\Psi(\mathbf{r},t)$ into eqn (2.1.1) produces the energy eigenvalue equation,

$$\hat{\mathcal{H}}_A\psi_i(\mathbf{r}) = E_i\psi_i(\mathbf{r}). \tag{2.1.3}$$

In these equations, i labels the different solutions and $\mathbf{r}$ symbolizes the coordinates of all the particles that make up the atom.

The atomic states described by eqns (2.1.2) and (2.1.3) are known as *stationary states*. An atom in one of its stationary states has the property that the average value of any observable is independent of the time, provided that the operator representing the observable does not involve the time explicitly. A system put into one of the stationary states of its total Hamiltonian remains in that state for all subsequent times. It is assumed in the following calculations that the atomic energy eigenstates are known functions of $\mathbf{r}$.

For the calculation of the Einstein B coefficient, we are concerned with only two energy eigenstates, labelled $i = 1, 2$, whose eigenvalue equations are

$$\hat{\mathcal{H}}_A\psi_1(\mathbf{r}) = E_1\psi_1(\mathbf{r}) \quad \text{and} \quad \hat{\mathcal{H}}_A\psi_2(\mathbf{r}) = E_2\psi_2(\mathbf{r}) \tag{2.1.4}$$

The corresponding time-dependent wavefunctions are

$$\Psi_1(\mathbf{r},t) = \psi_1(\mathbf{r})\exp(-iE_1t/\hbar) \quad \text{and} \quad \Psi_2(\mathbf{r},t) = \psi_2(\mathbf{r})\exp(-iE_2t/\hbar). \tag{2.1.5}$$

The energy levels are as shown in Fig. 1.8, except that we now call the transition frequency ω_0,

$$\hbar\omega_0 = E_2 - E_1. \tag{2.1.6}$$

Consider the effect on the atom of an incident light beam, whose electric and magnetic fields vary with position and time. Their interaction with the atom causes it to have an additional electromagnetic energy, which can be represented by an addition $\hat{\mathcal{H}}_I$ to the Hamiltonian. The functional form of the interaction Hamiltonian is discussed shortly; for the moment we treat it as a general operator that depends on $\mathbf{r}$ and t. The total Hamiltonian is

$$\hat{\mathcal{H}} = \hat{\mathcal{H}}_A + \hat{\mathcal{H}}_I. \tag{2.1.7}$$

The wave equation (2.1.1) no longer has stationary-state solutions of the form given by eqn (2.1.2) as $\hat{\mathcal{H}}$ now depends explicitly on the time.

An expression for the Einstein coefficient B_{12} associated with the absorption of radiation is derived as follows. We assume that the atom is in state $\psi_1(\mathbf{r})$ at some instant of time. Owing to the presence of the term $\hat{\mathcal{H}}_I$ in the Hamiltonian, $\psi_1(\mathbf{r})$ is not a stationary state and there is a nonzero probability for the atom to be found in state $\psi_2(\mathbf{r})$ at a later time. This probability can be expressed as a rate of transition from state 1 to state 2, and hence related to the B coefficient. The required information is contained in the time-dependent wavefunction $\Psi(\mathbf{r},t)$ for the atom, obtained by solution of eqn (2.1.1).

If the frequency ω of the light is close to ω_0, only the two selected atomic states are involved in the radiative processes and, at any instant of time, the wavefunction must be a linear superposition

$$\Psi(\mathbf{r},t) = C_1(t)\Psi_1(\mathbf{r},t) + C_2(t)\Psi_2(\mathbf{r},t). \tag{2.1.8}$$

The combination of states varies with the time but we may require the state $\Psi(\mathbf{r},t)$ to be normalized at all times

$$\int dV |\Psi(\mathbf{r},t)|^2 = |C_1(t)|^2 + |C_2(t)|^2 = 1, \tag{2.1.9}$$

where eqn (2.1.5) has been used and the functions $\psi_1(\mathbf{r})$ and $\psi_2(\mathbf{r})$ are normalized and orthogonal. The coefficients C_1 and C_2 are independent of position.

Substitution of eqn (2.1.8) into eqn (2.1.1) provides an equation for the coefficients that is simplified by the use of eqns (2.1.4), (2.1.5) and (2.1.7) to

$$\hat{\mathcal{H}}_I(C_1\Psi_1 + C_2\Psi_2) = i\hbar\left(\Psi_1\frac{dC_1}{dt} + \Psi_2\frac{dC_2}{dt}\right), \tag{2.1.10}$$

where the space and time arguments are henceforth omitted from the various functions. Multiplication from the left by the complex conjugate Ψ_1^*, followed by integration over all space leads with the use of eqns (2.1.5) and (2.1.6) to

$$\langle 1|\hat{\mathcal{H}}_I|1\rangle C_1 + \exp(-i\omega_0 t)\langle 1|\hat{\mathcal{H}}_I|2\rangle C_2 = i\hbar\, dC_1/dt. \tag{2.1.11}$$

The usual Dirac notation is used for the matrix elements of the interaction Hamiltonian, for example

$$\langle 1|\hat{\mathcal{H}}_I|1\rangle = \int dV \psi_1^* \hat{\mathcal{H}}_I \psi_1. \tag{2.1.12}$$

Multiplication of eqn (2.1.10) from the left by Ψ_2^* similarly leads to

$$\exp(i\omega_0 t)\langle 2|\hat{\mathcal{H}}_I|1\rangle C_1 + \langle 2|\hat{\mathcal{H}}_I|2\rangle C_2 = i\hbar\, dC_2/dt. \tag{2.1.13}$$

Equations (2.1.11) and (2.1.13) contain no position dependence and the

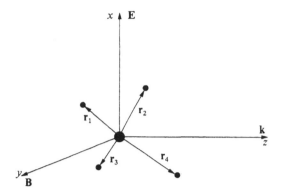

Fig. 2.1. Coordinate system for the atom and electromagnetic wave.

matrix elements are known functions of the time when the functional form of the interaction Hamiltonian is specified. The equations can in principle be solved for C_1 and C_2 as functions of the time, and the wavefunction Ψ is then determined by eqn (2.1.8).

2.2 Form of the interaction Hamiltonian

The complete form of the Hamiltonian for the interaction between the electromagnetic field and an atom is quite complicated, and a more detailed discussion is deferred to §4.8. However, a knowledge of the main features of the interaction is sufficient for calculation of the B coefficient.

Consider the atom illustrated in Fig. 2.1, which consists of a nucleus of charge Ze surrounded by Z electrons each of charge $-e$. The atom interacts with a polarized electromagnetic wave whose electric and magnetic fields are shown in the figure. It is convenient to take a travelling electromagnetic wave, instead of the standing-wave solutions given in eqns (1.1.3) to (1.1.5), with fields

$$\mathbf{E} = \mathbf{E}_0 \cos(kz - \omega t) \quad \text{and} \quad \mathbf{B} = \mathbf{B}_0 \cos(kz - \omega t). \tag{2.2.1}$$

The magnitudes of the electron radii are typified by the Bohr radius

$$a_{\mathrm{B}} = 4\pi\varepsilon_0 \hbar^2 / me^2 \approx 5 \times 10^{-11} \text{ m}, \tag{2.2.2}$$

where m is the electron mass. The radius is much smaller than the typical wavelength 6×10^{-7} m of an electromagnetic wave in the visible part of the frequency spectrum. For such frequencies,

$$ka_{\mathrm{B}} \ll 1, \tag{2.2.3}$$

and the spatial variations of the electric and magnetic fields across the dimensions of the atom are very small. It is then a good approximation to put $z = 0$, the coordinate of the atomic nucleus, in the cosines in eqn (2.2.1).

The total electric-dipole moment of the atom is written $-e\mathbf{D}$, where

$$\mathbf{D} = \sum_{\alpha=1}^{Z} \mathbf{r}_\alpha. \tag{2.2.4}$$

The main contribution to the interaction Hamiltonian arises from the potential energy of this electric dipole in the electric field of the light beam and we can write

$$\hat{\mathcal{H}}_{\mathrm{I}} = \hat{\mathcal{H}}_{\mathrm{ED}} = e\mathbf{D}.\mathbf{E}_0 \cos(\omega t). \tag{2.2.5}$$

The other contributions to $\hat{\mathcal{H}}_{\mathrm{I}}$ are considered in §4.8 and shown to be much smaller than the electric-dipole interaction $\hat{\mathcal{H}}_{\mathrm{ED}}$.

The interaction Hamiltonian is real and it has odd parity; that is, $\hat{\mathcal{H}}_{\mathrm{ED}}$ changes sign under inversion in the nucleus, when $\mathbf{r}_\alpha$ is replaced by $-\mathbf{r}_\alpha$. Its matrix elements diagonal in the atomic states must therefore vanish because the corresponding integrands, as defined in eqn (2.1.12), are odd functions of position,

$$\langle 1|\hat{\mathcal{H}}_{\mathrm{ED}}|1\rangle = \langle 2|\hat{\mathcal{H}}_{\mathrm{ED}}|2\rangle = 0. \tag{2.2.6}$$

Provided that the two atomic states have opposite parity, the matrix elements off-diagonal in the atomic states are not required to vanish, but they are related by

$$\langle 2|\hat{\mathcal{H}}_{\mathrm{ED}}|1\rangle = \langle 1|\hat{\mathcal{H}}_{\mathrm{ED}}|2\rangle^*. \tag{2.2.7}$$

As $\mathbf{E}_0$ points in the x direction in Fig. 2.1, the explicit form of the matrix element is

$$\langle 1|\hat{\mathcal{H}}_{\mathrm{ED}}|2\rangle = eE_0 X_{12} \cos(\omega t), \tag{2.2.8}$$

where

$$X_{12} = \int \mathrm{d}V \psi_1^* X \psi_2 ; \tag{2.2.9}$$

X is the x component of the atomic dipole moment $\mathbf{D}$, and the integration runs over all of the electron coordinates. It is convenient to use a shorthand notation for the numerical factor in the electric-dipole matrix element and we set

$$\hbar \mathcal{V} = eE_0 X_{12}, \tag{2.2.10}$$

where $\mathcal{V}$ has the dimensions of frequency. This is a complex quantity in general but it is real for transitions between bound states, where ψ_1 and ψ_2 are real wavefunctions. The calculations that follow refer to bound states, with $\mathcal{V}$ and X_{12} taken as real quantities.

The basic equations (2.1.11) and (2.1.13) are now simplified with the use of the results and notation of the previous paragraph to

$$\mathcal{V} \cos(\omega t) \exp(-i\omega_0 t) C_2 = i\, \mathrm{d}C_1/\mathrm{d}t \tag{2.2.11}$$

and

$$\mathcal{V}\cos(\omega t)\exp(i\omega_0 t)C_1 = i\,dC_2/dt. \tag{2.2.12}$$

These equations are exact within the restriction to a pair of atomic states.

Problem 2.1   Consider the transition from the 1S ground state of atomic hydrogen to the excited $2P_X$ state. Using the standard normalized wavefunctions for these states [1,2], prove that

$$\mathcal{V} = 2^{15/2}\,eE_0 a_B/3^5\,\hbar. \tag{2.2.13}$$

The transition frequency is

$$\omega_0 = 3\omega_R/4, \tag{2.2.14}$$

where the Rydberg frequency is defined as

$$\omega_R = me^4/32\pi^2\varepsilon_0^2\hbar^3 \approx 2\times10^{16}\ \mathrm{s}^{-1}. \tag{2.2.15}$$

Show that a field strength E_0 of about $2.5\times10^{11}\ \mathrm{Vm}^{-1}$, corresponding to a light-beam intensity of order $10^{20}\ \mathrm{Wm}^{-2}$, is needed to make $\mathcal{V}$ equal to ω_0.

For transitions in the visible region of the spectrum and for light beams of normally available intensities

$$\mathcal{V} \ll \omega_0. \tag{2.2.16}$$

Problem 2.2   Prove from eqns (2.2.11) and (2.2.12) that $|C_1|^2 + |C_2|^2$ does not change with time, thus ensuring that the normalization condition from eqn (2.1.9) remains valid throughout the time-dependent processes.

Problem 2.3   Consider the solution of eqns (2.2.11) and (2.2.12) for a constant applied electric field with $\omega = 0$. Show that C_2 satisfies

$$\frac{d^2 C_2}{dt^2} - i\omega_0\frac{dC_2}{dt} + \mathcal{V}^2 C_2 = 0, \tag{2.2.17}$$

and hence prove that the probability for the atom initially in its lower state at $t = 0$ to be found in its excited state at time $t > 0$ is

$$|C_2|^2 = \frac{4\mathcal{V}^2}{\omega_0^2 + 4\mathcal{V}^2}\sin^2\left\{\tfrac{1}{2}\left(\omega_0^2 + 4\mathcal{V}^2\right)^{1/2}t\right\}. \tag{2.2.18}$$

2.3 Expressions for the Einstein coefficients

The calculation of the Einstein B coefficient requires the solution of eqns (2.2.11) and (2.2.12) for conditions in which the frequency ω of the light beam is close to the transition frequency ω_0. The atom is again assumed to be in its lower state ψ_1 at time $t = 0$ and $|C_2|^2$ is the probability that the atom is excited after a time t. The corresponding excitation rate is the same quantity as the absorption rate defined in eqn (1.5.5), and we can make the identification

$$B_{12}\langle W(\omega_0)\rangle = |C_2|^2/t,\tag{2.3.1}$$

where $\langle W(\omega_0)\rangle$ is, as before, the mean energy density in the light beam per unit frequency range at frequency ω_0. A calculation of C_2 thus leads to the required quantum-mechanical expression for B_{12}.

Despite their simple forms, eqns (2.2.11) and (2.2.12) are quite difficult to solve in general, and it is necessary to look for approximate solutions. The inequality (2.2.16) suggests that it is useful to obtain the coefficients C_1 and C_2 as power series in $\mathcal{V}$. Their initial values are

$$C_1(0) = 1 \quad \text{and} \quad C_2(0) = 0,\tag{2.3.2}$$

and if these are substituted on the left-hand sides of eqns (2.2.11) and (2.2.12), integration with respect to time gives a first approximation to the time dependence of the coefficients. The results are

$$C_1(t) = 1\tag{2.3.3}$$

and

$$C_2(t) = \frac{\mathcal{V}}{2}\left\{\frac{1 - \exp[i(\omega_0 + \omega)t]}{\omega_0 + \omega} + \frac{1 - \exp[i(\omega_0 - \omega)t]}{\omega_0 - \omega}\right\}.\tag{2.3.4}$$

A second approximation is obtained by substituting these solutions on the left-hand sides of eqns (2.2.11) and (2.2.12) and integrating again; the improved solutions have the forms

$$C_1(t) = 1 + \mathcal{V}^2 \times (\text{function of } t)\tag{2.3.5}$$

and $C_2(t)$ is unchanged from eqn (2.3.4). A third approximation can now be obtained, and so on. A series in even powers of $\mathcal{V}$ is obtained for $C_1(t)$, while $C_2(t)$ is expressed as a series in odd powers of $\mathcal{V}$.

According to eqn (2.2.10), the series can equally well be regarded as expansions in powers of the electric field strength E_0. The energy in the travelling wave represented by eqn (2.2.1) is obtained by integration over the volume of a notional cavity, similar to eqn (1.2.3), with the electric and magnetic field strengths related by eqn (1.2.5). The energy density obtained by division of the electromagnetic energy by the cavity volume is

$$\tfrac{1}{2}\varepsilon_0 E_0^2 = \int d\omega \langle W(\omega) \rangle, \tag{2.3.6}$$

where the integral on the right-hand side is an alternative expression for the same energy density. Comparison of powers of E_0 on the two sides of eqn (2.3.1) shows that the required expression for B_{12} is obtained by evaluation of $C_2(t)$ to first order in E_0, or equivalently $\mathcal{V}$. The first approximation in eqn (2.3.4) is already adequate for this purpose. The Einstein theory of absorption and emission does not apply in situations where the terms of higher order in E_0 are important. The study of these higher-order terms forms the topic of nonlinear optics, which is considered in Chapter 9.

For light whose frequency ω is exactly equal to ω_0, eqn (2.3.4) reduces to

$$C_2(t) = -\left(i\mathcal{V}/2\omega_0\right)\left\{\sin(\omega_0 t)\exp(i\omega_0 t) + \omega_0 t\right\}. \tag{2.3.7}$$

It is shown later in the section that atomic transitions take place over time spans that are typically of order 10^{-7} s or longer, whereas the transition frequency from eqn (1.6.7) is typically of order 10^{15} s^{-1}. Thus the inequality

$$\omega_0 t \gg 1 \tag{2.3.8}$$

is very well satisfied for the times t of interest in the present calculation. In physical terms, many cycles of oscillation of the electromagnetic wave occur before there is a significant probability of atomic excitation. The second term in the bracket of eqn (2.3.7) is therefore much larger than the first, and a similar disparity holds more generally for the two terms in eqn (2.3.4) when ω is close, but not exactly equal, to ω_0.

With the first term neglected, eqn (2.3.4) gives

$$|C_2(t)|^2 = \mathcal{V}^2 \frac{\sin^2\left\{\tfrac{1}{2}(\omega_0 - \omega)t\right\}}{(\omega_0 - \omega)^2} \tag{2.3.9}$$

The excitation probability increases quadratically with the time when ω is exactly equal to ω_0

$$|C_2(t)|^2 = \tfrac{1}{4}\mathcal{V}^2 t^2, \tag{2.3.10}$$

but an oscillatory behaviour occurs when ω differs from ω_0. Figure 2.2 shows the dependence of the excitation probabilty on ω at a given time t. As time increases, the maximum in the curve moves upwards proportional to t^2 and the zeros of the function move in along the horizontal axis towards the origin.

The neglect of the first term in eqn (2.3.4) is known as the *rotating-wave approximation*. The term that is retained contains the difference of the phase angles ωt and $\omega_0 t$ for the light and the atomic transition, whereas the much smaller discarded term contains the sum of these phase angles.

Equation (2.3.9) gives the required solution for the probability of excitation of the atom into its upper level, but a further development is needed before this result

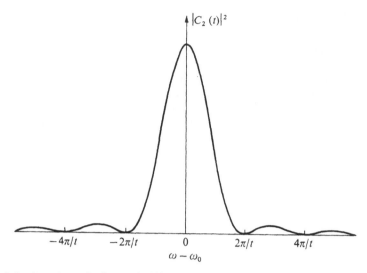

Fig. 2.2. Atomic excitation probability at time t as a function of the frequency ω of the light beam.

can be related to the measurable transition rate. We have assumed that the transition frequency ω_0 is a well-defined quantity with an exact value, but this does not correspond to the conditions in a practicable experiment, where there is always some uncertainty or statistical spread in the values of ω_0. Some forms of the associated lineshape function are considered in §§2.5 and 2.6. The Einstein theory applies to absorption and stimulated emission processes in which the energy in the light beam is distributed smoothly across the range $\Delta\omega$ of transition frequencies. Thus, if ω_0 is now interpreted as the centre of the distribution of transition frequencies, the excitation probability is obtained from eqn (2.3.9) as

$$|C_2(t)|^2 = \frac{2e^2 X_{12}^2}{\varepsilon_0 \hbar^2} \int_{\omega_0 - \frac{1}{2}\Delta\omega}^{\omega_0 + \frac{1}{2}\Delta\omega} d\omega \langle W(\omega) \rangle \frac{\sin^2\{\frac{1}{2}(\omega_0 - \omega)t\}}{(\omega_0 - \omega)^2}, \tag{2.3.11}$$

where the form of integration follows from eqn (2.3.6) with the use of eqn (2.2.10). It should be noted that this expression for the excitation probability has the form of an integral over the *probabilities* for the individual frequencies ω that make up the incident light beam. In general, the total excitation probability should be obtained as the square of an integral over ω of the excitation *amplitudes* $C_2(t)$. However, the form used in eqn (2.3.11) is correct when different frequency components in the light have randomly-distributed phases, so that the effects of interference between them average to zero.

   For a radiative energy density that varies little across the range of transition frequencies, the result (2.3.11) can be written

$$|C_2(t)|^2 = \left\{ 2e^2 X_{12}^2 \langle W(\omega_0) \rangle / \varepsilon_0 \hbar^2 \right\} \text{Int}(t), \tag{2.3.12}$$

where

$$\text{Int}(t) = \int_{\omega_0 - \frac{1}{2}\Delta\omega}^{\omega_0 + \frac{1}{2}\Delta\omega} d\omega \frac{\sin^2\left\{\frac{1}{2}(\omega_0 - \omega)t\right\}}{(\omega_0 - \omega)^2}. \tag{2.3.13}$$

The variation of this integral with t and $\Delta\omega$, shown in Fig. 2.3, has slight undulations associated with the zeros of the function illustrated in Fig. 2.2. For small $t\Delta\omega$, it is seen by reference to Fig. 2.2 that the integrand is almost constant with a value obtained from eqn (2.3.10),

$$\text{Int}(t) = \tfrac{1}{4}t^2\Delta\omega \quad (t\Delta\omega \ll 1). \tag{2.3.14}$$

For large $t\Delta\omega$, the integration covers almost all of the area under the curve in Fig. 2.2, and the limiting value is [3]

$$\text{Int}(t) = \tfrac{1}{2}\pi t \quad (t\Delta\omega \gg 1). \tag{2.3.15}$$

This limiting linear behaviour, illustrated by the dashed line in Fig. 2.3, is seen to provide a progressively better approximation for the larger values of $t\Delta\omega$.

In accordance with eqn (2.3.1) the Einstein B coefficient is associated with

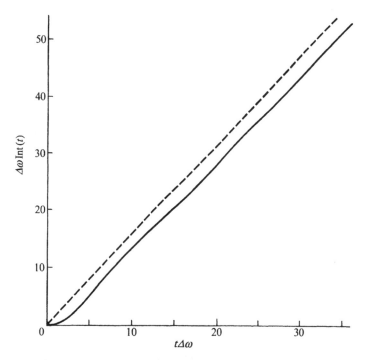

Fig. 2.3. The solid curve shows the time variation of the integral in eqn (2.3.13) while the dashed line shows the approximation of eqn (2.3.15).

absorption in which the probability of atomic excitation is proportional to the elapsed time t. This corresponds to the linear behaviour shown in eqn (2.3.15), whose insertion into eqn (2.3.12) gives

$$|C_2(t)|^2 = \pi e^2 X_{12}^2 \langle W(\omega_0) \rangle t / \varepsilon_0 \hbar^2. \tag{2.3.16}$$

The linear approximation embodied in eqn (2.3.16) clearly breaks down for times sufficiently long that the excitation probability becomes greater than unity, contrary to the normalization condition of eqn (2.1.9). However, the long-time behaviour of the atomic excitation can be determined in the manner of eqn (1.7.9) once the magnitude of the B coefficient is determined from eqn (2.3.16).

The calculation is completed by a generalization of the single-atom results to a gas of N identical atoms. With the light beam incident simultaneously on the atoms, all initially in their ground states, the mean numbers of atoms in the two states at time t are

$$N_1(t) = N|C_1(t)|^2 \quad \text{and} \quad N_2(t) = N|C_2(t)|^2. \tag{2.3.17}$$

The normalization condition of eqn (2.1.9) ensures conservation of the number of atoms. The electronic states of the atoms or molecules in the gas have random spatial orientations and it is useful to take account of these in the expression for the B coefficient. Let $\mathbf{e}$ be a unit vector in the direction of the electric field of the light beam, taken as the x axis in Fig. 2.1. The matrix element from eqn (2.2.9) can be written

$$X_{12} = \int dV \psi_1^* \mathbf{e}.\mathbf{D}\psi_2 = \mathbf{e}.\mathbf{D}_{12}, \tag{2.3.18}$$

where $\mathbf{D}$ is defined in eqn (2.2.4) and $\mathbf{D}_{12}$ is known as the *transition dipole moment*. This is a real vector for transitions between bound states. With the angle between $\mathbf{e}$ and $\mathbf{D}_{12}$ denoted by θ, an average over the random orientations of the transition dipole moment leads to the replacement of the squared matrix element in eqn (2.3.16) according to

$$X_{12}^2 \to \overline{X_{12}^2} = \overline{\cos^2 \theta} D_{12}^2 = \tfrac{1}{3} D_{12}^2, \tag{2.3.19}$$

where the overbar denotes the average over θ.

The Einstein B coefficient for absorption of light by the gas is now obtained by substitution of eqn (2.3.19) into eqn (2.3.16) and comparison with eqn (2.3.1) as

$$B_{12} = \pi e^2 D_{12}^2 / 3\varepsilon_0 \hbar^2. \tag{2.3.20}$$

The corresponding expression for the Einstein coefficient B_{21} that governs the rate of stimulated emission by the gas is obtained by a similar calculation in which the atom is assumed to be initially in its upper state ψ_2. The result is identical to eqn (2.3.20) if the two levels are nondegenerate, while the connection between the two B coefficients is given by eqn (1.5.12) for degenerate levels.

   The semiclassical method used here gives no prescription for calculating the Einstein A coefficient. The interaction Hamiltonian of eqn (2.2.5) vanishes in the absence of an incident light beam, when the atomic states become stationary. An atom in the upper state ψ_2 thus remains there indefinitely and the spontaneous emission process does not occur. This defect of the semiclassical theory is rectified in the fully-quantized theory of §4.10, where both the electromagnetic field and the atom are treated quantum-mechanically. In the meantime, the correct expression for the A coefficient is found with use of the general relations (1.5.12) and (1.5.13) as

$$A_{21} = \frac{1}{\tau_R} = \frac{\hbar \omega_0^3 g_1}{\pi^2 c^3 g_2} B_{12} = \frac{e^2 \omega_0^3 g_1 D_{12}^2}{3 \pi \varepsilon_0 \hbar c^3 g_2}, \tag{2.3.21}$$

where the radiative lifetime τ_R for the transition is defined in eqn (1.7.12). The A coefficients are readily calculated for states with known wavefunctions.

Problem 2.4   Show that the magnitude of the A coefficient for the transition 2P to 1S in atomic hydrogen is

$$A_{21} = 6 \times 10^8 \text{ s}^{-1}, \tag{2.3.22}$$

noting that the composite absorption rate for the corresponding upwards transition to the threefold degenerate 2P state has three times the value obtained from eqn (2.3.20).

The 2P to 1S transition frequency, given by eqns (2.2.14) and (2.2.15), lies in the ultraviolet region of the spectrum. The spontaneous emission rates for transitions in the visible region tend to be smaller on account of the ω_0^3 dependence of the A coefficient, and typical values are

$$A \approx 5 \times 10^6 \text{ s}^{-1} \quad \text{and} \quad \tau_R \approx 2 \times 10^{-7} \text{ s}, \tag{2.3.23}$$

where a representative visible angular frequency is given in eqn (1.6.7).

2.4 The Dirac delta-function and Fermi's golden rule

The method of calculation of the transition rate in the previous section has a wide applicability, beyond its use in the determination of the Einstein coefficients. The result of the calculation can be cast in a general form that facilitates these wider applications. The present section digresses from the development of the atom–radiation theory to extract the more general expressions for future use.

   Consider the *Dirac delta-function* defined by

$$\delta(\omega_0 - \omega) = \frac{2}{\pi} \underset{t \to \infty}{\text{Lt}} \frac{\sin^2 \{\tfrac{1}{2}(\omega_0 - \omega)t\}}{(\omega_0 - \omega)^2 t}. \tag{2.4.1}$$

According to eqn (2.3.15),

$$\int_{-\infty}^{\infty} d\omega\, \delta(\omega_0 - \omega) = 1. \tag{2.4.2}$$

The delta function is illustrated by the limit of the curve plotted in Fig. 2.2 for infinite t, where all the zeros crowd in to the origin and the value at $\omega = \omega_0$ tends to infinity. All of the area under the curve is concentrated at $\omega = \omega_0$ in the limit and the delta-function is zero everywhere else. Thus eqn (2.4.2) can be written more generally as

$$\int_{\omega_1}^{\omega_2} d\omega\, \delta(\omega_0 - \omega) = 1 \quad \text{provided} \quad \omega_1 < \omega_0 < \omega_2$$

$$= \tfrac{1}{2} \quad \text{if} \quad \omega_0 = \omega_1 \text{ or } \omega_2 \tag{2.4.3}$$

$$= 0 \quad \text{otherwise.}$$

The delta-function is highly singular and it occurs in physical quantities only when the expression includes an integration over the variable of the delta-function, ω in this case.

The definition in eqn (2.4.1) is used to prove the basic property of the delta-function. Let $f(\omega)$ be any function of ω that is non-singular at $\omega = \omega_0$, and consider the integral

$$\int_{\omega_1}^{\omega_2} d\omega\, f(\omega) \delta(\omega_0 - \omega) = \frac{2}{\pi} \operatorname*{Lt}_{t \to \infty} \int_{\omega_1}^{\omega_2} d\omega\, f(\omega) \frac{\sin^2\left\{\tfrac{1}{2}(\omega_0 - \omega)t\right\}}{(\omega_0 - \omega)^2 t}, \tag{2.4.4}$$

where $\omega_1 < \omega_0 < \omega_2$. A change in the variable of integration to $x = (\omega - \omega_0)t$ gives

$$\int_{\omega_1}^{\omega_2} d\omega\, f(\omega) \delta(\omega_0 - \omega) = \frac{2}{\pi} \operatorname*{Lt}_{t \to \infty} \int_{(\omega_1 - \omega_0)t}^{(\omega_2 - \omega_0)t} dx\, f\left(\frac{x}{t} + \omega_0\right) \frac{\sin^2(x/2)}{x^2}$$

$$= \frac{2}{\pi} f(\omega_0) \int_{-\infty}^{\infty} dx\, \frac{\sin^2(x/2)}{x^2} = f(\omega_0). \tag{2.4.5}$$

Inclusion of the delta-function in an integral thus picks out the value of the integrand at the point specified by the constant in the delta-function.

There are many alternative representations for the delta-function in addition to the form of limit given in eqn (2.4.1). The criterion for an acceptable representation is that eqn (2.4.5) should be satisfied, and each representation must be justified by evaluation of the integral on the left-hand side of eqn (2.4.4). Some examples are given in the following problem.

Problem 2.5   Prove that the following are acceptable representations of the Dirac delta-function

$$\delta(\omega_0 - \omega) = \frac{1}{\pi} \underset{\gamma \to 0}{\mathrm{Lt}} \, \mathrm{Im} \frac{1}{\omega_0 - \omega - i\gamma}$$

$$= \frac{1}{\pi} \underset{\gamma \to 0}{\mathrm{Lt}} \frac{\gamma}{(\omega_0 - \omega)^2 + \gamma^2}, \qquad (2.4.6)$$

$$\delta(\omega_0 - \omega) = \underset{\Delta \to 0}{\mathrm{Lt}} \left(2\pi\Delta^2\right)^{-1/2} \exp\left\{-(\omega_0 - \omega)^2 / 2\Delta^2\right\} \quad (2.4.7)$$

and

$$\delta(\omega_0 - \omega) = \underset{T \to \infty}{\mathrm{Lt}} \frac{\sin\{(\omega_0 - \omega)T\}}{\pi(\omega_0 - \omega)}, \qquad (2.4.8)$$

or

$$\delta(\omega_0 - \omega) = \frac{1}{2\pi} \int_{-\infty}^{\infty} dt \exp\{i(\omega_0 - \omega)t\}. \qquad (2.4.9)$$

Some useful manipulative properties of the delta-function are established by examination of the basic integral on the left-hand side of eqn (2.4.4). The most commonly used properties are given in the following problem.

Problem 2.6   Prove the following delta-function properties

$$\delta(\omega - \omega_0) = \delta(\omega_0 - \omega), \qquad (2.4.10)$$

$$\delta(\omega_0 - b\omega) = \frac{1}{|b|} \delta\left(\frac{\omega_0}{b} - \omega\right), \qquad (2.4.11)$$

and

$$\delta\{(\omega_1 - \omega)(\omega_2 - \omega)\} = \frac{\delta(\omega_1 - \omega) + \delta(\omega_2 - \omega)}{|\omega_1 - \omega_2|}, \qquad (2.4.12)$$

where b is a constant and, in each case, the equality means that the two sides of the equation produce the same result when multiplied by some function of ω and integrated.

The expression (2.3.20) for the B coefficient is derived from the limiting value in eqn (2.3.15) of the integral in eqn (2.3.13), which is valid for times t that are much larger than $1/\omega_0$ and $1/\Delta\omega$, the characteristic time spans that control the experimental observation of the absorption of light. In this large-time limit, it is appropriate to introduce the delta-function notation by eqn (2.4.1) and write eqn (2.3.9) in the form of an expression for the transition rate $1/\tau$,

$$1/\tau = |C_2(t)|^2 / t = \tfrac{1}{2} \pi V^2 \delta(\omega_0 - \omega). \qquad (2.4.13)$$

The expression is more compact in this form but it becomes physically significant only after integration over the range $\Delta\omega$ of transition frequencies, as in eqn (2.3.11), which is trivial to perform when eqn (2.4.5) is used.

This transition-rate formula can be applied to any process where there is an interaction Hamiltonian similar to eqn (2.2.5) and a continuous distribution of frequencies ω around the transition frequency ω_0. We note that the effect of the rotating-wave aproximation defined in §2.3 is to select a single term from the cosine in eqn (2.2.8), so that the transition matrix element is essentially

$$\langle 1|\hat{\mathcal{H}}_{\text{ED}}|2\rangle = \tfrac{1}{2}eE_0X_{12}\exp(-i\omega t) = \tfrac{1}{2}\hbar\mathcal{V}\exp(-i\omega t),\qquad(2.4.14)$$

with use of eqn (2.2.10). This replacement for $\mathcal{V}$ in eqn (2.4.13) gives

$$\frac{1}{\tau} = \frac{2\pi}{\hbar^2}\sum_f \left|\langle f|\hat{\mathcal{H}}_{\text{ED}}|i\rangle\right|^2\delta\!\left(\omega_f - \omega_i\right),\qquad(2.4.15)$$

where the formula is generalized to apply to the transition between an initial state $|i\rangle$ and a range of final states $|f\rangle$, with energies $\hbar\omega_i$ and $\hbar\omega_f$ respectively. The transition-rate formula of eqn (2.4.15) is known as *Fermi's golden rule*. It is again emphasized that the highly-singular delta-function can be physically meaningful only when its variable is integrated, and the summation over f should strictly be interpreted as an integral over a continuous range of final states $|f\rangle$. Alternatively, the rule takes the form

$$\frac{1}{\tau} = \frac{2\pi}{\hbar^2}\int d\omega_f \left|\langle f|\hat{\mathcal{H}}_{\text{ED}}|i\rangle\right|^2\delta\!\left(\omega_f - \omega_i\right)\qquad(2.4.16)$$

for transitions between a discrete initial state $|i\rangle$ and a continuous range of final states $|f\rangle$ with delta-function normalization

$$\langle f|f'\rangle = \delta\!\left(\omega_f - \omega_{f'}\right).\qquad(2.4.17)$$

An example of such states is given in §6.2.

2.5 Radiative broadening and linear susceptibility

The basic equations of motion for the atoms obtained from the time-dependent quantum mechanics of §§2.1 and 2.2 can be extended by inclusion of the effects of spontaneous emission. The more general theory provides an expression for the susceptibility of a gas of two-level atoms in the presence of radiative damping. The relation between susceptibility and attenuation coefficient then provides an expression for the radiatively-broadened absorption lineshape.

Consider again the perturbation of an atom by an electromagnetic wave whose frequency ω lies in the vicinity of a transition frequency ω_0. The interaction Hamiltonian given by eqn (2.2.5) represents the perturbation produced by the applied electric field

$$E(t) = E_0 \cos(\omega t) = \tfrac{1}{2} E_0 \{\exp(-i\omega t) + \exp(i\omega t)\}. \tag{2.5.1}$$

Application of this field to the gas produces a polarization $P(t)$ proportional to the electric field. The proportionality is determined by the linear susceptibility $\chi(\omega)$, such that the two-component field of eqn (2.5.1) generates a polarization

$$P(t) = \tfrac{1}{2} \varepsilon_0 E_0 \{\chi(\omega)\exp(-i\omega t) + \chi(-\omega)\exp(i\omega t)\}. \tag{2.5.2}$$

The method of calculation is to compute $P(t)$ quantum-mechanically and then to determine $\chi(\omega)$ by comparison of the resulting expression with eqn (2.5.2).

Suppose that the applied electric field points in the x direction, as in Fig. 2.1. The induced electric-dipole moment of the atom parallel to x is determined by the expectation value of X, the x component of $\mathbf{D}$ used in eqn (2.2.9). Thus at time t, when the atomic wavefunction is $\Psi(t)$, the electric-dipole moment is

$$d(t) = -\int dV \Psi^*(t) eX \Psi(t), \tag{2.5.3}$$

where the integration runs over all the coordinates of the Z electrons. The form of atomic wavefunction obtained from eqns (2.1.5) and (2.1.8) is

$$\Psi(t) = C_1(t)\psi_1 \exp(-iE_1 t/\hbar) + C_2(t)\psi_2 \exp(-iE_2 t/\hbar), \tag{2.5.4}$$

where the spatial dependence of the wavefunctions is suppressed. Substitution of eqn (2.5.4) into eqn (2.5.3) yields

$$d(t) = -e\{C_1^* C_2 X_{12} \exp(-i\omega_0 t) + C_2^* C_1 X_{21} \exp(i\omega_0 t)\}, \tag{2.5.5}$$

where X_{12} is defined in eqn (2.2.9) and the property

$$X_{11} = X_{22} = 0, \tag{2.5.6}$$

equivalent to eqn (2.2.6), is used. From the definition of X_{12},

$$X_{21} = X_{12}^* = X_{12} \tag{2.5.7}$$

for the real wavefunctions assumed here. Thus the dipole moment is also real, as expected physically.

The coefficients C_1 and C_2 needed for evaluation of eqn (2.5.5) are obtained by solution of their equations of motion (2.2.11) and (2.2.12). However, in the absence of any applied electromagnetic field, when $\mathcal{V} = 0$, these equations lead to constant values of C_1 and C_2, whatever the initial state of the atom. This is contrary to the requirement that an initially excited atom must decay to its ground state by spontaneous emission, with a mean radiative lifetime given by eqn (2.3.21). The deficiency is rectified by the addition of an extra term to the left-hand side of eqn (2.2.12) to represent the spontaneous emission. We take

$$\mathcal{V} \cos(\omega t)\exp(i\omega_0 t)C_1 - i\gamma_{sp} C_2 = i\, dC_2/dt \tag{2.5.8}$$

as the augmented equation of motion. In the absence of an applied electric field it can be integrated at once to give

$$C_2(t) = C_2(0)\exp(-\gamma_{sp}t).$$ (2.5.9)

For a gas of N similar atoms, the number excited at time t is given by eqns (2.3.17) and (2.5.9) to be

$$N_2(t) = N_2(0)\exp(-2\gamma_{sp}t).$$ (2.5.10)

Comparison with eqn (1.7.11) shows that the introduction of the additional term in eqn (2.5.8) gives the correct radiative decay rate if

$$2\gamma_{sp} = A_{21} = \frac{1}{\tau_R} = \frac{e^2\omega_0^3 D_{12}^2}{3\pi\varepsilon_0\hbar c^3},$$ (2.5.11)

where the form of the A coefficient for nondegenerate levels is taken from eqn (2.3.21). The other equation of motion, (2.2.11), is not needed for the present calculation and its modification to accommodate spontaneous emission is mentioned in §2.8.

The linear susceptibility is found by substitution of the solutions for the C coefficients into eqn (2.5.5). The method is similar to the determination of the Einstein B coefficient, in that solutions are needed for C_1 and C_2 correct to first order in $\mathcal{V}$ or E_0. Integration of eqn (2.5.8) with C_1 set equal to unity on the left-hand side leads to a solution for C_2 correct to first order in $\mathcal{V}$. Unlike the B coefficient, which refers to a nonequilibrium transfer of energy from radiation field to atom, the susceptibility refers to conditions in which the atoms have been subjected to the electric field for a sufficiently long time to achieve equilibrium with the radiation. It is therefore appropriate to take an indefinite integral of eqn (2.5.8), with the result

$$C_2(t) = -\tfrac{1}{2}\mathcal{V}\left\{\frac{\exp[i(\omega_0 + \omega)t]}{\omega_0 + \omega - i\gamma_{sp}} + \frac{\exp[i(\omega_0 - \omega)t]}{\omega_0 - \omega - i\gamma_{sp}}\right\}.$$ (2.5.12)

The normalization condition of eqn (2.1.9) shows that $C_1(t)$ differs from unity by a term of order $\mathcal{V}^2$. Thus to first order in $\mathcal{V}$, we must take

$$C_1(t) = 1.$$ (2.5.13)

The single-atom dipole moment is found by substitution of eqns (2.5.12) and (2.5.13) into eqn (2.5.5),

$$d(t) = \frac{e^2 X_{12}^2 E_0}{2\hbar}$$

$$\times\left\{\frac{\exp(i\omega t)}{\omega_0 + \omega - i\gamma_{sp}} + \frac{\exp(-i\omega t)}{\omega_0 - \omega - i\gamma_{sp}} + \frac{\exp(-i\omega t)}{\omega_0 + \omega + i\gamma_{sp}} + \frac{\exp(i\omega t)}{\omega_0 - \omega + i\gamma_{sp}}\right\},$$ (2.5.14)

where eqn (2.2.10) has also been used. The dipole moment of the single atom must now be related to the polarization of the gas. The random atomic orientations are taken into account by the same replacement (2.3.19) as made in the calculation of the B coefficient. Then $d(t)$ given by the modified eqn (2.5.14) represents the averaged dipole moment per atom at time t and the macroscopic polarization of the gas is

$$P(t) = Nd(t)/V, \tag{2.5.15}$$

where the justification for simple multiplication by N is discussed in §8.1.The linear susceptibility obtained by comparison with eqn (2.5.2) is

$$\chi(\omega) = \frac{Ne^2 D_{12}^2}{3\varepsilon_0 \hbar V} \left\{ \frac{1}{\omega_0 - \omega - i\gamma_{sp}} + \frac{1}{\omega_0 + \omega + i\gamma_{sp}} \right\}$$

$$= \frac{2\pi Nc^3}{\omega_0^3 V} \left\{ \frac{\gamma_{sp}}{\omega_0 - \omega - i\gamma_{sp}} + \frac{\gamma_{sp}}{\omega_0 + \omega + i\gamma_{sp}} \right\}, \tag{2.5.16}$$

where eqns (2.3.21) and (2.5.11) are used. The susceptibility at frequency $-\omega$ is given by a similar expression that satisfies

$$\chi(-\omega) = \chi^*(\omega). \tag{2.5.17}$$

The attenuation coefficient of the gas is related to the imaginary part of the susceptibility by

$$K(\omega) = \omega \chi''(\omega)/c\eta(\omega), \tag{2.5.18}$$

according to eqns (1.8.3) to (1.8.5). We consider a gas that is sufficiently dilute for the refractive index $\eta(\omega)$ to be close to unity at all frequencies. It is clear that the first term in the bracket of eqn (2.5.16) greatly exceeds the second term when ω is close to ω_0 and when γ_{sp} is very much smaller than ω_0. We therefore neglect the second term, thereby making once more the rotating-wave approximation introduced in eqn (2.3.9). Substitution of eqn (2.5.16) into eqn (2.5.18) then gives

$$K(\omega) = \frac{\pi Ne^2 \omega_0 D_{12}^2}{3\varepsilon_0 \hbar c V} \frac{\gamma_{sp}/\pi}{(\omega_0 - \omega)^2 + \gamma_{sp}^2} = \frac{2\pi^2 c^2 \gamma_{sp} N}{\omega_0^2 V} \frac{\gamma_{sp}/\pi}{(\omega_0 - \omega)^2 + \gamma_{sp}^2}, \tag{2.5.19}$$

where the narrowness of the absorption line allows ω to be replaced by ω_0 except in the final factor.

The frequency dependence of the attenuation coefficient is thus determined by a function of the form

$$F_L(\omega) = \frac{\gamma/\pi}{(\omega_0 - \omega)^2 + \gamma^2}, \tag{2.5.20}$$

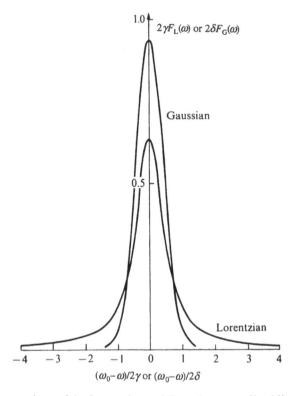

Fig. 2.4. Comparison of the Lorentzian and Gaussian normalized lineshapes. The functions are shown with the same width, equal to one unit on the horizontal scale.

known as the *Lorentzian lineshape* and illustrated in Fig. 2.4. This is the same function as used in the delta-function representation of eqn (2.4.6). The constant in the numerator is chosen to produce the normalization

$$\int d\omega F_L(\omega) = 1. \tag{2.5.21}$$

The full width of the Lorentzian line at half its maximum height (FWHM) is 2γ, and the radiative broadening produces a linewidth $2\gamma_{sp}$ equal to the spontaneous emission A coefficient for the transition. Thus the radiative linewidth of the 1S to 2P transition in hydrogen is obtained from eqn (2.3.22) as about 10^8 Hz in frequency (*not* angular frequency) units.

Radiative linewidths are generally very small and, in most experiments, the observed breadths of the atomic absorption lines are caused by other mechanisms, for example, the Doppler effect or atomic collisions, as discussed in §§2.6 and 2.9. However, these additional line-broadening contributions can always be reduced by methods such as cooling the gas or lowering its pressure. On the other hand, it is impossible to reduce the spontaneous emission rate of an atom in free space and the appropriate A coefficient gives the minimum value that can be achieved. This spontaneous emission linewidth is accordingly called the *natural width* of the spectral line.

The Lorentzian lineshape generally occurs for *homogeneous* broadening mechanisms, where each absorbing or emitting atom is identical to the rest. Thus in the radiative process, and in the collision broadening treated in §2.9, there is no experimental way of associating light of a certain frequency with a particular group of atoms. The width $\Delta\omega$ in these cases is a consequence of the finite mean length of time Δt for which the atomic state is undisturbed. According to the time–energy uncertainty relation [2], or the properties of Fourier transforms,

$$\Delta\omega\Delta t \geq 1, \tag{2.5.22}$$

in agreement with the explicit result for radiative broadening.

Problem 2.7   It is sometimes difficult to measure the width of a spectral line in the presence of background absorption from other transitions, and it may be easier to measure the separation between the two points of inflection in the lineshape. By what factor must this separation be multiplied to obtain the full width of the Lorentzian line at half maximum height?

2.6 Doppler broadening and composite lineshape

The atoms in a gas have a spread in their velocities that leads via the Doppler effect to an associated distribution in the frequencies at which they can absorb or emit light. Consider the absorption of light of frequency ω by an atom that is initially in its lower energy level E_1 and is moving with velocity $\mathbf{v}_1$. Absorption of a photon of energy $\hbar\omega$ promotes the atom into its excited energy level E_2. The photon carries a momentum $\hbar\mathbf{k}$, where $k = \omega/c$, and its absorption causes a recoil of the atom to a new velocity $\mathbf{v}_2$. The total momenta before and after the absorption must be equal,

$$M\mathbf{v}_1 + \hbar\mathbf{k} = M\mathbf{v}_2, \tag{2.6.1}$$

where M is the atomic mass. The energy conservation condition is

$$E_1 + \tfrac{1}{2}M v_1^2 + \hbar\omega = E_2 + \tfrac{1}{2}M v_2^2. \tag{2.6.2}$$

With the radiative broadening ignored for the present, a stationary atom can absorb only light of frequency ω_0 given by

$$\hbar\omega_0 = E_2 - E_1. \tag{2.6.3}$$

Elimination of $\mathbf{v}_2$ from eqn (2.6.2) with the help of eqns (2.6.1) and (2.6.3) gives

$$\hbar\omega_0 = \hbar\omega - \hbar\mathbf{v}_1.\mathbf{k} - \left(\hbar^2 k^2/2M\right), \tag{2.6.4}$$

and, if the incident light propagates parallel to the z axis,

$$\omega_0 = \omega - \left(\omega v_{1z}/c\right) - \left(\hbar\omega^2/2Mc^2\right). \tag{2.6.5}$$

Typical orders of magnitude for some quantities on the right-hand side of eqn (2.6.5) are

$$v_{1z}/c \approx 10^{-5} \quad \text{and} \quad \hbar\omega/2Mc^2 \approx 10^{-9}. \tag{2.6.6}$$

Thus ω differs only slightly from ω_0 and the final term on the right of eqn (2.6.5) can be neglected altogether, so that

$$\omega = \frac{\omega_0}{1 - \left(v_{1z}/c\right)} \approx \omega_0\left(1 + \frac{v_{1z}}{c}\right). \tag{2.6.7}$$

The light is absorbed only when its frequency ω differs from ω_0 by the Doppler shift appropriate to the initial atomic velocity.

It follows that the distribution of absorbed frequencies mirrors the distribution of atomic velocities in the gas. According to the Maxwellian velocity distribution [4], the relative probability that an atom in a gas at temperature T has the z component of its velocity between v_z and $v_z + \mathrm{d}v_z$ is

$$\exp\left\{-Mv_z^2/2k_{\mathrm{B}}T\right\}\mathrm{d}v_z = \exp\left\{-Mc^2\left(\omega_0 - \omega\right)^2/2\omega_0^2 k_{\mathrm{B}}T\right\}\left(c/\omega_0\right)\mathrm{d}\omega, \tag{2.6.8}$$

where eqn (2.6.7) is used. The expression on the right is the frequency distribution of the absorbed light, known as the *Gaussian lineshape*. The peak of the line is at $\omega = \omega_0$ and half its maximum strength occurs at frequencies that satisfy

$$\tfrac{1}{2} = \exp\left\{-Mc^2\left(\omega_0 - \omega\right)^2/2\omega_0^2 k_{\mathrm{B}}T\right\}. \tag{2.6.9}$$

The full width of the Doppler-broadened line at half its maximum height, or FWHM, is thus

$$2\delta = 2\omega_0\left(2k_{\mathrm{B}}T\ln 2/Mc^2\right)^{1/2}. \tag{2.6.10}$$

It is convenient to define the normalized form of the Gaussian lineshape function as

$$F_{\mathrm{G}}(\omega) = \left(2\pi\Delta^2\right)^{-1/2}\exp\left\{-\left(\omega_0 - \omega\right)^2/2\Delta^2\right\}, \tag{2.6.11}$$

with the property

$$\int \mathrm{d}\omega F_{\mathrm{G}}(\omega) = 1. \tag{2.6.12}$$

This is the same function as used in the delta-function representation of eqn

(2.4.7). The quantity Δ^2 is the variance of the Gaussian distribution, with

$$\Delta = \omega_0 \left(k_B T / Mc^2 \right)^{1/2} = \delta / (2 \ln 2)^{1/2} \approx \delta / 1.18. \qquad (2.6.13)$$

The Gaussian lineshape is illustrated in Fig. 2.4 together with the Lorentzian lineshape. Both lines are plotted with the same FWHM, that is $2\delta = 2\gamma$, and they enclose equal areas. It is seen that the Gaussian line is the more sharply peaked and it falls off very rapidly away from the central region. In contrast, the Lorentzian line has tails that extend some way from the region of the peak.

Problem 2.8   Determine the same multiplication factor for the Gaussian lineshape as considered in Problem 2.7 for the Lorentzian lineshape.

The Gaussian lineshape often occurs for *inhomogeneous* broadening mechanisms, where different atoms absorb or emit light at different frequencies because of a statistical spread in some parameter that determines the transition frequency. The atomic velocity is the relevant statistical parameter in Doppler broadening. The atoms that absorb at a particular frequency are, in principle, distinguishable from the rest by their component of velocity parallel to the absorbed light beam. Another example occurs for absorption by atoms embedded in crystals, where variations in the local strain can produce shifts in atomic transition frequency.

When the absorption frequencies are significantly affected by two or more line-broadening processes, it is necessary to determine the resulting composite lineshape. Consider the combination of two line-broadening mechanisms that individually generate normalized lineshape functions $F_1(\omega)$ and $F_2(\omega)$. The composite lineshape is

$$F(\omega) = \int_{-\infty}^{\infty} dv F_1(v) F_2(\omega + \omega_0 - v), \qquad (2.6.14)$$

where ω_0 is the common central frequency of the two distributions. In words, the integration associates with each frequency component in the shape F_1 a broadened distribution appropriate to the mechanism that generates the shape F_2. The integral is invariant under interchange of F_1 and F_2.

Any number of line-broadening mechanisms can be combined by repeated applications of eqn (2.6.14) and the final shape is independent of the order in which the contributions are combined. Specific examples of combined lineshapes are easier to evaluate in the temporal domain instead of the frequency domain of eqn (2.6.14), and the topic is reconsidered in §3.5. However, we give here the combination of the Lorentzian lineshape of eqn (2.5.20) with the Gaussian lineshape function of eqn (2.6.11), where

$$F(\omega) = \frac{\gamma}{2^{1/2} \pi^{3/2} \Delta} \int_{-\infty}^{\infty} dv \frac{\exp\left\{ -(v - \omega)^2 / 2\Delta^2 \right\}}{(\omega_0 - v)^2 + \gamma^2} = \frac{1}{(2\pi)^{1/2} \Delta} \mathrm{Re}\, w \left(\frac{\omega_0 - \omega + i\gamma}{2^{1/2} \Delta} \right),$$

$$(2.6.15)$$

and w is a form of complex error function [5]. There is no simple analytic expression for the linewidth in this case. The shape is named after Voigt and it is intermediate between the Lorentzian and Gaussian lineshapes, to which it reduces in the limits $\Delta \to 0$ and $\gamma \to 0$, respectively.

2.7 The optical Bloch equations

Equations (2.2.11) and (2.2.12), together with the definition of the coefficients in eqn (2.1.8), provide an exact description of the state of a two-level atom in interaction with an oscillatory electromagnetic field. The calculations in §2.3 are concerned with the solutions of these equations for a field that contains a smooth distribution of oscillation frequencies in the vicinity of the atomic transition. The solution is approximate in that only the terms of low order in $\mathcal{V}$ or E_0 are retained and the rotating-wave approximation is made in eqn (2.3.9).

We now look for more general solutions of eqns (2.2.11) and (2.2.12). The rotating-wave approximation is again made but we retain terms of all orders in $\mathcal{V}$ or E_0. Furthermore, the incident light is assumed to be monochromatic with an oscillatory electric field at the single frequency ω. The effect of a distribution of frequencies ω can then be found by an average of the results for monochromatic light, as is demonstrated in §§2.8 and 2.10.

In calculating the optical excitation of the atom, the quantity of interest is $|C_2|^2$ rather than the complex coefficient C_2. More generally, the expectation values of any observables for the atom in the state in eqn (2.1.8) depend upon products of the coefficients with complex conjugate coefficients. It is accordingly convenient to work directly with the bilinear products of the coefficients, and the four elements of the atomic *density matrix* ρ_{ij} are defined by

$$\rho_{11} = |C_1|^2 = N_1/N, \quad \rho_{22} = |C_2|^2 = N_2/N, \tag{2.7.1}$$

$$\rho_{12} = C_1 C_2^* \quad \text{and} \quad \rho_{21} = C_2 C_1^* = \rho_{12}^*. \tag{2.7.2}$$

The off-diagonal elements, or *coherences*, in eqn (2.7.2) are generally complex but the diagonal elements, or *populations*, are clearly real and from eqn (2.1.9) they satisfy

$$\rho_{11} + \rho_{22} = 1. \tag{2.7.3}$$

For a collection of N atoms, the relations of the diagonal elements in eqn (2.7.1) with the average numbers in the two levels are obtained from eqn (2.3.17).

Equations of motion for the density matrix elements are easily found. Thus

$$d\rho_{ij}/dt = C_i \left(dC_j^*/dt \right) + \left(dC_i/dt \right) C_j^*, \tag{2.7.4}$$

and with substitution from eqns (2.2.11) and (2.2.12), it follows that

$$d\rho_{22}/dt = -d\rho_{11}/dt = -i\mathcal{V}\cos(\omega t)\left\{ \exp(i\omega_0 t)\rho_{12} - \exp(-i\omega_0 t)\rho_{21} \right\} \tag{2.7.5}$$

and

$$dp_{12}/dt = dp_{21}^*/dt = i\mathcal{V}\cos(\omega t)\exp(-i\omega_0 t)(\rho_{11} - \rho_{22}).$$ (2.7.6)

These are exact equations for the density matrix elements.

The solution of the equations is greatly facilitated if the rotating-wave approximation is made. With the same justification as explained in connection with eqns (2.3.4) and (2.3.7), the effects of the terms that oscillate at frequency $\omega_0 + \omega$ are negligible compared to the effects of the terms that oscillate at frequency $\omega_0 - \omega$, when ω is close to ω_0. Removal of the rapidly-oscillating terms from eqns (2.7.5) and (2.7.6) produces the equations of motion

$$dp_{22}/dt = -dp_{11}/dt = -\tfrac{1}{2}i\mathcal{V}\left\{\exp\left[i(\omega_0 - \omega)t\right]\rho_{12} - \exp\left[-i(\omega_0 - \omega)t\right]\rho_{21}\right\}$$ (2.7.7)

and

$$dp_{12}/dt = dp_{21}^*/dt = \tfrac{1}{2}i\mathcal{V}\exp\left\{-i(\omega_0 - \omega)t\right\}(\rho_{11} - \rho_{22}).$$ (2.7.8)

These are known as the *optical Bloch equations*. They are similar to equations derived by Bloch to describe the motion of a spin in an oscillatory magnetic field. The quantum mechanics of the two-level atom considered here is formally identical to that of a spin system and there are many analogies between the influences of oscillatory fields on the two systems [6].

The optical Bloch equations can be solved without further approximation but, before doing so, it is convenient to remove the oscillatory factors by means of the substitutions

$$\tilde{\rho}_{12} = \exp\left\{i(\omega_0 - \omega)t\right\}\rho_{12} \quad \text{and} \quad \tilde{\rho}_{21} = \exp\left\{-i(\omega_0 - \omega)t\right\}\rho_{21}.$$ (2.7.9)

Eqns (2.7.7) and (2.7.8) then become

$$d\tilde{\rho}_{22}/dt = -d\tilde{\rho}_{11}/dt = -\tfrac{1}{2}i\mathcal{V}(\tilde{\rho}_{12} - \tilde{\rho}_{21})$$ (2.7.10)

and

$$d\tilde{\rho}_{12}/dt = d\tilde{\rho}_{21}^*/dt = \tfrac{1}{2}i\mathcal{V}(\tilde{\rho}_{11} - \tilde{\rho}_{22}) + i(\omega_0 - \omega)\tilde{\rho}_{12},$$ (2.7.11)

where $\tilde{\rho}_{ii} = \rho_{ii}$, the tildes on the diagonal elements being introduced for notational uniformity.

We take trial solutions of the form

$$\tilde{\rho}_{ij}(t) = \tilde{\rho}_{ij}(0)\exp(\lambda t).$$ (2.7.12)

The four equations that result from substitution into eqns (2.7.10) and (2.7.11) are written in matrix form as

$$
\begin{bmatrix}
-\lambda & 0 & \tfrac{1}{2}i\mathcal{V} & -\tfrac{1}{2}i\mathcal{V} \\
0 & -\lambda & -\tfrac{1}{2}i\mathcal{V} & \tfrac{1}{2}i\mathcal{V} \\
\tfrac{1}{2}i\mathcal{V} & -\tfrac{1}{2}i\mathcal{V} & i(\omega_0-\omega)-\lambda & 0 \\
-\tfrac{1}{2}i\mathcal{V} & \tfrac{1}{2}i\mathcal{V} & 0 & -i(\omega_0-\omega)-\lambda
\end{bmatrix}
\begin{bmatrix}
\tilde{\rho}_{11}(0) \\
\tilde{\rho}_{22}(0) \\
\tilde{\rho}_{12}(0) \\
\tilde{\rho}_{21}(0)
\end{bmatrix} = 0. \qquad (2.7.13)
$$

The solutions for λ are obtained from the 4×4 determinant of the coefficients, which yields the quartic equation

$$
\lambda^2\left\{\lambda^2 + (\omega_0-\omega)^2 + \mathcal{V}^2\right\} = 0. \qquad (2.7.14)
$$

The distinct roots are

$$
\lambda_1 = 0, \quad \lambda_2 = i\Omega \quad \text{and} \quad \lambda_3 = -i\Omega, \qquad (2.7.15)
$$

where

$$
\Omega = \left\{(\omega_0-\omega)^2 + \mathcal{V}^2\right\}^{1/2}. \qquad (2.7.16)
$$

The general solultions for the density matrix elements thus have the forms

$$
\tilde{\rho}_{ij}(t) = \tilde{\rho}_{ij}^{(1)} + \tilde{\rho}_{ij}^{(2)}\exp(i\Omega t) + \tilde{\rho}_{ij}^{(3)}\exp(-i\Omega t), \qquad (2.7.17)
$$

where the values of the coefficients are determined by the initial values of the density matrix elements for a given problem and by the initial values of their first and second derivatives obtained from the optical Bloch equations.

The presence of $\mathcal{V}^2$ in the roots λ_2 and λ_3 produces changes in the frequencies of the coupled system of atom and light beam from their uncoupled values of ω_0 and ω respectively. As $\mathcal{V}$ is proportional to the electric-field amplitude of the light, the shifts and splittings of the coupled frequencies are known as the *dynamic Stark effect*, by analogy with the shifts and splittings of atomic energy levels produced by the application of a static electric field. The dynamic Stark splitting produces important observable effects in resonance fluorescence spectra, as described in §8.4.

The solutions of the optical Bloch equations are quite lengthy for arbitrary initial conditions and we therefore limit the discussion to a simple special case.

Problem 2.9 Show that solution of eqns (2.7.10) and (2.7.11) for the initial conditions

$$
\tilde{\rho}_{22}(0) = 0 \quad \text{and} \quad \tilde{\rho}_{12}(0) = 0 \qquad (2.7.18)
$$

gives

$$
\tilde{\rho}_{22}(t) = (\mathcal{V}/\Omega)^2 \sin^2\left(\tfrac{1}{2}\Omega t\right) \qquad (2.7.19)
$$

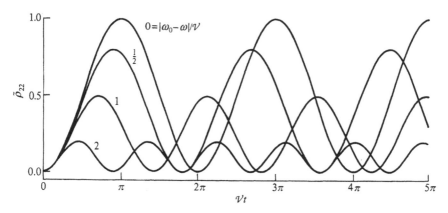

Fig. 2.5. Time dependences of the degree of atomic excitation $\tilde{\rho}_{22}$ for the ratios of detuning $|\omega_0 - \omega|$ to Rabi frequency $\mathcal{V}$ indicated against the curves.

and

$$\tilde{\rho}_{12}(t) = \left(\mathcal{V}/\Omega^2\right)\sin\left(\tfrac{1}{2}\Omega t\right)\left\{-(\omega_0 - \omega)\sin\left(\tfrac{1}{2}\Omega t\right) + i\Omega\cos\left(\tfrac{1}{2}\Omega t\right)\right\}.$$
(2.7.20)

The time dependence of the degree of excitation $\tilde{\rho}_{22}$ of the atom given by eqn (2.7.19) is illustrated in Fig. 2.5 for various values of the ratio of the detuning $|\omega_0 - \omega|$ of the incident light from the atomic transition to the transition matrix element $\mathcal{V}$. The degree of excitation oscillates between 0 and $(\mathcal{V}/\Omega)^2$ and, for detunings less than $\mathcal{V}$, the atom has a probability of excitation into its upper level greater than 0.5 for some values of t. This behaviour contrasts with that under broad-band illumination where, as discussed in §1.7, the excitation probability is always smaller than 0.5. For zero detuning, with $\omega_0 = \omega$, eqn (2.7.19) reduces to

$$\tilde{\rho}_{22}(t) = \sin^2\left(\tfrac{1}{2}\mathcal{V}t\right)$$
(2.7.21)

and the atom oscillates symmetrically between its ground and excited states. These are known as *Rabi oscillations* after the originator of similar solutions that occur in the problem of a spin subjected to an oscillatory magnetic field [7]. The oscillation frequency $\mathcal{V}$, obtained from eqns (2.2.10) and (2.3.18) as

$$\mathcal{V} = eE_0 X_{12}/\hbar = eE_0(\mathbf{e}.\mathbf{D}_{12})/\hbar,$$
(2.7.22)

is called the *Rabi frequency*.

It follows from the definitions of the density matrix elements in eqns (2.7.1) and (2.7.2) that

$$\rho_{11}\rho_{22} = \rho_{12}\rho_{21} = |\rho_{12}|^2,$$
(2.7.23)

and this relation is unaffected by the substitutions in eqn (2.7.9). It is clearly satisfied by the explicit solutions (2.7.19) and (2.7.20). However, the theory in the present section is valid only for two-level atoms in so-called *pure states* of the linear-superposition form given in eqn (2.1.8). The above solutions are also restricted to a monochromatic incident beam of light with a single frequency ω of oscillation. In practice, the sinusoidal behaviours of $\tilde{\rho}_{22}$ and $\tilde{\rho}_{12}$ can be observed experimentally with incident light whose frequency spread is small in comparison with the breadth of the distribution of atomic transition frequencies around ω_0. However, the processes that produce the broadening of the atomic transitions also modify the optical Bloch equations; they change the time dependences of the $\tilde{\rho}_{ij}$ and they remove the simple relation (2.7.23) between the different elements. These changes are considered in the following two sections.

2.8 Power broadening

The linear susceptibility of eqn (2.5.16) describes the response of the atoms correct to first order in the electric field E_0 of the incident light or in the Rabi frequency $\mathcal{V}$. Higher-order terms in the susceptibility are obtained by solution of the optical Bloch equations (2.7.10) and (2.7.11), but it is first necessary to generalize these equations by inclusion of the effects of spontaneous emission.

The Bloch equations with spontaneous emission included are obtained, as in eqns (2.7.4) to (2.7.11), from the equations of motion for the coefficients C_1 and C_2. However, eqn (2.5.8) rather than eqn (2.2.12) is now used for the rate of change of C_2, while eqn (2.2.11) is retained for C_1. The resulting equations are

$$d\tilde{\rho}_{22}/dt = -d\tilde{\rho}_{11}/dt = -\tfrac{1}{2}i\mathcal{V}\left(\tilde{\rho}_{12} - \tilde{\rho}_{21}\right) - 2\gamma_{\text{sp}}\tilde{\rho}_{22} \qquad (2.8.1)$$

and

$$d\tilde{\rho}_{12}/dt = d\tilde{\rho}_{21}^{*}/dt = \tfrac{1}{2}i\mathcal{V}\left(\tilde{\rho}_{11} - \tilde{\rho}_{22}\right) + \left[i\left(\omega_0 - \omega\right) - \gamma_{\text{sp}}\right]\tilde{\rho}_{12}. \qquad (2.8.2)$$

It is evident that this procedure for introducing γ_{sp} is not entirely justified because eqn (2.2.11) ought also to be modified by spontaneous emission. Indeed, the complete modification of eqn (2.2.11) needed to incorporate the damping is quite complicated and further changes to eqn (2.2.12) are required, beyond the form given in eqn (2.5.8) [8]. However, the forms of the optical Bloch equations (2.8.1) and (2.8.2) are the same as obtained by more rigorous derivations [9] and we do not give the modified forms of the fundamental equations for C_1 and C_2.

Spontaneous emission thus introduces damping terms into the optical Bloch equations. The solutions are no longer purely oscillatory, as in eqns (2.7.19) and (2.7.20), but the system now settles down into a steady state after a sufficiently long time has elapsed. With the time derivatives in eqns (2.8.1) and (2.8.2) all set equal to zero, the simultaneous equations for the steady-state density-matrix elements are readily solved to give

$$\tilde{\rho}_{22}(\infty) = \frac{\mathcal{V}^2/4}{\left(\omega_0 - \omega\right)^2 + \gamma_{\text{sp}}^2 + \left(\mathcal{V}^2/2\right)} \qquad (2.8.3)$$

and

$$\tilde{\rho}_{12}(\infty) = -\frac{(\mathcal{V}/2)\big(\omega_0 - \omega - i\gamma_{sp}\big)}{\big(\omega_0 - \omega\big)^2 + \gamma_{sp}^2 + \big(\mathcal{V}^2/2\big)}. \tag{2.8.4}$$

The presence of the term in $\mathcal{V}^2$ in the denominator of the excited-state population $\tilde{\rho}_{22}$ leads to saturation effects similar to those considered in §1.7. Thus for very intense monochromatic incident light, $\tilde{\rho}_{22}$ in eqn (2.8.3) tends to 1/2, the same value as obtained from eqn (1.7.4) for very intense broad-band incident light.

Consider the effect of these more general solutions for the steady-state behaviour on the expression for the susceptibility. The atomic dipole moment of eqn (2.5.5) is expressed in terms of density matrix elements with the use of eqns (2.7.2) and (2.7.9) as

$$d(t) = -e\big\{\tilde{\rho}_{21}X_{12}\exp(-i\omega t) + \tilde{\rho}_{12}X_{21}\exp(i\omega t)\big\}. \tag{2.8.5}$$

Thus with substitution of eqn (2.8.4) and its complex conjugate, the same procedure as in eqns (2.5.14) to (2.5.16) produces a susceptibility

$$\chi(\omega) = \frac{2\pi N c^3}{\omega_0^3 V} \frac{\gamma_{sp}\big(\omega_0 - \omega + i\gamma_{sp}\big)}{\big(\omega_0 - \omega\big)^2 + \gamma_{sp}^2 + \big(\mathcal{V}^2/2\big)}. \tag{2.8.6}$$

This is no longer a *linear* susceptibility because the field strength E_0 is contained in the quantity $\mathcal{V}$ that appears in the denominator. The expression (2.8.6) reduces to that obtained from the first term in the bracket of the second form of eqn (2.5.16) if $\mathcal{V}^2$ is neglected, but the susceptibility $\chi(\omega)$ otherwise contains additional contributions in all even powers of E_0. These contributions are related to successive components of the nonlinear susceptibility [10], which control the strengths of the varieties of higher-order process that occur in nonlinear optics, treated in Chapter 9.

The term in $\mathcal{V}^2$ in the denominator of eqn (2.8.6) reduces the rate of attenuation for very intense monochromatic incident light, similar to the saturation effect on the attenuation of broad-band incident light discussed in §1.8. The effect is equivalent to an increase in the effective linewidth of the transition from $2\gamma_{sp}$ to

$$\text{FWHM} = 2\big(\gamma_{sp}^2 + \tfrac{1}{2}\mathcal{V}^2\big)^{1/2}. \tag{2.8.7}$$

The additional contribution to the linewidth is known as *power* or *saturation broadening*.

The steady-state solutions in eqns (2.8.3) and (2.8.4) are independent of the initial conditions. For the variations of the atomic populations at shorter times, it is necessary to solve the Bloch equations (2.8.1) and (2.8.2) more generally, with insertion of the initial conditions. It is unfortunately not possible to write down the general explicit solutions of these equations and we therefore illustrate their forms by two special cases. Consider first the case of zero detuning with $\omega_0 = \omega$ and

$2\mathcal{V} \geq \gamma_{sp}$. The method of solution of eqns (2.8.1) and (2.8.2) is similar to that outlined in eqns (2.7.12) to (2.7.17), where the roots of the 4×4 determinant of the coefficients in the trial solution are now

$$\lambda_1 = 0, \quad \lambda_2 = -\gamma_{sp}, \quad \lambda_3 = -\tfrac{3}{2}\gamma_{sp} + i\lambda \quad \text{and} \quad \lambda_4 = -\tfrac{3}{2}\gamma_{sp} - i\lambda, \qquad (2.8.8)$$

and

$$\lambda = \left(\mathcal{V}^2 - \tfrac{1}{4}\gamma_{sp}^2\right)^{1/2}. \qquad (2.8.9)$$

The solution for the excited-state population with the same initial conditions as in eqn (2.7.18) is derived straightforwardly as

$$\tilde{\rho}_{22}(t) = \frac{\mathcal{V}^2/2}{2\gamma_{sp}^2 + \mathcal{V}^2}\left\{1 - \left[\cos(\lambda t) + \frac{3\gamma_{sp}}{2\lambda}\sin(\lambda t)\right]\exp\left(-\frac{3\gamma_{sp}t}{2}\right)\right\}. \qquad (2.8.10)$$

Figure 2.6 shows some examples of the time-dependent level of excitation obtained from eqn (2.8.10). The curve for zero radiative damping, $\gamma_{sp} = 0$, reproduces the pure Rabi oscillations shown for zero detuning in Fig. 2.5. The curves for $\gamma_{sp} \neq 0$ all tend to a steady-state limit given by the zero-detuning value of eqn (2.8.3). The oscillations are damped out more and more rapidly with increasing γ_{sp} and only a single maximum remains for $\gamma_{sp} = \mathcal{V}/3$. The incident light beam must thus be sufficiently strong for $\mathcal{V}$ to be much greater than $3\gamma_{sp}$ for significant oscillations in the level populations to be generated.

The second special case for which explicit solutions of the optical Bloch equations can be derived straightforwardly is that of a weak incident beam, where $\mathcal{V}$ is much smaller than the damping γ_{sp}. The roots of the determinant are now

$$\lambda_1 = 0, \quad \lambda_2 = -2\gamma_{sp}, \quad \lambda_3 = -\gamma_{sp} + i(\omega_0 - \omega) \quad \text{and} \quad \lambda_4 = -\gamma_{sp} - i(\omega_0 - \omega), \qquad (2.8.11)$$

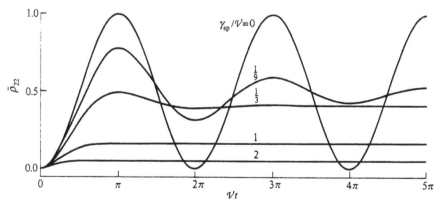

Fig. 2.6. Time dependences of the degree of atomic excitation $\tilde{\rho}_{22}$ for zero detuning and the ratios of radiative damping γ_{sp} to Rabi frequency $\mathcal{V}$ indicated against the curves.

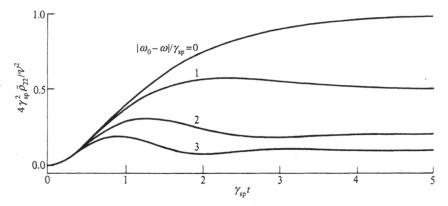

Fig. 2.7. Time dependences of the degree of atomic excitation $\tilde{\rho}_{22}$ for a weak incident beam and the ratios of detuning $|\omega_0 - \omega|$ to radiative damping γ_{sp} indicated against the curves.

and the solution for the excited state population with the initial conditions of eqn (2.7.18) is

$$\tilde{\rho}_{22}(t) = \frac{\mathcal{V}^2/4}{(\omega_0 - \omega)^2 + \gamma_{sp}^2}\left\{1 + \exp(-2\gamma_{sp}t) - 2\cos\left[(\omega_0 - \omega)t\right]\exp(-\gamma_{sp}t)\right\}.$$

$$(2.8.12)$$

Figure 2.7 shows the time dependence of the atomic excitation in the weak-beam limit for several values of the detuning. This result generalizes eqn (2.3.9), to which it reduces in the limit of zero γ_{sp}.

All of the curves in Figs. 2.5, 2.6 and 2.7 show the same initial quadratic dependence on time for $\mathcal{V}t \ll 1$. This property can be proved by a short-time expansion of the solutions of the optical Bloch equations. Thus with the initial conditions of eqn (2.7.18), it follows from eqns (2.8.1) and (2.8.2) that

$$\left.\frac{d\tilde{\rho}_{22}(t)}{dt}\right|_{t=0} = 0 \quad \text{and} \quad \left.\frac{d^2\tilde{\rho}_{22}(t)}{dt^2}\right|_{t=0} = \tfrac{1}{2}\mathcal{V}^2.$$

$$(2.8.13)$$

The first three terms of a Maclaurin series for $\tilde{\rho}_{22}(t)$ thus give

$$\tilde{\rho}_{22}(t) = \tfrac{1}{4}\mathcal{V}^2 t^2$$

$$(2.8.14)$$

irrespective of the values of $\omega_0 - \omega$ and γ_{sp}.

The broad-band result of the Einstein theory is retrieved by integration of eqn (2.8.12) over the radiatively-broadened absorption line. A simple contour integration gives

$$\int_{-\infty}^{\infty} d\omega\, \tilde{\rho}_{22}(t) = \left(\pi\mathcal{V}^2/4\gamma_{sp}\right)\left\{1 - \exp(-2\gamma_{sp}t)\right\}.$$

$$(2.8.15)$$

It follows from eqns (2.2.10), (2.3.6) and (2.3.20) that

$$\int_{-\infty}^{\infty} d\omega B \langle W(\omega) \rangle = \pi \mathcal{V}^2 / 2, \tag{2.8.16}$$

and the Einstein A coefficient is equal to $2\gamma_{sp}$, as in eqn (2.5.11). Eqn (2.8.15) thus agrees exactly with the weak-beam limit of the broad-band result (1.7.9) derived in the Einstein theory of absorption and emission.

It is readily verified that the steady-state solutions (2.8.3) and (2.8.4) satisfy the inequality

$$\tilde{\rho}_{11}(\infty)\tilde{\rho}_{22}(\infty) = \left[1 - \tilde{\rho}_{22}(\infty)\right]\tilde{\rho}_{22}(\infty) \geq \left|\tilde{\rho}_{12}(\infty)\right|^2, \tag{2.8.17}$$

where the equality occurs only for the trivial case of $\mathcal{V} = 0$. The existence of the steady state and the violation of the equality (2.7.23) are consequences of the introduction of radiative damping into the optical Bloch equations. The atom is no longer in a pure state of the form shown in eqn (2.1.8) but is described by a *statistical mixture*. It can be shown [2,9] that

$$\tilde{\rho}_{11}(t)\tilde{\rho}_{22}(t) \geq \left|\tilde{\rho}_{12}(t)\right|^2, \tag{2.8.18}$$

where the equality occurs only for pure states of the two-level atom. Statistical mixture states are considered in detail in §4.6, particularly their application to the radiation field.

2.9 Collision broadening

The treatment of the main line-broadening processes, which has so far covered the radiative, Doppler and power mechanisms, is completed by a treatment of the collision process. The dominant effects in the majority of experiments are the Doppler and collision broadening, and the latter process is used in §3.1 to model the typical features of a chaotic light source.

The theory of collision broadening is an extensive field of study [11] and we here consider just enough detail of the process to illustrate the nature of the mechanism and its incorporation into the optical Bloch equations. The occurrence of collisions between the atoms in a gas is a random process. According to the kinetic theory of gases [12], the probability $p(\tau)d\tau$ that an atom has a period of free flight between collisions lasting a length of time in the range τ to $\tau + d\tau$ is

$$p(\tau)d\tau = (1/\tau_0)\exp(-\tau/\tau_0)d\tau, \tag{2.9.1}$$

where the mean period τ_0 of free flight is given by

$$\frac{1}{\tau_0} = \frac{4d^2 N}{V}\left(\frac{\pi k_B T}{M}\right)^{1/2}. \tag{2.9.2}$$

Here d is the distance between the centres of the atoms during a collision and M is the atomic mass.

The effect of a collision on the atomic energy levels and wavefunctions is quite complicated. The energy levels are shifted during the collision by the forces of interaction between the colliding atoms, and the wavefunction becomes some linear combination of the wavefunctions of the unperturbed atom. It is permissible to ignore any absorption or emission of light that occurs within the duration of the collision, if this is sufficiently brief. The collisions then influence optical processes only via the changes in the atomic states from those preceding to those following the collision. We assume in what follows that the collision durations are indeed very short compared to the average time τ_0 of the periods of free flight during which the atomic states remain unchanged.

Two categories of collision can be distinguished. *Inelastic* collisions cause a change in the state of the atom from one energy level to another. Their effect is represented by an additional decay rate for the atomic level populations and they appear in the optical Bloch equations via an appropriate addition to the radiative decay rate. *Elastic* collisions leave the atom in the same energy level as before and their effect is limited to changes in the phase of the atomic wavefunction. Changes in the phases of C_1 and C_2 in eqn (2.1.8) affect the off-diagonal elements of the atomic density matrix in eqn (2.7.2) but the diagonal elements of eqn (2.7.1) are unaffected. The elastic phase-interruption collisions thus introduce an additional decay rate into the off-diagonal optical Bloch equation (2.8.2) but the diagonal equation (2.8.1) is unchanged. We restrict attention to the elastic variety of collision, which has the dominant line-broadening effect for a wide range of physical conditions.

The elastic off-diagonal decay rate is denoted γ_{coll}, and it is shown in §3.5 that

$$\gamma_{\text{coll}} = 1/\tau_0 . \tag{2.9.3}$$

The factor of 2 difference between this and the analogous relation (2.5.11) for the radiative broadening should be noted. With collisions included, the off-diagonal optical Bloch equation (2.8.2) becomes

$$d\tilde{\rho}_{12}/dt = d\tilde{\rho}_{21}^*/dt = \tfrac{1}{2}i\mathcal{V}(\tilde{\rho}_{11} - \tilde{\rho}_{22}) + [i(\omega_0 - \omega) - \gamma]\tilde{\rho}_{12} , \tag{2.9.4}$$

where

$$\gamma = \gamma_{\text{sp}} + \gamma_{\text{coll}} . \tag{2.9.5}$$

The expressions (2.8.3) and (2.8.4) for the steady-state density-matrix elements are generalized to

$$\tilde{\rho}_{22}(\infty) = \frac{(\gamma/4\gamma_{\text{sp}})\mathcal{V}^2}{(\omega_0 - \omega)^2 + \gamma^2 + (\gamma/2\gamma_{\text{sp}})\mathcal{V}^2} \tag{2.9.6}$$

and

$$\tilde{\rho}_{12}(\infty) = -\frac{(\mathcal{V}/2)(\omega_0 - \omega - i\gamma)}{(\omega_0 - \omega)^2 + \gamma^2 + (\gamma/2\gamma_{sp})\mathcal{V}^2}. \qquad (2.9.7)$$

It is straightforward to rederive the susceptibility, and the generalization of eqn (2.8.6) is

$$\chi(\omega) = \frac{2\pi N c^3}{\omega_0^3 V} \frac{\gamma_{sp}(\omega_0 - \omega + i\gamma)}{(\omega_0 - \omega)^2 + \gamma^2 + (\gamma/2\gamma_{sp})\mathcal{V}^2}. \qquad (2.9.8)$$

The linewidth of the atomic transition with inclusion of radiative, power, and collision broadening is thus

$$\text{FWHM} = 2\left\{ \gamma^2 + (\gamma/2\gamma_{sp})\mathcal{V}^2 \right\}^{1/2}, \qquad (2.9.9)$$

generalizing eqn (2.8.7).

Most of the subsequent discussion refers to fairly weak light beams where the power broadening is negligible and eqn (2.9.9) reduces to

$$\text{FWHM} = 2\gamma = 2\gamma_{sp} + 2\gamma_{coll}. \qquad (2.9.10)$$

The absorption line has a Lorentzian shape and its width is obtained from a simple sum of the radiative and collisional contributions. A typical value of the collision time at the gas density appropriate to a pressure of 10^5 Pa at room temperature is

$$\tau_0 \approx 3 \times 10^{-11} \text{ s.} \qquad (2.9.11)$$

The collision linewidth for this value is about 10^4 times the natural or radiative linewidth derived from the A coefficient magnitude of eqn (2.3.23).

The optical Bloch equations with inclusion of the homogeneous processes of radiative and collision broadening are given by the standard forms in eqns (2.8.1) and (2.9.4). The effects of addition of inhomogeneous broadening processes, which produce the composite lineshape of eqn (2.6.15), cannot be so directly included in the Bloch equations. We consider here only the Doppler broadening treated in §2.6, where it is necessary to include the frequency shifts associated with the atomic motions. For the subset of atoms moving with velocity **v**, the optical Bloch equations have their standard forms given above, except that the transition frequency ω_0 is replaced by the appropriate Doppler-shifted frequency. The solution of the Bloch equations for these atoms proceeds in the usual way and the results provide expressions for the quantities of interest for the atomic subset. The effects of the spread of atomic velocities are now included by averages over their Gaussian distribution. Such calculations must usually be performed numerically, as must the solution of the optical Bloch equations for homogeneous broadening with more general values of the parameters than those assumed in the special cases treated here. Examples of the solutions for both homogeneous and inhomogeneous processes are to be found in the literature [13].

Problem 2.10 Prove that the Doppler and collisional contributions to the linewidth of an atomic transition are equal at a gas density for which the volume per atom is close to λd^2, where λ is the optical wavelength of the transition.

2.10 Bloch equations and rate equations

We conclude the chapter with a more detailed discussion of the relation between the optical Bloch equations and the simpler rate equations for the atomic populations. The basic rate equation (1.5.7) describes a range of optical processes. It involves only the atomic populations N_1 and N_2, equivalent to $\tilde{\rho}_{11}$ and $\tilde{\rho}_{22}$ in the Bloch equations (2.8.1) and (2.9.4). The coherences, or off-diagonal density-matrix elements $\tilde{\rho}_{12}$ and $\tilde{\rho}_{21}$, are not included in the rate-equation (1.5.7).

The connections between the rate equations and the optical Bloch equations are most easily studied in the limit of low beam intensity, where the solution (1.7.9) of the rate equations gives

$$N_2(t) = \left(N\langle W \rangle / W_{\mathrm{s}}\right)\{1 - \exp(-At)\}. \tag{2.10.1}$$

For the Bloch equations, with the same initial conditions (2.7.18) as before, the solution of eqns (2.8.1) and (2.9.4) to lowest order in $\mathcal{V}$ gives the excited-state population as

$$\tilde{\rho}_{22}(t) = \tfrac{1}{4}\mathcal{V}^2 \left\{ \frac{\gamma/\gamma_{\mathrm{sp}}}{\left(\omega_0 - \omega\right)^2 + \gamma^2} + \frac{\left[\left(2\gamma_{\mathrm{sp}} - \gamma\right)/\gamma_{\mathrm{sp}}\right]\exp(-2\gamma_{\mathrm{sp}}t)}{\left(\omega_0 - \omega\right)^2 + \left(2\gamma_{\mathrm{sp}} - \gamma\right)^2} - 2\exp(-\gamma t) \times \right.$$

$$\left. \frac{\left[\left(\omega_0 - \omega\right)^2 + \gamma\left(2\gamma_{\mathrm{sp}} - \gamma\right)\right]\cos\left[\left(\omega_0 - \omega\right)t\right] + 2\left(\omega_0 - \omega\right)\left(\gamma - \gamma_{\mathrm{sp}}\right)\sin\left[\left(\omega_0 - \omega\right)t\right]}{\left[\left(\omega_0 - \omega\right)^2 + \gamma^2\right]\left[\left(\omega_0 - \omega\right)^2 + \left(2\gamma_{\mathrm{sp}} - \gamma\right)^2\right]} \right\}. \tag{2.10.2}$$

In the absence of collision broadening, where $\gamma = \gamma_{\mathrm{sp}}$, this reduces to the result (2.8.12) derived previously. The off-diagonal density-matrix element $\tilde{\rho}_{21}$ is not of interest for the present discussion but it is needed in §8.3 to lowest order in $\mathcal{V}$ and for an arbitrary initial value, when its form is

$$\tilde{\rho}_{21}(t) = \frac{\mathrm{i}\mathcal{V}/2}{\mathrm{i}\left(\omega_0 - \omega\right) + \gamma}\left(\exp\left\{-\left[\mathrm{i}\left(\omega_0 - \omega\right) + \gamma\right]t\right\} - 1\right)$$

$$+ \tilde{\rho}_{21}(0)\exp\left\{-\left[\mathrm{i}\left(\omega_0 - \omega\right) + \gamma\right]t\right\}. \tag{2.10.3}$$

These approximate solutions are valid when $\mathcal{V}^2 \ll \gamma\gamma_{\mathrm{sp}}$.

There are two distinct regimes in which the expression (2.10.2) for the excited-state population exhibits a time dependence similar to eqn (2.10.1). The first is that of broad-band incident light, for which the rate-equation method is immediately valid. For the Bloch-equation method, it is necessary to integrate over the solution in eqn (2.10.2) for conditions in which the frequency spread of the incident light exceeds the linewidth 2γ of the atomic transition.

Problem 2.11 Prove by contour integration of eqn (2.10.2) that

$$\int_{-\infty}^{\infty} d\omega \tilde{\rho}_{22}(t) = \left(\pi V^2 / 4\gamma_{sp}\right)\left\{1 - \exp(-2\gamma_{sp}t)\right\}, \qquad (2.10.4)$$

valid whether γ is greater or smaller than $2\gamma_{sp}$.

This is exactly the same as eqn (2.8.15), and it shows that the result applies for all values of the collision broadening. Its full equivalence to the rate-equation result (2.10.1) follows from the relation (2.8.16) between beam intensities. The broad-band limit is represented schematically on the left of Fig. 2.8.

The second regime is that of large collision broadening, where γ is much greater than γ_{sp}. Then for times t of order $1/\gamma_{sp}$, the inequality $t \gg 1/\gamma$ is satisfied, and the general solution (2.10.2) reduces to

$$\tilde{\rho}_{22}(t) = \frac{V^2 \gamma / 4\gamma_{sp}}{\left(\omega_0 - \omega\right)^2 + \gamma^2}\left\{1 - \exp(-2\gamma_{sp}t)\right\} \quad \left(\gamma \gg \gamma_{sp}\right). \qquad (2.10.5)$$

The time dependence is again the same as that in the rate-equation solution (2.10.1) but the prefactor is generalized to allow for a transition lineshape that is substantially broadened beyond the radiative width. The form in eqn (2.10.5) is the same as that obtained by solution of the rate equation

$$d\tilde{\rho}_{22}(t)/dt = R - 2\gamma_{sp}\tilde{\rho}_{22}(t), \qquad (2.10.6)$$

where

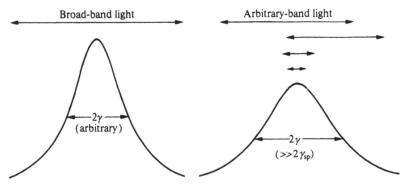

Fig. 2.8. Representations of the two regimes in which the optical Bloch equations reproduce the same atomic level populations as the Einstein rate equations.

$$R = \tfrac{1}{2}\mathcal{V}^2\gamma\Big/\big[(\omega_0 - \omega)^2 + \gamma^2\big].$$ (2.10.7)

The excited-state population in eqn (2.10.5) is thus effectively determined by the competition between an excitation or absorption process of rate R and radiative decay at the rate $2\gamma_{sp}$. The equivalence of the optical Bloch equation and rate equation solutions to order $\mathcal{V}^2$ in this second regime, shown schematically on the right of Fig. 2.8, holds for any bandwidth of the incident light.

The above discussion is based on the solution of the Bloch equations in the presence of collision broadening, but similar rate-equation limits can be established more generally when Doppler broadening is also present [13]. Thus the rate equations are valid in general when either (i) the bandwidth of the incident light exceeds the atomic transition linewidth or (ii) the combined collision and Doppler linewidth greatly exceeds the radiative linewidth of the transition. The rate equations involve only the atomic populations, while the optical Bloch equations also contain the atomic coherences.

References

[1]   Atkins, P.W. and Friedman, R.S., *Molecular Quantum Mechanics,* 3rd edn (Oxford University Press, Oxford, 1997).
[2]   Merzbacher, E., *Quantum Mechanics*, 3rd edn (Wiley, New York, 1998).
[3]   Dwight, H.B., *Tables of Integrals and Other Mathematical Data*, 4th edn (Macmillan, New York, 1961).
[4]   Landau, L.D. and Lifshitz, E.M., *Statistical Physics,* part 1, 3rd edn (Pergamon Press, Oxford, 1980) §29.
[5]   Abramowitz, M. and Stegun, I.A., *Handbook of Mathematical Functions* (Dover Publications, New York, 1965) chap. 7.
[6]   Allen, L. and Eberly, J.H., *Optical Resonance and Two-Level Atoms* (Dover Publications, New York, 1987).
[7]   Rabi, I.I., Space quantization in a gyrating magnetic field, *Phys. Rev.* **51**, 652–4 (1937).
[8]   Ackerhalt, J.R., Resonant laser excitation in a two-level model: a nonlinear Schrödinger equation?, *Opt. Lett.* **6**, 136–8 (1981).
[9]   Barnett, S.M. and Radmore, P.M., *Methods in Theoretical Quantum Optics* (Clarendon Press, Oxford, 1997).
[10] Butcher, P.N. and Cotter, D., *The Elements of Nonlinear Optics* (Cambridge University Press, Cambridge, 1990).
[11] Corney, A., *Atomic and Laser Spectroscopy* (Clarendon Press, Oxford, 1977).
[12] See §39 of ref [4].
[13] Smith, R.A., Excitation of transitions between atomic or molecular energy levels by monochromatic laser radiation, *Proc. R. Soc. Lond. A* **362**, 1–12 (1978): **362**, 13–25 (1978); **368**, 163–75 (1979).

3 Classical theory of optical fluctuations and coherence

The study of the classical theory of optical fluctuations and coherence is an important preliminary to the development of the corresponding quantum-mechanical theory. The classical theory is useful for a physical understanding of the various effects and many of the classical concepts carry over into the quantum theory. In addition, a knowledge of the classical theory is useful for identification of the nature of the specifically quantum properties that some light beams may exhibit. Thus it is shown in Chapter 5 that the classical and quantum theories yield the same predictions for chaotic light, but that there exist other kinds of light that elude any satisfactory classical description. The nature of such 'nonclassical' light is, of course, more sharply appreciated by contrast with the limitations on the possible kinds of light inherent in the classical theory.

The line-broadening mechanisms described in Chapter 2 produce the same lineshapes in both the absorption and emission spectra associated with a given atomic transition. In the present chapter we consider the characteristics of the emitted light generated by radiative transitions of excited atoms. The characteristics can in principle be measured by two different kinds of experiment. The first of these is ordinary spectroscopy, which measures the frequency distribution of the light and thus provides information on the nature and strengths of the line-broadening processes in the source.

Our main concern in the present chapter is with a second kind of experiment, which measures the time dependence of the amplitude or intensity of the light beam. It is shown that the line-broadening processes in the source cause the electric field and intensity of the beam to fluctuate around their mean values on time scales inversely proportional to the frequency differences contained in the spectrum of the light. These time-dependent fluctuation properties determine the results of optical interference experiments. The temporal fluctuations and the frequency spectrum are manifestations of the same physical properties of the radiating atoms that constitute the light source, but a knowledge of both aspects is needed to interpret the complete range of optical experiments.

The standard spectroscopic light source is the gas discharge lamp, where the different excited atoms emit their radiation independently of one another. The shape of the emission line is determined by the statistical spread in atomic velocities and the random occurrence of collisions. A conventional light source of this kind is designated as *chaotic*. The thermal cavity and the filament lamp are other examples of chaotic source. The light beams from any variety of chaotic source have similar statistical descriptions. An alternative type of light source is the laser, whose emitted light has quite different statistical properties. The properties of laser light are only briefly mentioned in the present chapter, a detailed treatment being deferred to §§7.3 and 7.4.

3.1 Models of chaotic light sources

The nature of the temporal fluctuations of chaotic light is most easily appreciated by considering a model source in which collision broadening predominates. We ignore radiative and Doppler broadening for the present and we suppose that the collisions are of the elastic phase-interruption variety that does not change the atomic state.

Consider a particular excited atom radiating light of frequency ω_0. A train of electromagnetic radiation emanates steadily from the atom until it suffers a collision. During a collision, the energy levels of the atom are shifted by amounts that depend on the severity of the collision, and the radiated wave train is interrupted for its duration. When the wave of frequency ω_0 is resumed after the collision, its characteristics are identical to those that it had prior to the collision, except that the phase of the wave is unrelated to the phase before the collision.

If the duration of the collision is sufficiently brief, it is permissible to ignore any radiation emitted during the collision, while the frequency is shifted from ω_0. The collision-broadening effect can then be adequately represented by a model in which each excited atom always radiates at frequency ω_0, but with random changes in the phase of the radiated wave each time a collision occurs. The apparent spread in the emitted frequencies arises because the wave is chopped into finite sections whose Fourier decompositions include frequencies other than ω_0.

The wave train radiated by a single atom is illustrated schematically in Fig. 3.1, which shows the variation of the electric field amplitude $E(t)$ at a fixed

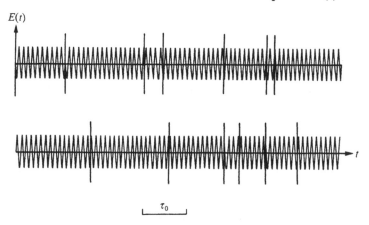

Fig. 3.1. The electric-field amplitude of the wave train radiated by a single atom. The vertical lines represent collisions separated by periods of free flight with the mean duration τ_0 indicated. The quantity $\omega_0 \tau_0$ is chosen unrealistically small in order to show the random phase changes caused by the collisions.

observation point as a function of the time. The occurrence of a collision is indicated by a vertical line accompanied by a random change in the phase of the wave. The periods of free flight in the figure are chosen in accordance with the probability distribution in eqn (2.9.1). The variation of the phase of the wave with time is shown in Fig. 3.2. In order to make Fig. 3.1 easy to draw, the quantity $\omega_0 \tau_0$ is given the small value of 60. More realistically, for visible light with the collision time given in eqn (2.9.11), we have

$$\omega_0 \tau_0 \approx 9 \times 10^4, \tag{3.1.1}$$

and the wave train radiated by an atom undergoes on average about 15,000 periods of oscillation between successive collisions.

The field amplitude of the wave illustrated in Figs. 3.1 and 3.2 is written in complex form as

$$E(t) = E_0 \exp\{-i\omega_0 t + i\varphi(t)\}. \tag{3.1.2}$$

As is shown in Fig. 3.2, the phase $\varphi(t)$ remains constant during periods of free flight but it changes abruptly each time a collision occurs. The amplitude E_0 and the frequency ω_0 are the same for every period. The total wave emitted by the collision-broadened source is represented by a sum of terms like eqn (3.1.2), one for each radiating atom. It is assumed that the waves emitted by different atoms are focused to form a plane parallel light beam and that the observed light has a fixed polarization, so that the electric fields can be added algebraically. The total electric field amplitude produced by a large number of radiating atoms is then

$$
\begin{aligned}
E(t) &= E_1(t) + E_2(t) + \ldots + E_\nu(t) \\
&= E_0 \exp(-i\omega_0 t)\{\exp(i\varphi_1(t)) + \exp(i\varphi_2(t)) + \ldots + \exp(i\varphi_\nu(t))\} \\
&= E_0 \exp(-i\omega_0 t) a(t) \exp(i\varphi(t)),
\end{aligned}
\tag{3.1.3}
$$

where the wave train from every atom is assigned the same amplitude E_0 and frequency ω_0, but the phases for the different atoms are completely unrelated.

The formal summation of the phase factors carried out in the final line of eqn (3.1.3) is illustrated in Fig. 3.3. The addition of unit vectors oriented in random directions is an example of a *random walk*, well known in the theory of stochastic processes. As the phase angles $\varphi_1, \varphi_2, \ldots, \varphi_\nu$ each have different random variations similar to that shown in Fig. 3.2, the end points of the walk, represented by the amplitude $a(t)$ and phase $\varphi(t)$, are different at different instants of time. The real electric field obtained from eqn (3.1.3) consists of a carrier wave of frequency ω_0 subjected to random amplitude and phase modulation. The Fourier decomposition of the modulated wave contains frequencies spread about ω_0 in a manner governed by the collision-broadened lineshape.

It is not possible in practice to resolve the oscillations in $E(t)$ that occur at the frequency of the carrier wave. A good experimental resolving time is of the order of 10^{-9} s, five or six orders of magnitude too long to detect oscillations at visible

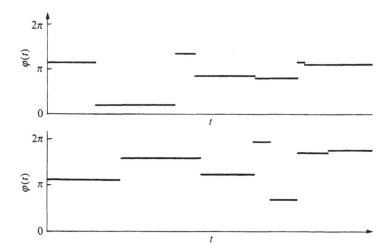

Fig. 3.2. Time dependence of the phase angle of the wave train illustrated in Fig. 3.1.

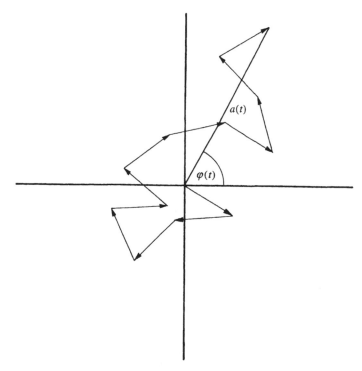

Fig. 3.3. Argand diagram showing the amplitude $a(t)$ and phase $\varphi(t)$ of the resultant vector formed by a random walk whose steps consist of a large number of unit vectors, each of which has a randomly-chosen phase angle.

frequencies. It is therefore appropriate for comparison with experiment to average the theoretical expressions over a cycle of oscillation of the carrier wave. The real electric field from eqn (3.1.3) has a zero cycle average. The beam intensity is given by the Poynting vector of eqn (1.8.1), with a cycle-average of the form given in eqn (1.10.9). The electric field is taken as the real part of the expression in eqn (3.1.3), and the average of the intensity over a cycle of the oscillation at frequency ω_0 is

$$\bar{I}(t) = \tfrac{1}{2}\varepsilon_0 c |E(t)|^2 = \tfrac{1}{2}\varepsilon_0 c E_0^2 a(t)^2. \tag{3.1.4}$$

The cycle average is again denoted by an overbar and the angle-bracket notation is reserved for statistical or longer-time averages used later in the chapter. The intensity $\bar{I}(t)$ retains the time dependence from the random amplitude modulation $a(t)$.

The variation of $\bar{I}(t)$ shown in Fig. 3.4, known as a *time series*, illustrates the type of fluctuation that occurs in the beam intensity. The figure is constructed by a computer simulation of a collision-broadened light source [1], in which the summation in eqn (3.1.3) is carried out numerically for a large number ν of atoms [2]. The dynamics of the phase-interruption process are represented by an appropriate random time-development of the variables used in the computer model. The light has the Lorentzian frequency distribution of eqn (2.5.20) with γ for a collision-broadened source set equal to γ_{coll}. The time scale in Fig. 3.4 is expressed in units of the *coherence time* τ_c of the light, which is determined by the dominant line-broadening processes in the source. For the collision broadening assumed here

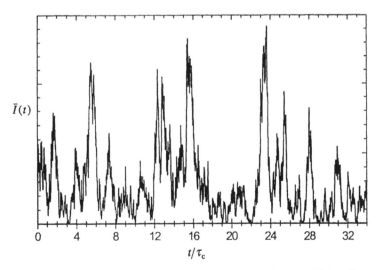

Fig. 3.4. Time series of the cycle-averaged intensity for a collision-broadened chaotic light beam obtained from a computer simulation [1]. The time scale is normalized by the coherence time τ_c, taken equal to the mean time between collisions. (After [1])

$$\tau_c = \tau_0 = 1/\gamma_{coll}, \qquad\qquad\qquad\qquad\qquad\qquad (3.1.5)$$

where τ_0 is the mean time between collisions given by eqn (2.9.2) and its relation to γ_{coll} is established in §3.5.

It is seen that gross fluctuations in the intensity occur over time spans of order τ_c but that additional fluctuations occur on shorter time scales to give a characteristic spiky structure to the time series. When the computer simulation is carried out with higher time-resolution than that used in Fig. 3.4, it is found that there are additional fluctuations with progressively shorter time scales. The existence of fluctuations with a wide range of time scales, extending in principle to arbitrarily short intervals, is a characteristic of the so-called *fractal* structures. The occurrence of such a fractal character in the intensity fluctuations of chaotic light emitted by a collision-broadened light source is a consequence of the significant strength in the wings of the Lorentzian lineshape. Thus, as the intensity fluctuations result from interference between the different frequencies in the light, there is a gross contribution resulting from the frequencies in the central component of the line, spread over a range of order $\gamma_{coll} = 1/\tau_c$, but there are also significant contributions resulting from frequencies well outside the line centre. The latter are responsible for the spiky structures seen in Fig. 3.4 and in the corresponding time series with higher time-resolutions.

Detailed measurements of the time series $\bar{I}(t)$ for the light from a collision-broadened source are quite difficult because of the short coherence times τ_c, typically of order 10^{-10} to 10^{-11} s, but similar fluctuations occur on the longer time scales, typically 10^{-3} to 10^{-4} s, that are found in laser light scattered from macroscopic particles in Brownian motion. Thus time series very similar to Fig. 3.4 are measured on the light scattered by a suspension of polystyrene spheres in water [3]. The experimental parameters are such that the thermal distribution of particle velocities is randomized by collisions, with a mean free path smaller than the optical wavelength. In these conditions the particle motion has the nature of a random walk and the suspension is a slowed-down analogue of a radiating gas, with the sharp laser frequency replacing the atomic transition frequency. The scattered light has a Lorentzian frequency spectrum and the associated fractal nature of the intensity fluctuations is verified over three orders of magnitude of the fluctuation time scales [4].

The above remarks apply specifically to the temporal fluctuations in the light emitted by a source in which collision broadening predominates, but fluctuations of similar forms occur in the presence of other line-broadening processes that give rise to Lorentzian frequency distributions, for example radiative broadening. Figure 3.5, however, shows the rather different form of time series for the cycle-averaged intensity $\bar{I}(t)$ that occurs in the light from a source dominated by Doppler broadening. The figure is again produced by means of a computer simulation [1] with the frequencies of the radiation distributed according to the Gaussian function of eqn (2.6.11). The spiky fractal nature of the time series for the Lorentzian-broadened light shown in Fig. 3.4 is removed, and much smoother fluctuations now occur on a single time scale determined by the coherence time. For the Gaussian spectrum of eqn (2.6.11), the coherence time is conveniently

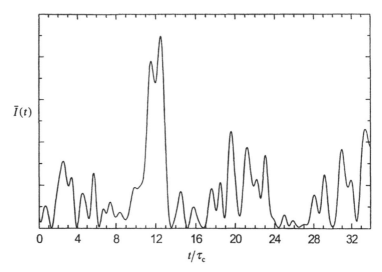

Fig. 3.5. Time series of the cycle-averaged intensity for a Doppler-broadened chaotic light beam obtained from a computer simulation [1]. The time scale is normalized by the coherence time τ_c given by eqn (3.1.6). (After [1])

defined as

$$\tau_c = \pi^{1/2}/\Delta . \tag{3.1.6}$$

and the time axis in Fig. 3.5 is normalized by this value of τ_c. The absence of fluctuations on shorter time scales is attributed to the lack of any significant wings on the Gaussian lineshape function, as is illustrated in Fig. 2.4. The much more rapid cutoff of the Gaussian function removes the higher frequency-difference components that are responsible for the fractal character of Lorentzian light.

The path length

$$\lambda_c = c\tau_c \tag{3.1.7}$$

associated with the coherence time is known as the *coherence length*. The graphs of the cycle-averaged intensity as a function of time at a fixed point in the beam shown in Figs. 3.4 and 3.5 can equally-well be regarded as graphs of the intensity as a function of distance z along the beam at a fixed instant of time. The horizontal axis can then be relabelled z/λ_c and the temporal and spatial aspects of the beam fluctuations are related by the simple scaling factor c. The beams considered here always have their coherence lengths much longer than the wavelength and their coherence times much longer than the period of oscillation.

3.2 The lossless optical beam-splitter

The temporal fluctuation properties of light beams are measured by optical interference experiments. The Mach–Zehnder interferometer analysed in the following section provides a simple example of first-order interference, while the Brown–Twiss interferometer treated in §3.8 is the paradigm of a second-order interference experiment. The central components in both of these experiments are optical beam splitters and it is useful to review their basic properties as a preparation for the discussions of interferometers. Beam splitters also play important roles in studies of the quantum aspects of light and the classical treatment that follows is recast in quantum-mechanical terms in §5.7.

Figure 3.6 shows a schematic representation of an optical beam-splitter with a notation for the electric fields of light beams in the two input and two output arms. The four beams are assumed to have a common linear polarization. The beam splitter is surrounded by free space and it is free of any loss processes that could remove energy from the light beams. The output fields are related to the input fields by linear relations of the forms

$$E_3 = \mathcal{R}_{31}E_1 + \mathcal{T}_{32}E_2 \quad \text{and} \quad E_4 = \mathcal{T}_{41}E_1 + \mathcal{R}_{42}E_2, \tag{3.2.1}$$

with an obvious notation for the beam-splitter reflection and transmission coefficients. The coefficients are generally complex quantities and they depend on the optical frequency. Their values here are those appropriate to the frequency of the input fields, which are assumed to be monochromatic. The relations are written in matrix form as

$$\begin{bmatrix} E_3 \\ E_4 \end{bmatrix} = \begin{bmatrix} \mathcal{R}_{31} & \mathcal{T}_{32} \\ \mathcal{T}_{41} & \mathcal{R}_{42} \end{bmatrix} \begin{bmatrix} E_1 \\ E_2 \end{bmatrix}, \tag{3.2.2}$$

where the 2×2 array is known as the *beam-splitter matrix*.

Important properties of the coefficients in the beam-splitter matrix are derived by considering the energy conservation between input and output arms [5]. Thus with both inputs illuminated, the total energy flow in the output arms is equal to the total energy flow in the input arms when

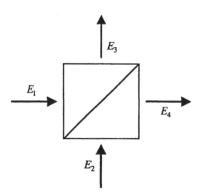

Fig. 3.6. Representation of a lossless beam-splitter showing the notation for electric field amplitudes of the input and output beams.

$$|E_3|^2 + |E_4|^2 = |E_1|^2 + |E_2|^2 \tag{3.2.3}$$

and this relation is satisfied for all input fields E_1 and E_2 if

$$|\mathcal{R}_{31}|^2 + |\mathcal{T}_{41}|^2 = |\mathcal{R}_{42}|^2 + |\mathcal{T}_{32}|^2 = 1 \quad \text{and} \quad \mathcal{R}_{31}\mathcal{T}_{32}^* + \mathcal{T}_{41}\mathcal{R}_{42}^* = 0. \tag{3.2.4}$$

The coefficients are separated into amplitude and phase factors as

$$\mathcal{R}_{31} = |\mathcal{R}_{31}|\exp(i\phi_{31}), \tag{3.2.5}$$

and similarly for the other coefficients. It then follows from the second part of eqn (3.2.4) that the phase angles are constrained by the relation

$$\phi_{31} + \phi_{42} - \phi_{32} - \phi_{41} = \pm\pi \tag{3.2.6}$$

and the amplitudes by

$$|\mathcal{R}_{31}|/|\mathcal{T}_{41}| = |\mathcal{R}_{42}|/|\mathcal{T}_{32}|. \tag{3.2.7}$$

This relation in conjunction with the first part of eqn (3.2.4) shows that the two reflection coefficients and the two transmission coefficients must have equal magnitudes,

$$|\mathcal{R}_{31}| = |\mathcal{R}_{42}| \equiv |\mathcal{R}| \quad \text{and} \quad |\mathcal{T}_{32}| = |\mathcal{T}_{41}| \equiv |\mathcal{T}|. \tag{3.2.8}$$

Equations (3.2.6) and (3.2.8) provide general contraints on the beam-splitter reflection and transmission coefficients.

These constraints ensure that the beam-splitter matrix is unitary, that is, the inverse of the matrix equals its complex-conjugate transpose. The relation of input fields to output fields, inverse to eqn (3.2.2), is thus

$$\begin{bmatrix} E_1 \\ E_2 \end{bmatrix} = \begin{bmatrix} \mathcal{R}_{31}^* & \mathcal{T}_{41}^* \\ \mathcal{T}_{32}^* & \mathcal{R}_{42}^* \end{bmatrix} \begin{bmatrix} E_3 \\ E_4 \end{bmatrix}. \tag{3.2.9}$$

The unitarity of the beam-splitter matrix is a direct consequence of the conservation of energy between the input and output arms.

The general structure of the beam-splitter matrix is sometimes simplified by additional assumptions on the forms of the reflection and transmission coefficients. Thus the coefficients can all be taken real, with $\phi_{31} = \phi_{32} = \phi_{41} = 0$ and $\phi_{42} = \pi$, when

$$\mathcal{R}_{31} = -\mathcal{R}_{42} = |\mathcal{R}| \quad \text{and} \quad \mathcal{T}_{32} = \mathcal{T}_{41} = |\mathcal{T}|. \tag{3.2.10}$$

Alternatively the coefficients can be taken symmetrical, with $\phi_{31} = \phi_{42} = \phi_{\mathcal{R}}$ and $\phi_{32} = \phi_{41} = \phi_{\mathcal{T}}$, when

$$\mathcal{R}_{31} = \mathcal{R}_{42} \equiv \mathcal{R} = |\mathcal{R}|\exp(i\phi_{\mathcal{R}}) \quad \text{and} \quad \mathcal{T}_{32} = \mathcal{T}_{41} \equiv \mathcal{T} = |\mathcal{T}|\exp(i\phi_{\mathcal{T}}). \quad (3.2.11)$$

The relations (3.2.4) reduce in this case to

$$|\mathcal{R}|^2 + |\mathcal{T}|^2 = 1 \quad \text{and} \quad \mathcal{R}\mathcal{T}^* + \mathcal{T}\mathcal{R}^* = 0, \quad \text{or} \quad \phi_{\mathcal{R}} - \phi_{\mathcal{T}} = \pm\pi/2. \quad (3.2.12)$$

This symmetrical case is assumed in all subsequent derivations. As a further specialization, it is sometimes convenient to consider a 50:50 beam splitter with reflection and transmission coefficients of equal magnitude,

$$|\mathcal{R}| = |\mathcal{T}| = 1/\sqrt{2} \quad \text{with} \quad \phi_{\mathcal{R}} - \phi_{\mathcal{T}} = \pi/2. \quad (3.2.13)$$

3.3 The Mach–Zehnder interferometer

The treatment of the fringes produced in a Mach–Zehnder interferometer provides a simple illustration of some general principles that are common to the whole class of interference experiments. The experiment is here analysed in sufficient detail to determine the conditions under which the field fluctuations associated with the chaotic properties of the light source affect the visibility of the interference fringes. Figure 3.7 shows a schematic representation of the experiment, which is equivalent to a Michelson interferometer, but is simpler to illustrate. Chaotic light from a point source is rendered parallel by means of a lens and then falls on the first beam splitter via arm 1. No light is incident via arm 2. The model experiment ignores all complications arising from the departure of the incident beam from a perfect plane parallel form, although the inclusion of such effects is essential for a rigorous treatment of the interference.

Let $E(t)$ be the complex electric field of the incident light beam in arm 1 at the

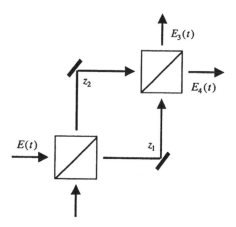

Fig. 3.7. Representation of a Mach–Zehnder interferometer showing the notation for the input and output field amplitudes and for the internal path lengths.

entrance to the first beam splitter. Both beam splitters are assumed to be symmetrical, with reflection and transmission coefficients $\mathcal{R}$ and $\mathcal{T}$ that satisfy the relations (3.2.12). Chaotic light contains a spread of frequencies, with spectra whose forms are discussed in Chapter 2, but it is assumed here that $\mathcal{R}$ and $\mathcal{T}$ are constant for the frequency components in the incident light beam. The fields in the vertical and horizontal output arms of the first beam splitter are thus $\mathcal{R}E(t)$ and $\mathcal{T}E(t)$ respectively, where the transit times of the light through the beam splitters are ignored. The two internal paths of the interferometer generally have different lengths z_1 and z_2, as shown in Fig. 3.7, and the fields that arrive in the two input arms of the second beam splitter are delayed by different times z_1/c and z_2/c. We consider the output at the exit from the second beam splitter to the arm labelled 4 in Fig. 3.7, where the electric field is

$$E_4(t) = \mathcal{R}\mathcal{T}E(t_1) + \mathcal{T}\mathcal{R}E(t_2), \tag{3.3.1}$$

with

$$t_1 = t - (z_1/c) \quad \text{and} \quad t_2 = t - (z_2/c). \tag{3.3.2}$$

The intensity of the output light, averaged over a cycle of oscillation, is

$$\begin{aligned}\bar{I}_4(t) &= \tfrac{1}{2}\varepsilon_0 c |E_4(t)|^2 \\ &= \tfrac{1}{2}\varepsilon_0 c |\mathcal{R}|^2 |\mathcal{T}|^2 \left\{ |E(t_1)|^2 + |E(t_2)|^2 + 2\,\mathrm{Re}\left[E^*(t_1)E(t_2) \right] \right\}.\end{aligned} \tag{3.3.3}$$

The intensity average $\bar{I}_4(t)$ has a similar nature to that defined in eqn (3.1.4).

The fringes in the Mach–Zehnder interferometer are normally observed by eye or recorded by photographic plates. Even with the use of photodetectors, the resolving time is normally much longer than τ_c, the coherence time of the chaotic light. To compare theory with experiment it is necessary to average $\bar{I}_4(t)$ over an observation period T much longer than τ_c. This averaged intensity is denoted $\langle \bar{I}_4(t) \rangle$, and it is obtained from eqn (3.3.3) as

$$\langle \bar{I}_4(t) \rangle = \tfrac{1}{2}\varepsilon_0 c |\mathcal{R}|^2 |\mathcal{T}|^2 \left\{ \langle |E(t_1)|^2 \rangle + \langle |E(t_2)|^2 \rangle + 2\,\mathrm{Re}\langle E^*(t_1)E(t_2) \rangle \right\}. \tag{3.3.4}$$

It is seen that the output intensity consists of three contributions. The first two terms represent the intensities produced by each internal path through the interferometer in the absence of the other. These two terms do not give rise to any interference effects. The fringes arise from the term that involves the *first-order correlation function* for the fields at times t_1 and t_2, which is defined more explicitly as

$$\langle E^*(t_1)E(t_2) \rangle = \frac{1}{T}\int_T dt_1\, E^*(t_1)E(t_2). \tag{3.3.5}$$

Note that there is only a single time variable in the integrand, as t_2 differs from t_1 by a fixed amount obtained from eqn (3.3.2). The averages in the first two terms in the large bracket of eqn (3.3.4) are given by the same expression (3.3.5) with the two times set equal.

The nature of the field correlation function and hence of the interference fringes depends on the kind of light incident on the interferometer. If the statistical properties of the light source are *stationary*, that is, the influences that govern the fluctuations do not change with time, then the average in eqn (3.3.5) does not depend on the particular starting time of the period T, provided that T is much longer than the characteristic time scale of the fluctuations. The time averaging in eqn (3.3.5) then samples all of the electric-field values allowed by the statistical properties of the source with their appropriate relative probabilities, and the result is independent of the magnitude of T. Thus, although the experimental effects of the first-order correlation function are manifested by a *time* averaging, as on the right of eqn (3.3.5), the function is calculated by a *statistical* average over all values of the fields at times t_1 and t_2. The result does not of course depend on the time t_1, and the correlation is a function only of the time delay $t_2 - t_1$ between the two field values. The equivalence of time averaging and statistical averaging is valid only for light beams whose fluctuations are produced by *ergodic* random processes [6]. Most of the derivations in the present chapter are concerned with ergodic beams of chaotic light, with time series of effectively infinite duration whose forms are similar to those shown in Figs. 3.4 and 3.5.

The above remarks do not apply to a *nonstationary* light source, for example one which is active for only a limited time so that it emits light in the form of an optical pulse. The measured intensity of such a light beam clearly depends on the time of observation and the electric-field correlation is a function of both t_1 and t_2. Several varieties of pulsed light are considered in detail in Chapter 6 but the treatments here are restricted to stationary beams, except where indicated.

It is convenient to rewrite the definition (3.3.5) as

$$\left\langle E^*(t)E(t+\tau)\right\rangle = \frac{1}{T}\int_T \mathrm{d}t E^*(t)E(t+\tau). \tag{3.3.6}$$

An important property of this first-order correlation function for light with stationary statistics results from its dependence only on the time delay τ between the two field values. It follows that

$$\left\langle E^*(t)E(t-\tau)\right\rangle = \left\langle E^*(t+\tau)E(t)\right\rangle = \left\langle E^*(t)E(t+\tau)\right\rangle^* \tag{3.3.7}$$

and the real part of the correlation function is therefore the same at positive and negative τ. The *degree of first-order temporal coherence* of light with stationary statistics is defined as a normalized version of the first-order correlation function,

$$g^{(1)}(\tau) = \frac{\left\langle E^*(t)E(t+\tau)\right\rangle}{\left\langle E^*(t)E(t)\right\rangle}. \tag{3.3.8}$$

This definition allows the output intensity of the Mach–Zehnder interferometer given by eqn (3.3.4) to be written

$$\langle \bar{I}_4(t) \rangle = 2|\mathcal{R}|^2|\mathcal{T}|^2\langle \bar{I}(t)\rangle\{1 + \operatorname{Re} g^{(1)}(\tau)\},\tag{3.3.9}$$

where

$$\langle \bar{I}(t) \rangle = \tfrac{1}{2}\varepsilon_0 c\langle |E(t)|^2\rangle\tag{3.3.10}$$

is the averaged input intensity and

$$\tau = (z_1 - z_2)/c.\tag{3.3.11}$$

It follows from eqns (3.3.7) and (3.3.8) that

$$g^{(1)}(-\tau) = g^{(1)}(\tau)^*,\tag{3.3.12}$$

and the intensity given by eqn (3.3.9) is symmetrical with respect to changes in the sign of τ. The detailed evaluation of the Mach–Zehnder fringe pattern thus requires a knowledge of the degree of first-order coherence.

3.4 Degree of first-order coherence

The model of a collision-broadened light source described in §3.1 is here used to calculate the first-order electric-field correlation function of the light and hence its degree of first-order coherence. Thus with use of the ergodic nature of the light and the form of electric field given in the middle line of eqn (3.1.3), the time average in the definition of eqn (3.3.6) is interpreted as equivalent to a statistical average and the required function is

$$\langle E^*(t)E(t+\tau)\rangle = E_0^2\exp(-i\omega_0\tau)$$
$$\times\langle\{\exp(-i\varphi_1(t)) + \exp(-i\varphi_2(t)) + ... + \exp(-i\varphi_\nu(t))\}\tag{3.4.1}$$
$$\times\{\exp(i\varphi_1(t+\tau)) + \exp(i\varphi_2(t+\tau)) + ... + \exp(i\varphi_\nu(t+\tau))\}\rangle.$$

In multiplying out the large brackets, the phase angles of the wave trains from different atoms have different random values and the cross terms give a zero average contribution. The remaining terms give

$$\langle E^*(t)E(t+\tau)\rangle = E_0^2\exp(-i\omega_0\tau)\sum_{i=1}^{\nu}\langle\exp[i(\varphi_i(t+\tau)-\varphi_i(t))]\rangle\tag{3.4.2}$$
$$= \nu\langle E_i^*(t)E_i(t+\tau)\rangle$$

since all ν radiating atoms are equivalent.

The correlation function for the beam as a whole is thus determined by the single-atom contributions. Now the phase angle of each wave train jumps to a random value after its source atom suffers a collision, subsequently producing a zero average contribution to the field correlation. Thus the single-atom correlation function in eqn (3.4.2) is proportional to the probability that the atom has a period of free flight longer than τ and, with use of the probability distribution from eqn (2.9.1), we can put

$$\left\langle E_i^*(t)E_i(t+\tau)\right\rangle = E_0^2 \exp(-i\omega_0\tau)\left\langle \exp[i(\varphi_i(t+\tau)-\varphi_i(t))]\right\rangle$$

$$= E_0^2 \exp(-i\omega_0\tau)\int_\tau^\infty d\tau' p(\tau') \tag{3.4.3}$$

$$= E_0^2 \exp\{-i\omega_0\tau-(\tau/\tau_0)\}.$$

The correlation function (3.4.1) thus becomes

$$\left\langle E^*(t)E(t+\tau)\right\rangle = vE_0^2 \exp\{-i\omega_0\tau-(\tau/\tau_0)\} \tag{3.4.4}$$

and the normalized correlation, or degree of first-order coherence, from eqn (3.3.8) is

$$g^{(1)}(\tau) = \exp\{-i\omega_0\tau-(|\tau|/\tau_0)\} = \exp\{-i\omega_0\tau-\gamma_{coll}|\tau|\}. \tag{3.4.5}$$

The above calculation assumes a positive value for the time delay τ but its final result is generalized to both negative and positive values by use of the symmetry property in eqn (3.3.12).

The degree of first-order coherence of the light from a source in which both collision and radiative broadening are important is calculated by a generalization of the model. The effect of the radiative damping on the correlation function of eqn (3.4.1) is an additional multiplicative factor of $\exp(-\gamma_{sp}|\tau|)$. The end result is the same as eqn (3.4.5), except that the collision time τ_0 is replaced by a coherence time

$$\tau_c = 1/\gamma, \tag{3.4.6}$$

where γ is the total damping rate defined in eqn (2.9.5). The degree of first-order coherence is thus written more generally as

$$g^{(1)}(\tau) = \exp\{-i\omega_0\tau-(|\tau|/\tau_c)\} = \exp\{-i\omega_0\tau-\gamma|\tau|\}. \tag{3.4.7}$$

The form of the modulus of this function is illustrated in Fig. 3.8. The existence of the finite slope at the origin is associated with the spiky structure shown in the time series of Fig. 3.4. Thus even over short times τ, the fractal character of the fluctuations produces significant deviations in the field amplitude and the intensity from their initial values, leading to the sharp fall-off seen in Fig. 3.8.

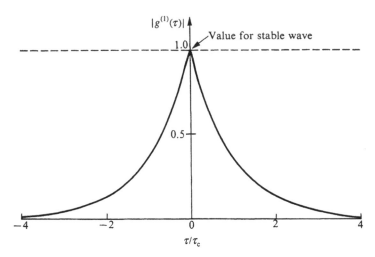

Fig. 3.8. The modulus of the degree of first-order coherence of chaotic light with a Lorentzian frequency spectrum.

The degree of first-order coherence is also readily calculated for a chaotic light source in which Doppler broadening is the main cause of the spread in emission frequencies. The emission lineshape is the same as that in absorption, given by eqn (2.6.11). If all other causes of line broadening are ignored, the total electric field with a given polarization in a plane-parallel radiated light beam at a fixed observation point can be written

$$E(t) = E_0 \sum_{i=1}^{v} \exp(-i\omega_i t + i\varphi_i), \tag{3.4.8}$$

where E_0 and φ_i are the fixed amplitude and phase of the wave train radiated by the ith atom. Different atoms have different frequencies ω_i of radiation that are Doppler shifted from ω_0 by amounts determined by the atomic velocities. The first-order correlation function is thus

$$\langle E^*(t)E(t+\tau) \rangle = E_0^2 \sum_{i,j=1}^{v} \langle \exp\{i\omega_i t - i\varphi_i - i\omega_j(t+\tau) + i\varphi_j\} \rangle. \tag{3.4.9}$$

The phase angles have fixed values but they are randomly distributed, and the contributions for $i \neq j$ average to zero, leaving

$$\langle E^*(t)E(t+\tau) \rangle = E_0^2 \sum_{i=1}^{v} \exp(-i\omega_i \tau). \tag{3.4.10}$$

The sum that remains is converted to an integral over the Gaussian distribution of Doppler-shifted frequencies given in eqn (2.6.11),

$$\left\langle E^*(t)E(t+\tau)\right\rangle$$

$$= vE_0^2\left(2\pi\Delta^2\right)^{-1/2}\int_0^\infty d\omega_i\exp(-i\omega_i\tau)\exp\left\{-(\omega_0-\omega_i)^2/2\Delta^2\right\} \qquad (3.4.11)$$

$$= vE_0^2\exp\left(-i\omega_0\tau-\tfrac{1}{2}\Delta^2\tau^2\right).$$

The degree of first-order coherence for Doppler-broadened chaotic light, defined in accordance with eqn (3.3.8), is therefore

$$g^{(1)}(\tau)=\exp\left(-i\omega_0\tau-\tfrac{1}{2}\Delta^2\tau^2\right)=\exp\left\{-i\omega_0\tau-\tfrac{\pi}{4}(\tau/\tau_c)^2\right\}, \qquad (3.4.12)$$

where the coherence time τ_c is defined in eqn (3.1.6). The modulus of this function is illustrated in Fig. 3.9. Its more rounded shape for small values of the time delay, compared to the degree of first-order coherence for light with a Lorentzian spectrum shown in Fig. 3.8, reflects the smoother form of the time series of the intensity fluctuations of light with a Gaussian spectrum, as shown in Fig. 3.5. Thus, over times τ much shorter than τ_c, the field amplitude and intensity deviate from their initial values by insignificant amounts, corresponding to the slow fall-off seen in Fig. 3.9.

The expressions for the degrees of first-order coherence of chaotic light with Lorentzian and Gaussian spectra given in eqns (3.4.7) and (3.4.12) respectively satisfy

$$\tau_c=\int_{-\infty}^\infty d\tau\left|g^{(1)}(\tau)\right|^2, \qquad (3.4.13)$$

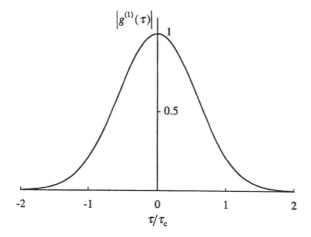

Fig. 3.9. The modulus of the degree of first-order coherence of chaotic light with a Gaussian frequency spectrum.

and this is sometimes taken as the definition of the coherence time for chaotic light with a known form of $g^{(1)}(\tau)$ [6].

The degree of first-order coherence for a stationary light beam introduced by eqns (3.3.6) and (3.3.8) is expressed in terms of the correlation between the fields in the beam at the same position but at different times. The definition of $g^{(1)}(\tau)$ is generalized for the correlation between the fields at two different space–time points (z_1, t_1) and (z_2, t_2) in the plane parallel light beam as

$$g^{(1)}(z_1, t_1; z_2, t_2) = \frac{\left\langle E^*(z_1, t_1) E(z_2, t_2) \right\rangle}{\left[\left\langle |E(z_1, t_1)|^2 \right\rangle \left\langle |E(z_2, t_2)|^2 \right\rangle \right]^{1/2}}, \tag{3.4.14}$$

where this form of definition also applies to light with nonstationary statistical properties, when the two factors in the denominator are generally different. The light at two space–time points has the following designations in terms of the value of the degree of first-order coherence.

$$\text{For } \left| g^{(1)}(z_1, t_1; z_2, t_2) \right| \begin{cases} = 1 \\ = 0 \\ \neq 0 \text{ or } 1 \end{cases} \quad \text{the light is} \begin{cases} \text{first–order coherent} \\ \text{incoherent} \\ \text{partially coherent.} \end{cases} \tag{3.4.15}$$

As the fluctuations in a parallel light beam in free space propagate at velocity c without any change in their form, the expressions for the degree of first-order coherence of stationary light derived above are simply converted by a redefinition of the delay time τ as

$$\tau = t_2 - t_1 - \frac{z_2 - z_1}{c}. \tag{3.4.16}$$

The expression for the degree of first-order coherence in eqn (3.3.8) remains valid with this replacement for τ. It is obvious from its definition that

$$g^{(1)}(0) = 1, \tag{3.4.17}$$

and the light remains approximately first-order coherent for delay times such that $\tau \ll \tau_c$. The field correlations in any kind of chaotic light must vanish for times τ much longer than τ_c, and as

$$\langle E(t) \rangle = 0, \tag{3.4.18}$$

it follows that the degree of first-order coherence must have a limiting value

$$g^{(1)}(\tau) \to 0 \quad \text{for} \quad \tau \gg \tau_c. \tag{3.4.19}$$

The light at such well separated space–time points is incoherent. The degrees of first-order coherence for chaotic light with Lorentzian and Gaussian spectra, shown in Figs. 3.8 and 3.9 respectively, illustrate these properties.

The coherence properties of chaotic light can be contrasted with those of the classical wave of stable amplitude and phase that is often assumed in theoretical treatments of optical experiments. For such a *deterministic* wave, propagated in the z direction with wavevector $k = \omega_0/c$, the electric field is

$$E(z,t) = E_0 \exp(ikz - i\omega_0 t + i\varphi), \tag{3.4.20}$$

where the amplitude and phase are fixed quantities, in contrast to the chaotic field of eqn (3.1.3). Figure 3.10 shows the variation of the electric field with time at a fixed observation point. The first-order correlation function is determined without any statistical averaging as

$$\langle E^*(z_1,t_1)E(z_2,t_2) \rangle = E_0^2 \exp(-i\omega_0 \tau), \tag{3.4.21}$$

with τ given by eqn (3.4.16). It follows that

$$g^{(1)}(\tau) = \exp(-i\omega_0 \tau) \tag{3.4.22}$$

and the light is first-order coherent at all space–time points. The expression (3.4.22) is the same as that obtained by letting the coherence times τ_c tend to infinity in the results for chaotic light given in eqns (3.4.7) and (3.4.12). Note that the property expressed by eqn (3.4.19) strictly refers to chaotic light and it does not apply to the classical stable wave. As is discussed in §7.4, the beam from a single-mode laser can approximate a classical stable wave of the kind represented in Fig. 3.10.

Problem 3.1   Consider a parallel light beam whose field contains a large number of contributions similar to the stable wave of Fig. 3.10, all with the

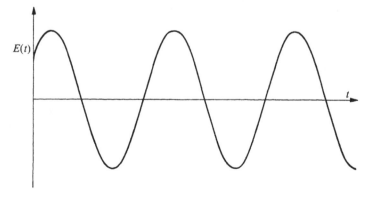

Fig. 3.10. Variation of electric field with time at a fixed observation point for a classical stable wave.

same frequency and wavevector but with a random distribution of phase angles. Prove that the beam is first-order coherent at any pair of space–time points.

Problem 3.2   Consider the beam of light produced by excitation of two stable waves, where the electric field is

$$E(z,t) = E_1 \exp(ik_1 z - i\omega_1 t) + E_2 \exp(ik_2 z - i\omega_2 t). \quad (3.4.23)$$

Prove that the light is first-order coherent at all pairs of points.

Problem 3.3   Consider a beam of light produced by excitation of two waves as in eqn (3.4.23), but where both exhibit random amplitudes and phases. If the average intensity is equally divided between the two waves, prove that

$$\left|g^{(1)}(\tau)\right| = \left|\cos\left\{\tfrac{1}{2}(\omega_1 - \omega_2)\tau\right\}\right|. \quad (3.4.24)$$

Note that the property in eqn (3.4.19) does not apply to any of the field excitations considered in these three problems.

Problem 3.4   Consider light from a source that simultaneously has collision and Doppler broadening. Prove that the degree of first-order coherence is

$$g^{(1)}(\tau) = \exp\left\{-i\omega_0 \tau - \gamma_{\text{coll}}|\tau| - \tfrac{1}{2}\Delta^2 \tau^2\right\}. \quad (3.4.25)$$

3.5 Interference fringes and frequency spectra

With the properties of the degree of first-order coherence now evaluated, we may return to the theory of the Mach–Zehnder interferometer outlined in §3.3. The output intensity, given by eqns (3.3.9) and (3.3.10), can be written in the more explicit form

$$\langle \bar{I}_4(t) \rangle = 2|\mathcal{R}|^2 |\mathcal{T}|^2 \langle \bar{I}(t) \rangle \left\{1 + \exp\left[-|z_1 - z_2|/c\tau_c\right] \cos\left[\omega_0(z_1 - z_2)/c\right]\right\}, \quad (3.5.1)$$

where the light is assumed to have Lorentzian broadening with a degree of first-order coherence given by eqn (3.4.7). The variations of the second term in the large bracket with relative path length $z_1 - z_2$ correspond to the interference fringes, whose visibility is conveniently defined as the modulation amplitude,

$$\text{visibility} = \exp\left[-|z_1 - z_2|/c\tau_c\right] = \left|g^{(1)}(\tau)\right|. \quad (3.5.2)$$

The chaotic nature of the light thus causes the fringe visibility to fall from its unit value for equal paths at a rate determined by the coherence length λ_c, defined in eqn (3.1.7), and the fringes disappear altogether for z_1 sufficiently different from z_2. However, for a narrow emission line, very many fringes are generated by the cosine term in eqn (3.5.1) before the exponential fall-off renders them invisible. For the parameter values assumed in eqn (3.1.1), there are of order 10^4 sharp fringes centred on the equal-path configuration and the interference is only slightly affected by the chaotic nature of the light. In this regime of first-order coherence of the light in the two internal paths of the interferometer, the output intensity of eqn (3.5.1) is approximately

$$\langle \bar{I}_4(t) \rangle = 4|\mathcal{R}|^2|\mathcal{T}|^2 \langle \bar{I}(t) \rangle \cos^2\left[\omega_0(z_1 - z_2)/2c\right]. \tag{3.5.3}$$

The most important cause of fringe blurring in a practical Mach–Zehnder interferometer is often the departure of the incident light beam from plane parallel form, arising for example from the finite extent of the source perpendicular to the beam axis. Such transverse effects belong to the theory of *spatial* coherence in contrast to the *temporal* coherence considered here.

The degree of first-order temporal coherence defined by eqns (3.3.8) or (3.4.14) is nevertheless important in describing the ultimate potential of plane parallel light of given statistical properties for the formation of interference fringes. The latter definition can be generalized to three-dimensional form as

$$g^{(1)}(\mathbf{r}_1,t_1;\mathbf{r}_2,t_2) = \frac{\langle E^*(\mathbf{r}_1,t_1)E(\mathbf{r}_2,t_2) \rangle}{\left[\langle |E(\mathbf{r}_1,t_1)|^2 \rangle \langle |E(\mathbf{r}_2,t_2)|^2 \rangle\right]^{1/2}} \tag{3.5.4}$$

and the coherence of the light at the space–time points $(\mathbf{r}_1,t_1)$ and $(\mathbf{r}_2,t_2)$ has similar designations to those given in eqn (3.4.15). For light of stationary statistical properties, the averages in the denominator of eqn (3.5.4) are independent of t_1 and t_2, while the average in the numerator is a function of the spatial positions and the time difference $t_2 - t_1$. The magnitude of the degree of first-order coherence determines the visibility of the interference fringes that could ideally be produced by superposition of light from the two space–time points, and the definition in eqn (3.5.2) is generalized to

$$\text{visibility} = \left|g^{(1)}(\mathbf{r}_1,t_1;\mathbf{r}_2,t_2)\right|. \tag{3.5.5}$$

These three-dimensional expressions are required for many practical interference experiments but their more complicated forms obscure some of the physical principles that govern optical coherence and fluctuations. We restrict the treatment that follows to plane parallel light beams where inclusion of one spatial dimension is adequate.

It is clear from the above discussions that the degree of first-order temporal coherence and the frequency spectrum are different aspects of the same physical processes that govern the emission of light by the active material in the light source. The electric fields in the time and frequency domains are related by the restricted Fourier transform

$$E_T(\omega) = \frac{1}{\sqrt{2\pi}} \int_T dt E(t) \exp(i\omega t), \qquad (3.5.6)$$

where T denotes a range of integration, similar to that in the definition of the first-order correlation function in eqn (3.3.6). The time T is taken to be much longer than the coherence time τ_c of the light and, with this assumption, the cycle-averaged intensity of the light of frequency ω is determined by the *power spectral density*, defined as

$$f(\omega) = \frac{|E_T(\omega)|^2}{T} = \frac{1}{2\pi T} \int_T dt \int_T dt' E^*(t) E(t') \exp[-i\omega(t-t')]. \qquad (3.5.7)$$

As is discussed in §3.3, the terms on the right-hand side that generate the first-order correlation function depend only on the time difference $\tau = t' - t$ for stationary light beams. Thus, with a change of variable from t' to τ and use of the definition (3.3.6), the power spectral density takes the form

$$f(\omega) = \frac{1}{2\pi} \int_{-\infty}^{\infty} d\tau \langle E^*(t) E(t+\tau) \rangle \exp(i\omega\tau), \qquad (3.5.8)$$

where the insertion of infinite limits on the integral is valid for $T \gg \tau_c$. Use of the delta-function representation from eqn (2.4.9) gives

$$\int_{-\infty}^{\infty} d\omega f(\omega) = \langle E^*(t) E(t) \rangle. \qquad (3.5.9)$$

Division of eqn (3.5.8) by eqn (3.5.9) and use of the definition of the degree of first-order temporal coherence in eqn (3.3.8) gives an expression for the normalized spectrum as

$$F(\omega) = \frac{1}{2\pi} \int_{-\infty}^{\infty} d\tau g^{(1)}(\tau) \exp(i\omega\tau). \qquad (3.5.10)$$

This relation between the spectrum of the light and its degree of first-order coherence is a form of the *Wiener–Khintchine theorem*. It gives a formal connection between the results of spectroscopic experiments and the results of measurements of the time-dependent fluctuation properties of light. With the use of eqn (3.3.12), the relation can be recast as

$$F(\omega) = \frac{1}{\pi} \text{Re} \int_0^{\infty} d\tau g^{(1)}(\tau) \exp(i\omega\tau), \qquad (3.5.11)$$

which involves only integration over positive τ.

Problem 3.5   Use the Wiener–Khintchine theorem to obtain the frequency spectrum of the excitation described in Problem 3.3.

Problem 3.6   Prove that the normalized Lorentzian and Gaussian lineshape functions of eqns (2.5.20) and (2.6.11) respectively are correctly generated by the Wiener–Khintchine theorem with use of the appropriate degrees of first-order coherence. More generally, show that the expression (2.6.15) for the composite lineshape, with γ replaced by γ_{coll}, is obtained with use of the degree of first-order coherence given in eqn (3.4.25).

The relation (2.9.3) between the collisional decay rate, as it appears in the broadened Lorentzian lineshape, and the mean time between collisions is justified by the result of this problem.

The expression (2.6.14) for the composite lineshape in the presence of two distinct line-broadening processes takes a useful form when the frequency distributions in the integrand are replaced by the Wiener–Khintchine spectra from eqn (3.5.10), to give

$$F(\omega) = \frac{1}{2\pi} \int_{-\infty}^{\infty} d\tau g_1^{(1)}(\tau) g_2^{(1)}(\tau) \exp\{i(\omega + \omega_0)\tau\}. \tag{3.5.12}$$

Here $g_1^{(1)}(\tau)$ and $g_2^{(1)}(\tau)$ are the degrees of first-order coherence associated with light beams subjected to the two line-broadening processes separately. This form shows that the composite lineshape is essentially the Fourier transform of the product of the appropriate degrees of coherence, in agreement with the composite degrees of coherence in eqn (3.4.7) for the combination of radiative and collision broadening and in eqn (3.4.25) for the combination of collision and Doppler broadening. It also follows from eqn (3.5.12) that the composite lineshape for the combination of two processes that each generate Gaussian lineshapes is another Gaussian whose variance is the sum of the variances of the individual lineshapes.

3.6 Intensity fluctuations of chaotic light

The properties of chaotic light derived so far concern the electric-field correlation function and its role in the determination of the fringes in a Mach–Zehnder interferometer and other first-order interference experiments. The second main topic of the chapter concerns the direct measurement of intensity fluctuations of the kind shown in Figs. 3.4 and 3.5. We describe the observation of higher-order interference effects that depend on the correlations of two *intensities* rather than the correlations of two *fields*.

As a preliminary to the discussion of intensity interference experiments, the present section considers the statistical properties of the intensity fluctuations in chaotic light. The fluctuations in the cycle-averaged intensity are too rapid for direct observation in many cases, and what is measured is some average of the fluctuations over a detector response time. We suppose initially however that an ideal detector is available, with response time much shorter than the coherence time τ_c, that can take effectively instantaneous measurements of the intensity.

The average of a large number of values of the cycle-averaged intensity $\bar{I}(t)$ measured over a period of time very much longer than τ_c is easy to calculate. With the long observation period and the assumption of instantaneous intensity measurements, the time average can be replaced by a statistical average over the distribution of phase angles, denoted by angle brackets, as explained in §3.3. The long-time average intensity in a plane parallel light beam from eqns (3.1.3) and (3.1.4) is

$$\bar{I} \equiv \langle \bar{I}(t) \rangle = \tfrac{1}{2}\varepsilon_0 c E_0^2 \langle |\exp(i\varphi_1(t)) + \exp(i\varphi_2(t)) + \ldots + \exp(i\varphi_v(t))|^2 \rangle$$
$$= \tfrac{1}{2}\varepsilon_0 c E_0^2 v, \tag{3.6.1}$$

as the cross-terms between the phase factors for different radiating atoms give zero average contributions.

The average values of higher powers of the cycle-averaged intensity are calculated in a similar fashion. Thus the mean square intensity is

$$\langle \bar{I}(t)^2 \rangle = \tfrac{1}{4}\varepsilon_0^2 c^2 E_0^4 \langle |\exp(i\varphi_1(t)) + \exp(i\varphi_2(t)) + \ldots + \exp(i\varphi_v(t))|^4 \rangle. \tag{3.6.2}$$

In taking the average of the fourth power of the sum of phase factors, only the terms in which each factor is multiplied by its complex conjugate are non-zero. These contributions give

$$\langle \bar{I}(t)^2 \rangle = \tfrac{1}{4}\varepsilon_0^2 c^2 E_0^4 \left\{ \sum_i \langle |\exp(i\varphi_i(t))|^4 \rangle + \sum_{i>j} \langle |2\exp[i(\varphi_i(t) + \varphi_j(t))]|^2 \rangle \right\}$$
$$= \tfrac{1}{4}\varepsilon_0^2 c^2 E_0^4 \{ v + 2v(v-1) \}, \tag{3.6.3}$$

where the first and second terms are the contributions of the individual atoms and pairs of atoms, respectively. The second summation counts the number of distinct pairs and the factor of 2 is the number of occurrences of the sum of phase angles shown in the exponent for a given pair i, j. Thus in terms of the average intensity from eqn (3.6.1), the mean-square intensity is

$$\langle \bar{I}(t)^2 \rangle = \left(2 - \frac{1}{v} \right) \bar{I}^2. \tag{3.6.4}$$

The number v of radiating atoms is normally very large and eqn (3.6.4) can be written

$$\langle \bar{I}(t)^2 \rangle = 2\bar{I}^2 \quad (v \gg 1), \tag{3.6.5}$$

to a very good approximation. Note that this expression results from the pair term in eqn (3.6.3).

The variance of the cycle-averaged intensity is

$$(\Delta I)^2 = \left\langle \bar{I}(t)^2 \right\rangle - \left\langle \bar{I}(t) \right\rangle^2 = \bar{I}^2. \tag{3.6.6}$$

The size of fluctuation, characterized by ΔI, is thus equal to the mean value $\bar{I}$, as is qualitatively evident in Figs. 3.4 and 3.5. A similar result was found in eqns (1.4.7) or (1.4.8) for the fluctuations in the number of thermally excited photons in a single cavity mode for $\langle n \rangle \gg 1$.

The averages of powers of $\bar{I}(t)$ higher than the second are quite complicated in general, but relatively simple results are obtained if the number of radiating atoms is assumed to be very large. The dominant contribution to the average of the rth power of the intensity then comes from the terms that involve the square moduli of products of the phase factors of r distinct atoms, as in the second-moment calculation in eqn (3.6.3). Thus, approximately,

$$\left\langle \bar{I}(t)^r \right\rangle = \left(\tfrac{1}{2} \varepsilon_0 c E_0^2 \right)^r \sum_{i>j>k>\dots} \left\langle \left| r! \exp\left[i \left(\varphi_i(t) + \varphi_j(t) + \varphi_k(t) + \dots \right) \right] \right|^2 \right\rangle$$

$$= \left(\tfrac{1}{2} \varepsilon_0 c E_0^2 \right)^r {}^{\nu}C_r (r!)^2, \tag{3.6.7}$$

where the summation runs over all the different sets of r atoms and the factor of $r!$ is the number of occurrences of the sum of r phase angles shown in the exponent for a given selection of r atoms. Then with ν assumed to be very much larger than r,

$$\left\langle \bar{I}(t)^r \right\rangle = r! \bar{I}^r \quad (\nu \gg r), \tag{3.6.8}$$

where the average intensity is taken from eqn (3.6.1).

This result for the rth moment of the intensity fluctuation distribution, although derived for the collision-broadened source, is in fact valid for any kind of chaotic light. The occurrence of the $r!$ factor in eqn (3.6.8) is a universal characteristic of chaotic light. The similarity to the rth *factorial* moment of the thermal photon distribution in eqn (1.4.4) should be noted.

Provided that the number of radiating atoms is very large, it is possible to calculate not only the moments, as above, but also the explicit form of the probability distribution for the cycle-averaged intensity. The first step is the determination of the statistical distribution of the values of $a(t)$. It is seen from eqn (3.1.3) and Fig. 3.3 that $a(t)$ is the distance from the origin in an Argand diagram after ν steps of unit length in random directions specified by the angles $\varphi_1(t)$, $\varphi_2(t)$, ..., $\varphi_\nu(t)$. Let $p[a(t)]$ be the probability that the end point of this random walk lies in unit area around the point specified by coordinates $a(t)$ and $\varphi(t)$ in the plane of Fig. 3.3. The result given by standard random-walk theory [6] for $\nu \gg 1$ is

$$p[a(t)] = (1/\pi v)\exp\!\left(-a(t)^2/v\right),$$ (3.6.9)

being independent of $\varphi(t)$, as expected from the physical nature of the problem. The probability is normalized,

$$\int_0^\infty da(t) \int_0^{2\pi} d\varphi(t) a(t) p[a(t)] = 1.$$ (3.6.10)

The distribution in eqn (3.6.9) is a Gaussian function and, in accordance with eqn (3.1.3), the corresponding distribution for the modulus of the electric field amplitude $|E(t)|$ is obtained upon multiplication of $a(t)$ by E_0. The probability is greatest at $|E(t)| = 0$ and electric-field variance, obtained by comparison with the standard Gaussian form in eqn (2.6.11), is

$$\left(\Delta|E(t)|\right)^2 = \tfrac{1}{2} E_0^2 v.$$ (3.6.11)

The field distribution remains independent of $\varphi(t)$ and some of its contours are shown in Fig. 3.11, where the shaded disc has a diameter equal to the square root of the variance. Similar distributions occur for all chaotic light beams, and the light from a chaotic source is sometimes called Gaussian light.

It is important not to confuse the Gaussian distribution of the field amplitude $|E(t)|$ with the distribution of frequencies in the beam, which is a Lorentzian function for the collision-broadening mechanism assumed here. Such light is commonly referred to as *Gaussian–Lorentzian*, where 'Gaussian' refers to the amplitude statistics and 'Lorentzian' to the frequency spectrum. Chaotic light with a Gaussian spectrum is correspondingly referred to as *Gaussian–Gaussian*.

The probability distribution for $a(t)$ in eqn (3.6.9) is converted to a probab-

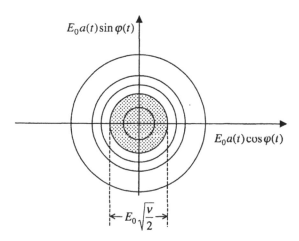

Fig. 3.11. Contours of the probability distribution for the amplitude and phase of the electric field of a chaotic light beam.

ility distribution for $\bar{I}(t)$ with the use of eqn (3.1.4). Thus with the expression (3.6.1) for $\bar{I}$, the probability that an instantaneous measurement of the cycle-averaged intensity yields a value between $\bar{I}(t)$ and $\bar{I}(t) + d\bar{I}(t)$ is $p[\bar{I}(t)]d\bar{I}(t)$, where

$$p[\bar{I}(t)] = \left(1/\bar{I}\right)\exp\left(-\bar{I}(t)/\bar{I}\right). \tag{3.6.12}$$

The normalization of the distribution is easily verified. Its form shows that the most probable value of $\bar{I}(t)$ is always zero, and this feature is qualitatively evident from Figs. 3.4 and 3.5. Like the amplitude distribution of eqn (3.6.9), the intensity distribution of eqn (3.6.12) is valid for all kinds of chaotic light, irrespective of its frequency spectrum. The moments of the intensity distribution, given by

$$\left\langle \bar{I}(t)^r \right\rangle = \left(1/\bar{I}\right)\int_0^\infty d\bar{I}(t)\bar{I}(t)^r \exp\left(-\bar{I}(t)/\bar{I}\right) = r!\bar{I}^r, \tag{3.6.13}$$

agree with the result in eqn (3.6.8) obtained previously.

The random-walk probability distribution refers to the end points of a large number of walks that all begin at the origin, and each such point defines an amplitude a and a phase φ for a light beam. The collection of light beams with all possible amplitudes and phases forms a statistical ensemble of the kind described in §1.4. Each light beam in the ensemble has fixed a and φ, and the intensity distribution of eqn (3.6.12) is rigorously valid for the ensemble of fixed-amplitude and fixed-phase beams. Its application to determine time averages of a long series of instantaneous intensity measurements on a single beam of chaotic light relies once more on the ergodic statistics of the beam. All of the above results refer to idealized experiments that measure the instantaneous intensity. The effects of a finite detector response time, particularly on the observed second moment of the intensity fluctuations, are discussed in §3.8.

A contrast to the results for chaotic light is provided by the classical stable wave of Fig. 3.10. There is no need to employ statistics in this case, as the cycle-averaged intensity is constant. The result analogous to eqn (3.6.6) is

$$(\Delta I)^2 = 0 \tag{3.6.14}$$

and there are no intensity fluctuations.

3.7 Degree of second-order coherence

The intensity-fluctuation properties of chaotic light described in the previous section refer to averages of intensity readings taken at single instants of time. We now consider two-time measurements in which many pairs of readings of the cycle-averaged intensity are taken with a fixed time delay τ. The readings are again taken at a fixed point in space and only a single polarization is measured.

The average of the product of each pair of readings is the intensity correlation function of the light, analogous to the electric-field correlation function defined in eqn (3.3.6). The measurement of intensity correlations is described in §3.8; we first consider the theory of the correlation function.

It is convenient to work with a normalized form of the correlation function called the *degree of second-order temporal coherence,*

$$g^{(2)}(\tau) = \frac{\langle \bar{I}(t)\bar{I}(t+\tau)\rangle}{\bar{I}^2} = \frac{\langle E^*(t)E^*(t+\tau)E(t+\tau)E(t)\rangle}{\langle E^*(t)E(t)\rangle^2}, \tag{3.7.1}$$

where $\bar{I}$ is the long-time average intensity, defined as in eqn (3.6.1), and the order of field factors in the second-order electric-field correlation function follows a convention. The light beam is again assumed to have stationary statistical properties. It is clear from the symmetry of the definition that the property analogous to that in eqn (3.3.12) is

$$g^{(2)}(-\tau) = g^{(2)}(\tau), \tag{3.7.2}$$

and calculations need only be made for positive τ. It is assumed that the measurements satisfy the conditions for equivalence of statistical and time averages.

We have seen that the magnitude of the degree of first-order coherence takes values in the range 0 to 1. The allowed range of values of the degree of second-order coherence is controlled by Cauchy's inequality, which applies to any pair of real numbers. Thus two measurements of the intensity at times t_1 and t_2 must satisfy

$$2\bar{I}(t_1)\bar{I}(t_2) \le \bar{I}(t_1)^2 + \bar{I}(t_2)^2. \tag{3.7.3}$$

By applying this inequality to the cross terms, it is easy to show that

$$\left\{ \frac{\bar{I}(t_1)+\bar{I}(t_2)+...+\bar{I}(t_N)}{N} \right\}^2 \le \frac{\bar{I}(t_1)^2 + \bar{I}(t_2)^2 +...+\bar{I}(t_N)^2}{N} \tag{3.7.4}$$

for the results of N measurements of the intensity. Thus in terms of statistical averages,

$$\bar{I}^2 \equiv \langle \bar{I}(t)\rangle^2 \le \langle \bar{I}(t)^2\rangle, \tag{3.7.5}$$

and the degree of second-order coherence for zero time delay from eqn (3.7.1) satisfies

$$1 \le g^{(2)}(0). \tag{3.7.6}$$

It is not possible to establish any upper limit, and the complete range of allowed values is

$$1 \le g^{(2)}(0) \le \infty. \tag{3.7.7}$$

The derivations of these inequalities assume a stationary light beam but they also apply to a series of measurements on nonstationary light. For the extreme example of a single optical pulse, it is clear that measurements at a fixed observation point must produce quite different results that depend on the locations of the pulse at the times of measurement. There is no equivalence between time and statistical averaging in this case. However, the N measurements of the intensity envisaged in eqn (3.7.4) can in principle be made at the same time but on the members of an ensemble of N realizations of the same optical pulse. The inequalities in eqns (3.7.6) and (3.7.7) continue to apply with this interpretation of the averaging, and they can be taken as general properties of all kinds of classical light.

The above proof cannot be extended to nonzero time delays, and the only restriction on the degree of second-order coherence then results from the essentially positive nature of the intensity, which gives

$$0 \le g^{(2)}(\tau) \le \infty \quad \tau \ne 0. \tag{3.7.8}$$

There is, however, another important property that follows from the inequality

$$\left\{ \bar{I}(t_1)\bar{I}(t_1 + \tau) + \ldots + \bar{I}(t_N)\bar{I}(t_N + \tau) \right\}^2$$

$$\le \left\{ \bar{I}(t_1)^2 + \ldots + \bar{I}(t_N)^2 \right\} \left\{ \bar{I}(t_1 + \tau)^2 + \ldots + \bar{I}(t_N + \tau)^2 \right\}, \tag{3.7.9}$$

which is also readily established with the use of eqn (3.7.3). The two summations on the right are equal for a sufficiently long and numerous series of measurements, and the square root of eqn (3.7.9) then produces the result

$$\left\langle \bar{I}(t)\bar{I}(t + \tau) \right\rangle \le \left\langle \bar{I}(t)^2 \right\rangle, \tag{3.7.10}$$

or

$$g^{(2)}(\tau) \le g^{(2)}(0). \tag{3.7.11}$$

The degree of second-order coherence can therefore never exceed its value for zero time delay. The inequality again applies both to stationary and nonstationary light beams in the classical theory.

These general properties of the degree of second-order coherence are illustrated by the example of chaotic light. We suppose, as in the models used in §3.4 for both collision and Doppler-broadened light, that the electric field of the plane parallel light beam is made up of independent contributions from the different radiating atoms i,

$$E(t) = \sum_{i=1}^{v} E_i(t).$$

(3.7.12)

The second-order electric-field correlation function that appears in eqn (3.7.1) is thus,

$$\langle E^*(t)E^*(t+\tau)E(t+\tau)E(t)\rangle = \sum_{i=1}^{v}\langle E_i^*(t)E_i^*(t+\tau)E_i(t+\tau)E_i(t)\rangle$$

$$+ \sum_{i\neq j}\{\langle E_i^*(t)E_j^*(t+\tau)E_j(t+\tau)E_i(t)\rangle + \langle E_i^*(t)E_j^*(t+\tau)E_i(t+\tau)E_j(t)\rangle\},$$

(3.7.13)

where only those terms are retained in which the field from each atom is multiplied by its complex conjugate. All other terms vanish because of the random relative phases of the waves from the different atoms. Thus taking account of the equivalence of the contributions from each atom

$$\langle E^*(t)E^*(t+\tau)E(t+\tau)E(t)\rangle = v\langle E_i^*(t)E_i^*(t+\tau)E_i(t+\tau)E_i(t)\rangle$$

$$+ v(v-1)\{\langle E_i^*(t)E_i(t)\rangle^2 + |\langle E_i^*(t)E_i(t+\tau)\rangle|^2\}.$$

(3.7.14)

If the number of radiating atoms is now assumed to be very large, the contributions that involve pairs of atoms greatly exceed the single-atom contributions and, to a good approximation,

$$\langle E^*(t)E^*(t+\tau)E(t+\tau)E(t)\rangle = v^2\{\langle E_i^*(t)E_i(t)\rangle^2 + |\langle E_i^*(t)E_i(t+\tau)\rangle|^2\}.$$

(3.7.15)

The corresponding result for the first-order electric-field correlation function is given in eqn (3.4.2). Thus with the definitions in eqns (3.3.8) and (3.7.1) of the degrees of first- and second-order coherence, eqns (3.4.2) and (3.7.15) give

$$g^{(2)}(\tau) = 1 + |g^{(1)}(\tau)|^2 \quad (v \gg 1).$$

(3.7.16)

It follows from the limiting values of the degree of first-order coherence given in eqns (3.4.17) and (3.4.19) that

$$g^{(2)}(0) = 2$$

(3.7.17)

and

$$g^{(2)}(\tau) \rightarrow 1 \quad \text{for} \quad \tau \gg \tau_c.$$

(3.7.18)

These important relations hold for all varieties of chaotic light.

The degree of second-order coherence of Gaussian–Lorentzian light is found by substitution of the degree of first-order coherence from eqn (3.4.7) into eqn (3.7.16) as

$$g^{(2)}(\tau) = 1 + \exp(-2\gamma|\tau|) = 1 + \exp(-2|\tau|/\tau_c). \qquad (3.7.19)$$

The corresponding expression for Gaussian–Gaussian light emitted by a Doppler-broadened source is obtained with the use of eqn (3.4.12) as

$$g^{(2)}(\tau) = 1 + \exp(-\Delta^2\tau^2) = 1 + \exp\left\{-\pi(\tau/\tau_c)^2\right\}. \qquad (3.7.20)$$

These degrees of second-order coherence for chaotic light are illustrated in Fig. 3.12. It is evident that the general inequalities (3.7.6) and (3.7.11) are satisfied and both kinds of chaotic light have the limiting values of second-order coherence given in eqns (3.7.17) and (3.7.18).

The peak in the degree of second-order coherence of chaotic light for $\tau < \tau_c$ is a manifestation of the kinds of intensity fluctuation shown in Figs. 3.4 and 3.5. For such small delay times, the two intensity measurements to be correlated in the degree of second-order coherence often fall within the same fluctuation peak to give an enhanced contribution. For longer delay-times, $\tau > \tau_c$, the two intensities tend to be uncorrelated and the degree of second-order coherence is close to unity.

These properties of chaotic light contrast with those of the classical wave of stable amplitude and phase, whose electric field variation is represented by eqn (3.4.20). It is easily shown by substitution of the field into the definition of eqn (3.7.1) that

$$g^{(2)}(\tau) = 1, \qquad (3.7.21)$$

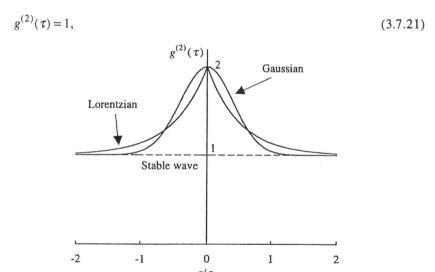

Fig. 3.12. Degrees of second-order coherence of chaotic light having Gaussian and Lorentzian frequency distributions with coherence time τ_c. The dashed line shows the constant unit second-order coherence of a classical stable wave.

independent of the time delay τ. This stable-wave result is indicated by the horizontal dashed line in Fig. 3.12.

The degrees of second-order coherence calculated above assume a stationary, polarized, plane-parallel beam of light and a common observation point. The definition (3.7.1) can be generalized, analogous to eqn (3.5.4), to cover non-stationary optical fields with a three-dimensional spatial dependence. The degree of second-order temporal coherence between the light at space–time points $(\mathbf{r}_1, t_1)$ and $(\mathbf{r}_2, t_2)$ is defined as

$$g^{(2)}(\mathbf{r}_1, t_1; \mathbf{r}_2, t_2) = \frac{\left\langle E^*(\mathbf{r}_1, t_1) E^*(\mathbf{r}_2, t_2) E(\mathbf{r}_2, t_2) E(\mathbf{r}_1, t_1) \right\rangle}{\left\langle \left| E(\mathbf{r}_1, t_1) \right|^2 \right\rangle \left\langle \left| E(\mathbf{r}_2, t_2) \right|^2 \right\rangle}, \tag{3.7.22}$$

where the angle brackets again denote ensemble averages. Note that this function is a special case of a more general second-order coherence in which the four fields in the correlation function are evaluated at four different space–time points, but we do not consider this further generalization. The light at points $(\mathbf{r}_1, t_1)$ and $(\mathbf{r}_2, t_2)$ is said to be second-order coherent if simultaneously

$$\left| g^{(1)}(\mathbf{r}_1, t_1; \mathbf{r}_2, t_2) \right| = 1 \quad \text{and} \quad g^{(2)}(\mathbf{r}_1, t_1; \mathbf{r}_2, t_2) = 1. \tag{3.7.23}$$

The inclusion of spatial dependence in the degree of second-order coherence is here restricted to stationary plane-polarized parallel beams where only the single z coordinate is needed. The spatial dependence of the field is then incorporated as in the degree of first-order coherence defined by eqn (3.4.14) and it is not difficult to verify that

$$g^{(2)}(z_1, t_1; z_2, t_2) = g^{(2)}(\tau), \tag{3.7.24}$$

where τ is defined in eqn (3.4.16). With this interpretation of τ, the expressions already derived for the degree of second-order coherence apply to spatially separated points in the parallel beam.

It is evident from the discussion of eqn (3.4.14) that chaotic light is always first-order coherent for space–time points sufficiently close together. This is illustrated by the approach of the degrees of first-order coherence of the chaotic light in Figs. 3.8 and 3.9 towards unity for small values of τ/τ_c. However, the small τ limit produces a degree of second-order coherence equal to 2, according to eqn (3.7.17) and as illustrated in Fig. 3.12. It is not possible for chaotic light to be second-order coherent in accordance with the definition in eqn (3.7.23) for any choice of space–time points. By contrast, eqns (3.4.22) and (3.7.21) show that the classical stable wave is second-order coherent at all space–time points.

Problem 3.7 Consider the single-mode chaotic light beam, defined in Problem 3.1 as a randomly-phased superposition of a large number of stable waves. Prove from first principles that

$$g^{(2)}(\tau) = 2.   \qquad (3.7.25)$$

The coherence time is infinite in this case and eqn (3.7.18) does not apply.

The degrees of second-order coherence derived above refer to measurements on a single beam of light. The theory is readily generalized to refer to measurements on two distinct beams, denoted by subscript labels a and b. The beams may differ, for example, in their polarizations, their directions of propagation, or their non-overlapping frequency distributions, such that they can be distinguished experimentally. Consider measurements made on the two beams at a common spatial position but at different times. Then, analogous to eqn (3.7.1), there are now four degrees of second-order coherence

$$g_{l,m}^{(2)}(t_1,t_2) = \frac{\langle \bar{I}_l(t_1)\bar{I}_m(t_2)\rangle}{\langle \bar{I}_l(t_1)\rangle\langle \bar{I}_m(t_2)\rangle} = \frac{\langle E_l^*(t_1)E_m^*(t_2)E_m(t_2)E_l(t_1)\rangle}{\langle E_l^*(t_1)E_l(t_1)\rangle\langle E_m^*(t_2)E_m(t_2)\rangle}$$

$l,m = a,a$ or b,b: *intrabeam* degrees of coherence   (3.7.26)

$l,m = a,b$ or b,a: *interbeam* degrees of coherence.

The values of these degrees of second-order coherence for two beams that have stationary and ergodic statistical properties are subject to a restriction obtained from an inequality similar to (3.7.9), but for products of the intensities of beams a and b, as

$$\left.\begin{array}{c} \left[g_{a,b}^{(2)}(t_1,t_2)\right]^2 \\[2mm] \left[g_{b,a}^{(2)}(t_2,t_1)\right]^2 \end{array}\right\} \leq g_{a,a}^{(2)}(t_1,t_1)g_{b,b}^{(2)}(t_2,t_2). \qquad (3.7.27)$$

The relation can also be derived for statistical averages on an ensemble of pairs of nonstationary light beams, and the inequality (3.7.27) represents another general classical property of light. Some examples of nonstationary light beams are treated in §6.7.

A simple special case of these degrees of second-order coherence occurs for measurements that do not distinguish between the contributions of the two beams.

Problem 3.8   Consider the light beam formed by superposition of two independent stationary beams, labelled a and b, with a total cycle-averaged intensity

$$\bar{I}(t) = \bar{I}_a(t) + \bar{I}_b(t). \qquad (3.7.28)$$

Show that the overall degree of second-order coherence for a measurement that does not distinguish the two beams is

$$g^{(2)}(\tau) = \frac{\bar{I}_a^2 g_{a,a}^{(2)}(\tau) + \bar{I}_b^2 g_{b,b}^{(2)}(\tau) + 2\bar{I}_a \bar{I}_b}{\left(\bar{I}_a + \bar{I}_b\right)^2}, \tag{3.7.29}$$

where eqn (3.7.1) is used and $\tau = t_2 - t_1$.

This result can be applied to an unpolarized light beam where the mode labels a and b refer to two orthogonal polarizations. With equal intensities and degrees of second-order coherence in the two components, eqn (3.7.29) reduces to

$$g^{(2)}(\tau) = \tfrac{1}{2}\left\{1 + g_{a,a}^{(2)}(\tau)\right\}. \tag{3.7.30}$$

For unpolarized chaotic light, the degree of second-order coherence has the limiting values of 3/2 for $\tau = 0$ and 1 for $\tau = \infty$.

The concept of optical coherence can be generalized to higher orders, where the degrees of first- and second-order coherence defined in eqns (3.5.4) and (3.7.22) appear as the first two members of an infinite hierarchy. Thus the general degree of rth-order coherence has a correlation of the fields at $2r$ space–time points in the numerator and a denominator similar to that in eqn (3.5.4), but with a string of $2r$ factors inside the square root. A less general degree of rth order coherence has the corrrelation of $2r$ fields in the numerator, but at only r different space–time points, with a string of r factors in the denominator, similar to the structure of the degree of second-order coherence in eqn (3.7.22). We consider only this latter form, which essentially involves the optical intensities at r space–time points and their correlation.

Two simple results can be written down with the use of previous discussion. Thus for the classical stable wave, the same arguments as used to derive the degree of second-order coherence in eqn (3.7.21) show that

$$g^{(r)}\left(\mathbf{r}_1, t_1; \mathbf{r}_2, t_2; \ldots; \mathbf{r}_r, t_r\right) = 1 \quad \text{for all } r, \tag{3.7.31}$$

where the notation for the degree of rth-order coherence is an obvious extension of that used for $r = 2$ in eqn (3.7.22). By analogy with eqn (3.7.23), the classical stable wave can be said to be coherent in all orders. There is no similarly simple result for a chaotic light beam, but the very special case where all the space–time points are the same is easily obtained. Thus for a polarized beam propagated parallel to the z axis, the correlation function is the same as that evaluated in eqn (3.6.8), and

$$g^{(r)}\left(z_1, t_1; z_1, t_1; \ldots; z_1, t_1\right) = r!, \tag{3.7.32}$$

with the second-order result in eqn (3.7.17) as a special case.

3.8 The Brown–Twiss interferometer

The results of all the classical interference experiments, typified by the Mach–

Zehnder interferometer, are controlled by the correlation of two electric-field amplitudes, conveniently expressed by the degree of first-order coherence of the light. The correlation of two optical intensities, conveniently expressed in terms of the degree of second-order coherence, was first measured by Hanbury Brown and Twiss [7,8]. Their experiment typifies all subsequent measurements of degrees of second-order coherence. Such measurements have a particular significance for the correspondence between the classical and quantum theories of light, which is explored in §5.10. We here consider the classical theory of the experiment as a preliminary to the more general discussion of §§5.8 and 5.9 in the framework of quantum theory.

The apparatus is shown schematically in Fig. 3.13. Light from a mercury arc is filtered to retain only the 435.8 nm emission line and the beam is divided into two equal portions by a 50:50 beam splitter, as specified by eqn (3.2.13). The intensity of each portion is measured by a photomultiplier detector, the principles of which are outlined in §§3.9 and 4.11, and the outputs of these detectors, D_3 and D_4, are multiplied together in the correlator. The integrated value of the product over a long period of observation provides a measurement of the intensity fluctuations. We assume here an idealized arrangement in which the detectors are symmetrically placed with respect to the beam splitter. They thus measure the intensities in the beams at equal linear distances z from the light source.

According to classical theory, the beam splitter divides the incident cycle-averaged intensity $\bar{I}_1(z,t)$ into two identical beams in arms 3 and 4 of the experiment, where we use the notation of Fig. 3.6 for the input and output arms of the beam splitter. In an obvious notation,

$$\bar{I}_3(z,t) = \bar{I}_4(z,t) = \tfrac{1}{2}\bar{I}_1(z,t),\tag{3.8.1}$$

and the long-time average intensities in the two outputs are

$$\bar{I}_3 = \bar{I}_4 = \tfrac{1}{2}\bar{I}_1.\tag{3.8.2}$$

Thus with the degrees of second-order coherence subscripted as in eqn (3.7.26) to

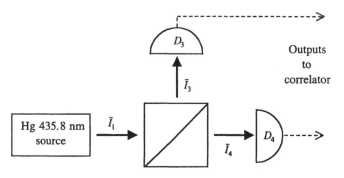

Fig. 3.13. Arrangement of the main components in the Brown–Twiss intensity interference experiment.

indicate the correlated input or output beams, we have

$$g_{3,4}^{(2)}(\tau) = \frac{\langle \bar{I}_3(t)\bar{I}_4(t+\tau)\rangle}{\bar{I}_3\bar{I}_4} = \frac{\langle \bar{I}_1(t)\bar{I}_1(t+\tau)\rangle}{\bar{I}_1^2} = g_{1,1}^{(2)}(\tau),$$ (3.8.3)

where the z dependence is removed. It is seen that the normalized cross correlation between the two output beams provides a measurement of the degree of second-order coherence of the input beam.

With ideal detectors that make instantaneous measurements of the intensities at times t_3 and t_4, the correlator in the experiment is designed to form the average

$$\langle [\bar{I}_3(z,t_3) - \bar{I}_3][\bar{I}_4(z,t_4) - \bar{I}_4]\rangle = \tfrac{1}{4}\{\langle \bar{I}_1(z,t_3)\bar{I}_1(z,t_4)\rangle - \bar{I}_1^2\}.$$ (3.8.4)

The normalized version of this Brown–Twiss correlation is readily expressed in terms of the degree of second-order coherence of the incident light as

$$\frac{\langle [\bar{I}_3(z,t_3) - \bar{I}_3][\bar{I}_4(z,t_4) - \bar{I}_4]\rangle}{\bar{I}_3\bar{I}_4} = g_{1,1}^{(2)}(\tau) - 1,$$ (3.8.5)

where $\tau = t_4 - t_3$. Apart from the normalization factor, the experiment of Hanbury Brown and Twiss provides a direct measurement of the deviation from unity of the degree of second-order coherence. The results can be envisaged in terms of the examples shown in Fig. 3.12. Only chaotic light was available at the time of the original experiments; in this case the correlation in eqn (3.8.5) equals unity for $\tau \ll \tau_c$ but falls to zero for $\tau \gg \tau_c$. Similar experiments have since been made on light that closely corresponds to the classical stable wave; in this case the Brown–Twiss correlation vanishes for all values of τ.

For Gaussian–Lorentzian chaotic light, the correlation given by eqns (3.7.19) and (3.8.5) is

$$g_{1,1}^{(2)}(\tau) - 1 = \exp(-2|t_4 - t_3|/\tau_c) = \exp(-2|\tau|/\tau_c).$$ (3.8.6)

However, experiments are never able to make instantaneous measurements of the beam intensity and the expected correlation is modified accordingly. Practical detectors always have some minimum resolving time, or integration time T, such that the recorded intensity is a mean over the period T. Consider a Brown–Twiss interferometer where the 'instantaneous' intensities are measured simultaneously by two detectors of equal integration time T. If the additional average over the detector resolving time is indicated by further angle brackets, similar to the notation used in §3.3, the predicted result is

$$\langle g_{1,1}^{(2)}(0)\rangle - 1 = \frac{1}{T^2}\int_0^T dt_3 \int_0^T dt_4 \exp(-2|t_4 - t_3|/\tau_c)$$ (3.8.7)

$$= (\tau_c^2/2T^2)\{\exp(-2T/\tau_c) - 1 + (2T/\tau_c)\}.$$

For a very short integration time, $T \ll \tau_c$, the right-hand side of eqn (3.8.7) reduces to unity, in agreement with the result of eqn (3.8.6) for simultaneous measurement of the instantaneous intensities. However, in the opposite limit of a very long integration time, $T \gg \tau_c$, the right-hand side of eqn (3.8.7) is approximately equal to τ_c/T, and the observed magnitude of the intensity fluctuations is greatly reduced. The original experiments had a reduction factor of order 10^5, and their success depended upon a processing of the measured intensities to obtain the correlation in eqn (3.8.5) directly.

The intensity interference effects described above are governed by the second-order *temporal* coherence of the light. The temporal coherence controls the fluctuation and correlation properties of the fields and intensities in a direction parallel to the beam, and it is determined by the dynamics of the light-emitting atoms in the source. However, similar to the Mach–Zehnder interferometer, the practical observation of the intensity correlations is also affected by the spatial coherence of the light, associated with the finite dimensions of the source and the detectors. The spatial coherence controls mainly the correlation properties of the light in directions perpendicular to the beam. The original purpose of the Brown–Twiss interferometer was the measurement of the angular diameters of stars by observation of spatial coherence effects [8]. Thus one of the detectors was mounted on an adjustable transverse slide to allow measurements to be made of the fall-off in correlation with increase in the lateral separation of the two detectors. The spatial coherence is not of great interest here as it is not so fundamental a property of the light as the temporal coherence and we consider it no further [9].

Irrespective of any transverse spatial dependence in the observed correlations, use of the double-detector arrangement of the Brown–Twiss interferometer is beneficial even for spatially-coincident measurements because of detector recovery effects. Detectors suffer a period of paralysis following a measurement, known as the *deadtime*. A pair of closely-spaced, or effectively coincident, measurements is therefore more effectively made with the use of different detectors for the two observations.

3.9 Semiclassical theory of optical detection

The various properties of light beams derived in the present chapter are somewhat abstract, in that they concern the electric fields and intensities of the light, without any serious consideration of how these properties are measured. The observation, or integration, times associated with measurements are mentioned in previous sections but there is no description of the methods by which the basic variables associated with the light are converted into experimental readings. Almost all practical devices measure the intensity of a light beam rather than its electric field. Their operation depends on the absorption of a portion of the beam, whose energy is converted to a detectable form. The photographic plate, the phototube, and the bolometer all belong to this category.

It is convenient to base the discussion on the photomultiplier, or phototube, which essentially converts optical intensity into electrical current pulses. The phototube relies on the photoelectric effect, in which absorption of a photon by an atom excites an electron from a bound state to a free state, in which it can leave the atom altogether. The phototube is designed so that the freed electron produces more free electrons in a cascade process until the electron current is sufficiently large to be detectable. The current pulse generated by a single ionization event may contain up to 10^6 electrons, with careful design of the phototube. The detection of each pulse corresponds to a *photocount*, and the measured quantity is the *photocount rate*. Thus in the semiclassical theory, where the atom is treated quantum-mechanically but the electromagnetic field is treated classically, the phototube has the effect of converting the continuous cycle-averaged classical intensity $\bar{I}(t)$ of the beam into a succession of discrete photocounts. The quantum theory of photodetection is considered in §§4.11 and 6.10.

The atomic ionization rate resembles the upwards transition rate between bound states, treated in §2.3, in being proportional to the optical energy density or intensity. The number m of photocounts produced in an integration time T thus provides a measure of the optical intensity. A photocount experiment consists of a large number, for example 10^5, of successive measurements of the numbers of photocounts in integration times of the same length T. The results are expressible as a probability distribution $P_m(T)$ for the occurrence of m photocounts during an observation time T. It is assumed throughout the derivations below that the measured light beams are stationary. The aim of the calculations is to determine the relation of the measured statistical distribution $P_m(T)$ of photocounts to the statistical properties of the light beam.

For the present account of the semiclassical theory, it is only necessary to make the reasonable assumption that the probability $p(t)$ per unit time of a photoionization at time t is proportional to the cycle-averaged intensity $\bar{I}(t)$ at that time. The probability that the light beam causes a single photoionization, registered as a single photocount, during the time interval between t and $t + \mathrm{d}t$ then has the form

$$p(t)\mathrm{d}t = \xi \bar{I}(t)\mathrm{d}t, \qquad\qquad (3.9.1)$$

where $\mathrm{d}t$ is assumed sufficiently large for the transition probability theory of eqn (2.3.16) to be valid, but sufficiently small for the probability of more than one photoionization to be negligible. The constant of proportionality ξ represents the efficiency of the detector, including the matrix elements for the photoelectric transition and factors that depend on the construction of the phototube.

Consider the period of time that extends from t to $t + T$ and let $P_m(t, T)$ be the probability that m photocounts occur in this time interval. The required distribution $P_m(T)$ is determined by a subsequent average over a large number of different starting times t. Let $t + t'$ be a time that lies within the interval between t and $t + T$, as shown in Fig. 3.14, and let $\mathrm{d}t'$ be a short period of time similar to that used in eqn (3.9.1). The probability that m photocounts occur between t and $t + t' + \mathrm{d}t'$ is by definition

$$P_m(t, t' + \mathrm{d}t'). \qquad\qquad (3.9.2)$$

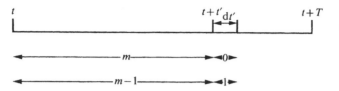

Fig. 3.14. Time intervals used in the calculation of $P_m(t,T)$, showing the two ways in which m photocounts can occur in the period between t and $t + t' + dt'$.

There are, however, two distinct ways in which m photocounts can occur during the given time interval. Either

(1) m photocounts occur between t and $t + t'$ and none in the interval dt', with total probability

$$P_m(t,t')[1 - p(t + t')dt'],\qquad(3.9.3)$$

or (2) $m - 1$ photocounts occur between t and $t + t'$ and one in the interval dt', with total probability

$$P_{m-1}(t,t')p(t + t')dt'.\qquad(3.9.4)$$

The probability for more than one photocount during the time interval dt' is negligible by hypothesis.

The two ways of computing the probability that m photocounts occur between t and $t + t' + dt'$, either by eqn (3.9.2) or by eqns (3.9.3) and (3.9.4), must be equivalent, so that

$$P_m(t, t' + dt') = P_m(t,t')[1 - p(t + t')dt'] + P_{m-1}(t,t')p(t + t')dt'.\qquad(3.9.5)$$

The standard definition of differentiation and the insertion of eqn (3.9.1) converts this last equation to

$$dP_m(t,t')/dt' = \xi\bar{I}(t + t')[P_{m-1}(t,t') - P_m(t,t')],\qquad(3.9.6)$$

which applies for general m except that the first term on the right is absent for $m = 0$, when

$$dP_0(t,t')/dt' = -\xi\bar{I}(t + t')P_0(t,t').\qquad(3.9.7)$$

The photocount probabilities for the different values of m are thus related by a chain of differential equations.

The chain of equations is solved by recursion, beginning with eqn (3.9.7), subject to the boundary condition that there can certainly be no photocounts during a zero time interval, so that

$$P_0(t,0) = 1 \quad \text{and} \quad P_m(t,0) = 0 \quad \text{for} \quad m > 0.\qquad(3.9.8)$$

The integration of eqn (3.9.7) with this boundary condition gives

$$P_0(t,T) = \exp\left[-\xi \bar{I}(t,T)T\right],$$ (3.9.9)

where $\bar{I}(t,T)$ is the mean intensity that falls on the phototube during the period from t to $t+T$,

$$\bar{I}(t,T) = \frac{1}{T}\int_t^{t+T} dt'' \bar{I}(t'').$$ (3.9.10)

The remaining elements of the distribution are determined from eqn (3.9.6), beginning with $m = 1$ and proceeding to the higher values of m.

Problem 3.9   Prove by induction that the general solution of eqn (3.9.6), subject to the boundary conditions in eqn (3.9.8), is

$$P_m(t,T) = \frac{\left[\xi \bar{I}(t,T)T\right]^m}{m!}\exp\left[-\xi \bar{I}(t,T)T\right].$$ (3.9.11)

This probability gives the distribution of the numbers of photocount readings in a series of experiments that observe a light beam with the same mean intensity $\bar{I}(t,T)$. Real experiments measure the distribution of photocount numbers in a series of successive periods of duration T, so that the observed distribution is a time average of eqn (3.9.11) over the fluctuations in the light beam. Thus for stationary and ergodic light, the required photocount distribution $P_m(T)$ is obtained as an average over the successive starting times t or equivalently as a statistical average over the intensity fluctuations of the expression in eqn (3.9.11)

$$P_m(T) = \left\langle \frac{\left[\xi \bar{I}(t,T)T\right]^m}{m!}\exp\left[-\xi \bar{I}(t,T)T\right]\right\rangle.$$ (3.9.12)

This result is known as the Mandel formula [10,11]. It is readily verified, with the usual Maclaurin expansion of the exponential, that the distribution is normalized, and the mean number of photocounts is also easily calculated as

$$\langle m \rangle = \sum_m m P_m(T) = \left\langle \xi \bar{I}(t,T)T\right\rangle.$$ (3.9.13)

The statistical average in eqn (3.9.12) is difficult to perform in general but there are a few simple special cases. The simplest case of all is that of the classical stable wave of Fig. 3.10, where the cycle-averaged intensity has the fixed value $\bar{I}$, independent of the time. There is no need to perform the averaging in this case and the distribution in eqn (3.9.12) reduces to

$$P_m(T) = \frac{\langle m \rangle^m}{m!} e^{-\langle m \rangle}, \tag{3.9.14}$$

with a mean number of photocounts

$$\langle m \rangle = \xi \bar{I} T. \tag{3.9.15}$$

This is a Poisson distribution and its form is illustrated in Fig. 5.3 for three values of the mean count, where n in the figure is here replaced by m. The photocount statistics are the same as those found for the arrival of raindrops in a steady downpour or for the emission of particles during radioactive decay of a long-lived isotope. The Poisson photocount distribution in eqn (3.9.14) is measurable at the level of an undergraduate experiment [12] using the coherent light beam from a laser source.

The Poisson form of photocount distribution also occurs for an experiment on a light beam with fluctuating intensity, in which the integration time T is much longer than the characteristic time scale of the fluctuations. The mean intensity in the measurement period, defined in eqn (3.9.10), is then independent of the starting time t, and there is again no need to perform the average in eqn (3.9.12). The mean number of photocounts is again given by eqn (3.9.15), where $\bar{I}$ is now the long-time average intensity. The Poisson distribution thus applies to chaotic light when T is much longer than the coherence time τ_c, so that the intensity fluctuations are averaged out by the duration of each measurement.

The photocount distribution is also simply derived for chaotic light in the opposite extreme of an integration time that is much shorter than the coherence time. The instantaneous intensity $\bar{I}(t)$ in this case is essentially constant over the period T and eqn (3.9.10) becomes

$$\bar{I}(t,T) = \bar{I}(t) \quad \text{for} \quad T \ll \tau_c. \tag{3.9.16}$$

The probability distribution for the instantaneous intensity of chaotic light is given by eqn (3.6.12). Thus for a stationary ergodic beam of chaotic light, the average in eqn (3.9.12) is evaluated as

$$P_m(T) = \frac{1}{\bar{I}} \int d\bar{I}(t) \exp\left[-\frac{\bar{I}(t)}{\bar{I}}\right] \frac{\left[\xi\bar{I}(t)T\right]^m}{m!} \exp\left[-\xi\bar{I}(t)T\right] = \frac{\langle m \rangle^m}{\left(1 + \langle m \rangle\right)^{1+m}}, \tag{3.9.17}$$

where $\langle m \rangle$ is defined by eqn (3.9.15). The geometric photocount distribution for chaotic light with short integration times is therefore similar to that of the photon distribution in a single mode of a thermal source given by eqn (1.4.2) and illustrated in Fig. 1.7. However, the photocount distribution in eqn (3.9.17) is not restricted to the excitation of a single mode but applies generally to any chaotic light. The photocount distribution in eqn (3.9.17) is also measurable in the undergraduate laboratory [12].

The form of the photocount distribution for chaotic light is much more difficult to calculate for integration times that are neither long nor short compared to the coherence time. Numerical evaluations [13] for increasing integration times, with a fixed coherence time, provide a succession of distributions that change continuously from the geometric form of eqn (3.9.17) for $T \ll \tau_c$ to the Poisson form of eqn (3.9.14) for $T \gg \tau_c$. The coherence time for chaotic light can be determined by fitting the measured photocount distribution to the numerical results.

Although the complete photocount distribution for chaotic light requires a numerical calculation for general values of T, it is not difficult to evaluate the size of the photocount fluctuations. Thus the second moment of the distribution given by eqn (3.9.12) is

$$\langle m^2 \rangle = \sum_m m^2 P_m(T) = \langle \xi \bar{I}(t,T)T \rangle + \langle \left(\xi \bar{I}(t,T)T \right)^2 \rangle = \langle m \rangle + \langle \left(\xi \bar{I}(t,T)T \right)^2 \rangle,$$

(3.9.18)

where the mean value $\langle m \rangle$ is given by eqn (3.9.15). The variance of the photocount distribution is therefore

$$(\Delta m)^2 = \langle m \rangle + \xi^2 T^2 \left[\langle \bar{I}(t,T)^2 \rangle - \bar{I}^2 \right].$$

(3.9.19)

The first contribution to the variance is the same for all kinds of light. It is known as the *shot noise* in the detection process and it results from the discrete nature of the photoelectric ionization. The second term is known as the *excess noise* and it clearly depends on the nature of the light. The intensity variance that occurs in this term measures the *wave fluctuations* in the light beam.

The simplest example of the photocount variance occurs for the classical stable wave, where the cycle-averaged intensity has the fixed value of $\bar{I}$ and the second term in eqn (3.9.19) vanishes, to give

$$(\Delta m)^2 = \langle m \rangle.$$

(3.9.20)

The equality of the mean and the variance is a standard property of the Poisson distribution in eqn (3.9.14).

A second straightforward example is provided by Gaussian–Lorentzian chaotic light, where the integration needed for the wave-fluctuation variance in eqn (3.9.19) is already evaluated in eqn (3.8.7). Thus, with use of the mean count from eqn (3.9.15), the photocount variance is

$$(\Delta m)^2 = \langle m \rangle + \langle m \rangle^2 \left(\tau_c^2 / 2T^2 \right) \left\{ \exp(-2T/\tau_c) - 1 + (2T/\tau_c) \right\}.$$

(3.9.21)

For short integration times, the variance reduces to

$$(\Delta m)^2 = \langle m \rangle + \langle m \rangle^2 \quad \text{for} \quad T \ll \tau_c.$$

(3.9.22)

Thus all chaotic light beams produce the same photocount variance in this limit, and the expression is the same as that found in eqn (1.4.7) for the photon-number variance in a single thermally-excited mode. In the opposite limit of long integration times, the variance reduces to

$$(\Delta m)^2 = \langle m \rangle + \langle m \rangle^2 (\tau_c/T) \quad \text{for} \quad T \gg \tau_c \qquad (3.9.23)$$

and the Poisson result of eqn (3.9.20) is retrieved when the second term can be neglected.

The results of the present section show how measurements of the continuous time dependence of the classical cycle-averaged intensity by a phototube produce information in the form of discrete photocounts, governed by a digital probability distribution. The semiclassical theory given here contrasts with the photodetection theory of the quantized electromagnetic field given in §6.10, where the field intensity is itself expressed in terms of discrete photons. It is found that the semiclassical results for the stable wave and for chaotic light survive in the fully-quantized theory but the more general quantum theory also embraces the photocount distributions for light beams not describable by the classical theory.

References

[1]   Squire, A., University of Essex PhD thesis (unpublished); Bain, A.J. and Squire, A., Super-Gaussian field statistics in broadband laser emission, *Opt. Comm.* **135**, 157–63 (1997).

[2]   Vannucci, G. and Teich, M.C., Computer simulation of superposed coherent and chaotic radiation, *Appl. Opt.* **19**, 548–53 (1980).

[3]   Harris, M., Light-field fluctuations in space and time, *Contemp. Phys.* **36**, 215–33 (1995).

[4]   Harris, M., Pearson, G.N., Hill, C.A. and Vaughan, J.M., The fractal character of Gaussian-Lorentzian light, *Opt. Comm.* **116**, 15–9 (1995).

[5]   Ou, Z.Y. and Mandel, L., Derivation of reciprocity relations for a beam splitter from energy balance, *Am. J. Phys.* **57**, 66–7 (1989).

[6]   Goodman, J.W., *Statistical Optics* (Wiley, New York, 1985).

[7]   Brown, R.H. and Twiss, R.Q., Correlation between photons in two coherent beams of light, *Nature* **177**, 27–9 (1956).

[8]   An interesting account of the original experiments and of their development for measurements of stellar diameters is given by Brown, R.H., *The Intensity Interferometer* (Taylor and Francis, London, 1974).

[9]   Original papers on all aspects of coherence theory are collected in Mandel, L. and Wolf, E., *Selected Papers on Coherence and Fluctuations of Light*, vols. 1 and 2 (Dover, New York, 1970).

[10]   Mandel, L., Fluctuations in photon beams and their correlations, *Proc. Phys. Soc.* **72**, 1037–48 (1958); Fluctuations of photon beams: the distribution of photoelectrons, *Proc. Phys. Soc.* **74**, 233–43 (1959). These papers are reprinted in ref [9].

[11]   Mandel, L. and Wolf, E., *Optical Coherence and Quantum Optics* (Cambridge University Press, Cambridge, 1995) chap. 9.
[12]   Koczyk, P., Wiewiór, P. and Radzewicz, C., Photon counting statistics – undergraduate experiment, *Am. J. Phys.* **64**, 240–5 (1996).
[13]   Jakeman, E. and Pike, E.R., The intensity-fluctuation distribution of Gaussian light, *J. Phys. A* **1**, 128–38 (1968).

4 Quantization of the radiation field

Quantization conditions on the energy of the radiation field are used to derive Planck's law in §1.3 but most calculations in the preceding chapters are based on the semiclassical theory, in which the radiation fields $\mathbf{E}$ and $\mathbf{B}$ are treated as classical variables while the atoms are treated quantum-mechanically. A more consistent theory must treat the whole system of radiation and atoms in quantum-mechanical terms, with the fields represented by operators $\hat{\mathbf{E}}$ and $\hat{\mathbf{B}}$. The semiclassical theory provides theoretical expressions for a wide range of quantities that agree with those obtained by a fully quantum-mechanical theory, but other examples of measured phenomena cannot be described semiclassically.

The present chapter is devoted to a description of the procedure by which quantum mechanics is applied to the electromagnetic field. We obtain expressions for the operators that represent the field observables and we examine some of their properties. The quantization introduces characteristic quantum-mechanical effects into the properties of the radiation field, for example field operators that do not commute with each other and thus represent incompatible measured quantities. The Hamiltonian of the coupled radiation–atomic system is expressed wholly in terms of quantum-mechanical operators. We show that the full quantum theory of the interaction between the radiation field and atoms produces the same transition rates for absorption and stimulated emission as does the semiclassical theory. The rate of spontaneous emission confirms that obtained from the phenomenological Einstein theory. The treatment of absorption of radiation by atoms also identifies the quantum-mechanical operator that represents a measurement of the energy or intensity of a light beam, and this has important consequences for the descriptions of the experiments that can be performed in practice.

Despite the many successes of the semiclassical theory, the existence of phenomena that lie outside its domain of application limits its usefulness and undermines confidence in its reliability. We adopt the views in subsequent calculations that quantum mechanics provides the best available predictions for measured quantities and that the results of semiclassical theory are reliable only when they agree with the quantum results. In terms of the coherence properties of light beams, there is a range of overlap between the two theories but the quantum field can have kinds of coherence not allowed by classical theory. The nature of the relation between the classical and quantum theories of the radiation field is further discussed in §5.10.

Much of the following material is concerned with general quantum-mechanical procedures and with the formal quantum mechanics of the harmonic oscillator and the electromagnetic field. Readers familiar with some of this material, or prepared to accept it without detailed discussion, may prefer to proceed to its quantum-optical applications covered in subsequent chapters.

4.1 Potential theory for the classical electromagnetic field

The quantum theory of the radiation field has many similarities with the classical theory. The field vectors in quantum theory must be taken as operators instead of the algebraic quantities of classical theory, but both theories are based on the familiar Maxwell equations. It is, of course, not possible to derive the quantum theory from the classical equations, but the transition to quantum mechanics can be accomplished most easily if the equations of classical electromagnetic theory are first put into a suitably suggestive form. The central step in the quantization is the replacement of a classical harmonic oscillator by the corresponding quantum-mechanical harmonic oscillator. The first task is to cast the classical equations in a form where the harmonic-oscillator dependence of the field variables is appropriate for the conversion to quantum mechanics.

The microscopic Maxwell equations for the electromagnetic fields in a medium are

$$\nabla \times \mathbf{E} = -\partial \mathbf{B}/\partial t, \tag{4.1.1}$$

$$\mu_0^{-1} \nabla \times \mathbf{B} = \varepsilon_0 (\partial \mathbf{E}/\partial t) + \mathbf{J}, \tag{4.1.2}$$

$$\varepsilon_0 \nabla.\mathbf{E} = \sigma, \tag{4.1.3}$$

$$\nabla.\mathbf{B} = 0, \tag{4.1.4}$$

where σ and $\mathbf{J}$ are the charge and current densities, respectively. The classical fields and densities are functions of position and time, for example $\mathbf{E} \equiv \mathbf{E}(\mathbf{r},t)$, but the shorter notation is used in the present section. The quantization is facilitated if the classical Maxwell equations are re-expressed in terms of the scalar and vector potentials, ϕ and $\mathbf{A}$ respectively. The fourth of the equations is satisfied identically if $\mathbf{B}$ is derived from a potential $\mathbf{A}$ according to

$$\mathbf{B} = \nabla \times \mathbf{A}. \tag{4.1.5}$$

Use of the identity

$$\nabla \times \nabla \phi \equiv 0, \tag{4.1.6}$$

where ϕ is a scalar function, then shows that the first Maxwell equation is satisfied identically if ϕ is defined by

$$\nabla \phi = -\mathbf{E} - (\partial \mathbf{A}/\partial t), \tag{4.1.7}$$

or equivalently

$$\mathbf{E} = -\nabla\phi - (\partial\mathbf{A}/\partial t). \tag{4.1.8}$$

Equations (4.1.5) and (4.1.8) enable the fields to be found if the scalar and vector potentials are known.

The potentials are determined, in principle, by substitution of the fields from eqns (4.1.5) and (4.1.8) into the remaining Maxwell equations (4.1.2) and (4.1.3), with the results

$$\nabla(\nabla.\mathbf{A}) - \nabla^2\mathbf{A} + \frac{1}{c^2}\frac{\partial}{\partial t}\nabla\phi + \frac{1}{c^2}\frac{\partial^2\mathbf{A}}{\partial t^2} = \mu_0\mathbf{J} \tag{4.1.9}$$

and

$$-\varepsilon_0\nabla^2\phi - \varepsilon_0\nabla.(\partial\mathbf{A}/\partial t) = \sigma, \tag{4.1.10}$$

where

$$c = (\varepsilon_0\mu_0)^{-1/2} \tag{4.1.11}$$

and the vector identity

$$\nabla\times\nabla\times\mathbf{A} \equiv \nabla(\nabla.\mathbf{A}) - \nabla^2\mathbf{A} \tag{4.1.12}$$

is used. Equations (4.1.9) and (4.1.10) are the *field equations*; they determine the fields that result from given distributions of charge σ and current $\mathbf{J}$.

The field equations are quite complicated owing to the mixtures of terms involving $\mathbf{A}$ and ϕ that occur on their left-hand sides. They are simplified by the imposition of an additional condition on the potentials. The fields obtained from eqns (4.1.5) and (4.1.8) are the same for pairs of potentials $\mathbf{A},\phi$ and $\mathbf{A}',\phi'$ related by a *gauge transformation* [1] defined by

$$\mathbf{A} = \mathbf{A}' - \nabla\Xi \tag{4.1.13}$$

and

$$\phi = \phi' + (\partial\Xi/\partial t), \tag{4.1.14}$$

where the *gauge function* Ξ is an arbitrary function of the position $\mathbf{r}$ and time t. A gauge for the electromagnetic field is specified by some condition on $\mathbf{A}$ and ϕ that can be realized by a gauge transformation from an arbitrary pair of solutions of the field equations for the given σ and $\mathbf{J}$.

The electromagnetic field is said to be in the *Coulomb gauge* when the vector potential satisfies the condition

$$\nabla.\mathbf{A} = 0. \tag{4.1.15}$$

It is always possible to transform to the Coulomb gauge from an arbitrary potential

$\mathbf{A}'$ by the use of a gauge function that satisfies

$$\nabla^2 \Xi = \nabla . \mathbf{A}'. \qquad (4.1.16)$$

The choice of Coulomb gauge simplifies the field equations (4.1.9) and (4.1.10), which become

$$-\nabla^2 \mathbf{A} + \frac{1}{c^2}\frac{\partial}{\partial t}\nabla\phi + \frac{1}{c^2}\frac{\partial^2 \mathbf{A}}{\partial t^2} = \mu_0 \mathbf{J} \qquad (4.1.17)$$

and

$$-\nabla^2 \phi = \sigma/\varepsilon_0. \qquad (4.1.18)$$

The scalar potential now satisfies Poisson's equation of electrostatics, whose solutions are well known [1].

The field equation (4.1.17) that determines $\mathbf{A}$ is still complicated by a term in ϕ, but this is removed by the following arguments. According to Helmholtz' theorem [2], any vector field can be written as a sum of two components, one of which has zero divergence and the other of which has zero curl. For the current density, the sum is written

$$\mathbf{J} = \mathbf{J}_T + \mathbf{J}_L, \qquad (4.1.19)$$

where

$$\nabla . \mathbf{J}_T = 0 \qquad (4.1.20)$$

and

$$\nabla \times \mathbf{J}_L = 0; \qquad (4.1.21)$$

$\mathbf{J}_T$ is called the *transverse* or *solenoidal* component and $\mathbf{J}_L$ is the *longitudinal* or *irrotational* component. Similar definitions and nomenclatures apply to the other field vectors. For example, it is evident from eqn (4.1.5) that the magnetic field $\mathbf{B}$ is determined solely by the transverse part $\mathbf{A}_T$ of the vector potential. It follows from eqn (4.1.6) that $\nabla\Xi$ is longitudinal for all gauge functions Ξ. The gauge transformation of eqn (4.1.13) thus affects only the longitudinal part $\mathbf{A}_L$ of the vector potential and the transverse part $\mathbf{A}_T$ is gauge-invariant, leaving the observable magnetic field unchanged. The Coulomb gauge definition in eqn (4.1.15) identifies $\mathbf{A}$ as wholly transverse, with the longitudinal part completely transformed away.

With the use of these definitions and the assumption of the Coulomb gauge condition (4.1.15), the complete field equation (4.1.17) is readily separated into its transverse and longitudinal parts as

$$-\nabla^2 \mathbf{A} + \frac{1}{c^2}\frac{\partial^2 \mathbf{A}}{\partial t^2} = \mu_0 \mathbf{J}_T \qquad (4.1.22)$$

and

$$\frac{1}{c^2}\frac{\partial}{\partial t}\nabla\phi = \mu_0 \mathbf{J}_L.$$ (4.1.23)

The vector potential is thus determined by the transverse part of the current density and eqn (4.1.22) has well-known formal solutions [1]. The scalar potential satisfies both eqns (4.1.18) and (4.1.23) and its elimination from the two gives

$$\nabla.\mathbf{J}_L = -\partial\sigma/\partial t,$$ (4.1.24)

which is the equation of charge conservation.

The electric vector can also be divided into transverse and longitudinal parts, where use of eqn (4.1.8) shows that the parts are identified in Coulomb gauge as

$$\mathbf{E}_T = -\partial\mathbf{A}/\partial t$$ (4.1.25)

and

$$\mathbf{E}_L = -\nabla\phi.$$ (4.1.26)

The magnetic vector $\mathbf{B}$ is entirely transverse according to the Maxwell equation (4.1.4). The great advantage of the Coulomb gauge for the radiation field and its interaction with charges and currents lies in the clean separations of the field equations and Maxwell's equations into two distinct sets.

In summary, the transverse equations are associated with the vector potential $\mathbf{A}$ as in eqn (4.1.25). The relevant field equation (4.1.22) has the form of a wave equation and Maxwell's equations (4.1.1) to (4.1.4) give

$$\nabla\times\mathbf{E}_T = -\partial\mathbf{B}/\partial t,$$ (4.1.27)

$$\mu_0^{-1}\nabla\times\mathbf{B} = \varepsilon_0(\partial\mathbf{E}/\partial t) + \mathbf{J}_T,$$ (4.1.28)

$$\nabla.\mathbf{E}_T = 0,$$ (4.1.29)

and

$$\nabla.\mathbf{B} = 0.$$ (4.1.30)

These transverse equations describe electromagnetic waves, which are influenced only by the transverse part of the current density.

The longitudinal equations, on the other hand, are associated with the scalar potential ϕ as in eqn (4.1.26) and the relevant field equation is (4.1.18). The Maxwell equation (4.1.3) gives

$$\nabla.\mathbf{E}_L = \sigma/\varepsilon_0$$ (4.1.31)

and the longitudinal current density is obtained from eqns (4.1.23) and (4.1.26) as

$$\mathbf{J}_L = -\varepsilon_0 \, \partial \mathbf{E}_L / \partial t. \tag{4.1.32}$$

The longitudinal equations describe the electric fields that arise from the charge density, as determined by the equations of electrostatics.

4.2 The free classical field

In the earlier part of the chapter, we consider the electromagnetic waves in a region of space where

$$\mathbf{J}_T = 0 \tag{4.2.1}$$

and hence from eqn (4.1.22)

$$-\nabla^2 \mathbf{A} + \frac{1}{c^2} \frac{\partial^2 \mathbf{A}}{\partial t^2} = 0. \tag{4.2.2}$$

The field in such a region of space is said to be free. The theory of the interaction of radiation with atoms requires solution of the more difficult problem where $\mathbf{J}_T$ represents the transverse current produced by the atomic electrons. The interaction of radiation with atoms is treated in §4.8.

The quantization of the electromagnetic field proceeds by the replacement of the classical vector potential $\mathbf{A}$ by a quantum-mechanical operator $\hat{\mathbf{A}}$. We put the classical theory into a form from which the transition to quantum mechanics is straightforward. Consider a cubic region of space of side L, similar to the cavity illustrated in Fig. 1.1. The cavity is now regarded merely as a region of space, without any real boundaries, known as the *quantization cavity*. Instead of the standing-wave solutions for the field given in eqns (1.1.3) to (1.1.5), we take running waves and subject them to periodic boundary conditions. Also, in contrast to the calculations of Chapter 1, we work with the vector potential instead of the electric field.

With these preliminaries, the vector potential is expanded as a sum of contributions from the modes of the cavity,

$$\mathbf{A}(\mathbf{r},t) = \sum_{\mathbf{k}} \sum_{\lambda=1,2} \mathbf{e}_{\mathbf{k}\lambda} A_{\mathbf{k}\lambda}(\mathbf{r},t), \tag{4.2.3}$$

where

$$A_{\mathbf{k}\lambda}(\mathbf{r},t) = A_{\mathbf{k}\lambda}(t)\exp(\mathrm{i}\mathbf{k}.\mathbf{r}) + A_{\mathbf{k}\lambda}^*(t)\exp(-\mathrm{i}\mathbf{k}.\mathbf{r}). \tag{4.2.4}$$

Here the components of the wavevector $\mathbf{k}$ take the values

$$k_x = 2\pi v_x/L, \quad k_y = 2\pi v_y/L, \quad k_z = 2\pi v_z/L, \tag{4.2.5}$$

with

$$v_x, v_y, v_z = 0, \pm 1, \pm 2, \pm 3, ...$$ (4.2.6)

The density of modes defined in this way is the same as for the modes specified by eqns (1.1.3) to (1.1.5). The $\mathbf{e}_{\mathbf{k}\lambda}$ in eqn (4.2.3) are unit polarization vectors and the Coulomb gauge condition of eqn (4.1.15) is satisfied if these are both transverse, with

$$\mathbf{e}_{\mathbf{k}\lambda}.\mathbf{k} = 0.$$ (4.2.7)

The polarizations are chosen to be perpendicular to each other with

$$\mathbf{e}_{\mathbf{k}1}.\mathbf{e}_{\mathbf{k}2} = 0.$$ (4.2.8)

The normalization and orthogonality of the two polarization vectors for each $\mathbf{k}$ are expressed in the single condition

$$\mathbf{e}_{\mathbf{k}\lambda}.\mathbf{e}_{\mathbf{k}\lambda'} = \delta_{\lambda,\lambda'},$$ (4.2.9)

where the Kronecker delta-function equals unity when its two indices are identical and vanishes otherwise

The modal components of the vector potential are independent and they separately obey the field equation (4.2.2),

$$k^2 A_{\mathbf{k}\lambda}(t) + \frac{1}{c^2}\frac{\partial^2 A_{\mathbf{k}\lambda}(t)}{\partial t^2} = 0.$$ (4.2.10)

The modal coefficients, and their complex conjugates, thus satisfy the simple-harmonic equation of motion,

$$\frac{\partial^2 A_{\mathbf{k}\lambda}(t)}{\partial t^2} + \omega_k^2 A_{\mathbf{k}\lambda}(t) = 0,$$ (4.2.11)

where it is convenient to subscript the mode angular frequency as

$$\omega_k = ck.$$ (4.2.12)

The electromagnetic field is quantized by conversion of the classical harmonic oscillator to a quantum-mechanical counterpart. The nature of the conversion is suggested by the form of the field energy, which we now evaluate.

The solution of eqn (4.2.11) is taken in the form

$$A_{\mathbf{k}\lambda}(t) = A_{\mathbf{k}\lambda} \exp(-i\omega_k t),$$ (4.2.13)

and the modal contribution to the vector potential from eqn (4.2.4) becomes

$$A_{\mathbf{k}\lambda}(\mathbf{r},t) = A_{\mathbf{k}\lambda}\exp(-i\omega_k t + i\mathbf{k}.\mathbf{r}) + A_{\mathbf{k}\lambda}^*\exp(i\omega_k t - i\mathbf{k}.\mathbf{r}). \tag{4.2.14}$$

The complete vector potential is obtained by substitution of this expression into eqn (4.2.3). The corresponding complete transverse electric field is obtained from eqn (4.1.25) as

$$\mathbf{E}_T(\mathbf{r},t) = \sum_{\mathbf{k}}\sum_{\lambda=1,2}\mathbf{e}_{\mathbf{k}\lambda}E_{\mathbf{k}\lambda}(\mathbf{r},t), \tag{4.2.15}$$

where

$$E_{\mathbf{k}\lambda}(\mathbf{r},t) = i\omega_k\{A_{\mathbf{k}\lambda}\exp(-i\omega_k t + i\mathbf{k}.\mathbf{r}) - A_{\mathbf{k}\lambda}^*\exp(i\omega_k t - i\mathbf{k}.\mathbf{r})\} \tag{4.2.16}$$

and the magnetic field is obtained from eqn (4.1.5) as

$$\mathbf{B}(\mathbf{r},t) = \sum_{\mathbf{k}}\sum_{\lambda=1,2}\frac{\mathbf{k}\times\mathbf{e}_{\mathbf{k}\lambda}}{k}B_{\mathbf{k}\lambda}(\mathbf{r},t), \tag{4.2.17}$$

where

$$B_{\mathbf{k}\lambda}(\mathbf{r},t) = ik\{A_{\mathbf{k}\lambda}\exp(-i\omega_k t + i\mathbf{k}.\mathbf{r}) - A_{\mathbf{k}\lambda}^*\exp(i\omega_k t - i\mathbf{k}.\mathbf{r})\}. \tag{4.2.18}$$

The ratio of these electric and magnetic field amplitudes equals the velocity of light, as expected for a free electromagnetic wave.

The total energy of the electromagnetic radiation field in the cavity is

$$\mathcal{E}_R = \tfrac{1}{2}\int_{\text{cavity}} dV[\varepsilon_0\mathbf{E}_T(\mathbf{r},t).\mathbf{E}_T(\mathbf{r},t) + \mu_0^{-1}\mathbf{B}(\mathbf{r},t).\mathbf{B}(\mathbf{r},t)], \tag{4.2.19}$$

and this is evaluated by substitution of the fields from eqns (4.2.15) and (4.2.17). The spatial integrals that occur have the forms

$$\int_{\text{cavity}} dV\exp[\pm i(\mathbf{k}-\mathbf{k}').\mathbf{r}] = V\delta_{\mathbf{k},\mathbf{k}'} \tag{4.2.20}$$

and

$$\int_{\text{cavity}} dV\exp[\pm i(\mathbf{k}+\mathbf{k}').\mathbf{r}] = V\delta_{\mathbf{k},-\mathbf{k}'}, \tag{4.2.21}$$

where $V = L^3$. The energy from eqn (4.2.19) thus becomes

$$\mathcal{E}_R = \tfrac{1}{2} \sum_{\mathbf{k}} \sum_{\lambda,\lambda'} V \Big\{ \big(A_{\mathbf{k}\lambda} A^*_{\mathbf{k}\lambda'} + A^*_{\mathbf{k}\lambda} A_{\mathbf{k}\lambda'} \big)$$

$$\times \big(\varepsilon_0 \omega_k^2 \mathbf{e}_{\mathbf{k}\lambda} \cdot \mathbf{e}_{\mathbf{k}\lambda'} + \mu_0^{-1} \mathbf{k} \times \mathbf{e}_{\mathbf{k}\lambda} \cdot \mathbf{k} \times \mathbf{e}_{\mathbf{k}\lambda'} \big)$$

$$- \big(A_{\mathbf{k}\lambda} A_{-\mathbf{k}\lambda'} e^{-2i\omega_k t} + A^*_{\mathbf{k}\lambda} A^*_{-\mathbf{k}\lambda'} e^{2i\omega_k t} \big)$$

$$\times \big(\varepsilon_0 \omega_k^2 \mathbf{e}_{\mathbf{k}\lambda} \cdot \mathbf{e}_{-\mathbf{k}\lambda'} - \mu_0^{-1} \mathbf{k} \times \mathbf{e}_{\mathbf{k}\lambda} \cdot \mathbf{k} \times \mathbf{e}_{-\mathbf{k}\lambda'} \big) \Big\}, \quad (4.2.22)$$

as the modes $\mathbf{k}\lambda$ and $-\mathbf{k}\lambda$ have the same frequency ω_k. The factors that include the mode polarization vectors simplify considerably [3], as the usual properties of the vector product lead to

$$\mathbf{k} \times \mathbf{e}_{\mathbf{k}\lambda} \cdot \mathbf{k} \times \mathbf{e}_{\pm\mathbf{k}\lambda'} = k^2 \mathbf{e}_{\mathbf{k}\lambda} \cdot \mathbf{e}_{\pm\mathbf{k}\lambda'}. \quad (4.2.23)$$

Then with the help of eqns (4.1.11), (4.2.9) and (4.2.12), it is found that the time-dependent terms in eqn (4.2.22) vanish, as a result of cancellation between the relevant parts of the electric and magnetic field energies. The total radiative energy thus reduces to a sum of time-independent contributions from the individual modes,

$$\mathcal{E}_R = \sum_{\mathbf{k}} \sum_{\lambda} \mathcal{E}_{\mathbf{k}\lambda} \quad (4.2.24)$$

with

$$\mathcal{E}_{\mathbf{k}\lambda} = \varepsilon_0 V \omega_k^2 \big(A_{\mathbf{k}\lambda} A^*_{\mathbf{k}\lambda} + A^*_{\mathbf{k}\lambda} A_{\mathbf{k}\lambda} \big). \quad (4.2.25)$$

The mode coefficients $A_{\mathbf{k}\lambda}$ and $A^*_{\mathbf{k}\lambda}$ are left in the orders in which they occur in the squares of the field amplitudes, although these classical coefficients commute and they could be combined in a single term.

4.3 The quantum-mechanical harmonic oscillator

The quantum-mechanical Hamiltonian for a one-dimensional harmonic oscillator is

$$\hat{\mathcal{H}} = \frac{\hat{p}^2}{2m} + \tfrac{1}{2} m \omega^2 \hat{q}^2, \quad (4.3.1)$$

where the position operator $\hat{q}$ and the momentum operator $\hat{p}$ obey the usual commutation relation,

$$[\hat{q}, \hat{p}] = i\hbar. \quad (4.3.2)$$

It is convenient to replace $\hat{q}$ and $\hat{p}$ by a pair of dimensionless operators defined as

$$\hat{a} = (2m\hbar\omega)^{-1/2} (m\omega\hat{q} + i\hat{p}) \quad (4.3.3)$$

and

$$\hat{a}^\dagger = (2m\hbar\omega)^{-1/2}(m\omega\hat{q} - i\hat{p}),$$

(4.3.4)

or conversely

$$\hat{q} = (\hbar/2m\omega)^{1/2}\left(\hat{a}^\dagger + \hat{a}\right)$$

(4.3.5)

and

$$\hat{p} = i(m\hbar\omega/2)^{1/2}\left(\hat{a}^\dagger - \hat{a}\right).$$

(4.3.6)

The operators $\hat{a}$ and $\hat{a}^\dagger$ are called, respectively, the *destruction* and *creation* operators for the harmonic oscillator. They have simple properties and they are extremely useful in calculations, although they do not represent observable features of the oscillator.

It follows from eqns (4.3.3) and (4.3.4) that

$$\hat{a}\hat{a}^\dagger = (2m\hbar\omega)^{-1}\left(\hat{p}^2 + m^2\omega^2\hat{q}^2 - im\omega\hat{q}\hat{p} + im\omega\hat{p}\hat{q}\right)$$
$$= (\hbar\omega)^{-1}\left(\hat{\mathcal{H}} + \tfrac{1}{2}\hbar\omega\right)$$

(4.3.7)

and

$$\hat{a}^\dagger\hat{a} = (\hbar\omega)^{-1}\left(\hat{\mathcal{H}} - \tfrac{1}{2}\hbar\omega\right),$$

(4.3.8)

where eqns (4.3.1) and (4.3.2) are used. The difference of these expressions provides the commutator of the creation and destruction operators as

$$\left[\hat{a}, \hat{a}^\dagger\right] = \hat{a}\hat{a}^\dagger - \hat{a}^\dagger\hat{a} = 1,$$

(4.3.9)

while their sum leads to an expression for the Hamiltonian in the form

$$\hat{\mathcal{H}} = \tfrac{1}{2}\hbar\omega\left(\hat{a}\hat{a}^\dagger + \hat{a}^\dagger\hat{a}\right) = \hbar\omega\left(\hat{a}^\dagger\hat{a} + \tfrac{1}{2}\right).$$

(4.3.10)

The theory of the quantum-mechanical harmonic oscillator is developed equivalently in terms of the commutation relation and Hamiltonian given by eqns (4.3.2) and (4.3.1) or by eqns (4.3.9) and (4.3.10).

The energy eigenvalues are easily determined from the form of theory based on the creation and destruction operators. Thus let $|n\rangle$ be an energy eigenstate with eigenvalue E_n. The eigenvalue equation is

$$\hat{\mathcal{H}}|n\rangle = \hbar\omega\left(\hat{a}^\dagger\hat{a} + \tfrac{1}{2}\right)|n\rangle = E_n|n\rangle.$$

(4.3.11)

Multiply both sides from the left by $\hat{a}^\dagger$,

$$\hbar\omega\left(\hat{a}^\dagger\hat{a}^\dagger\hat{a} + \tfrac{1}{2}\hat{a}^\dagger\right)|n\rangle = E_n\hat{a}^\dagger|n\rangle. \tag{4.3.12}$$

Use of the commutator given by eqn (4.3.9) on the first term on the left yields

$$\hbar\omega\left(\hat{a}^\dagger\hat{a}\hat{a}^\dagger - \hat{a}^\dagger + \tfrac{1}{2}\hat{a}^\dagger\right)|n\rangle = E_n\hat{a}^\dagger|n\rangle, \tag{4.3.13}$$

which is rearranged as

$$\hbar\omega\left(\hat{a}^\dagger\hat{a} + \tfrac{1}{2}\right)\hat{a}^\dagger|n\rangle = \hat{\mathcal{H}}\hat{a}^\dagger|n\rangle = \left(E_n + \hbar\omega\right)\hat{a}^\dagger|n\rangle. \tag{4.3.14}$$

This last equation has the form of an energy eigenvalue equation. It shows that the state $\hat{a}^\dagger|n\rangle$ is an eigenstate of the harmonic oscillator with eigenvalue $E_n + \hbar\omega$. We denote the new eigenstate and eigenvalue as

$$|n+1\rangle = \hat{a}^\dagger|n\rangle \tag{4.3.15}$$

and

$$E_{n+1} = E_n + \hbar\omega, \tag{4.3.16}$$

and eqn (4.3.14) is now written

$$\hat{\mathcal{H}}|n+1\rangle = E_{n+1}|n+1\rangle. \tag{4.3.17}$$

These results show that, given a harmonic-oscillator energy level E_n, there exists another level higher than the first by an amount $\hbar\omega$. The energy levels thus form an equally-spaced ladder, which is illustrated in Fig. 4.1. As in classical mechanics, there is no restriction on the maximum energy of the oscillator, and the ladder extends upwards to infinity.

The lower end of the ladder is investigated by multiplication of the energy eigenvalue equation (4.3.11) from the left by $\hat{a}$. Manipulations similar to those that lead from eqn (4.3.12) to eqn (4.3.14) give

$$\hat{\mathcal{H}}\hat{a}|n\rangle = \left(E_n - \hbar\omega\right)\hat{a}|n\rangle. \tag{4.3.18}$$

The state $\left(\hat{a}|n\rangle\right)$ is thus an energy eigenstate with eigenvalue $E_n - \hbar\omega$, denoted by

$$|n-1\rangle = \hat{a}|n\rangle \tag{4.3.19}$$

and

$$E_{n-1} = E_n - \hbar\omega, \tag{4.3.20}$$

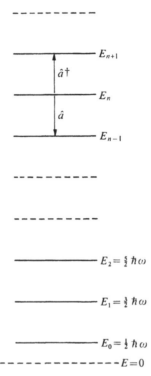

Fig. 4.1. Energy-level diagram for the quantum-mechanical harmonic oscillator, showing the roles of the creation operator $\hat{a}^\dagger$ and the destruction operator $\hat{a}$ in respectively adding and subtracting a quantum $\hbar\omega$ to or from the energy.

so that eqn (4.3.18) becomes

$$\hat{\mathcal{H}}|n-1\rangle = E_{n-1}|n\rangle. \qquad (4.3.21)$$

The ladder of energy levels thus extends downwards with equal steps $\hbar\omega$. However, the ladder must have a lower end because the oscillator kinetic and potential energies are positive quantities, and the eigenvalues are not allowed to go negative. Let $|0\rangle$ be the ground state with energy E_0, for which eqn (4.3.18) gives

$$\hat{\mathcal{H}}\,\hat{a}|0\rangle = (E_0 - \hbar\omega)\hat{a}|0\rangle. \qquad (4.3.22)$$

As, by hypothesis, there is no eigenstate of lower energy than the ground state, the only solution of eqn (4.3.22) consistent with this physical interpretation is

$$\hat{a}|0\rangle = 0. \qquad (4.3.23)$$

This *ground-state* or *vacuum-state condition* is used to determine E_0, as the ground-state form of the eigenvalue equation (4.3.11) is

$$\hat{\mathcal{H}}|0\rangle = \tfrac{1}{2}\hbar\omega|0\rangle = E_0|0\rangle. \tag{4.3.24}$$

Thus

$$E_0 = \tfrac{1}{2}\hbar\omega, \tag{4.3.25}$$

and it follows from eqn (4.3.16) that

$$E_n = \left(n + \tfrac{1}{2}\right)\hbar\omega, \quad n = 0, 1, 2, \dots, \tag{4.3.26}$$

the usual result for the quantum harmonic oscillator.

Figure 4.1 shows the energy-level scheme and indicates the roles of $\hat{a}$ and $\hat{a}^\dagger$ in, respectively, destroying or creating a quantum $\hbar\omega$ in the oscillator's excitation energy, and thus causing a step down or up the ladder. The states $|n\rangle$ are simultaneous eigenstates of the Hamiltonian $\hat{\mathcal{H}}$ and the *number operator* $\hat{n}$ defined as

$$\hat{n} = \hat{a}^\dagger\hat{a}, \tag{4.3.27}$$

where it is evident from eqns (4.3.11) and (4.3.26) that

$$\hat{n}|n\rangle = n|n\rangle. \tag{4.3.28}$$

The states $|n\rangle$ are called the *number states* of the harmonic oscillator.

The derivation so far has not paid any attention to normalization of the eigenstates, expressed by the conditions

$$\langle n-1|n-1\rangle = \langle n|n\rangle = \langle n+1|n+1\rangle = 1. \tag{4.3.29}$$

Additional factors appear in the relations between different eigenstates when the states are normalized and, for example, eqn (4.3.19) is generalized to

$$C_n|n-1\rangle = \hat{a}|n\rangle. \tag{4.3.30}$$

Thus with the use of eqn (4.3.29), the constant C_n is determined by taking the products of both sides of eqn (4.3.30) with their Hermitian conjugates, leading to

$$|C_n|^2 = n. \tag{4.3.31}$$

The phase of the normalization constant is conventionally taken to be zero, and eqn (4.3.30) becomes

$$\hat{a}|n\rangle = n^{1/2}|n-1\rangle. \tag{4.3.32}$$

A similar analysis for eqn (4.3.15) produces the result,

$$\hat{a}^\dagger|n\rangle = (n+1)^{1/2}|n+1\rangle. \tag{4.3.33}$$

Equations (4.3.32) and (4.3.33) are always used in preference to eqns (4.3.15) and (4.3.19) as it is advantageous to work with normalized eigenstates. Note that the ground-state condition (4.3.23) is included as a special case of the general result (4.3.32). The different energy eigenstates of the harmonic oscillator are orthogonal and it follows that the only nonvanishing matrix elements of the operators $\hat{a}$ and $\hat{a}^\dagger$ are those of the forms

$$\langle n-1|\hat{a}|n\rangle = n^{1/2} \quad \text{and} \quad \langle n+1|\hat{a}^\dagger|n\rangle = (n+1)^{1/2}. \tag{4.3.34}$$

Every Hermitian operator $\hat{O}$ has matrix elements that satisfy the condition

$$\langle i|\hat{O}|j\rangle = \langle j|\hat{O}|i\rangle^*. \tag{4.3.35}$$

It is clear that $\hat{a}$ and $\hat{a}^\dagger$ are not Hermitian operators and, according to the general principles of quantum mechanics, they cannot represent observable quantities [4]. Their property of destroying or creating a quantum of energy is, however, easily appreciated in a physical sense. The simple structures of their nonzero matrix elements also greatly facilitate calculations. It is however sometimes more convenient to work with dimensionless forms of the position and momentum operators used in the initial expression (4.3.1) for the harmonic-oscillator Hamiltonian. The *quadrature operators* are defined by

$$\hat{X} = (m\omega/2\hbar)^{1/2}\hat{q} = \tfrac{1}{2}\left(\hat{a}^\dagger + \hat{a}\right) \quad \text{and} \quad \hat{Y} = (2m\hbar\omega)^{-1/2}\hat{p} = \tfrac{1}{2}i\left(\hat{a}^\dagger - \hat{a}\right) \tag{4.3.36}$$

with the inverse relations

$$\hat{a} = \hat{X} + i\hat{Y} \quad \text{and} \quad \hat{a}^\dagger = \hat{X} - i\hat{Y}. \tag{4.3.37}$$

The operators $\hat{X}$ and $\hat{Y}$ do satisfy the Hermitian operator condition (4.3.35). It is easily shown that the Hamiltonian of eqn (4.3.1) is expressed as

$$\hat{\mathcal{H}} = \hbar\omega\left(\hat{X}^2 + \hat{Y}^2\right) \tag{4.3.38}$$

and that the quadrature-operator commutation relation is

$$\left[\hat{X}, \hat{Y}\right] = i/2. \tag{4.3.39}$$

The commutator leads by the standard procedure [4] to a dimensionless form of the Heisenberg position–momentum uncertainty relation

$$(\Delta X)^2(\Delta Y)^2 \geq 1/16. \tag{4.3.40}$$

The calculations that follow sometimes use the pair of operators $\hat{a}$ and $\hat{a}^\dagger$, and sometimes the pair $\hat{X}$ and $\hat{Y}$, whichever is the more convenient.

Problem 4.1   Prove that

$$(n+3)(n+2)\langle n|\left(\hat{a}^\dagger\right)^3 \hat{a}^4 \hat{a}^\dagger|n\rangle = (n-1)(n-2)\langle n|\hat{a}^3\left(\hat{a}^\dagger\right)^4 \hat{a}|n\rangle.$$

(4.3.41)

Problem 4.2   Prove the commutators

$$\left[\hat{a},\left(\hat{a}^\dagger\right)^2\right] = 2\hat{a}^\dagger \quad \text{and} \quad \left[\hat{a}^2,\hat{a}^\dagger\right] = 2\hat{a},$$

(4.3.42)

and in general where n is a positive integer

$$\left[\hat{a},\left(\hat{a}^\dagger\right)^n\right] = n\left(\hat{a}^\dagger\right)^{n-1} \quad \text{and} \quad \left[\hat{a}^n,\hat{a}^\dagger\right] = n\hat{a}^{n-1}.$$

(4.3.43)

Hence show that

$$\left[\hat{a},\exp\left(\beta\hat{a}^\dagger\right)\right] = \beta\exp\left(\beta\hat{a}^\dagger\right),$$

(4.3.44)

where the operator $\exp\left(\beta\hat{a}^\dagger\right)$ is defined by its Maclaurin series in powers of $\beta\hat{a}^\dagger$.

Problem 4.3   Prove that the nth excited state of the oscillator can be expressed in terms of the ground state according to

$$|n\rangle = \hat{N}(n)|0\rangle,$$

(4.3.45)

where

$$\hat{N}(n) = \left(\hat{a}^\dagger\right)^n \big/ (n!)^{1/2}$$

(4.3.46)

has the nature of a creation operator for the number state as a whole.

4.4 Quantization of the electromagnetic field

The electromagnetic field is quantized by the association of a quantum-mechanical harmonic oscillator with each mode $\mathbf{k}\lambda$ of the radiation field in the quantization cavity defined in §4.2. The modes to which the quantum-mechanical operators refer are indicated by $\mathbf{k}\lambda$ subscripts. Thus the destruction and creation operator relations (4.3.32) and (4.3.33) for cavity mode $\mathbf{k}\lambda$ take the forms

$$\hat{a}_{\mathbf{k}\lambda}|n_{\mathbf{k}\lambda}\rangle = n_{\mathbf{k}\lambda}^{1/2}|n_{\mathbf{k}\lambda} - 1\rangle$$

(4.4.1)

and

$$\hat{a}_{k\lambda}^{\dagger}|n_{k\lambda}\rangle = (n_{k\lambda}+1)^{1/2}|n_{k\lambda}+1\rangle, \tag{4.4.2}$$

with the physical interpretation that the operators respectively destroy and create one photon of energy $\hbar\omega_k$ in mode $k\lambda$. Note that the angular frequency of the photon depends only on the magnitude of its wavevector and it is independent of the mode polarization specified by the index $\lambda = 1,2$. The number $n_{k\lambda}$ of photons excited in the cavity mode is given by the eigenvalue of the appropriate photon-number operator

$$\hat{n}_{k\lambda} = \hat{a}_{k\lambda}^{\dagger}\hat{a}_{k\lambda}, \tag{4.4.3}$$

with the eigenvalue relation

$$\hat{n}_{k\lambda}|n_{k\lambda}\rangle = \hat{a}_{k\lambda}^{\dagger}\hat{a}_{k\lambda}|n_{k\lambda}\rangle = n_{k\lambda}|n_{k\lambda}\rangle \quad n_{k\lambda} = 0, 1, 2, ..., \tag{4.4.4}$$

similar to eqns (4.3.27) and (4.3.28). The orthonormal eigenstates $|n_{k\lambda}\rangle$ are known as the *photon-number states* or *Fock states* of the electromagnetic field.

A number state of the total electromagnetic field in the cavity is specified by a string of photon numbers, one for each of the allowed modes, with the convention that the mode labels are arranged in some particular order as, for example, in Fig. 1.4. The different cavity modes are independent, and their associated operators commute, so that the basic commutation relation (4.3.9) is generalized to

$$\left[\hat{a}_{k\lambda}, \hat{a}_{k'\lambda'}^{\dagger}\right] = \delta_{k,k'}\delta_{\lambda,\lambda'}. \tag{4.4.5}$$

The state of the total field is written as a product of the states of the individual modes

$$|n_{k_1 1}, n_{k_1 2}, n_{k_2 1}, n_{k_2 2}, ...\rangle = |n_{k_1 1}\rangle|n_{k_1 2}\rangle|n_{k_2 1}\rangle|n_{k_2 2}\rangle\cdots = |\{n_{k\lambda}\}\rangle. \tag{4.4.6}$$

The notation on the far right is a short-hand for the somewhat cumbersome notations to the left, where $\{n_{k\lambda}\}$ denotes the complete set of numbers that specify the excitation levels of all the harmonic oscillators associated with the cavity modes. There are always, of course, infinitely many such oscillators. The multimode number states defined in eqn (4.4.6) form a complete set of states for the electromagnetic field in the cavity when every member of the set $\{n_{k\lambda}\}$ runs over the values of zero and all the positive integers. The action of a specific mode operator on the multimode state picks out the corresponding single-mode state in the product so that, for example, the effects of the destruction and creation operators can be evaluated by the use of eqns (4.4.1) and (4.4.2).

The Hamiltonian of the radiation field is obtained by summation of the harmonic-oscillator contributions similar to eqn (4.3.10) as

$$\hat{\mathcal{H}}_R = \sum_k \sum_\lambda \hat{\mathcal{H}}_{k\lambda}, \tag{4.4.7}$$

where

$$\hat{\mathcal{H}}_{k\lambda} = \tfrac{1}{2}\hbar\omega_k\left(\hat{a}_{k\lambda}\hat{a}_{k\lambda}^\dagger + \hat{a}_{k\lambda}^\dagger\hat{a}_{k\lambda}\right). \tag{4.4.8}$$

Comparison with the expression (4.2.25) for the classical cycle-averaged energy suggests that the conversions from the classical vector-potential amplitudes to the quantum-mechanical mode operators should be

$$A_{k\lambda} \to \left(\hbar/2\varepsilon_0 V\omega_k\right)^{1/2}\hat{a}_{k\lambda} \quad \text{and} \quad A_{k\lambda}^* \to \left(\hbar/2\varepsilon_0 V\omega_k\right)^{1/2}\hat{a}_{k\lambda}^\dagger. \tag{4.4.9}$$

With these substitutions, the classical vector potential of eqn (4.2.3) is converted to the operator

$$\hat{\mathbf{A}}(\mathbf{r},t) = \sum_k \sum_{\lambda=1,2}\mathbf{e}_{k\lambda}\hat{A}_{k\lambda}(\mathbf{r},t), \tag{4.4.10}$$

where

$$\hat{A}_{k\lambda}(\mathbf{r},t) = \left(\hbar/2\varepsilon_0 V\omega_k\right)^{1/2}\left\{\hat{a}_{k\lambda}\exp(-i\omega_k t + i\mathbf{k}\cdot\mathbf{r}) + \hat{a}_{k\lambda}^\dagger\exp(i\omega_k t - i\mathbf{k}\cdot\mathbf{r})\right\} \tag{4.4.11}$$

is the contribution to the magnitude of the vector-potential operator from cavity mode $k\lambda$.

The corresponding results for the electric and magnetic field operators are obtained from the operator versions of eqns (4.1.25) and (4.1.5) or by conversion of eqns (4.2.15) to (4.2.18). The expressions for the operators are simplified by defining a phase angle for the mode waveforms by

$$\chi_k(\mathbf{r},t) = \omega_k t - \mathbf{k}\cdot\mathbf{r} - \tfrac{\pi}{2}. \tag{4.4.12}$$

The complete electric-field operator is conveniently separated as

$$\hat{\mathbf{E}}_T(\mathbf{r},t) = \hat{\mathbf{E}}_T^+(\mathbf{r},t) + \hat{\mathbf{E}}_T^-(\mathbf{r},t), \tag{4.4.13}$$

where the two contributions

$$\hat{\mathbf{E}}_T^+(\mathbf{r},t) = \sum_k \sum_\lambda \mathbf{e}_{k\lambda}\left(\hbar\omega_k/2\varepsilon_0 V\right)^{1/2}\hat{a}_{k\lambda}\exp[-i\chi_k(\mathbf{r},t)] \tag{4.4.14}$$

and

$$\hat{\mathbf{E}}_T^-(\mathbf{r},t) = \sum_k \sum_\lambda \mathbf{e}_{k\lambda}\left(\hbar\omega_k/2\varepsilon_0 V\right)^{1/2}\hat{a}_{k\lambda}^\dagger\exp[i\chi_k(\mathbf{r},t)] \tag{4.4.15}$$

are known respectively as the *positive* and *negative frequency parts* of the electric-field operator. These names are somewhat counter-intuitive, as the plus sign is associated with the destruction operator components and the minus sign with the creation operator components, but this is nevertheless the conventional nomenclature. The magnetic field operator is written in the analogous form

$$\hat{\mathbf{B}}(\mathbf{r},t) = \hat{\mathbf{B}}^+(\mathbf{r},t) + \hat{\mathbf{B}}^-(\mathbf{r},t), \tag{4.4.16}$$

where

$$\hat{\mathbf{B}}^+(\mathbf{r},t) = \sum_{\mathbf{k}}\sum_{\lambda} \mathbf{k} \times \mathbf{e}_{\mathbf{k}\lambda}(\hbar/2\varepsilon_0\omega_k V)^{1/2} \hat{a}_{\mathbf{k}\lambda} \exp[-i\chi_{\mathbf{k}}(\mathbf{r},t)] \tag{4.4.17}$$

and

$$\hat{\mathbf{B}}^-(\mathbf{r},t) = \sum_{\mathbf{k}}\sum_{\lambda} \mathbf{k} \times \mathbf{e}_{\mathbf{k}\lambda}(\hbar/2\varepsilon_0\omega_k V)^{1/2} \hat{a}_{\mathbf{k}\lambda}^{\dagger} \exp[i\chi_{\mathbf{k}}(\mathbf{r},t)]. \tag{4.4.18}$$

The complete electric and magnetic field operators in eqns (4.4.13) and (4.4.16) are Hermitian, in accordance with the condition in eqn (4.3.35), and they represent the observable electromagnetic fields in the cavity.

Alternative expressions for the field operators are obtained with the use of the quadrature operators defined in eqns (4.3.36) and (4.3.37). Thus with insertion of the mode subscripts as before, the commutator in eqn (4.3.39) is generalized to

$$[\hat{X}_{\mathbf{k}\lambda}, \hat{Y}_{\mathbf{k}'\lambda'}] = (i/2)\delta_{\mathbf{k},\mathbf{k}'}\delta_{\lambda,\lambda'}. \tag{4.4.19}$$

The electric field operator from eqns (4.4.13) to (4.4.15) takes the form

$$\hat{\mathbf{E}}_T(\mathbf{r},t) = \sum_{\mathbf{k}}\sum_{\lambda} \mathbf{e}_{\mathbf{k}\lambda}(2\hbar\omega_k/\varepsilon_0 V)^{1/2}\{\hat{X}_{\mathbf{k}\lambda} \cos[\chi_{\mathbf{k}}(\mathbf{r},t)] + \hat{Y}_{\mathbf{k}\lambda} \sin[\chi_{\mathbf{k}}(\mathbf{r},t)]\} \tag{4.4.20}$$

and the magnetic field operator from eqns (4.4.16) to (4.4.18) is given by a similar expression. The form of eqn (4.4.20) shows that the operators $\hat{X}_{\mathbf{k}\lambda}$ and $\hat{Y}_{\mathbf{k}\lambda}$ are associated with the cosine and sine quadratures of the mode phase angle, in accordance with the naming of these operators.

The electromagnetic radiation Hamilonian is written in a form analogous to the classical energy of eqn (4.2.19) as

$$\hat{\mathcal{H}}_R = \tfrac{1}{2} \int_{\text{cavity}} \mathrm{d}V \left[\varepsilon_0 \hat{\mathbf{E}}_T(\mathbf{r},t).\hat{\mathbf{E}}_T(\mathbf{r},t) + \mu_0^{-1}\hat{\mathbf{B}}(\mathbf{r},t).\hat{\mathbf{B}}(\mathbf{r},t) \right], \tag{4.4.21}$$

where the field operators are given by eqns (4.4.13) to (4.4.18). The integral is evaluated by the same steps as used in the derivation of eqn (4.2.25) and the result is identical to that given by eqns (4.4.7) and (4.4.8). The final result can be written in the equivalent forms

$$\hat{\mathcal{H}}_R = \sum_k \sum_\lambda \hbar\omega_k\left(\hat{a}^\dagger_{k\lambda}\hat{a}_{k\lambda} + \tfrac{1}{2}\right) = \sum_k \sum_\lambda \hbar\omega_k\left(\hat{X}^2_{k\lambda} + \hat{Y}^2_{k\lambda}\right), \tag{4.4.22}$$

where the commutators in eqns (4.4.5) and (4.4.19) are used. The energy eigen-value relation for the multimode number state defined in eqn (4.4.6) is

$$\hat{\mathcal{H}}_R\big|\{n_{k\lambda}\}\big\rangle = \sum_k \sum_\lambda \hbar\omega_k\left(n_{k\lambda} + \tfrac{1}{2}\right)\big|\{n_{k\lambda}\}\big\rangle. \tag{4.4.23}$$

The ground state of the electromagnetic field in which no photons are excited in any of the field modes, that is

$$n_{k\lambda} = 0 \quad \text{for all } k \text{ and } \lambda, \tag{4.4.24}$$

is called the *vacuum state* of the field. The vacuum state is denoted $|\{0\}\rangle$, and it satisfies ground-state conditions of the form of eqn (4.3.23) for every mode,

$$\hat{a}_{k\lambda}|\{0\}\rangle = 0 \quad \text{for all } k \text{ and } \lambda. \tag{4.4.25}$$

The eigenvalue equation (4.4.23) takes the form

$$\hat{\mathcal{H}}_R|\{0\}\rangle = \tfrac{1}{2}\sum_k \sum_\lambda \hbar\omega_k|\{0\}\rangle \equiv \mathcal{E}_0|\{0\}\rangle \tag{4.4.26}$$

for the vacuum state, where the ground-state energy,

$$\mathcal{E}_0 = \tfrac{1}{2}\sum_k \sum_\lambda \hbar\omega_k = \sum_k \hbar\omega_k, \tag{4.4.27}$$

is known as the *zero-point energy* or *vacuum energy*. This contribution has no analogue in the classical theory. The eigenvalue relation (4.4.23) can be written as

$$\hat{\mathcal{H}}_R\big|\{n_{k\lambda}\}\big\rangle = \left(\mathcal{E}_R + \mathcal{E}_0\right)\big|\{n_{k\lambda}\}\big\rangle, \tag{4.4.28}$$

where

$$\mathcal{E}_R = \sum_k \sum_\lambda \hbar\omega_k n_{k\lambda} \tag{4.4.29}$$

is the excitation energy of the electromagnetic field above its zero-point value, analogous to the classical energy of eqn (4.2.24).

The allowed frequencies ω_k have no upper bound and the zero-point energy $\mathcal{E}_0$ of eqn (4.4.27) is accordingly infinite, an awkward feature of the quantized electromagnetic field. It is shown in §4.11 that only the part $\mathcal{E}_R$ contributes to the energy of the electromagnetic field as determined by measurements of the intensity of a beam of light. However, both parts are associated with measurable forces that act on material bodies. This and other features of the zero-point energy are discussed in §6.12.

Problem 4.4   Prove the commutation relation

$$\left[\hat{\mathcal{H}}_R, \hat{\mathbf{A}}(\mathbf{r},t)\right] = i\hbar \hat{\mathbf{E}}_T(\mathbf{r},t).$$ (4.4.30)

4.5 Canonical commutation relation

The quantum theory of the electromagnetic field is rigorously derived by quantization of the classical Hamiltonian, which is itself derived from the classical Lagrangian with the use of appropriate generalized coordinates and their canonically conjugate momenta [5]. When applied to the free field, this procedure provides a more rigorous derivation of the same quantum Hamiltonian and the associated field operators as obtained in the previous section by an informal method based essentially on plausibility arguments. The formal quantization is well described in several monographs [6,7], which provide detailed accounts of the canonical procedure, and this material is not repeated here.

However, an important aspect of the formal results is the nature of the commutator of the transverse vector potential operator $\hat{\mathbf{A}}(\mathbf{r},t)$ with its canonically conjugate momentum, $-\varepsilon_0 \hat{\mathbf{E}}_T(\mathbf{r},t)$ in the Coulomb gauge [6–8]. This canonical commutator is needed for subsequent calculations and it can be derived from the forms of the field operators derived in the previous section. The equal-time commutator between Cartesian components of the vector potential and transverse electric field operators is obtained from eqns (4.4.10) to (4.4.15) as

$$\left[\hat{A}_i(\mathbf{r},t), -\varepsilon_0 \hat{E}_{Tj}(\mathbf{r}',t)\right] = \frac{i\hbar}{2V} \sum_{\mathbf{k}} \sum_{\lambda} e_{\mathbf{k}\lambda i} e_{\mathbf{k}\lambda j} \left\{ \exp[i\mathbf{k}.(\mathbf{r}-\mathbf{r}')] + \exp[-i\mathbf{k}.(\mathbf{r}-\mathbf{r}')] \right\},$$ (4.5.1)

where the basic commutator in eqn (4.4.5) is used.

In order to proceed further in the evaluation of the canonical commutator, it is necessary to carry out the summations over wavevector and polarization. Let k, θ and ϕ be the polar coordinates of the wavevector $\mathbf{k}$, so that its Cartesian components are

$$\mathbf{k} = k(\sin\theta\cos\phi, \sin\theta\sin\phi, \cos\theta).$$ (4.5.2)

Possible choices of polarization vectors that satisfy eqns (4.2.7) to (4.2.9) include

$$\mathbf{e}_{\mathbf{k}1} = (\sin\phi, -\cos\phi, 0)$$ (4.5.3)

and

$$\mathbf{e}_{\mathbf{k}2} = (\cos\theta\cos\phi, \cos\theta\sin\phi, -\sin\theta).$$ (4.5.4)

It is easily verified that

$$e_{\mathbf{k}1i}e_{\mathbf{k}1j} + e_{\mathbf{k}2i}e_{\mathbf{k}2j} = \delta_{ij} - \left(k_i k_j / k^2\right), \tag{4.5.5}$$

where the Kronecker delta has its usual properties. This result holds generally for any choice of orthogonal polarization vectors.

It is convenient to convert the summation over $\mathbf{k}$ in eqn (4.5.1) to an integration. The conversion is given by eqn (1.1.11), except that the function to be integrated is not now isotropic in wavevector space, with the two polarizations explicitly taken into account in eqn (4.5.5). The generalized conversion is

$$\sum_{\mathbf{k}} \rightarrow \left[V/(2\pi)^3\right]\!\int\! d\mathbf{k}, \tag{4.5.6}$$

where the integration runs over all of three-dimensional wavevector space. Thus, with the use of eqn (4.5.5) and noting that the two exponential terms in eqn (4.5.1) make the same contribution to the purely imaginary commutator, we find

$$\left[\hat{A}_i(\mathbf{r},t), -\varepsilon_0 \hat{E}_{Tj}(\mathbf{r}',t)\right] = \frac{i\hbar}{(2\pi)^3}\int d\mathbf{k}\left(\delta_{ij} - \frac{k_i k_j}{k^2}\right)\exp[i\mathbf{k}.(\mathbf{r}-\mathbf{r}')]. \tag{4.5.7}$$

The integral can be written as a variant of the Dirac delta-function, known as the *transverse delta-function* [6,7,9], and defined as

$$\delta_{Tij}(\mathbf{r}-\mathbf{r}') = \frac{1}{(2\pi)^3}\int d\mathbf{k}\left(\delta_{ij} - \frac{k_i k_j}{k^2}\right)\exp[i\mathbf{k}.(\mathbf{r}-\mathbf{r}')]. \tag{4.5.8}$$

It enables the canonical commutator to be written in the final form

$$\left[\hat{A}_i(\mathbf{r},t), -\varepsilon_0 \hat{E}_{Tj}(\mathbf{r}',t)\right] = i\hbar\delta_{Tij}(\mathbf{r}-\mathbf{r}'). \tag{4.5.9}$$

The role of the transverse delta-function is understood by evaluating its effect on a general vector-field operator. We first need to consider the division of an operator into transverse and longitudinal parts, in accordance with eqns (4.1.19) to (4.1.21), in greater detail. Thus, let $\hat{\mathbf{V}}(\mathbf{r})$ be any vector-field operator expressed in terms of its Fourier transform $\hat{\mathbf{V}}(\mathbf{k})$ by

$$\hat{\mathbf{V}}(\mathbf{r}) = \frac{1}{(2\pi)^{3/2}}\int d\mathbf{k}\hat{\mathbf{V}}(\mathbf{k})\exp(i\mathbf{k}.\mathbf{r}). \tag{4.5.10}$$

By the ordinary expansion of the triple vector product, we can write

$$\hat{\mathbf{V}}(\mathbf{k}) = \hat{\mathbf{V}}_T(\mathbf{k}) + \hat{\mathbf{V}}_L(\mathbf{k}), \tag{4.5.11}$$

where

$$\hat{\mathbf{V}}_T(\mathbf{k}) = \left\{\mathbf{k} \times \hat{\mathbf{V}}(\mathbf{k})\right\} \times \mathbf{k}/k^2 \tag{4.5.12}$$

and

$$\hat{\mathbf{V}}_L(\mathbf{k}) = \left\{\mathbf{k}.\hat{\mathbf{V}}(\mathbf{k})\right\}\mathbf{k}/k^2 . \tag{4.5.13}$$

These two functions are the Fourier transforms of the transverse and longitudinal parts of the vector field $\hat{\mathbf{V}}(\mathbf{r})$, and we can write

$$\hat{\mathbf{V}}(\mathbf{r}) = \hat{\mathbf{V}}_T(\mathbf{r}) + \hat{\mathbf{V}}_L(\mathbf{r}) \tag{4.5.14}$$

where

$$\hat{\mathbf{V}}_T(\mathbf{r}) = \frac{1}{(2\pi)^{3/2}} \int d\mathbf{k} \frac{\left\{\mathbf{k} \times \hat{\mathbf{V}}(\mathbf{k})\right\} \times \mathbf{k}}{k^2} \exp(i\mathbf{k}.\mathbf{r}) \tag{4.5.15}$$

and

$$\hat{\mathbf{V}}_L(\mathbf{r}) = \frac{1}{(2\pi)^{3/2}} \int d\mathbf{k} \frac{\left\{\mathbf{k}.\hat{\mathbf{V}}(\mathbf{k})\right\}\mathbf{k}}{k^2} \exp(i\mathbf{k}.\mathbf{r}). \tag{4.5.16}$$

It is readily verified that these are indeed the transverse and longitudinal parts, as they clearly satisfy

$$\nabla.\hat{\mathbf{V}}_T(\mathbf{r}) = 0 \tag{4.5.17}$$

and

$$\nabla \times \hat{\mathbf{V}}_L(\mathbf{r}) = 0, \tag{4.5.18}$$

similar to eqns (4.1.20) and (4.1.21).

Now consider the integral

$$\sum_j \int d\mathbf{r}' \delta_{Tij}(\mathbf{r} - \mathbf{r}')\hat{V}_j(\mathbf{r}') = \frac{1}{(2\pi)^{3/2}} \sum_j \int d\mathbf{k} \left(\delta_{ij} - \frac{k_i k_j}{k^2}\right)\hat{V}_j(\mathbf{k})\exp(i\mathbf{k}.\mathbf{r}), \tag{4.5.19}$$

where eqns (4.5.8) and (4.5.10) are used. The right-hand side of eqn (4.5.19) is straightforwardly shown to be identical to the ith Cartesian component of the right-hand side of eqn (4.5.15), so that

$$\sum_j \int d\mathbf{r}' \delta_{Tij}(\mathbf{r} - \mathbf{r}')\hat{V}_j(\mathbf{r}') = \hat{V}_{Ti}(\mathbf{r}). \tag{4.5.20}$$

The transverse delta-function thus extracts the transverse part of the vector-field operator at the position specified. A similar procedure shows that the *longitudinal delta-function*, defined as [6,9]

$$\delta_{Lij}(\mathbf{r}-\mathbf{r}') = \frac{1}{(2\pi)^3}\int d\mathbf{k}\,\frac{k_i k_j}{k^2}\exp[i\mathbf{k}.(\mathbf{r}-\mathbf{r}')], \tag{4.5.21}$$

performs the analogous task of extracting the longitudinal part of the vector field,

$$\sum_j \int d\mathbf{r}' \delta_{Lij}(\mathbf{r}-\mathbf{r}')\hat{V}_j(\mathbf{r}') = \hat{V}_{Li}(\mathbf{r}). \tag{4.5.22}$$

It is clear from their definitions that the sum of the transverse delta-function from eqn (4.5.8) and the longitudinal delta-function from eqn (4.5.21) is a product of Kronecker and Dirac delta-functions

$$\delta_{Tij}(\mathbf{r}-\mathbf{r}') + \delta_{Lij}(\mathbf{r}-\mathbf{r}') = \delta_{ij}\delta(\mathbf{r}-\mathbf{r}'). \tag{4.5.23}$$

The commutators of other combinations of field operators are readily evaluated with the use of their forms given in eqns (4.4.10) to (4.4.18).

Problem 4.5   Prove the commutation relations

$$\begin{aligned}
\left[\hat{E}_{Ti}(\mathbf{r},t),\hat{E}_{Tj}(\mathbf{r}',t)\right] &= \left[\hat{B}_i(\mathbf{r},t),\hat{B}_j(\mathbf{r}',t)\right] \\
&= \left[\hat{A}_i(\mathbf{r},t),\hat{A}_j(\mathbf{r}',t)\right] \\
&= \left[\hat{B}_i(\mathbf{r},t),\hat{A}_j(\mathbf{r}',t)\right] = 0.
\end{aligned} \tag{4.5.24}$$

Problem 4.6   Prove that

$$\left[\hat{E}_{Ti}(\mathbf{r},t),\hat{B}_j(\mathbf{r}',t)\right] = \frac{\hbar}{(2\pi)^3\varepsilon_0}\int d\mathbf{k}\,k_h\exp[i\mathbf{k}.(\mathbf{r}-\mathbf{r}')], \tag{4.5.25}$$

where *ijh* are a cyclic permutation of *xyz*. The sign of the result is changed for a non-cyclic permutation and the commutator vanishes for $i = j$.

Problem 4.7   If $\hat{\mathbf{V}}(\mathbf{r})$ and $\hat{\mathbf{W}}(\mathbf{r})$ are any vector-field operators, prove that

$$\int d\mathbf{r}\,\hat{\mathbf{V}}_T(\mathbf{r}).\hat{\mathbf{W}}_L(\mathbf{r}) = 0. \tag{4.5.26}$$

Problem 4.8   If

$$\hat{U} = \exp\left\{\frac{i}{\hbar}\int d\mathbf{r}\,\hat{\mathbf{V}}(\mathbf{r}).\hat{\mathbf{A}}(\mathbf{r},t)\right\}, \tag{4.5.27}$$

where $\hat{\mathbf{V}}(\mathbf{r})$ is any vector field that commutes with the transverse electric field and vector potential operators, prove that

$$\hat{U}^{-1}\hat{\mathbf{E}}_T(\mathbf{r},t)\hat{U} = \hat{\mathbf{E}}_T(\mathbf{r},t) - (1/\varepsilon_0)\hat{\mathbf{V}}_T(\mathbf{r}), \qquad (4.5.28)$$

where it may be useful to express the exponential in eqn (4.5.27) as a Maclaurin series in powers of the exponent.

4.6 Pure states and statistical mixtures

The multimode number states defined in eqn (4.4.6) form a complete set of basis states for the electromagnetic field. A general pure state of the electromagnetic field is expressible as a superposition of these basis states of the form

$$|\text{pure state}\rangle = \sum_{\{n_{\mathbf{k}\lambda}\}} c(\{n_{\mathbf{k}\lambda}\})|\{n_{\mathbf{k}\lambda}\}\rangle, \qquad (4.6.1)$$

where the summation runs in general over all sets $\{n_{\mathbf{k}\lambda}\}$ of photon numbers $n_{\mathbf{k}\lambda} = 0, 1, 2,\ldots$, one for each mode $\mathbf{k}\lambda$ in the quantization cavity. The full generality of this state is rarely used, but there are important pure states of the radiation field in which the superposition is restricted to the number states of a single mode, for example the coherent states of §5.3 and the squeezed states of §§5.5 and 5.6, or to excitations of two modes, for example the photon pair states of §6.7. Superposition states have the common feature that the contribution of each basis state $|\{n_{\mathbf{k}\lambda}\}\rangle$ has a specified amplitude and a definite phase, relative to the overall phase angle of the complete superposition, embodied in the complex coefficient $c(\{n_{\mathbf{k}\lambda}\})$.

However, not all excitations of the radiation field are expressible as linear superpositions of the basis states defined in eqn (4.4.6), and they cannot be represented by *pure states* of the form of eqn (4.6.1). Some analogies can be drawn between the descriptions of light beams in the classical and quantum theories. Thus, while the electric field of the classical stable wave illustrated in Fig. 3.10 is specified by a definite amplitude and phase, the electric field of the light beam from a chaotic source is shown in §3.6 to be specifiable only in terms of probabilities for the field to have particular values of amplitude and phase. A similar situation occurs in quantum mechanics, where the nature of a chaotic light source excludes the possibility of a description of the emitted field in terms of pure states with specified phase relations between the different photon-number states in the output beam. When all that can be specified is a set of probabilities that the radiation field is found in a range of states, each corresponding to one of a complete set of basis states, then the state of the field is a *statistical mixture*. Statistical distributions are introduced into quantum mechanics by means of the *density operator*. The matrix elements of the density operator for a two-level atom are used in the treatment of the optical Bloch equations in §§2.7 to 2.10, together with some discussion of pure states and statistical mixtures, but more detailed properties are needed for applications to the radiation field.

Consider a cavity electromagnetic field for which there is a known probability P_R that the field is in state $|R\rangle$. Here R is a label that runs over a set of basis pure states sufficient to describe the field. The multimode number states defined in eqn (4.4.6) are a possible set $|R\rangle$, but other complete sets of basis states can be formed from suitable linear superpositions of the $|\{n_{\mathbf{k}\lambda}\}\rangle$. The probability distribution is assumed to be normalized, with

$$\sum_R P_R = 1. \tag{4.6.2}$$

Consider some observable O that is represented by a quantum-mechanical operator $\hat{O}$. The average value of the observable for the pure state $|R\rangle$ is $\langle R|\hat{O}|R\rangle$, and hence its average value for the statistical mixture specified by P_R is the *ensemble average*

$$\langle \hat{O} \rangle = \sum_R P_R \langle R|\hat{O}|R\rangle. \tag{4.6.3}$$

This is the basic expression for predictions of measurements on statistical mixtures of the radiation field.

It is convenient to cast eqn (4.6.3) into a different form that is easier to work with and provides more elegant expressions for the ensemble averages. Let $|S\rangle$ be any complete basis set of pure states for the field considered, where the label S specifies different states in the set. According to the closure theorem [4],

$$\sum_S |S\rangle\langle S| = 1 \tag{4.6.4}$$

and insertion of this unit quantity after the operator $\hat{O}$ in eqn (4.6.3) gives

$$\langle \hat{O} \rangle = \sum_R P_R \sum_S \langle R|\hat{O}|S\rangle\langle S|R\rangle = \sum_R \sum_S P_R \langle S|R\rangle\langle R|\hat{O}|S\rangle. \tag{4.6.5}$$

The density operator $\hat{\rho}$ is defined as [3,4]

$$\hat{\rho} = \sum_R P_R |R\rangle\langle R| \tag{4.6.6}$$

and the average value of O from eqn (4.6.5) is written

$$\langle \hat{O} \rangle = \sum_S \langle S|\hat{\rho}\hat{O}|S\rangle. \tag{4.6.7}$$

The density operator contains exactly the same information as the probability distribution P_R, and $\hat{\rho}$ is determined once the P_R are specified for a given set of basis states $|R\rangle$.

It is evident from the above derivation that the ensemble average evaluated by means of eqn (4.6.7) has the same value for any complete set $|S\rangle$ of basis states. The result can thus be written in the general form

$$\langle \hat{O} \rangle = \text{Trace}\{\hat{\rho}\hat{O}\} \equiv \text{Tr}\{\hat{\rho}\hat{O}\}, \tag{4.6.8}$$

where the *trace* of an operator is the invariant sum of its diagonal matrix elements for any complete set of states. The result (4.6.8) is entirely equivalent to the basic expression (4.6.3) but the use of the density operator simplifies the expressions for ensemble averages and the manipulations in their evaluation. A special case of eqn (4.6.8), where $\hat{O}$ is taken to be the unit number, gives the normalization condition

$$\text{Tr}\{\hat{\rho}\} = 1. \tag{4.6.9}$$

The result (2.7.3) is an example of eqn (4.6.9) for the density operator of a two-level atom.

The probability distribution P_R contains the physical information about a field excitation needed to evaluate the ensemble averages. In contrast to the arbitrariness of the states $|S\rangle$ used to evaluate the trace in eqn (4.6.8), it is important to make a careful choice of the complete set of pure states $|R\rangle$ used in defining the density operator. The choice must be made in such a way as to preserve all the available information about the state of the system. Thus, it should be noted that the matrix elements of $\hat{\rho}$ itself have different properties for different complete sets of states. For the defining set $|R\rangle$ used in the construction of $\hat{\rho}$, only the diagonal matrix elements are non-zero,

$$\langle R'|\hat{\rho}|R''\rangle = \sum_R P_R \langle R'|R\rangle\langle R|R''\rangle = P_{R'}\delta_{R',R''}. \tag{4.6.10}$$

However, a typical matrix element for some other complete set of states $|S\rangle$ is

$$\langle S|\hat{\rho}|S'\rangle = \sum_R P_R \langle S|R\rangle\langle R|S'\rangle, \tag{4.6.11}$$

and there are no general constraints on the $|S\rangle$ and $|S'\rangle$ for which the right-hand side is non-zero. For any statistical mixture state of the radiation field, the appropriate set of states $|R\rangle$ in eqn (4.6.6) is that for which the corresponding P_R provide all information on the state. These considerations are illustrated by specific examples given later in this and subsequent chapters.

A pure state is a special case of a statistical mixture in which one of the probabilities P_R equals unity and all the remaining elements of the distribution vanish. The density operator from eqn (4.6.6) is

$$\hat{\rho} = |R\rangle\langle R|. \tag{4.6.12}$$

and the radiation field is definitely in a specific pure state $|R\rangle$. The statistical description becomes redundant but the concept of the density operator remains valid. A property that holds only for a pure-state density operator is

$$\hat{\rho}^2 = \hat{\rho}, \tag{4.6.13}$$

which is easily proved from eqn (4.6.12). It is often convenient to use the density-operator formalism in general expressions for quantum-mechanical averages so that the results cover both pure states and statistical mixtures.

An important example of the statistical mixture is provided by the thermal excitation of photons in a single mode of a cavity maintained at temperature T. This is the system treated in §1.3 in the derivation of Planck's law, and the probability $P(n)$ that n photons are excited is given by eqns (1.3.2) and (1.3.5). Thus according to eqn (4.6.6), the density operator based on the number states is

$$\hat{\rho} = \sum_n P(n)|n\rangle\langle n| = \{1 - \exp(-\hbar\omega/k_B T)\}\sum_n \exp(-n\hbar\omega/k_B T)|n\rangle\langle n|. \tag{4.6.14}$$

The number states are the correct basis for the density operator in this case because the thermal distribution gives information only on the probabilities of finding a system in its various energy eigenstates, which are the same as the states $|n\rangle$. The density operator of eqn (4.6.14) has only diagonal number-state matrix elements.

Problem 4.9   Determine the mean number of thermally-excited photons by the use of eqn (4.6.8) in the form

$$\langle n \rangle = \text{Tr}\{\hat{\rho}\hat{a}^\dagger\hat{a}\} \tag{4.6.15}$$

and hence show that the density operator of eqn (4.6.14) can be re-expressed as

$$\hat{\rho} = \sum_n \frac{\langle n \rangle^n}{(1 + \langle n \rangle)^{1+n}}|n\rangle\langle n|. \tag{4.6.16}$$

The calculation is a quantum version of the classical procedure in eqn (1.3.6).

Problem 4.10   Show that the density operator of eqn (4.6.14) for single-mode thermal light can be written in the equivalent form

$$\hat{\rho} = \{1 - \exp(-\hbar\omega/k_B T)\}\exp(-\hbar\omega\hat{a}^\dagger\hat{a}/k_B T), \tag{4.6.17}$$

where the exponential is defined by its usual power-series expansion.

Problem 4.11   Prove that the density operator of single-mode thermal light satisfies

$$\text{Tr}\{\hat{\rho}^2\} = \frac{1}{1+2\langle n \rangle}.$$
$$\tag{4.6.18}$$

This is equal to unity only for $\langle n \rangle = 0$, corresponding to the vacuum state, which is a pure state in accordance with eqns (4.6.9) and (4.6.13).

Now consider the thermal excitation of all the cavity modes. The different field modes are independent and their combined density operator is a product of the contributions of the individual modes. Thus with use of the product states defined in eqn (4.4.6), the density operator is expressed as

$$\hat{\rho} = \sum_{\{n_{k\lambda}\}} P(\{n_{k\lambda}\}) |\{n_{k\lambda}\}\rangle\langle\{n_{k\lambda}\}|,$$
$$\tag{4.6.19}$$

where the summation runs over all sets $\{n_{k\lambda}\}$ of the photon numbers $n_{k\lambda}$ for all the modes $k\lambda$. The grand probability distribution is obtained by multiplying together factors similar to that in eqn (4.6.16), one for each mode,

$$P(\{n_{k\lambda}\}) = \prod_k \prod_\lambda \frac{\langle n_{k\lambda} \rangle^{n_{k\lambda}}}{(1+\langle n_{k\lambda} \rangle)^{1+n_{k\lambda}}}.$$
$$\tag{4.6.20}$$

The mean photon numbers are related to the mode frequency ω_k and the temperature by

$$\langle n_{k\lambda} \rangle = \{\exp(\hbar\omega_k/k_B T) - 1\}^{-1}$$
$$\tag{4.6.21}$$

and, with this expression, the density operator contains essentially the same information as the results of the thermal probability calculation employed in the derivation of Planck's radiation law in §1.3.

The density operator given by eqns (4.6.19) and (4.6.20) applies, however, not only to the thermal photon distribution but also to a wide range of excitations in which the statistical properties of the light are suitably random. For example, it is shown in §7.2 that the light generated by a source in which the atoms are kept at an excitation level different from that in thermal equilibrium has the same form of photon distribution as thermal radiation. More generally, this form of density operator applies to all varieties of chaotic light, and it provides the tool by which the classical derivations of Chapter 3 are recast in quantum-mechanical form. The spectral distributions of the radiation in these cases differ from that in thermal equilibrium, and the mean photon numbers are determined by the nature of the random field excitation. Thus eqn (4.6.21) is not generally applicable and, for example, in a beam of chaotic light with a Lorentzian frequency distribution it is replaced by an appropriate Lorentzian function of ω_k.

Problem 4.12 For the density operator of eqns (4.6.19) and (4.6.20), prove the normalization requirement of eqn (4.6.9) explicitly and show that the total mean number of photons is

$$\langle n \rangle = \sum_{\mathbf{k}} \sum_{\lambda} \mathrm{Tr}\left\{ \hat{\rho} \hat{a}^{\dagger}_{\mathbf{k}\lambda} \hat{a}_{\mathbf{k}\lambda} \right\} = \sum_{\mathbf{k}} \sum_{\lambda} \langle n_{\mathbf{k}\lambda} \rangle . \tag{4.6.22}$$

4.7 Time-development of quantum-optical systems

A given quantum-mechanical system can in general be described in several *pictures*, or *representations*, which differ in their treatment of the time dependence. The extreme examples are the *Schrödinger picture*, where all the operators are time-independent and the time dependence is carried entirely by the wavefunctions, and the *Heisenberg picture*, where the wavefunctions are independent of the time and the time dependence is carried by the operators. The electromagnetic field operators derived in §4.4 are in the Heisenberg picture, and these provide the closest quantum-mechanical approach to the dynamical variables of the classical theory. It is, however, sometimes convenient to use the Schrödinger picture for other calculations, for example the theory of the interaction of the quantized field with atoms covered in the following section. The present section gives a brief compilation of the main results in the two pictures [10,11].

The *interaction picture* has properties intermediate between those of the Schrödinger and Heisenberg pictures, with the time dependence shared between operators and wavefunctions. The main features of the interaction picture are outlined in §4.9. Applications of all three pictures occur in the remaining chapters.

(1) *The Schrödinger picture.* The operators $\hat{O}_S$ that represent observables in the Schrödinger picture are all time-independent,

$$d\hat{O}_S / dt = 0. \tag{4.7.1}$$

The expressions for electromagnetic field operators derived in §4.4 are converted to the Schrödinger picture by simple removal of their time dependences. The electric field operator is thus obtained from eqns (4.4.12) to (4.4.15) as

$$\hat{\mathbf{E}}_T(\mathbf{r}) = i \sum_{\mathbf{k}} \sum_{\lambda} \mathbf{e}_{\mathbf{k}\lambda} \left(\hbar \omega_k / 2\varepsilon_0 V \right)^{1/2} \left\{ \hat{a}_{\mathbf{k}\lambda} \exp(i\mathbf{k}.\mathbf{r}) - \hat{a}^{\dagger}_{\mathbf{k}\lambda} \exp(-i\mathbf{k}.\mathbf{r}) \right\}. \tag{4.7.2}$$

The time-dependent Schrödinger wavefunction $\Psi_S(t)$ of the atom–radiation system satisfies eqn (2.1.1) in the form

$$\hat{\mathcal{H}} \Psi_S(t) = i\hbar d\Psi_S(t)/dt, \tag{4.7.3}$$

where the Hamiltonian is independent of the time. The formal solution is

$$\Psi_S(t) = \exp\left(-i\hat{\mathcal{H}}t/\hbar\right)\Psi_S(0),$$ (4.7.4)

where the exponential quantity is known as the *time-development operator*. The expectation value of a Schrödinger-picture operator for a system in this pure state is

$$\left\langle \hat{O}(t) \right\rangle_S = \left\langle \Psi_S(t) \middle| \hat{O}_S \middle| \Psi_S(t) \right\rangle.$$ (4.7.5)

For a statistical mixture in the Schrödinger picture, with a probability P_S that the system is in state Ψ_S at time $t = 0$, the density operator at a general time t is defined in accordance with eqn (4.6.6) as

$$\begin{aligned}
\hat{\rho}_S(t) &= \sum_S P_S \left| \Psi_S(t) \right\rangle \left\langle \Psi_S(t) \right| \\
&= \sum_S P_S \exp\left(-i\hat{\mathcal{H}}t/\hbar\right) \left| \Psi_S(0) \right\rangle \left\langle \Psi_S(0) \middle| \exp\left(i\hat{\mathcal{H}}t/\hbar\right).
\end{aligned}$$ (4.7.6)

The density operator *does* vary with time in the Schrödinger picture, essentially because it is a form of representation of the wavefunction, and it is easily shown by differentiation of eqn (4.7.6) that

$$i\hbar \, d\hat{\rho}_S(t)/dt = \left[\hat{\mathcal{H}}, \hat{\rho}_S(t)\right].$$ (4.7.7)

The expectation value of an operator for a statistical mixture is

$$\left\langle \hat{O}(t) \right\rangle_S = \mathrm{Tr}\left\{\hat{\rho}_S(t)\hat{O}_S\right\},$$ (4.7.8)

similar to eqn (4.6.8).

(2) *The Heisenberg picture.* The Heisenberg wavefunctions are related to the Schrödinger wavefunctions by

$$\Psi_H = \exp\left(i\hat{\mathcal{H}}t/\hbar\right)\Psi_S(t) = \Psi_S(0),$$ (4.7.9)

where $\hat{\mathcal{H}}$ is the time-independent Hamiltonian. The Heisenberg-picture wavefunction is clearly independent of the time, with

$$d\Psi_H/dt = 0.$$ (4.7.10)

The time-dependent Heisenberg operators are defined in terms of the time-independent Schrödinger operators by

$$\hat{O}_H(t) = \exp\left(i\hat{\mathcal{H}}t/\hbar\right)\hat{O}_S \exp\left(-i\hat{\mathcal{H}}t/\hbar\right),$$ (4.7.11)

and their equation of motion is obtained by simple differentiation as

$$i\hbar \, d\hat{O}_{\mathrm{H}}(t)/dt = \left[\hat{O}_{\mathrm{H}}(t), \hat{\mathcal{H}}\right].$$ (4.7.12)

The commutation properties of the operators are unchanged by the transformation in eqn (4.7.11), even though the operators themselves vary with the time. Unlike most other operators, the total Hamiltonian remains independent of the time in the Heisenberg picture,

$$\hat{\mathcal{H}}(t) = \hat{\mathcal{H}}.$$ (4.7.13)

The formalism for the quantized electromagnetic field presented in §§4.4 and 4.5 is all in the Heisenberg picture.

Calculated results for measurable quantities are the same in both pictures. Thus the Heisenberg expectation value

$$\left\langle \hat{O}(t) \right\rangle_{\mathrm{H}} = \left\langle \varPsi_{\mathrm{H}} \middle| \hat{O}_{\mathrm{H}}(t) \middle| \varPsi_{\mathrm{H}} \right\rangle$$ (4.7.14)

is seen, with the use of eqns (4.7.9) and (4.7.11), to be identical to the Schrödinger expectation value in eqn (4.7.5). For statistical mixtures, the Heisenberg density operator is defined in the manner of eqn (4.7.11) as

$$\hat{\rho}_{\mathrm{H}} = \exp\!\left(i\hat{\mathcal{H}}t/\hbar\right)\hat{\rho}_{\mathrm{S}}(t)\exp\!\left(-i\hat{\mathcal{H}}t/\hbar\right).$$ (4.7.15)

It follows from eqn (4.7.6) that this is another exception to the usual time dependence of operators in the Heisenberg picture, again because the density operator is associated with the wavefunction, now independent of the time. The Heisenberg expectation value for a statistical mixture, given by

$$\left\langle \hat{O}(t) \right\rangle_{\mathrm{H}} = \mathrm{Tr}\!\left\{\hat{\rho}_{\mathrm{H}}\hat{O}_{\mathrm{H}}(t)\right\},$$ (4.7.16)

is seen with the use of eqns (4.7.11) and (4.7.15) to be identical to the Schrödinger value in eqn (4.7.8), as the trace of a product of operators in unchanged by cyclic permutation.

4.8 Interaction of the quantized field with atoms

The derivations in the present chapter, from §4.2 onwards, are exclusively concerned with the properties of the free radiation field, that is, the field in a region of space where there is no matter to interact with the radiation. However, all experiments on the generation and detection of light involve the interactions of electromagnetic waves with matter. Thus, although our main interest is in the properties of the quantized electromagnetic field itself, it is necessary to study its interaction with atoms in order to appreciate what kinds of field excitation can be produced and what features of the excitations can be measured in practice. Indeed, the differences between the classical and quantum theories are largely accounted for by their different descriptions of the emission and absorption of light. These differences can be understood in terms of the simplest form of coupling between light and atoms.

The electric-dipole interaction is already treated in §2.2 and our purpose here is to give more detail of its formal derivation and its incorporation into the quantum theory of the radiation field. The derivation begins with a consideration of the classical theory of the atom–radiation interaction. The classical fields are again functions of position and time but only the former is shown explicitly as the $\mathbf{r}$ dependences of the fields are important in determining the forms of the contributions to the interaction.

The classical theory of §4.1 covers the interaction of the electromagnetic field with the atom illustrated in Fig. 2.1 if the charge and current densities in Maxwell's equations (4.1.1) to (4.1.4) are given by

$$\sigma(\mathbf{r}) = -e\sum_{\alpha} \delta(\mathbf{r} - \mathbf{r}_{\alpha}) + Ze\delta(\mathbf{r}) \tag{4.8.1}$$

and

$$\mathbf{J}(\mathbf{r}) = -e\sum_{\alpha} \dot{\mathbf{r}}_{\alpha} \delta(\mathbf{r} - \mathbf{r}_{\alpha}), \tag{4.8.2}$$

where α runs over the Z electrons of charge $-e$. The positions $\mathbf{r}_{\alpha}$ of the point charges are specified by three-dimensional delta-functions, formed from products of the three Dirac delta-functions that specify the individual Cartesian coordinates of the charges. The relatively massive atomic nucleus is assumed to be fixed at the origin of coordinates and a small contribution to the charge density from its motion is ignored in eqn (4.8.2).

If the Coulomb gauge specified by eqn (4.1.15) is adopted, the scalar potential obtained by solution of Poisson's equation (4.1.18) for the charge density of eqn (4.8.1) is

$$\phi(\mathbf{r}) = \frac{1}{4\pi\varepsilon_0} \left\{ -\sum_{\alpha} \frac{e}{|\mathbf{r} - \mathbf{r}_{\alpha}|} + \frac{Ze}{r} \right\}. \tag{4.8.3}$$

This is the usual Coulomb potential of the atomic electrons and nucleus.

The interaction between the atom and the electromagnetic field is conveniently treated in terms of the polarization $\mathbf{P}(\mathbf{r})$ and magnetization $\mathbf{M}(\mathbf{r})$, which are associated with the atomic charges by the usual relations [1]

$$\sigma(\mathbf{r}) = -\nabla.\mathbf{P}(\mathbf{r}) \tag{4.8.4}$$

and

$$\mathbf{J}(\mathbf{r}) = \dot{\mathbf{P}}(\mathbf{r}) + \nabla \times \mathbf{M}(\mathbf{r}). \tag{4.8.5}$$

Only the longitudinal part of the polarization contributes to the charge density and comparison of eqns (4.1.31) and (4.8.4) gives

$$\mathbf{P}_L(\mathbf{r}) = -\varepsilon_0 \mathbf{E}_L(\mathbf{r}). \tag{4.8.6}$$

It is straightforward, but tedious, to verify that eqns (4.8.4) and (4.8.5) have solutions for the atomic polarization and magnetization in the integral forms [6,8]

$$\mathbf{P}(\mathbf{r}) = -e \sum_\alpha \mathbf{r}_\alpha \int_0^1 \mathrm{d}\zeta \delta(\mathbf{r} - \zeta \mathbf{r}_\alpha) \tag{4.8.7}$$

and

$$\mathbf{M}(\mathbf{r}) = -e \sum_\alpha \mathbf{r}_\alpha \times \dot{\mathbf{r}}_\alpha \int_0^1 \mathrm{d}\zeta \zeta \delta(\mathbf{r} - \zeta \mathbf{r}_\alpha), \tag{4.8.8}$$

and these forms are convenient for the calculations that follow.

The potential energy of the atom in the transverse electric field $\mathbf{E}_T(\mathbf{r})$ is simply

$$V_E = -\int \mathrm{d}\mathbf{r} \mathbf{P}(\mathbf{r}) . \mathbf{E}_T(\mathbf{r}) = e \sum_\alpha \int_0^1 \mathrm{d}\zeta \, \mathbf{r}_\alpha . \mathbf{E}_T(\zeta \mathbf{r}_\alpha), \tag{4.8.9}$$

where the complete expression for the polarization from eqn (4.8.7) is inserted but, in accordance with eqn (4.5.26), only its transverse part contributes to the integral. The time dependences of the electromagnetic fields are not explicitly shown in the present section. A Taylor expansion of the transverse electric vector gives

$$
\begin{aligned}
V_E &= e \sum_\alpha \int_0^1 \mathrm{d}\zeta \left\{ 1 + \zeta \mathbf{r}_\alpha . \nabla + \frac{1}{2!} (\zeta \mathbf{r}_\alpha . \nabla)^2 + \dots \right\} \mathbf{r}_\alpha . \mathbf{E}_T(0) \\
&= e \sum_\alpha \left\{ 1 + \frac{1}{2!} \mathbf{r}_\alpha . \nabla + \frac{1}{3!} (\mathbf{r}_\alpha . \nabla)^2 + \dots \right\} \mathbf{r}_\alpha . \mathbf{E}_T(0).
\end{aligned}
\tag{4.8.10}
$$

The gradient operator in these expressions acts only on the spatial coordinate of the transverse electric field, which is set equal to zero after the differentiations are carried out. This expansion expresses the electrical potential energy in terms of the *multipole moments* of the atomic charge distribution. The first term is equivalent to the electric-dipole interaction given in eqn (2.2.5), with the atomic dipole moment

$$-e\mathbf{D} = -\sum_\alpha e \mathbf{r}_\alpha \tag{4.8.11}$$

and $\mathbf{D}$ as defined in eqn (2.2.4). The second term contains the product of the electric quadrupole moment

$$Q = -\tfrac{1}{2}\sum_{\alpha} e\mathbf{r}_\alpha \mathbf{r}_\alpha \qquad (4.8.12)$$

with the electric field gradients. The subsequent terms are the contributions of the third- and higher-order multipoles.

The potential energy of the atom in the magnetic field $\mathbf{B(r)}$ is similarly derived as

$$V_M = -\int d\mathbf{r}\,\mathbf{M(r).B(r)} = \frac{e}{m}\sum_\alpha \left\{\frac{1}{2!} + \frac{2}{3!}\mathbf{r}_\alpha.\nabla + \frac{3}{4!}(\mathbf{r}_\alpha.\nabla)^2 + ...\right\}\mathbf{l}_\alpha.\mathbf{B}(0), \qquad (4.8.13)$$

where

$$\mathbf{l}_\alpha = m\mathbf{r}_\alpha \times \dot{\mathbf{r}}_\alpha \qquad (4.8.14)$$

is the orbital angular momentum of electron α about the nucleus. This expansion expresses the magnetic potential energy in terms of the magnetic multipole moments of the moving atomic charges. The first term is the energy of the magnetic dipole moment

$$-e\mathbf{D}_M = -\tfrac{1}{2}\sum_\alpha \frac{e}{m}\mathbf{l}_\alpha \qquad (4.8.15)$$

in the magnetic field $\mathbf{B}(0)$. The subsequent terms are the contributions of the magnetic quadrupole and higher-order magnetic multipole moments.

The above expressions refer to an atom whose nucleus lies at the origin of coordinates. For a nucleus located at a general position $\mathbf{R}$, the electron coordinates $\mathbf{r}_\alpha$ are replaced by $\mathbf{r}_\alpha - \mathbf{R}$ and the electric and magnetic fields are evaluated at position $\mathbf{R}$.

The orders of magnitude of the different contributions to the potential energies V_E and V_M are compared by means of rough estimates of the quantities involved. Thus the electron coordinate $\mathbf{r}_\alpha$ relative to the atomic nucleus is taken to have the order of magnitude of the Bohr radius a_B from eqn (2.2.2), the electron angular momentum $\mathbf{l}_\alpha$ to have magnitude $\hbar$, the field gradient $\nabla E_T(0)$ to have magnitude $(\omega/c)E_T(0)$, and ω is taken as the visible frequency given in eqn (1.6.7). Then the magnitudes of the first two terms in the potential energy of eqn (4.8.10) are

$$e\mathbf{D}.\mathbf{E}_T(0) \sim E(0)4\pi\varepsilon_0\hbar^2/me \qquad (4.8.16)$$

for the electric-dipole contribution and

$$-\nabla.\mathbf{Q}.\mathbf{E}_T(0) \sim E_T(0)3e\hbar/16mc \qquad (4.8.17)$$

for the electric-quadrupole contribution. The leading term in the potential energy of eqn (4.8.13) is the magnetic-dipole contribution

$$e\mathbf{D}_M \cdot \mathbf{B}(0) \sim B(0)e\hbar/2m \sim E_T(0)e\hbar/2mc. \tag{4.8.18}$$

It is seen that the electric-quadrupole and magnetic-dipole potential energies have similar magnitudes, but they are smaller than the electric-dipole energy by the order of a dimensionless factor

$$e^2/4\pi\varepsilon_0\hbar c \approx 1/137, \tag{4.8.19}$$

known as the *fine structure constant*. Subsequent terms in the expansions of the electric and magnetic potential energies given in eqns (4.8.10) and (4.8.13) are smaller than the electric-dipole contribution by a quantity of the second or some higher order in the fine structure constant.

The discussion so far is concerned with the classical theory of the coupling of the atomic charge and current densities to the electromagnetic field. For the conversion to quantum mechanics, it is convenient to use field operators in the Schrödinger picture, as in the electric field operator of eqn (4.7.2). The quantum-mechanical Hamiltonian for the coupled system of radiation field and atom is derived by quantization of the classical Hamiltonian, which is again derived from the classical Lagrangian with the use of appropriate generalized coordinates and their conjugate momenta. It is well known [6,7] that the Hamiltonian is similar to a sum of the uncoupled Hamiltonians of the atom and the fields, except that the momentum operator conjugate to the charge position is replaced according to

$$\hat{\mathbf{p}}_\alpha \rightarrow \hat{\mathbf{p}}_\alpha + e\hat{\mathbf{A}}(\mathbf{r}_\alpha). \tag{4.8.20}$$

The complete Hamiltonian in the Coulomb gauge is thus

$$\hat{\mathcal{H}}' = \frac{1}{2m}\sum_\alpha\left[\hat{\mathbf{p}}_\alpha + e\hat{\mathbf{A}}(\mathbf{r}_\alpha)\right]^2 + \tfrac{1}{2}\int d\mathbf{r}\sigma(\mathbf{r})\phi(\mathbf{r}) + \tfrac{1}{2}\int d\mathbf{r}\left[\varepsilon_0\hat{\mathbf{E}}_T(\mathbf{r})^2 + \mu_0^{-1}\hat{\mathbf{B}}(\mathbf{r})^2\right], \tag{4.8.21}$$

which is known as the *minimal-coupling* form of the Hamiltonian. Its first term includes the kinetic energies of the atomic charges and additional terms

$$\hat{\mathcal{H}}_I = (e/m)\sum_\alpha\hat{\mathbf{A}}(\mathbf{r}_\alpha)\cdot\hat{\mathbf{p}}_\alpha + \left(e^2/2m\right)\sum_\alpha\hat{\mathbf{A}}(\mathbf{r}_\alpha)^2 \tag{4.8.22}$$

that represent the interaction between the atom and the electromagnetic radiation. The second term in the Hamiltonian is the electrostatic energy of the various charges that constitute the atom and it does not involve any quantum operators of the transverse radiation field. The final term is the transverse field energy, identical to the form of the free-field Hamiltonian given in eqn (4.4.21).

The interaction Hamiltonian of eqn (4.8.22), known as the **p.A** form, is not very convenient for the present calculation. The electromagnetic field appears via the transverse vector potential, which is not directly observable, and the expression does not take advantage of the small variation of the electromagnetic field of visible light across the dimensions of the atom. There is no obvious connection with the classical electric and magnetic potential energies given by eqns (4.8.10) and (4.8.13). The Hamilonian is converted to a more convenient and transparent form, known as the Power–Zienau–Woolley Hamiltonian [12,13], by a unitary transformation. Define an operator

$$\hat{U} = \exp\left\{\frac{i}{\hbar}\int d\mathbf{r}\hat{\mathbf{P}}(\mathbf{r}).\hat{\mathbf{A}}(\mathbf{r})\right\}, \tag{4.8.23}$$

which takes the form

$$\hat{U} = \exp\left\{-\frac{ie}{\hbar}\sum_\alpha\int_0^1 d\zeta\,\mathbf{r}_\alpha.\hat{\mathbf{A}}(\zeta\mathbf{r}_\alpha)\right\} \tag{4.8.24}$$

after substitution of the solution for the atomic polarization given in eqn (4.8.7). The transformed Hamiltonian and wavefunctions are given by

$$\hat{\mathcal{H}} = \hat{U}^{-1}\hat{\mathcal{H}}'\hat{U} \quad \text{and} \quad \psi = \hat{U}^{-1}\psi', \tag{4.8.25}$$

and the physical predictions that follow from the transformed theory are the same as from the original theory. The interaction between the atom and the radiation in the transformed Hamiltonian is expressed as series in the electric and magnetic multipole moments, but we derive here only the leading electric-dipole term in the interaction.

The transformation in eqn (4.8.25) is applied term-by-term to the Hamiltonian $\hat{\mathcal{H}}'$ of eqn (4.8.21) and it produces a new Hamiltonian expressed in terms of the electric and magnetic field operators rather than the vector potential operator. Thus the transformation of the **A.p** term in the interaction Hamiltonian of eqn (4.8.22) produces the quantum analogue of the magnetic potential energy of eqn (4.8.13) and the $\mathbf{A}^2$ term produces a nonlinear magnetic contribution. The second term in the Hamiltonian of eqn (4.8.21) is unchanged by the transformation. The quantum analogue of the electrical potential energy of eqn (4.8.9) arises from transformation of the electric field part of the radiative energy in the final term of the Hamiltonian (4.8.21). We have shown in eqns (4.8.16) to (4.8.18) that the electric-dipole contribution dominates the contributions from the higher-order electric multipoles and from all of the magnetic multipoles. We accordingly neglect the transformed interaction Hamiltonian obtained from eqn (4.8.22), which contains only magnetic contributions. The first two terms from eqn (4.8.21) then provide the atomic Hamiltonian

$$\hat{\mathcal{H}}_A = \frac{1}{2m}\sum_\alpha\hat{\mathbf{p}}_\alpha^2 + \tfrac{1}{2}\int d\mathbf{r}\sigma(\mathbf{r})\phi(\mathbf{r}). \tag{4.8.26}$$

It is verified by substitution of eqns (4.8.1) and (4.8.3) that the second term in eqn (4.8.26) includes the usual Coulomb interaction energies between all of the atomic charges. It also includes the infinite self-energies of the Coulomb interactions of the charges with themselves, and these contributions must be removed from the Hamiltonian.

The transformed electric field is obtained with the use of eqn (4.5.28) as

$$\hat{U}^{-1}\hat{\mathbf{E}}_T(\mathbf{r})\hat{U} = \hat{\mathbf{E}}_T(\mathbf{r}) - (1/\varepsilon_0)\mathbf{P}_T(\mathbf{r}), \tag{4.8.27}$$

and the final term of the Hamiltonian of eqn (4.8.21) is transformed into

$$\tfrac{1}{2}\int d\mathbf{r}\left[\varepsilon_0\hat{\mathbf{E}}_T(\mathbf{r})^2 + \mu_0^{-1}\hat{\mathbf{B}}(\mathbf{r})^2\right] - \int d\mathbf{r}\mathbf{P}(\mathbf{r}).\hat{\mathbf{E}}_T(\mathbf{r}) + (1/2\varepsilon_0)\int d\mathbf{r}\mathbf{P}_T(\mathbf{r})^2. \tag{4.8.28}$$

The first term is the same as the free-field Hamiltonian of eqn (4.4.21),

$$\hat{\mathcal{H}}_R = \tfrac{1}{2}\int d\mathbf{r}\left[\varepsilon_0\hat{\mathbf{E}}_T(\mathbf{r})^2 + \mu_0^{-1}\hat{\mathbf{B}}(\mathbf{r})^2\right]. \tag{4.8.29}$$

The second term is the quantized version of the electrical potential energy of eqn (4.8.9), and the theorem of eqn (4.5.26) is used to substitute the complete polarization field for its transverse part. The final term in eqn (4.8.28) is a function only of the atomic variables; it produces changes in the atomic energy levels but it does not contribute to the atom–field interaction and we neglect it here.

We make the *electric-dipole approximation* by neglecting all except the dominant contribution obtained from the first term in the expansion of the electrical potential energy in eqn (4.8.28), similar to the expansion in eqn (4.8.10). The resulting electric-dipole interaction Hamiltonian is

$$\hat{\mathcal{H}}_{ED} = e\sum_\alpha \mathbf{r}_\alpha.\hat{\mathbf{E}}_T(0) = e\mathbf{D}.\hat{\mathbf{E}}_T(0), \tag{4.8.30}$$

where **D** is defined in eqn (2.2.4) or (4.8.11). This interaction Hamiltonian, known as the **r.E** form, is equivalent to that derived more qualitatively in eqn (2.2.5). The complete Hamiltonian is a sum of the three contributions from eqns (4.8.26), (4.8.29) and (4.8.30),

$$\hat{\mathcal{H}} = \hat{\mathcal{H}}_A + \hat{\mathcal{H}}_R + \hat{\mathcal{H}}_{ED}. \tag{4.8.31}$$

The more complete Power–Zienau–Woolley, or multipolar, Hamiltonian is derived by retaining terms in the expansion of the electrical potential energy beyond the first and by keeping the magnetic potential energy and the nonlinear terms obtained from transformation of the **p.A** Hamiltonian of eqn (4.8.22). However, the electric-dipole term of eqn (4.8.30) is sufficient for our subsequent derivations. The predominance of the electric-dipole interaction in causing atomic transitions, including those involved in the photodetection process, ensures a more important role for the electric-field operator $\hat{\mathbf{E}}_T$, compared to that of the magnetic-field operator $\hat{\mathbf{B}}$, in the calculations that follow. It is emphasized that the **p.A** form of the interaction Hamiltonian produces the same results as the **r.E** form, provided that the corresponding approximations are made, and both forms play useful roles in appropriate calculations. The **r.E** form is more convenient for the calulations that follow.

4.9 Second quantization of the atomic Hamiltonian

The radiative part of the Hamiltonian in eqn (4.8.31) is expressed in terms of the photon creation and destruction operators when the Schrödinger-picture field operators $\hat{\mathbf{E}}_T(0)$ and $\hat{\mathbf{B}}(0)$ are inserted. It is also convenient to introduce creation and destruction operators for the atomic part of the Hamiltonian. The entire Hamiltonian is then expressed as products of algebraic factors with creation and destruction operators. It is straightforward to apply this form of the theory to higher-order radiative processes where multiple interactions of light and atoms occur.

Consider the atomic Hamiltonian of eqn (4.8.26) and let $|i\rangle$ be an energy eigenstate with eigenvalue $\hbar\omega_i$,

$$\hat{\mathcal{H}}_A|i\rangle = \hbar\omega_i|i\rangle. \tag{4.9.1}$$

According to the closure theorem, similar to eqn (4.6.4),

$$\sum_i |i\rangle\langle i| = 1, \tag{4.9.2}$$

where the sum runs over all the eigenstates of $\hat{\mathcal{H}}_A$, both bound and free. Use of the closure theorem twice establishes the identity

$$\hat{\mathcal{H}}_A = \sum_i |i\rangle\langle i|\hat{\mathcal{H}}_A \sum_j |j\rangle\langle j|. \tag{4.9.3}$$

However, it follows from eqn (4.9.1) and the orthonormality of the energy eigenstates that

$$\langle i|\hat{\mathcal{H}}_A|j\rangle = \hbar\omega_i\delta_{ij}, \tag{4.9.4}$$

and eqn (4.9.3) reduces to

$$\hat{\mathcal{H}}_A = \sum_i \hbar\omega_i|i\rangle\langle i|. \tag{4.9.5}$$

The transformation to the form in eqn (4.9.5) is known as *second quantization* of the atomic Hamiltonian. In this terminology, the normal quantum-mechanical procedure for determining the energy eigenstates and eigenvalues is regarded as first quantization of the motion of the atomic electrons. The second-quantized Hamiltonian is a useful form for calculations where the atom interacts with some other physical system, such as the radiation field.

Now consider the effect of a general combination $|i\rangle\langle j|$ applied to some arbitrary atomic eigenstate $|l\rangle$. As the atomic eigenstates are orthonormal,

$$|i\rangle\langle j|l\rangle = |i\rangle\delta_{jl},\tag{4.9.6}$$

and $|i\rangle\langle j|$ applied to an atomic eigenstate $|l\rangle$ thus changes the state to $|i\rangle$ if the original state $|l\rangle$ is the same as $|j\rangle$, but gives zero otherwise. In other words, $|i\rangle\langle j|$ is a *transition* or *projection* operator whose application to an atom in state $|j\rangle$ removes it from that state and projects it into state $|i\rangle$. We can say that $|i\rangle\langle j|$ destroys atomic state $|j\rangle$ and creates atomic state $|i\rangle$. Note that the electrons themselves cannot be created or destroyed in interactions that take place at photon energies in the visible region of the spectrum, but they can only be shifted from one state to another.

It remains to express the electric-dipole interaction Hamiltonian in terms of the atomic creation and destruction operators. According to eqn (4.8.30), the coordinates of the electrons enter $\hat{\mathcal{H}}_{ED}$ only in the vector quantity $\hat{\mathbf{D}}$, here written with a circumflex to emphasize its operator nature. Use of the closure theorem of eqn (4.9.2) twice establishes the identity

$$\hat{\mathbf{D}} = \sum_i |i\rangle\langle i|\hat{\mathbf{D}}\sum_j |j\rangle\langle j| = \sum_{i,j} \mathbf{D}_{ij}|i\rangle\langle j|,\tag{4.9.7}$$

where

$$\mathbf{D}_{ij} = \langle i|\hat{\mathbf{D}}|j\rangle.\tag{4.9.8}$$

The terms in eqn (4.9.7) with $i = j$ all vanish, as in eqn (2.2.6), because they involve diagonal matrix elements of the odd-parity operator $\hat{\mathbf{D}}$. The result (4.9.5) for the second-quantized Hamiltonian is a special case of the more general expression (4.9.7), which holds when the operator is not diagonal in the energy eigenstates.

The electric-dipole interaction of eqn (4.8.30) is now second-quantized by substitution of eqn (4.9.7) for $\hat{\mathbf{D}}$ and insertion of $\hat{\mathbf{E}}_T(0)$ from eqn (4.7.2). However, for slightly greater generality, we take the atomic nucleus to be placed at position $\mathbf{R}$ rather than at the origin of coordinates. The electric-dipole interaction is then

$$\hat{\mathcal{H}}_{ED} = e\hat{\mathbf{D}}.\hat{\mathbf{E}}_T(\mathbf{R}) = ie\sum_k\sum_\lambda\sum_{i,j}(\hbar\omega_k/2\varepsilon_0 V)^{1/2}\mathbf{e}_{k\lambda}.\mathbf{D}_{ij}$$
$$\times\left\{\hat{a}_{k\lambda}\exp(i\mathbf{k}.\mathbf{R}) - \hat{a}_{k\lambda}^\dagger\exp(-i\mathbf{k}.\mathbf{R})\right\}|i\rangle\langle j|.\tag{4.9.9}$$

The complete Hamiltonian of eqn (4.8.31) for the atom–radiation system is now in second-quantized form, with its component parts given by eqns (4.9.5), (4.4.22) and (4.9.9). All the operator properties of the Hamiltonian are contained in the creation, destruction and transition operators, and the remaining factors are ordinary numbers, variables or vector quantities.

Many calculations are concerned with radiative processes that involve only two atomic states, $|1\rangle$ and $|2\rangle$ say, with energies $\hbar\omega_1$ and $\hbar\omega_2$. We take state $|1\rangle$ to be the ground state and state $|2\rangle$ to be an excited state, and we define the transition frequency as

$$\omega_0 = \omega_2 - \omega_1. \tag{4.9.10}$$

It is convenient to introduce a shorthand notation for the relevant transition operators of the effectively two-level atom, and we define

$$\hat{\pi}^\dagger = |2\rangle\langle 1| \quad \text{and} \quad \hat{\pi} = |1\rangle\langle 2|. \tag{4.9.11}$$

Thus $\hat{\pi}^\dagger$ shifts the atom from its ground state to its excited state and $\hat{\pi}$ produces the reverse transition, with

$$\hat{\pi}^\dagger|1\rangle = |2\rangle \quad \text{and} \quad \hat{\pi}|2\rangle = |1\rangle \tag{4.9.12}$$

and

$$\hat{\pi}^\dagger|2\rangle = 0 \quad \text{and} \quad \hat{\pi}|1\rangle = 0. \tag{4.9.13}$$

Other properties of the transition operators are easily obtained from their definitions. Thus

$$\hat{\pi}^\dagger\hat{\pi} = |2\rangle\langle 1|1\rangle\langle 2| = |2\rangle\langle 2| \tag{4.9.14}$$

and similarly

$$\hat{\pi}\hat{\pi}^\dagger = |1\rangle\langle 1|. \tag{4.9.15}$$

These operators represent the probabilities of finding the atom in its excited state $|2\rangle$ and its ground state $|1\rangle$ respectively. The closure theorem of eqn (4.9.2) for a two-level atom gives

$$|1\rangle\langle 1| + |2\rangle\langle 2| = \hat{\pi}^\dagger\hat{\pi} + \pi\hat{\pi}^\dagger = 1, \tag{4.9.16}$$

equivalent to the statement that the atom must be in one or the other of its two states. The squares of the transition operators vanish, as

$$\hat{\pi}^\dagger\hat{\pi}^\dagger = |2\rangle\langle 1|2\rangle\langle 1| = 0, \tag{4.9.17}$$

and similarly

$$\hat{\pi}\hat{\pi} = 0. \tag{4.9.18}$$

If the zero of energy is taken at the level of the ground state $|1\rangle$, the atomic Hamitonian of eqn (4.9.5) reduces to

$$\hat{\mathcal{H}}_A = \hbar\omega_0|2\rangle\langle 2| = \hbar\omega_0\hat{\pi}^\dagger\hat{\pi}. \tag{4.9.19}$$

The electric-dipole Hamiltonian can be similarly re-expressed. The wavefunctions of states $|1\rangle$ and $|2\rangle$ are again assumed to be real, so that $\mathbf{D}_{21}$ and $\mathbf{D}_{12}$ are equal real vectors, and the second-quantized form of $\hat{\mathbf{D}}$ from eqn (4.9.7) reduces to

$$\hat{\mathbf{D}} = \mathbf{D}_{12}\left(\hat{\pi}^\dagger + \hat{\pi}\right). \tag{4.9.20}$$

The electric-dipole Hamiltonian from eqn (4.9.9) thus becomes

$$\hat{\mathcal{H}}_{ED} = i\sum_{\mathbf{k}}\sum_{\lambda} \hbar g_{\mathbf{k}\lambda}\left\{\hat{a}_{\mathbf{k}\lambda}\exp(i\mathbf{k}.\mathbf{R}) - \hat{a}_{\mathbf{k}\lambda}^\dagger\exp(-i\mathbf{k}.\mathbf{R})\right\}\left(\hat{\pi}^\dagger + \hat{\pi}\right), \tag{4.9.21}$$

where

$$g_{\mathbf{k}\lambda} = e(\omega_k/2\varepsilon_0\hbar V)^{1/2}\mathbf{e}_{\mathbf{k}\lambda}.\mathbf{D}_{12} \tag{4.9.22}$$

is a real quantity with the dimensions of frequency.

It is useful to employ pictorial representations of the various kinds of inter-action process included in $\hat{\mathcal{H}}_{ED}$. Four distinct terms occur when the brackets in eqn (4.9.21) are multiplied out, and each of them produces a definite final state of the atom–radiation system when applied to an appropriate initial state. These four types of term are represented by the diagrams of Fig. 4.2. The initial state is represented on the right of each diagram, and the final state into which $\hat{\mathcal{H}}_{ED}$ converts it is represented on the left. Thus diagram (b) represents the absorption of a photon $\mathbf{k}\lambda$ accompanied by excitation of the atom from state $|1\rangle$ to state $|2\rangle$, while the inverse process of photon emission is represented in diagram (c). The events represented by diagrams (a) and (d) do not correspond to allowed absorption and emission processes as it is not possible for the final states to have the same energies as the initial states. Neglect of these contributions produces the same rotating-wave approximation as made in the treatment of absorption and emission processes in §2.3, when eqn (4.9.21) reduces to

$$\hat{\mathcal{H}}_{ED} = i\sum_{\mathbf{k}}\sum_{\lambda} \hbar g_{\mathbf{k}\lambda}\left\{\hat{\pi}^\dagger\hat{a}_{\mathbf{k}\lambda}\exp(i\mathbf{k}.\mathbf{R}) - \hat{a}_{\mathbf{k}\lambda}^\dagger\hat{\pi}\exp(-i\mathbf{k}.\mathbf{R})\right\}. \tag{4.9.23}$$

It is shown in Chapters 8 and 9 that the non-energy-conserving terms in $\hat{\mathcal{H}}_{ED}$ make important contributions to higher-order radiative processes, where energy is conserved in the final state but not in some of the intermediate states. The full form of interaction Hamiltonian from eqn (4.9.21) must be used for such processes.

The electric-dipole interaction Hamiltonians given in eqns (4.9.9), (4.9.21)

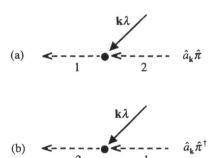

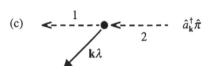

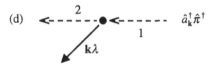

Fig. 4.2. Diagrammatic representations of the four kinds of term in the electric-dipole interaction. The continuous lines represent photons and the broken lines atomic states. A line whose arrow is directed towards (away from) the interaction point corresponds to a photon or atomic state that is destroyed (created) in the interaction.

and (4.9.23) are all in the Schrödinger picture. The corresponding Hamiltonians in the Heisenberg picture have the same general forms but the various creation and destruction operators now depend on the time. The Heisenberg form of the complete Hamiltonian in the rotating-wave approximation is

$$\hat{\mathcal{H}} = \hbar\omega_0\hat{\pi}^\dagger(t)\hat{\pi}(t) + \sum_k\sum_\lambda \hbar\omega_k\left(\hat{a}_{k\lambda}^\dagger(t)\hat{a}_{k\lambda}(t) + \tfrac{1}{2}\right)$$

$$+ i\sum_k\sum_\lambda \hbar g_{k\lambda}\left\{\hat{\pi}^\dagger(t)\hat{a}_{k\lambda}(t)\exp(ik.R) - \hat{a}_{k\lambda}^\dagger(t)\hat{\pi}(t)\exp(-ik.R)\right\}, \tag{4.9.24}$$

where the time dependences of the operators are determined by solution of the appropriate Heisenberg equation of motion from eqn (4.7.12) with substitution of this Hamiltonian. The complete Hamiltonian is independent of the time, as in eqn (4.7.13), but its individual terms *do* depend on the time.

In addition to the Schrödinger and Heisenberg pictures, it is advantageous in some calculations to the use the interaction picture, whose main features are given below in a format similar to that of §4.7. This third and final picture has wave-functions and operators that all vary with the time. Let the total Schrödinger Hamiltonian be written as

$$\hat{\mathcal{H}} = \hat{\mathcal{H}}_0 + \hat{\mathcal{H}}_{ED}, \tag{4.9.25}$$

where

$$\hat{\mathcal{H}}_0 = \hat{\mathcal{H}}_A + \hat{\mathcal{H}}_R = \hbar\omega_0 \hat{\pi}^\dagger \hat{\pi} + \sum_k \sum_\lambda \hbar\omega_k \left(\hat{a}_{k\lambda}^\dagger \hat{a}_{k\lambda} + \tfrac{1}{2} \right) \tag{4.9.26}$$

is the uncoupled Hamiltonian of the atom and field components. The interaction-picture wavefunction is related to the Schrödinger wavefunction by

$$\Psi_I(t) = \exp\left(i\hat{\mathcal{H}}_0 t/\hbar\right)\Psi_S(t) = \exp\left(i\hat{\mathcal{H}}_0 t/\hbar\right)\exp\left(-i\hat{\mathcal{H}} t/\hbar\right)\Psi_S(0), \tag{4.9.27}$$

where the exponentials on the right cannot be combined as different contributions to the Hamiltonian do not commute. The equation of motion obtained by differentiation of this relation is

$$i\hbar\, d\Psi_I(t)/dt = \hat{\mathcal{H}}_{ED}(t)\Psi_I(t), \tag{4.9.28}$$

where interaction-picture time-dependent operators, like $\hat{\mathcal{H}}_{ED}(t)$, are defined by

$$\hat{O}_I(t) = \exp\left(i\hat{\mathcal{H}}_0 t/\hbar\right)\hat{O}_S \exp\left(-i\hat{\mathcal{H}}_0 t/\hbar\right). \tag{4.9.29}$$

The operator equation of motion, obtained by differentiation of this definition, is

$$i\hbar\, d\hat{O}_I(t)/dt = \left[\hat{O}_I(t), \hat{\mathcal{H}}_0\right]. \tag{4.9.30}$$

According to eqn (4.9.29), the uncoupled part of the Hamiltonian is time independent,

$$\hat{\mathcal{H}}_0(t) = \hat{\mathcal{H}}_0, \tag{4.9.31}$$

but the interaction part varies with the time. The time dependences of the atomic and field operators are readily obtained from eqns (4.9.26) and (4.9.30) as

$$\hat{\pi}(t) = \hat{\pi}\exp\left(-i\omega_0 t\right), \quad \hat{\pi}^\dagger(t) = \hat{\pi}^\dagger \exp\left(i\omega_0 t\right) \tag{4.9.32}$$

and

$$\hat{a}_k(t) = \hat{a}_k \exp\left(-i\omega_k t\right), \quad \hat{a}_k^\dagger(t) = \hat{a}_k^\dagger \exp\left(i\omega_k t\right). \tag{4.9.33}$$

Substitution of these time-dependent operators into eqn (4.9.26) verifies the property in eqn (4.9.31), while substitution into the electric-dipole Hamiltonian obtained from eqn (4.9.23) gives

$$\hat{\mathcal{H}}_{ED}(t) = i\sum_{k}\sum_{\lambda}\hbar g_{k\lambda}\left\{\hat{\pi}^{\dagger}\hat{a}_{k\lambda}\exp\left[i(\omega_0 - \omega_k)t + i\mathbf{k}.\mathbf{R}\right]\right.$$
$$\left. -\hat{a}_{k\lambda}^{\dagger}\hat{\pi}\exp\left[-i(\omega_0 - \omega_k)t - i\mathbf{k}.\mathbf{R}\right]\right\}. \tag{4.9.34}$$

The electromagnetic-field operators of §4.4, in the absence of atom–radiation interaction, already have the correct time dependences for the interaction picture.

Calculated results for measurable quantities are again the same in the interaction picture as in the other pictures. Thus the interaction expectation value

$$\left\langle\hat{O}(t)\right\rangle_{I} = \left\langle\Psi_{I}(t)\middle|\hat{O}_{I}(t)\middle|\Psi_{I}(t)\right\rangle \tag{4.9.35}$$

gives identical results to eqns (4.7.5) and (4.7.14). The interaction density operator is defined in the manner of eqn (4.9.29) as

$$\hat{\rho}_{I}(t) = \exp\left(i\hat{\mathcal{H}}_{0}t/\hbar\right)\hat{\rho}_{S}(t)\exp\left(-i\hat{\mathcal{H}}_{0}t/\hbar\right), \tag{4.9.36}$$

and its equation of motion is obtained with the use of eqn (4.7.7) as

$$i\hbar\, d\hat{\rho}_{I}(t)/dt = \left[\hat{\mathcal{H}}_{ED}(t), \hat{\rho}_{I}(t)\right]. \tag{4.9.37}$$

The equation of motion for the density operator once again differs from that of other operators, given by eqn (4.9.30) in the interaction picture. The interaction expectation value for a statistical mixture, given by

$$\left\langle\hat{O}(t)\right\rangle_{I} = \mathrm{Tr}\left\{\hat{\rho}_{I}(t)\hat{O}_{I}(t)\right\}, \tag{4.9.38}$$

is identical to the Schrödinger value in eqn (4.7.8) and the Heisenberg value in eqn (4.7.16).

4.10 Photon absorption and emission rates

As a first application of the second-quantized interaction Hamiltonian, we use it to calculate the rates of emission and absorption of photons by an atom that can make transitions between states $|1\rangle$ and $|2\rangle$. The problem is the same as that treated by semiclassical radiation theory in §2.3. However, the latter theory is incapable of treating spontaneous emission in a rigorous manner and the radiative lifetime of an excited atom is derived in eqn (2.3.21) via the relations between the Einstein A and B coefficients. The direct calculation carried out below includes spontaneous emission automatically and it provides a more rigorous justification of the phenomenological Einstein theory.

We denote the state of the combined atom–radiation system, with $n_{\mathbf{k}\lambda}$ photons present in mode $\mathbf{k}\lambda$ and the atom in state $|1\rangle$, by the ket $|n_{\mathbf{k}\lambda},1\rangle$. A similar notation is used for the other combination states. It is a simple matter to write down the matrix elements of $\hat{\mathcal{H}}_{\text{ED}}$, given by eqn (4.9.34) in the interaction picture, for the photon absorption and emission processes of diagrams (b) and (c) in Fig. 4.2, respectively,

$$\langle n_{\mathbf{k}\lambda} - 1,2|\hat{\mathcal{H}}_{\text{ED}}|n_{\mathbf{k}\lambda},1\rangle = i\hbar g_{\mathbf{k}\lambda} \exp\left[i(\omega_0 - \omega_k)t + i\mathbf{k}.\mathbf{R}\right]n_{\mathbf{k}\lambda}^{1/2} \qquad (4.10.1)$$

and

$$\langle n_{\mathbf{k}\lambda} + 1,1|\hat{\mathcal{H}}_{\text{ED}}|n_{\mathbf{k}\lambda},2\rangle = -i\hbar g_{\mathbf{k}\lambda} \exp\left[-i(\omega_0 - \omega_k)t - i\mathbf{k}.\mathbf{R}\right]\left(n_{\mathbf{k}\lambda} + 1\right)^{1/2}, \quad (4.10.2)$$

where eqns (4.4.1), (4.4.2), (4.9.12) and (4.9.13) are used. Note that only the destruction operator, or positive frequency part of the electric field operator, contributes to the absorption matrix element of eqn (4.10.1) and only the creation operator, or negative frequency part, contributes to the emission matrix element of eqn (4.10.2). In working with the second-quantized interaction Hamiltonian it is essential for the initial state of the system to be on the right of all matrix elements and the final state on the left.

The atomic contribution to the above matrix elements is the quantity $\mathbf{D}_{12}$ included in $g_{\mathbf{k}\lambda}$ defined by eqn (4.9.22). It is emphasized that the present theory is valid only when $\mathbf{D}_{12}$ is nonvanishing, that is when the transitions between states $|1\rangle$ and $|2\rangle$ are *electric-dipole allowed*. In cases where $\mathbf{D}_{12}$ vanishes, the electric-dipole approximation cannot be made, and it is necessary to retain higher-order terms in the interaction Hamiltonian considered in §4.8. It may be that the matrix elements of the electric-quadrupole or magnetic-dipole moments defined in eqns (4.8.12) or (4.8.15) are nonzero for the atomic states concerned. Thus, while electric-dipole transitions require the wavefunctions of the initial and final atomic states to have opposite parity, wavefunctions of the same parity are required for both electric-quadrupole and magnetic-dipole transitions. In these transitions, the appropriate forms of the atom–radiation interaction are determined by extensions of the theory of §4.8 and the matrix elements in eqns (4.10.1) and (4.10.2) for absorption and emission are modified. The modifications are straightforward but we do not discuss them here, and the theory that follows assumes the applicability of the electric-dipole approximation.

The matrix elements in eqns (4.10.1) and (4.10.2) are used to calculate the rates of absorption and emission of photons with the use of the transition-rate theory from §2.4. The theory of §§2.1 to 2.3 is presented in a kind of semi-classical interaction picture, where the quantum atomic states have a Schrödinger-like time dependence shown in eqn (2.1.5) but the classical radiation fields have a Heisenberg-like time dependence shown in eqn (2.2.1). However, as we now show, the absorption and emission rates obtained in this hybrid theory are the same as those found by use of the standard interaction picure of §4.9, where the time dependences of the atomic and radiative parts of the system are treated on the same footing.

Consider the absorption process first, where the transition rate is obtained by substitution of the matrix element from eqn (4.10.1) into Fermi's golden rule from eqn (2.4.15). However, the transition-rate theory is applicable only for broadband illumination of the atom where the energy density $\langle W(\omega_k) \rangle$ of the radiation varies slowly for frequencies ω_k in the vicinity of ω_0. Thus, analogous to eqn (2.3.6), the field excitation energy in the single mode $k\lambda$ must be replaced by an integration over frequency according to

$$n_{k\lambda} \hbar \omega_k \to V \int d\omega_k \langle W(\omega_k) \rangle. \tag{4.10.3}$$

Inclusion of this integation in the Fermi golden rule leads to a transition rate

$$\frac{1}{\tau} = 2\pi V \int d\omega_k \frac{\overline{g_{k\lambda}^2}}{\hbar \omega_k} \langle W(\omega_k) \rangle \delta(\omega_k - \omega_0), \tag{4.10.4}$$

where the overbar denotes the usual average over the random orientations of $\mathbf{D}_{12}$, included in the coefficient $g_{k\lambda}$, as derived in eqn (2.3.19). The result for the Einstein B coefficient, obtained by procedures similar to those of §2.3, is

$$B_{12} = \pi e^2 D_{12}^2 / 3\varepsilon_0 \hbar^2, \tag{4.10.5}$$

in agreement with the expression (2.3.20) derived by semiclassical radiation theory.

The photon emission process is treated in a similar way, with the matrix element now given by eqn (4.10.2). The square modulus of the emission matrix element that occurs in Fermi's golden rule is the same as that for absorption, except that the factor $n_{k\lambda}$ is replaced by $n_{k\lambda} + 1$, and there are thus two distinct contributions to the emission rate. The part linear in $n_{k\lambda}$ gives a rate proportional to the number of photons $k\lambda$ present initially and it leads to a rate of emission equal to the rate of photon absorption. The equality of these two rates, proved here quantum-mechanically, is equivalent to the equality of the Einstein coefficients B_{12} and B_{21} for a pair of nondegenerate states, derived indirectly in eqn (1.5.12). Note how the stimulated emission process arises naturally with use of the second-quantized interaction Hamiltonian. Futhermore, it is evident from the matrix element of eqn (4.9.2) that the photon emitted in the stimulated process has the same wavevector $\mathbf{k}$ and polarization label λ as those of the $n_{k\lambda}$ photons present initially. These properties of the stimulated emission are stated without proof in the discussion of Fig. 1.9.

The contribution to the emission rate that is independent of $n_{k\lambda}$ corresponds to spontaneous emission, as the transition to the lower level occurs even in the complete absence of any radiation, when the radiation field is initially in its vacuum state $|\{0\}\rangle$ defined in eqns (4.4.24) and (4.4.25). Note that the factor $n_{k\lambda} + 1$, which governs the ratio of the stimulated and spontaneous emission rates for a particular photon $k\lambda$, is similar to the factor $\langle n \rangle + 1$, which controls the ratio of the two rates in eqn (1.6.2) in the presence of a mean number $\langle n \rangle$ of thermally excited photons per cavity mode. In the quantum theory of light, the dependences of the absorption and emission rates on the number of photons arise solely from the basic properties in eqns (4.3.32) and (4.3.33) for the subtraction and addition of a quantum of energy to the eigenstates of the simple harmonic oscillator. Spontaneous emission can in principle occur into all modes of the cavity, unlike the stimulated emission which must always match the mode of the stimulating photons.

The total spontaneous emission rate is obtained by summation of the spontaneous contribution over all modes. The radiative lifetime is thus given by

$$1/\tau_R = 2\pi \sum_{\mathbf{k}} \sum_{\lambda} g_{\mathbf{k}\lambda}^2 \delta(\omega_k - \omega_0). \tag{4.10.6}$$

This formula is a special case of Fermi's golden rule from eqn (2.4.15), where the final states correspond to the various photons $\mathbf{k}\lambda$ that can be emitted as the atom decays to state $|1\rangle$. The calculation is completed by conversion of the sum over $\mathbf{k}$ in eqn (4.10.6) to an integral, as in eqn (4.5.6). We take polar coordinates with the wavevector and polarizations given by eqns (4.5.2) to (4.5.4). The coordinates are chosen so that $\mathbf{D}_{12}$ lies in the zx plane with orientation Θ to the z axis,

$$\mathbf{D}_{12} = D_{12}(\sin\Theta, 0, \cos\Theta). \tag{4.10.7}$$

Problem 4.13 Prove that

$$\sum_{\mathbf{k}} \sum_{\lambda} (\mathbf{e}_{\mathbf{k}\lambda} \cdot \mathbf{D}_{12})^2 = \frac{V D_{12}^2}{3\pi^2} \int_0^\infty dk\, k^2, \tag{4.10.8}$$

independent of the dipole orientation Θ. Hence show that

$$\sum_{\mathbf{k}} \sum_{\lambda} g_{\mathbf{k}\lambda}^2 = \frac{e^2 D_{12}^2}{6\pi^2 \varepsilon_0 \hbar c^3} \int d\omega_k\, \omega_k^3. \tag{4.10.9}$$

Insertion of this result into eqn (4.10.6) produces

$$1/\tau_R \equiv 2\gamma_{sp} \equiv A_{21} = e^2 \omega_0^3 D_{12}^2 / 3\pi\varepsilon_0 \hbar c^3. \tag{4.10.10}$$

This agrees with the result for the Einstein A coefficient obtained in eqn (2.3.21) when the two levels are nondegenerate and it replicates the relation (2.5.11) between the different ways of expressing the radiative decay rate.

The rates of absorption and stimulated emission can be expressed in terms of the Rabi frequency $\mathcal{V}$, defined by eqn (2.7.22). The squared field E_0^2 in mode $\mathbf{k}\lambda$ is converted to the photon number in the mode with the use of eqns (2.3.6) and (4.10.3). The expression for $\mathcal{V}$ then gives

$$\mathcal{V}^2 = \frac{2e^2 \omega_k n_{\mathbf{k}\lambda}}{\varepsilon_0 \hbar V} (\mathbf{e}_{\mathbf{k}\lambda} \cdot \mathbf{D}_{12})^2 = 4g_{\mathbf{k}\lambda}^2 n_{\mathbf{k}\lambda}, \tag{4.10.11}$$

where the form on the right is derived with the use of eqn (4.9.22). The Rabi frequency thus determines the rates of absorption and stimulated emission produced by the $n_{k\lambda}$ photons in mode $k\lambda$. An average over the random atomic orientations gives

$$\overline{\mathcal{V}^2} = \frac{2e^2\omega_k n_{k\lambda}}{3\varepsilon_0\hbar V}\overline{D_{12}^2} = 4\overline{g_{k\lambda}^2}n_{k\lambda}. \tag{4.10.12}$$

The Rabi frequency $\mathcal{V}$ expressed by eqn (4.10.11) vanishes in the absence of an initial photon excitation and, as discussed in §§2.7 and 2.8, it is the interaction with an incident light beam that generates the Rabi oscillations in the atomic populations. However, oscillations in the populations still occur in the absence of any incident beam and it is useful to define a *vacuum Rabi frequency* $\mathcal{V}_{\text{vac}}$ by

$$\mathcal{V}_{\text{vac}}^2 = \frac{2e^2\omega_k}{\varepsilon_0\hbar V}\left(\mathbf{e}_{k\lambda}.\mathbf{D}_{12}\right)^2 = 4g_{k\lambda}^2. \tag{4.10.13}$$

This frequency governs the oscillations in population of an initially excited atom coupled to an initially-unexcited, or vacuum, field. Unlike the multimode spontaneous emission process, the vacuum Rabi oscillations occur in situations where the atom interacts with a single-mode field confined to a well-defined volume V by a suitable optical cavity. The system of a two-level atom coupled to a single cavity mode, whether excited by incident light or not, is known as the *Jaynes–Cummings model*; it is discussed no further here (see [14] for a review).

It is important to distinguish between the different formalisms that can be used to calculate the emission of light by an excited atom. In the *fully-quantized* theory used to derive eqn (4.10.10), where both the electromagnetic field and the atom are treated quantum-mechanically, spontaneous emission is a consequence of the coupling of the atomic transition to the vacuum electromagnetic field. The resulting decay rate, or Einstein A coefficient, depends on both the atomic transition frequency ω_0 and the dipole matrix element D_{12}. The expression (4.10.10) agrees very well with measured decay rates and it appears to be universally valid.

In the *semiclassical* theory, where the atom is treated quantum-mechanically but the electromagnetic field is treated classically, spontaneous emission does not occur in the absence of any incident radiation, and the atom remains in its excited state. The absence of a classical electromagnetic field nullifies the strength of the coupling between the atomic transition and the field, so that the emission rate vanishes. The Einstein B coefficients for the rates of absorption and stimulated emission can be calculated by semiclassical theory, as in §2.3, and the Einstein A coefficient is then deduced from the relation between the two obtained by comparison of the radiation equilibrium state with that given by Planck's law. However, this procedure essentially introduces the full quantum theory by a back door, and the result cannot strictly be regarded as a derivation from the semiclassical theory.

In the *classical* theory, where both the atom and the electromagnetic field are treated classically, spontaneous emission is equivalent to the radiation by an oscillating dipole moment. The fields of the emitted radiation, proportional to the acceleration of the dipole charges, are routinely calculated in textbooks on electromagnetic theory [1]. The rate of decay of the dipole energy caused by the radiation is

$$(1/\tau_R)_{classical} = e^2\omega_0^2/6\pi\varepsilon_0 mc^3,$$ (4.10.14)

where m is the dipole mass and the frequency ω_0 is identified with that of the dipole oscillation. The phenomenon of spontaneous emission thus occurs in both the fully classical and fully quantum theories, but the atomic decay rates are different. In particular, the classical rate does not depend on the dipole moment and the theory provides no valid means of calculating the rates of specific atomic transitions.

4.11 The photon intensity operator

The semiclassical theory of optical detection, presented in §3.9, assumes a photoionization rate proportional to the intensity of the light beam. The phototube thus measures the rate of atomic ionizations, which is in turn proportional to the rate of photon arrivals at the tube and hence to the beam intensity. In the present section we consider the ionization process in a little more detail in order to understand the nature of intensity measurement. We determine the form of the operator that represents the observable intensity of a light beam.

Let us consider the rate at which a particular atom in a phototube is ionized by a beam of photons. The theory of the photoelectric effect is similar to the calculation of the photon absorption rate given in the previous section, except that the final atomic state is one in which the electron is detached from the atom. Suppose that the atom is initially in state $|1\rangle$ and that the ionization occurs by absorption of a photon $k\lambda$, of which there are initially $n_{k\lambda}$. It is convenient to use the Heisenberg picture, with time-dependent field operators analogous to the field variables of the classical theory. The relevant matrix element of the electric-dipole interaction Hamiltonian from eqn (4.9.9) is

$$\langle f|\hat{\mathcal{H}}_{ED}|i\rangle = e\langle \mathbf{q}|\hat{\mathbf{D}}|1\rangle\cdot\langle n_{k\lambda}-1|\hat{E}_T^+(\mathbf{R},t)|n_{k\lambda}\rangle,$$ (4.11.1)

where $\mathbf{q}$ denotes the wavevector of the electron freed from the ionized atom and the field operators are given by eqns (4.4.13) to (4.4.15). The matrix element is factorized into its atomic and radiative parts and only the contributing positive frequency, or destruction operator, part is retained in the latter. The time dependence of the atomic dipole moment is not shown explicitly.

The photoelectric transition rate is now calculated by insertion of the matrix element into the Fermi golden rule of eqn (2.4.15). The delta function in this expression imposes the energy conservation condition

$$\hbar q^2/2m = \omega_k + \omega_1,$$ (4.11.2)

where it should be borne in mind that $\hbar\omega_1$ is negative for an atomic bound state. The sum over final states in eqn (2.4.15) is represented by an integration over the wavevectors $\mathbf{q}$ of the freed electron allowed by energy conservation. The integration depends upon the wavefunctions for the bound and free states of the atom and its result is governed only by the atomic factor in the matrix element of eqn (4.11.1). This atomic part of the expression for the transition rate is not needed for the present discussion.

The radiative part of the matrix element contributes independently of the atomic part and it enters the transition rate via the factor

$$\left|\left\langle n_{\mathbf{k}\lambda} - 1\left|\hat{E}_T^+(\mathbf{R},t)\right|n_{\mathbf{k}\lambda}\right\rangle\right|^2 = \left\langle n_{\mathbf{k}\lambda}\left|\hat{E}_T^-(\mathbf{R},t)\right|n_{\mathbf{k}\lambda} - 1\right\rangle\left\langle n_{\mathbf{k}\lambda} - 1\left|\hat{E}_T^+(\mathbf{R},t)\right|n_{\mathbf{k}\lambda}\right\rangle$$

$$= \left\langle n_{\mathbf{k}\lambda}\left|\hat{E}_T^-(\mathbf{R},t)\hat{E}_T^+(\mathbf{R},t)\right|n_{\mathbf{k}\lambda}\right\rangle,$$

(4.11.3)

where the final step is readily verifed by evaluation of the right-hand sides with and without the operator $\left|n_{\mathbf{k}\lambda} - 1\right\rangle\left\langle n_{\mathbf{k}\lambda} - 1\right|$ in place. The matrix element in eqn (4.11.1) assumes an initial number state for the photons in mode $\mathbf{k}\lambda$ but the results are easily generalized to obtain the photoelectric transition rate for an incident beam that corresponds to a superposition or a statistical mixture of number states. Let the beam in its initial state be specified by a probability $P(n_{\mathbf{k}\lambda})$ associated with the state $\left|n_{\mathbf{k}\lambda}\right\rangle$. The expression for the transition rate remains valid except that the electromagnetic field factor in eqn (4.11.3) is generalized to

$$\sum_{n_{\mathbf{k}\lambda}} P(n_{\mathbf{k}\lambda})\left\langle n_{\mathbf{k}\lambda}\left|\hat{E}_T^-(\mathbf{R},t)\hat{E}_T^+(\mathbf{R},t)\right|n_{\mathbf{k}\lambda}\right\rangle = \text{Tr}\left\{\hat{\rho}\hat{E}_T^-(\mathbf{R},t)\hat{E}_T^+(\mathbf{R},t)\right\}, \qquad (4.11.4)$$

where the electromagnetic density operator $\hat{\rho}$ is used to evaluate the ensemble averages in accordance with the procedure of eqns (4.6.3) to (4.6.8). For a multi-mode incident beam, the trace in eqn (4.11.4) is evaluated with a multimode density operator of form similar to that in eqn (4.6.19).

The operator $\hat{E}_T^-(\mathbf{R},t)\hat{E}_T^+(\mathbf{R},t)$ in eqn (4.11.4) represents the observable intensity of the electromagnetic field as measured by the current that the beam of radiation induces in a phototube. The derivation of the radiative factor in the transition rate is free of any serious assumptions about the nature of the light beam or its coupling to the atoms, except that the validity of the electric-dipole approximation is assumed. It also relies on the applicability of standard transition-rate theory, together with use of the rotating-wave approximation in restricting the interaction processes to the kinds shown in parts (b) and (c) of Fig. 5.2. The radiative factor is not changed by a more detailed modelling of the construction of the phototube or of the dynamics of the excitation of the photocurrent. Analysis of other types of intensity-measuring device leads to the same conclusion that their response is proportional to the expectation value in eqn (4.11.4).

The intensity of a light beam in classical electromagnetic theory is determined by the Poynting vector [1], whose form is given in eqn (1.8.1). The corresponding quantum-optical intensity operator is defined as

$$\hat{\mathbf{I}}(\mathbf{R},t) = \varepsilon_0 c^2 \left\{ \hat{\mathbf{E}}_T^-(\mathbf{R},t) \times \hat{\mathbf{B}}^+(\mathbf{R},t) - \hat{\mathbf{B}}^-(\mathbf{R},t) \times \hat{\mathbf{E}}_T^+(\mathbf{R},t) \right\}, \tag{4.11.5}$$

where the second term on the right, including the minus sign, is the Hermitian conjugate of the first, and the electric and magnetic field operators are defined by eqns (4.4.13) to (4.4.18). Photoelectric measurements of the intensity are usually made on light beams that are at least approximately parallel. We accordingly assume that, for light beams described by multimode excitations of the electromagnetic field, all of the wavevectors $\mathbf{k}$ involved point in a common direction.

Problem 4.14   Prove that, for a polarized parallel light beam, the intensity operator defined by eqn (4.11.5) is equivalent to an operator of magnitude

$$\hat{I}(\mathbf{R},t) = 2\varepsilon_0 c \hat{E}_T^-(\mathbf{R},t) \hat{E}_T^+(\mathbf{R},t) \tag{4.11.6}$$

oriented parallel to the beam. The electric-field operators are given by eqns (4.4.14) and (4.4.15) with a single polarization and with the $\mathbf{k}$ summations restricted to parallel wavevectors.

The definition of the intensity operator adopted in eqn (4.11.5) thus agrees with the form of observed intensity derived in eqn (4.11.4) by consideration of the photoelectric effect.

The expectation value of the intensity operator simplifies for incident light whose density operator has no off-diagonal number-state matrix elements, for example chaotic light described by eqns (4.6.19) and (4.6.20). The intensity for a given polarization λ obtained with the use of eqn (4.11.6) is independent of position and time in this case, with the value

$$\langle \hat{I} \rangle = (c/V) \sum_k \hbar \omega_k \langle n_{\mathbf{k}\lambda} \rangle = (c/V) \langle \mathcal{E}_R \rangle, \tag{4.11.7}$$

where $\mathcal{E}_R$ is the excitation energy of the electromagnetic field above its zero-point value, as defined in eqn (4.4.29), but here restricted to a single polarization and to wavevectors parallel to the beam direction.

The nature of the intensity-measuring process, which relies upon the absorption of light, or the destruction of photons, leads to the definition (4.11.5) or (4.11.6) of the intensity operator. The destruction operators lie to the right of the creation operators in these expressions, which are said to be *normally-ordered*. The expectation value of the operator product $\hat{E}_T^- \hat{E}_T^+$ is proportional, as in eqn (4.11.7), to the mean numbers of photons in the modes or to the mode excitation energies. The detector response is *not* proportional to the expectation value of $\hat{E}_T \hat{E}_T$, which is evaluated in the derivation of eqn (4.4.22) from eqn (4.4.21) and which includes a zero-point contribution from every mode $\mathbf{k}\lambda$ of the radiation field, whether or not the beam contains any such photons. It is therefore asserted that the zero-point energy $\mathcal{E}_0$ of eqn (4.4.27) makes no contribution to the observable radiative intensity, which is determined only by the excitation energy $\mathcal{E}_R$ of eqn (4.4.29).

It is shown in the previous section that the semiclassical and fully-quantized theories give the same predictions for the rate of absorption by atoms. Thus the semiclassical theory of absorption, which remains valid for the excitation of an atom to an ionized state, also gives a transition rate proportional to the energy of the incident light. Indeed, semiclassical theory accounts for all the main features of the photoelectric effect, including the Einstein relation, which is essentially eqn (4.11.2). The form of this relation does not 'prove' the need for quantization of the radiation field [15], contrary to statements sometimes made.

4.12 Quantum degrees of first and second-order coherence

The Mach–Zehnder and Brown–Twiss interferometers are treated more generally by quantum versions of the classical theories of interference given in Chapter 3. Such quantum theories are covered in more detail in Chapters 5 and 6. However, the measurements in both kinds of interferometer involve optical intensities, and the main changes introduced by the quantum theory result from the form of the quantum-mechanical intensity operator given by eqn (4.11.5), or more simply by eqn (4.11.6). In the present section we outline the ways in which the quantum theories of interference lead to quantum versions of the classical degrees of coherence [16]. It is assumed throughout that the measurements detect a single linear polarization so that the fields can be written as scalar operators.

The quantum theory of the Mach–Zehnder interferometer closely parallels the classical theory of §3.3. The main change is the replacement of classical electric-field variables by quantized Heisenberg-picture electric-field operators, which need to be put in normal $\hat{E}_T^- \hat{E}_T^+$ order in expressions for measured intensities. The quantum version of eqn (3.3.4) is accordingly

$$\langle \hat{I}_4(t) \rangle = \tfrac{1}{2} \varepsilon_0 c |\mathcal{R}|^2 |\mathcal{T}|^2 \left\{ \langle \hat{E}_T^-(t_1) \hat{E}_T^+(t_1) \rangle + \langle \hat{E}_T^-(t_2) \hat{E}_T^+(t_2) \rangle \right.$$
$$\left. + 2 \operatorname{Re} \langle \hat{E}_T^-(t_1) \hat{E}_T^+(t_2) \rangle \right\}. \tag{4.12.1}$$

The angle brackets now represent quantum-mechanical expectation values calculated in accordance with eqn (4.6.8), for example

$$\langle \hat{E}_T^-(t_1) \hat{E}_T^+(t_2) \rangle = \operatorname{Tr} \left\{ \hat{\rho} \hat{E}_T^-(t_1) \hat{E}_T^+(t_2) \right\}, \tag{4.12.2}$$

where $\hat{\rho}$ is the density operator of the light beam.

Considerations similar to those given in §§3.4 and 3.5 now lead to a definition of the quantum degree of first-order coherence, analogous to the classical definition in eqn (3.5.4), as

$$g^{(1)}(\mathbf{r}_1, t_1; \mathbf{r}_2, t_2) = \frac{\langle \hat{E}_T^-(\mathbf{r}_1, t_1) \hat{E}_T^+(\mathbf{r}_2, t_2) \rangle}{\left[\langle \hat{E}_T^-(\mathbf{r}_1, t_1) \hat{E}_T^+(\mathbf{r}_1, t_1) \rangle \langle \hat{E}_T^-(\mathbf{r}_2, t_2) \hat{E}_T^+(\mathbf{r}_2, t_2) \rangle \right]^{1/2}}. \tag{4.12.3}$$

This quantity retains its physical significance as a measure of the ability of light at the space–time points $(\mathbf{r}_1,t_1)$ and $(\mathbf{r}_2,t_2)$ to form interference fringes when superposed. For the plane parallel light beams assumed in the previous two sections, this general three-dimensional definition reduces to a quantum analogue of eqn (3.4.14) as

$$g^{(1)}(z_1,t_1;z_2,t_2) = \frac{\left\langle \hat{E}_T^-(z_1,t_1)\hat{E}_T^+(z_2,t_2)\right\rangle}{\left[\left\langle \hat{E}_T^-(z_1,t_1)\hat{E}_T^+(z_1,t_1)\right\rangle\left\langle \hat{E}_T^-(z_2,t_2)\hat{E}_T^+(z_2,t_2)\right\rangle\right]^{1/2}}. \qquad (4.12.4)$$

For light beams that are also stationary, as defined in §3.3, the degree of first-order coherence takes the simple form

$$g^{(1)}(\tau) = \frac{\left\langle \hat{E}_T^-(t)\hat{E}_T^+(t+\tau)\right\rangle}{\left\langle \hat{E}_T^-(t)\hat{E}_T^+(t)\right\rangle}, \qquad (4.12.5)$$

where τ is defined in eqn (3.4.16). The designations of the kinds of light given in eqn (3.4.15) remain in force for the quantum definition of the degree of first-order coherence. This definition allows the Mach–Zehnder output intensity from eqn (4.12.1) to be put in a compact form, similar to eqn (3.3.9), as

$$\left\langle \hat{I}_4(t)\right\rangle = 2|\mathcal{R}|^2|\mathcal{T}|^2\left\langle \hat{I}(t)\right\rangle\left\{1 + \operatorname{Re} g^{(1)}(\tau)\right\}, \qquad (4.12.6)$$

where τ is defined with $t_1 = t_2$ as in eqn (3.3.11).

The quantum theory of the Brown–Twiss interferometer parallels the classical theory of §3.8 in a similar fashion. The experiment measures the correlation of light intensities at two space–time points. With the intensities determined by two photodetectors, the correlation is proportional to the transition rate for a joint absorption of photons at the two points. A simple extension of the photoelectric theory of §4.11 shows that the transition amplitude is proportional to a matrix element of $\hat{E}_T^+(\mathbf{r}_2,t_2)\hat{E}_T^+(\mathbf{r}_1,t_1)$, and the transition rate involves the square of the matrix element, analogous to eqn (4.11.3). These considerations lead to the definition of the quantum-mechanical degree of second-order coherence as

$$g^{(2)}(\mathbf{r}_1,t_1;\mathbf{r}_2,t_2) = \frac{\left\langle \hat{E}_T^-(\mathbf{r}_1,t_1)\hat{E}_T^-(\mathbf{r}_2,t_2)\hat{E}_T^+(\mathbf{r}_2,t_2)\hat{E}_T^+(\mathbf{r}_1,t_1)\right\rangle}{\left\langle \hat{E}_T^-(\mathbf{r}_1,t_1)\hat{E}_T^+(\mathbf{r}_1,t_1)\right\rangle\left\langle \hat{E}_T^-(\mathbf{r}_2,t_2)\hat{E}_T^+(\mathbf{r}_2,t_2)\right\rangle}, \qquad (4.12.7)$$

where the angle-bracket ensemble averages are again evaluated with the use of the density operator similar to eqn (4.12.2). The physical significance of the quantum coherence is the same as that of its classical counterpart, and in particular the condition for second-order coherence at two space–time points is the same as eqn (3.7.23).

The general three-dimensional definition simplifies for a plane parallel light beam by the replacement of $\mathbf{r}$ by z, to give

$$g^{(2)}(z_1,t_1;z_2,t_2) = \frac{\left\langle \hat{E}_T^-(z_1,t_1)\hat{E}_T^-(z_2,t_2)\hat{E}_T^+(z_2,t_2)\hat{E}_T^+(z_1,t_1)\right\rangle}{\left\langle \hat{E}_T^-(z_1,t_1)\hat{E}_T^+(z_1,t_1)\right\rangle\left\langle \hat{E}_T^-(z_2,t_2)\hat{E}_T^+(z_2,t_2)\right\rangle},\qquad (4.12.8)$$

and this form of the degree of second-order coherence is used for most subsequent applications of the theory. The space–time coordinates enter the coherence only in the combination τ defined in eqn (3.4.16). If the light beam is also stationary, the degree of second-order coherence can be written as

$$g^{(2)}(\tau) = \frac{\left\langle \hat{E}_T^-(t)\hat{E}_T^-(t+\tau)\hat{E}_T^+(t+\tau)\hat{E}_T^+(t)\right\rangle}{\left\langle \hat{E}_T^-(t)\hat{E}_T^+(t)\right\rangle^2},\qquad (4.12.9)$$

analogous to the classical expression in eqn (3.7.1). The above expressions for the degree of second-order coherence include the field operators of a single light beam. These intrabeam expressions are converted to cover measurements of the interbeam coherence of a pair of light beams when the two intensity operators $\hat{E}_T^-\hat{E}_T^+$ in the numerators and denominators refer one each to the fields of the two beams, analogous to the classical definition in eqn (3.7.26). Examples of inter-beam degrees of second-order coherence are given in §§6.7, 6.9, 8.5 and 8.6.

The quantities in the numerators of eqns (4.12.7) to (4.12.9) are expectation values of products of operators with their Hermitian conjugates. These must be positive [16] and it follows that

$$0 \le g^{(2)}(\tau) \le \infty,\qquad (4.12.10)$$

similar to the classical range of eqn (3.7.8). It is not, however, possible to prove that the quantum degree of second-order coherence satisfies either of the classical inequalities (3.7.7) or (3.7.11). The resulting differences between the classical and quantum theories are discussed in §§5.10, 7.9, 8.3 and 8.4.

References

[1]   Jackson, J.D., *Classical Electrodynamics*, 3rd edn (Wiley, New York, 1999).

[2]   Arfken, G.B. and Weber, H.J., *Mathematical Methods for Physicists*, 4th edn (Academic Press, San Diego, 1995).

[3]   Louisell, W.H., *Quantum Statistical Properties of Radiation* (Wiley–Interscience, New York, 1973).

[4]   Bransden, B.H. and Joachain, O.J., *Introduction to Quantum Mechanics* (Longman, Harlow, 1989); Atkins, P.W. and Friedman, R.S., *Molecular Quantum Mechanics*, 3rd edn (Oxford University Press, Oxford, 1997).

[5]   Goldstein, H., *Classical Mechanics* (Addison–Wesley, Reading MA, 1980).

[6]   Craig, D.P. and Thirunamachandran, T., *Molecular Quantum Electro-dynamics* (Academic Press, London, 1984).

[7]   Cohen-Tannoudji, C., Dupont–Roc, J. and Grynberg, G., *Photons and Atoms: Introduction to Quantum Electrodynamics* (Wiley–Interscience, New York, 1989).

[8]   Babiker, M. and Loudon, R., Derivation of the Power–Zienau–Woolley Hamiltonian in quantum electrodynamics by gauge transformation, *Proc. R. Soc. Lond. A* **385**, 439–60 (1983).

[9]   Power, E.A., *Introductory Quantum Electrodynamics* (Longmans, London, 1964).

[10]   Bransden, B.H. and Joachain, O.J., *Introduction to Quantum Mechanics* (Longman, Harlow, 1989); Merzbacher, E., *Quantum Mechanics*, 3rd edn (Wiley, New York, 1998).

[11]   Barnett, S.M. and Radmore, P.M., *Methods in Theoretical Quantum Optics* (Clarendon Press, Oxford, 1997).

[12]   Power, E.A. and Zienau, S., Coulomb gauge in non-relativistic quantum electrodynamics and the shape of spectral lines, *Phil. Trans. R. Soc. Lond. A* **251**, 427–54 (1959).

[13]   Woolley, R.G., Molecular quantum electrodynamics, *Proc. R. Soc. Lond. A* **321**, 557–72 (1971).

[14]   Shore, B.W. and Knight, P.L., The Jaynes–Cummings model, *J. Mod. Opt.* **40**, 1195–238 (1993).

[15]   Scully, M.O. and Sargent, M., The concept of the photon, *Physics Today* **25**, 38–47 (March 1972).

[16]   Glauber, R.J., The quantum theory of optical coherence, *Phys. Rev.* **130,** 2529–39 (1963), Coherent and incoherent states of the radiation field, *Phys. Rev.* **131**, 2766–88 (1963).

5 Single-mode quantum optics

The theory of the electromagnetic field quantization developed in Chapter 4 simplifies considerably for applications to many of the key experiments in quantum optics. Thus many experiments use plane parallel light beams whose transverse intensity profiles are not important for the measured quantities. Again, it is often sufficient in interpreting the data to consider the light beams as exciting a single mode of the field. The present chapter begins with the specialization of the quantization to single-mode excitations. The quantum theory is illustrated by its application to several varieties of single-mode state and the main features of number states, coherent states, chaotic statistical mixtures and squeezed states are derived.

The optical beam splitter is an important component in the Mach–Zehnder and Brown–Twiss interferometers, as discussed in Chapter 3 within the framework of classical theory. The corresponding quantum theories of these experiments require a quantum-mechanical formulation of the beam splitter theory and this is derived in the present chapter. It is found that the predictions of the quantum theory are sometimes the same but sometimes differ from those of the classical theory. Indeed, the use of beam splitters is crucial in many of the experiments that demonstrate the specific quantum nature of light. The quantum theory of the beam splitter is used to rederive the Mach–Zehnder interference fringes and the Brown–Twiss correlation. It is convenient to express the results of these experiments in terms of the quantum degrees of coherence defined in §4.12.

A remarkable feature of the theory of light is the large measure of agreement between the predictions of classical and quantum theory, despite the fundamental differences between the two approaches. Thus, it turns out that the classical and quantum theories predict identical interference effects and associated degrees of coherence for experiments that use coherent or chaotic light, or light with intermediate coherence properties. In these examples, the quantum theory provides different conceptual descriptions of the experiments, but the electromagnetic field quantization has little impact on the observable phenomena.

Experiments that do show phenomena at variance with the predictions of classical theory generally rely on the use of *nonclassical* light beams whose fluctuation and coherence properties cannot be described in classical terms. The results of many such experiments can be expressed in terms of components of the degree of second-order coherence, which plays a crucial role in the distinctions between light beams that can or cannot be described by classical models. In the present and later chapters we pay particular attention to the sources of nonclassical light and to the outcomes of the quantum treatment that are quite different from the predictions of the classical theory.

5.1 Single-mode field operators

The electromagnetic field is assumed to excite a single travelling-wave mode with a given wavevector $\mathbf{k}$ and a polarization direction specified by the label λ. The forms of the electric and magnetic field operators for the given mode in the interaction picture, or in the free-field Heisenberg picture, are obtained from eqns (4.4.13) to (4.4.18). We give particular attention to the electric-field operator because of its important role in the electric-dipole interaction and in the definitions of the degrees of coherence. The direction of propagation is taken to be the z axis and, with the mode subscripts $\mathbf{k}$ and λ removed, the scalar electric-field operator for the given direction of linear polarization is written

$$\hat{E}(\chi) = \hat{E}^+(\chi) + \hat{E}^-(\chi) = \left(\hbar\omega/2\varepsilon_0 V\right)^{1/2}\left\{\hat{a}e^{-i\chi} + \hat{a}^\dagger e^{i\chi}\right\}, \qquad (5.1.1)$$

where the positive and negative frequency parts of the field operator correspond respectively to the two terms on the right-hand side. The explicit dependence on z and t is buried in the phase angle χ defined in eqn (4.4.12), which simplifies to

$$\chi = \omega t - kz - \tfrac{\pi}{2}, \qquad (5.1.2)$$

where $k = \omega/c$ in free space. It is convenient to write the electric-field operator in a yet simpler form by removal of the awkward square-root factor from eqn (5.1.1). Thus with the convention that the electric field is measured in units of $2\left(\hbar\omega/2\varepsilon_0 V\right)^{1/2}$, the operator reduces to

$$\hat{E}(\chi) = \hat{E}^+(\chi) + \hat{E}^-(\chi) = \tfrac{1}{2}\hat{a}e^{-i\chi} + \tfrac{1}{2}\hat{a}^\dagger e^{i\chi} = \hat{X}\cos\chi + \hat{Y}\sin\chi, \qquad (5.1.3)$$

where the quadrature operators for the selected mode $\mathbf{k}\lambda$ are as defined in eqns (4.3.36) and (4.3.37), and the final form on the right is the simplified version of eqn (4.4.20).

We show in §6.11 that the electric-field operator of eqn (5.1.3) represents a quantity related to the signal field measured in homodyne detection, where the phase angle χ is determined by the phase of the local oscillator. It has the property that the fields measured at different phase angles χ do not generally commute. Thus with the use of either form of field operator in eqn (5.1.3) and the commutation relations (4.3.9) or (4.3.39), it is easily shown that

$$\left[\hat{E}(\chi_1), \hat{E}(\chi_2)\right] = -\tfrac{i}{2}\sin(\chi_1 - \chi_2), \qquad (5.1.4)$$

where the different phase angles may result from different measurement times or positions, or from different local-oscillator phase angles in homodyne detection. It is emphasized that the complete field operators at equal times continue to commute, as shown in eqn (4.5.24), and that the lack of commutation shown in eqn (5.1.4) applies only to measurements of the electric field that are sensitive to a single travelling-wave mode.

The uncertainty relation that results from eqn (5.1.4) by the standard procedure [1] is

$$\Delta E(\chi_1)\Delta E(\chi_2) \geq \tfrac{1}{4}|\sin(\chi_1 - \chi_2)|, \tag{5.1.5}$$

where the variance of the electric field is defined by

$$(\Delta E(\chi))^2 = \left\langle \left(\hat{E}(\chi)\right)^2 \right\rangle - \left\langle \hat{E}(\chi) \right\rangle^2. \tag{5.1.6}$$

It follows from eqn (5.1.5) that measurements of the values of the fields at phase angles χ that differ by integer multiples of π can in principle be made precisely, but that measurements for other pairs of phase angles must suffer intrinsic quantum-mechanical uncertainties. The size of the uncertainty ΔE depends on the nature of the field excitation. The mean and the variance of the field for the vacuum state $|0\rangle$ are easily calculated as

$$\left\langle \hat{E}(\chi) \right\rangle = 0 \tag{5.1.7}$$

and

$$(\Delta E(\chi))^2 = \tfrac{1}{4}. \tag{5.1.8}$$

Calculations for specific kinds of light beam are presented in subsequent sections.

The electric-field uncertainty is independent of the measurement phase angle χ for the vacuum state and for many important varieties of field excitation, for example the single-mode number state of §5.2 and the single-mode coherent state of §5.3. In these examples eqn (5.1.5) must be satisfied with the maximal value on the right-hand side, so that

$$(\Delta E(\chi))^2 \geq \tfrac{1}{4} \tag{5.1.9}$$

for excitations with phase-independent uncertainty. However, there exist other kinds of light for which the uncertainty $\Delta E(\chi)$ varies quite strongly with χ, for example the squeezed light considered in §§5.5 and 5.6. In such examples, the uncertainty relation (5.1.5) allows ΔE to be smaller than 1/2 at some phase angles χ_1 provided that ΔE is larger than 1/2 at other angles χ_2.

The *coherent signal* S carried by a light beam can be defined as the expectation value of the field operator in eqn (5.1.3),

$$S = \langle E(\chi) \rangle. \tag{5.1.10}$$

It is clear from the form of the field operator and from the matrix elements of the destruction and creation operators in eqn (4.3.34) that nonvanishing coherent signals occur only for states that contain superpositions of photon-number states with values of n differing by unity. The field uncertainty, controlled by the relation (5.1.5), represents *noise* on the optical signal. The magnitude $\mathcal{N}$ of the noise is conventionally defined by

$$\mathcal{N} = \left(\Delta E(\chi)\right)^2, \tag{5.1.11}$$

and the *signal-to-noise ratio* is defined as

$$\text{SNR} = \frac{S^2}{\mathcal{N}} = \frac{\langle E(\chi)\rangle^2}{\left(\Delta E(\chi)\right)^2}. \tag{5.1.12}$$

This ratio is closely related to the photocount signal-to-noise ratio in homodyne detection, treated in §6.11. It provides a measure of the information-carrying capacity of a light beam for this form of detection. The signal-to-noise ratio is evaluated for the range of field excitations considered in the following sections. It is generally a function of the measurement phase angle, as the signal varies with χ even for excitations that have phase-independent field uncertainty. It is emphasized that other definitions of signal-to-noise ratio apply for other forms of signal detection, for example the direct detection treated in §6.10, but these are not treated here.

The definition of the quantum degree of first-order coherence simplifies considerably for the single-mode light beam assumed in the present chapter. Thus, with insertion of the forms of the positive and negative frequency parts of the field operators from eqn (5.1.3), the degree of first-order coherence from eqn (4.12.4) reduces to

$$g^{(1)}(z_1,t_1;z_2,t_2) = g^{(1)}(\tau) = g^{(1)}(\chi_1;\chi_2) = \exp\{i(\chi_1 - \chi_2)\}, \tag{5.1.13}$$

where $\tau = \chi_2 - \chi_1$ in accordance with eqns (3.4.16) and (5.1.2). Note that the renormalization of the field operators between eqns (5.1.1) and (5.1.3) makes no difference to the value of the degree of first-order coherence. In accordance with eqn (3.4.15) it follows that any plane parallel single-mode light beam is first-order coherent for all pairs of space–time points.

The definition of the degree of second-order coherence also simplifies for single-mode light, where the phase-angle notation from eqn (5.1.2) and the field operators from eqn (5.1.3) convert the degree of second-order coherence in eqn (4.12.8) to the form

$$g^{(2)}(\tau) = g^{(2)}(\chi_1;\chi_2) = \left\langle \hat{a}^{\dagger}\hat{a}^{\dagger}\hat{a}\hat{a}\right\rangle / \left\langle \hat{a}^{\dagger}\hat{a}\right\rangle^2 \tag{5.1.14}$$

This expression is independent of position and time but, unlike the single-mode degree of first-order coherence in eqn (5.1.13), it does depend on the nature of the light beam. With the usual properties (4.3.9) and (4.3.27) of the destruction and creation operators, the degree of coherence is expressed in terms of the mean and mean-square photon numbers as

$$g^{(2)}(\tau) = \frac{\langle n(n-1) \rangle}{\langle n \rangle^2} = \frac{\langle n^2 \rangle - \langle n \rangle}{\langle n \rangle^2} = 1 + \frac{(\Delta n)^2 - \langle n \rangle}{\langle n \rangle^2}. \tag{5.1.15}$$

The photon-number variance must be a non-negative quantity,

$$0 \le (\Delta n)^2 \equiv \langle n^2 \rangle - \langle n \rangle^2. \tag{5.1.16}$$

It follows that the degree of second-order coherence for any single-mode field excitation must satisfy the inequality

$$1 - \frac{1}{\langle n \rangle} \le g^{(2)}(\tau) \quad \text{for } \langle n \rangle \ge 1, \tag{5.1.17}$$

and the lower limit is 0 for $\langle n \rangle < 1$. The lower limit in eqn (5.1.17) is consistent with the zero value derived more generally in eqn (4.12.10) but it provides a more restrictive range of values that depends on the mean photon-number excitation.

The degree of second-order coherence is evaluated in subsequent sections for a variety of states and the results are all independent of τ for the single-mode excitations considered in the present chapter. Time and space dependent degrees of second-order coherence occur for light beams that excite two or more modes, and these are treated in Chapter 6.

5.2 Number states

The photon number states are easy to comprehend and they are the basic states of the quantum theory of light. They form a complete set for the states of a single mode and they are easy to manipulate in calculations of their quantum-optical properties. They are the obvious basis for the Planck black-body distribution and, more generally, for other varieties of chaotic light. They are, however, less easy to generate experimentally and it is necessary to treat more complicated excitations for realistic interpretations of most quantum-optical experiments. Nevertheless, the photon number states form the natural starting point for a treatment of single-mode light.

Several properties of the single-mode number states $|n\rangle$ are derived in §4.3. The eigenvalue property (4.3.28) implies that there is no uncertainty in the photon number, whose variance therefore vanishes,

$$(\Delta n)^2 = 0. \tag{5.2.1}$$

More generally, the expectation value of an arbitrary power of the photon number is simply

$$\langle n^r \rangle = n^r. \tag{5.2.2}$$

The degree of second-order coherence for the number state $|n\rangle$ thus follows trivially from eqn (5.1.15) as

$$g^{(2)}(\tau) = 1 - \frac{1}{n} \quad \text{for} \quad n \geq 1, \tag{5.2.3}$$

and it has the minimum value allowed by the inequality (5.1.17). The classical inequality of eqn (3.7.6) for $\tau = 0$ is violated. The degree of second-order coherence for the vacuum number state $|0\rangle$ is not determined by the expressions in eqn (5.1.15) as both their numerators and denominators vanish.

The energy eigenvalue relation can be written in a variety of forms from eqns (4.3.11), (4.3.26) and (4.3.38),

$$\hat{\mathcal{H}}|n\rangle = \hbar\omega\left(\hat{a}^\dagger\hat{a} + \tfrac{1}{2}\right)|n\rangle = \hbar\omega\left(\hat{X}^2 + \hat{Y}^2\right)|n\rangle = \hbar\omega\left(n + \tfrac{1}{2}\right)|n\rangle. \tag{5.2.4}$$

The number state thus has the quadrature-operator eigenvalue property

$$\left(\hat{X}^2 + \hat{Y}^2\right)|n\rangle = \left(n + \tfrac{1}{2}\right)|n\rangle. \tag{5.2.5}$$

The quadrature-operator expectation values are

$$\langle n|\hat{X}|n\rangle = \langle n|\hat{Y}|n\rangle = 0 \tag{5.2.6}$$

and

$$(\varDelta X)^2 = (\varDelta Y)^2 = \tfrac{1}{2}\left(n + \tfrac{1}{2}\right). \tag{5.2.7}$$

The number states thus have identical properties for the $\hat{X}$ and $\hat{Y}$ quadrature operators. Their variances have the smallest values allowed by eqn (4.3.40) only for the vacuum state with $n = 0$. Thus the state $|0\rangle$ is an example of a *quadrature minimum-uncertainty state*. Figure 5.1 shows a qualitative representation of the

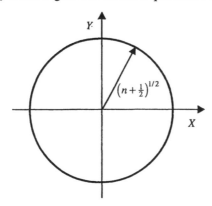

Fig. 5.1. Representation of the quadrature-operator expectation values for the photon number state.

quadrature properties (5.2.5) to (5.2.7) of the number state, where the two axes represent the quadrature variables. The heavy circle shows the eigenvalue given in eqn (5.2.5).

The expectation values of the electric-field operator in eqn (5.1.3) are also easily determined. The coherent signal or the mean field vanishes, in accordance with eqn (5.2.6),

$$S = \langle n|\hat{E}(\chi)|n\rangle = 0,\tag{5.2.8}$$

and the single-mode number state is thus unable to carry information in a form suitable for homodyne detection. The field variance, or the noise, is

$$\mathcal{N} = \left(\Delta E(\chi)\right)^2 = \langle n|\left(\hat{E}(\chi)\right)^2|n\rangle = \tfrac{1}{2}\left(n + \tfrac{1}{2}\right).\tag{5.2.9}$$

The phase-independent variance satisfies the uncertainty relation (5.1.9) and the signal-to-noise ratio defined in eqn (5.1.12) vanishes. The field variance has the smallest value from eqn (5.1.8) for the vacuum state $|0\rangle$ but the noise is larger than the minimum for the other number states.

Figure 5.2 shows a representation of the field properties of the single-mode number state. The vertical axis represents the electric field at some fixed point as a function of the time. The field oscillates like a sine wave of known frequency ω. The expression (5.2.9) shows that the wave amplitude is

$$E_0 = \left(n + \tfrac{1}{2}\right)^{1/2} \quad \text{in units of} \quad 2\left(\hbar\omega/2\varepsilon_0 V\right)^{1/2};\tag{5.2.10}$$

in these units, of order 10^{-2} Vm^{-1} for a mode in a typical gas-laser cavity, the electric field of each photon is represented by a vector of unit length and random orientation. However, the lateral position of the wave along the horizontal axis is undetermined owing to a complete uncertainty in the phase angle. This is indicated

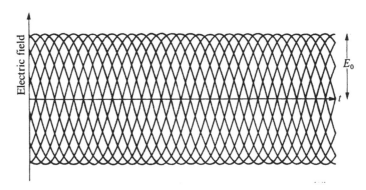

Fig. 5.2. Representation of the electric-field variation in a single cavity mode excited to a number state. The sine waves should more accurately form a horizontal continuum. The amplitude E_0 is defined in eqn (5.2.10).

in the figure by the inclusion of several waves, all of the same amplitude and frequency, but with their nodes progressively displaced along the axis. More accurately, the horizontal positions of the waves form a continuum, and the field at any time can take a continuous range of values between $-E_0$ and E_0.

Pictorial representations of quantum-mechanical states must generally be treated with caution. Thus Fig. 5.2 reproduces the expectation values given in eqns (5.2.8) and (5.2.9) but the upper and lower cut-offs of the field distribution are shown as too sharp. The figure provides an accurate representation only in the limit $n \gg 1$.

Problem 5.1 Evaluate the expectation value $\langle n|\left(\hat{E}(\chi)\right)^4|n\rangle$ and show that it exceeds the corresponding average for the sine waves of Fig. 5.2 by an amount 3/32. As this is independent of n, the importance of the discrepancy diminishes for large n.

The field variation of the single-mode number state shown in Fig. 5.2 is in stark contrast with that of the classical wave of stable amplitude and phase defined by eqn (3.4.20) and illustrated in Fig. 3.10. Thus, while the number state has a fairly well defined amplitude E_0 given by eqn (5.2.10), it shows no vestige of the phase angle φ of the classical wave form. Other kinds of quantum field excitation, for example the coherent state of §5.3, have amplitudes that are less well defined but phase angles that are better defined than those of the number state. It is natural to enquire how the phase properties of excitations are represented in the quantum theory of light.

The identification of a Hermitian operator that represents the phase, in conformity with its physical significance, is a subtle problem that requires careful discussion [2,3]. However, it is possible to establish the form of the probability distribution $P(\varphi)$ for the phase angle quite straightforwardly. The classical phase φ in eqn (3.4.20) can take any value, but changes in φ by integer multiples of 2π make no difference to the variation of the electric field. In other words, the phase angle φ is defined *modulo* 2π and it can thus be restricted to any convenient 2π range, for example 0 to 2π. The same features of the phase are expected to apply to quantum field excitations. It is therefore appropriate to take

$$\int_0^{2\pi} d\varphi \, P(\varphi) = 1. \tag{5.2.11}$$

as the normalization condition for the quantum-phase probability-distribution.

The state $|n\rangle$ of well-defined photon number has a completely random phase and the quantum phase is expected to be complementary to the photon number, similar to the position and momentum in quantum mechanics. Conversely, a state of well-defined phase should have a completely random photon-number distribution. It is convenient to replace the continuous range of phase angles from 0 to 2π by $r + 1$ equally-spaced discrete values φ_m, defined as

$$\varphi_m = \frac{2\pi m}{r+1}, \quad m = 0,1,....,r, \tag{5.2.12}$$

where r is a large number. The state associated with phase angle φ_m, known as the *phase state*, is the superposition of photon number states defined by

$$|\varphi_m\rangle = (r+1)^{-1/2} \sum_{n=0}^{r} e^{in\varphi_m} |n\rangle. \tag{5.2.13}$$

The highest photon number r eventually tends to infinity, but the limit cannot be taken at this stage in the calculation and for the present r is merely assumed to be much larger than the important photon-number constituents in a given field excitation. The coefficients in the superposition of eqn (5.2.13) for the different n all have equal amplitudes. It follows that the phase state $|\varphi_m\rangle$ does indeed have a uniform, or random, photon-number distribution over the lowest $r+1$ number states.

Problem 5.2   Prove that the phase states satisfy the orthonormality condition

$$\langle\varphi_m|\varphi_{m'}\rangle = \delta_{mm'} \tag{5.2.14}$$

and the completeness condition

$$\sum_{m=0}^{r} |\varphi_m\rangle\langle\varphi_m| = \sum_{n=0}^{r} |n\rangle\langle n|. \tag{5.2.15}$$

. This last relation shows that the $|\varphi_m\rangle$ span the same $r+1$ dimensional state space as the lowest $r+1$ number states.

The phase state also has the property

$$\exp(i\hat{n}\varphi_{m'})|\varphi_m\rangle = |\varphi_m + \varphi_{m'}\rangle = |\varphi_{m+m'}\rangle, \tag{5.2.16}$$

in accordance with the requirement that each operator in a complementary pair should generate shifts in the eigenvalues of the other [3]. The phase is again defined modulo 2π and, if $m + m'$ is greater than r on the right-hand side of eqn (5.2.16), an amount 2π should be subtracted from the composite phase to obtain one of the angles defined in eqn (5.2.12). More generally, it can be shown that the states $|\varphi_m\rangle$ defined by eqn (5.2.13) are eigenstates of the properly-defined Hermitian phase operator [2,3] and they are indeed the states of well-defined phase complementary to the number states $|n\rangle$.

An arbitrary single-mode pure state $|R\rangle$, defined as a superposition of the lowest $r+1$ number states, is expressed as

$$|R\rangle = \sum_{n=0}^{r} c(n)|n\rangle, \tag{5.2.17}$$

with the normalization condition

$$\sum_{n=0}^{r} |c(n)|^2 = 1. \tag{5.2.18}$$

The phase probability distribution of state $|R\rangle$ over the discrete angles φ_m is determined in the usual way by the overlap

$$P(\varphi_m) = |\langle \varphi_m | R\rangle|^2. \tag{5.2.19}$$

If r now tends to infinity, the separation $2\pi/(r+1)$ of adjacent phase angles φ_m defined by eqn (5.2.12) becomes infinitesimal and the discrete φ_m are replaced by a continuous phase variable φ. Thus, with allowance for the discrete spacing, the continuous phase distribution is given by

$$P(\varphi) = \underset{r \to \infty}{\mathrm{Lt}} \frac{r+1}{2\pi} |\langle \varphi_m | R\rangle|^2. \tag{5.2.20}$$

The definition is valid for all states for which the limit exists and it can be shown [2] that this condition includes all states that have finite expectation values of positive integer powers of the photon-number operator for $r \to \infty$. The phase state $|\varphi_m\rangle$ defined in eqn (5.2.13) does not satisfy the condition, but all of the physical electromagnetic field excitations ordinarily available in practical experiments do so.

The limit in eqn (5.2.20) is readily performed by substitution of $|R\rangle$ from eqn (5.2.17) and the phase state $|\varphi_m\rangle$ from eqn (5.2.13), with the result

$$P(\varphi) = \frac{1}{2\pi} \left| \sum_{n=0}^{\infty} c(n)\mathrm{e}^{-in\varphi} \right|^2. \tag{5.2.21}$$

It is readily verified with the use of eqn (5.2.18) that this phase probability distribution satisfies the normalization condition of eqn (5.2.11). For the number state $|n\rangle$ itself, eqn (5.2.21) reduces trivially to

$$P(\varphi) = 1/2\pi, \tag{5.2.22}$$

which confirms the expected uniform phase distribution of the number state.

Problem 5.3   Prove that the mean and the variance of the phase of the photon number state are respectively,

$$\langle \varphi \rangle = \pi \quad \text{and} \quad (\Delta \varphi)^2 = \pi^2/3 \tag{5.2.23}$$

when the range of phase angles is chosen to be 0 to 2π.

The theoretical simplicity of the number states contrasts with the experimental difficulties of generating them in the laboratory. The vacuum state $|0\rangle$ for a mode of frequency ω is of course easily produced, at least to a very good approximation, by taking an optical system at a temperature $T \ll \hbar\omega/k_B$, but this state is not of physical interest in the present context. The first excited state, or single-photon state, $|1\rangle$ is obtained experimentally [4] by first generating a photon pair state and then using the detection of one of the pair to isolate the second of the pair, which can be measured with a sufficient degree of confidence of its location in space and time. Further details of the experiment are given in §5.8, but fuller accounts of the natures of pair states and their use in the generation of single-photon states need the time-dependent continuous-mode theory developed in §§6.7 and 6.8. The pair states themselves are generated in cascade emission, treated in §8.6, and parametric down-conversion, treated in §9.4. They are important in a variety of quantum-optical experiments, which are discussed in these various sections. We do not consider the practical generation and use of the number states with more than two photons.

5.3 Coherent states

The most commonly found single-mode states correspond not to the individual number states but to linear superpositions of the states $|n\rangle$. There is of course a wide variety of possible superposition states but one kind, the coherent state [5], is of particular importance in the practical applications of the quantum theory of light. The coherent states, so called for reasons that will become apparent, are denoted by $|\alpha\rangle$. Their electric-field variation approaches that of the classical wave of stable amplitude and fixed phase shown in Fig. 3.10 in the limit of strong excitation. They are important, not only because they are one of the quantum-mechanical states whose properties most closely resemble those of a classical electromagnetic wave, but also because a single-mode laser operated well above threshold generates a coherent state excitation, as is discussed in §7.4.

Consider the properties of the state $|\alpha\rangle$ defined as the following linear superposition of the number states

$$|\alpha\rangle = \exp\left(-\tfrac{1}{2}|\alpha|^2\right)\sum_{n=0}^{\infty}\frac{\alpha^n}{(n!)^{1/2}}|n\rangle. \tag{5.3.1}$$

In this expression α is any complex number, and the coherent states so defined form a double continuum corresponding to the continuous ranges of values of the real and imaginary parts of α. It is easily verified that the state $|\alpha\rangle$ is normalized,

$$\langle\alpha|\alpha\rangle = \exp\left(-|\alpha|^2\right)\sum_n\frac{\alpha^{*n}\alpha^n}{n!} = 1. \tag{5.3.2}$$

Different coherent states are not however orthogonal as, for two different complex numbers α and β,

$$\langle\alpha|\beta\rangle = \exp\left(-\tfrac{1}{2}|\alpha|^2 - \tfrac{1}{2}|\beta|^2\right)\sum_n \frac{\alpha^{*n}\beta^n}{n!} = \exp\left(-\tfrac{1}{2}|\alpha|^2 - \tfrac{1}{2}|\beta|^2 + \alpha^*\beta\right). \quad (5.3.3)$$

Thus

$$\left|\langle\alpha|\beta\rangle\right|^2 = \exp\left(-|\alpha - \beta|^2\right). \quad (5.3.4)$$

It is apparent from the definition (5.3.1) that there are many more coherent states $|\alpha\rangle$ than there are states $|n\rangle$. The $|\alpha\rangle$ form an overcomplete set of states for the harmonic oscillator, and their lack of orthogonality is a consequence. Note, however, from eqn (5.3.4) that the states $|\alpha\rangle$ and $|\beta\rangle$ become approximately orthogonal if $|\alpha - \beta|$ is much greater than unity.

The coherent states are right eigenstates of the destruction operator, as follows from

$$\hat{a}|\alpha\rangle = \exp\left(-\tfrac{1}{2}|\alpha|^2\right)\sum_n \frac{\alpha^n}{(n!)^{1/2}} n^{1/2}|n-1\rangle = \alpha|\alpha\rangle. \quad (5.3.5)$$

The complex number α that labels the coherent state is thus the eigenvalue of the destruction operator. Note, however, that the state $|\alpha\rangle$ is not a right eigenstate of the creation operator as the summation over α analogous to that in eqn (5.3.5) cannot be rearranged to reproduce the coherent state from $\hat{a}^\dagger|\alpha\rangle$. Of course, the creation operator does satisfy the left-eigenvalue relation conjugate to eqn (5.3.5),

$$\langle\alpha|\hat{a}^\dagger = \langle\alpha|\alpha^*. \quad (5.3.6)$$

The simple eigenvalue relations in eqns (5.3.5) and (5.3.6) are very useful in the calculation of coherent-state expectation values of products of creation and destruction operators.

Problem 5.4   An alternative approach to the coherent state is to take the eigen-value equation (5.3.5) as its definition [5]. Show that the expansion (5.3.1) of the coherent state in number states can be derived from this alternative starting point. Prove that the creation operator $\hat{a}^\dagger$ has no right eigenstates.

The relations (4.3.45) and (4.3.46) are used to rewrite the definition (5.3.1) of the coherent state as

$$|\alpha\rangle = \exp\left(-\tfrac{1}{2}|\alpha|^2\right)\sum_n \frac{\left(\alpha\hat{a}^\dagger\right)^n}{n!}|0\rangle = \exp\left(\alpha\hat{a}^\dagger - \tfrac{1}{2}|\alpha|^2\right)|0\rangle. \quad (5.3.7)$$

Problem 5.5   The relation

$$\exp(\hat{c})\exp(\hat{d}) = \exp\left(\hat{c} + \hat{d} + \tfrac{1}{2}[\hat{c},\hat{d}]\right) \tag{5.3.8}$$

can be proved [3] for any pair of operators $\hat{c}$ and $\hat{d}$ that commute with their commutator, that is

$$\left[\hat{c},[\hat{c},\hat{d}]\right] = \left[\hat{d},[\hat{c},\hat{d}]\right] = 0. \tag{5.3.9}$$

With use of the vacuum-state condition of eqn (4.3.23), cast eqn (5.3.7) into the form

$$|\alpha\rangle = \exp\left(\alpha\hat{a}^\dagger - \alpha^*\hat{a}\right)|0\rangle. \tag{5.3.10}$$

This result is often written compactly as

$$|\alpha\rangle = \hat{D}(\alpha)|0\rangle, \tag{5.3.11}$$

where the *coherent-state displacement operator*, defined as

$$\hat{D}(\alpha) = \exp\left(\alpha\hat{a}^\dagger - \alpha^*\hat{a}\right), \tag{5.3.12}$$

is equivalent to a creation operator for the complete state, analogous to the number-state creation operator $\hat{N}(n)$ defined in eqn (4.3.46). It is seen by inspection that the displacement operator satisfies the conditions

$$\hat{D}^\dagger(\alpha)\hat{D}(\alpha) = \hat{D}(\alpha)\hat{D}^\dagger(\alpha) = 1 \tag{5.3.13}$$

for a unitary operator [1]. Its effect on the destruction operator is obtained with the use of eqn (4.3.44) as a displacement by an amount α,

$$\hat{D}^\dagger(\alpha)\hat{a}\hat{D}(\alpha) = \hat{a} + \alpha, \tag{5.3.14}$$

with the Hermitian conjugate relation

$$\hat{D}^\dagger(\alpha)\hat{a}^\dagger\hat{D}(\alpha) = \hat{a}^\dagger + \alpha^*. \tag{5.3.15}$$

The coherent-state expectation values for the number operator are readily obtained with use of the right and left eigenvalue properties (5.3.5) and (5.3.6),

$$\langle n\rangle = \langle\alpha|\hat{n}|\alpha\rangle = \langle\alpha|\hat{a}^\dagger\hat{a}|\alpha\rangle = |\alpha|^2. \tag{5.3.16}$$

For the second moment it is convenient to reorder the creation and destruction operators with the use of the commutation relation (4.3.9),

$$\hat{n}^2 = \hat{a}^\dagger \hat{a} \hat{a}^\dagger \hat{a} = \hat{a}^\dagger \left(\hat{a}^\dagger \hat{a} + 1\right)\hat{a} = \hat{a}^\dagger \hat{a}^\dagger \hat{a} \hat{a} + \hat{a}^\dagger \hat{a} =: \hat{a}^\dagger \hat{a} \hat{a}^\dagger \hat{a}: + \hat{a}^\dagger \hat{a}. \tag{5.3.17}$$

The terms with the destruction operators to the right of the creation operators are normally ordered and the :: notation on the right-hand side of eqn (5.3.17) means 'put the operators between the colons into normal order'. It is easy to evaluate the expectation values of normally-ordered operators for the coherent state with use of its eigenvalue properties. Thus

$$\left\langle n^2 \right\rangle = \left\langle \alpha | \hat{n}^2 | \alpha \right\rangle = |\alpha|^4 + |\alpha|^2 = \langle n \rangle^2 + \langle n \rangle, \tag{5.3.18}$$

and the photon-number variance is

$$(\Delta n)^2 = |\alpha|^2 = \langle n \rangle. \tag{5.3.19}$$

The fractional uncertainty in the number of photons in the coherent state is

$$\frac{\Delta n}{\langle n \rangle} = \frac{1}{|\alpha|} = \frac{1}{\sqrt{\langle n \rangle}}, \tag{5.3.20}$$

and this decreases with increasing values of the coherent state amplitude $|\alpha|$.

The photon-number variance of coherent light can be compared with the classical intensity variance of the stable wave, which is shown to vanish in eqn (3.6.14). The nonvanishing right-hand side of eqn (5.3.19) is a consequence of the particle-like aspects of light in the quantum theory. Equation (5.3.20) shows that the importance of these aspects diminishes with increasing values of the mean photon number $\langle n \rangle$.

The probability of finding n photons in the mode is obtained from the definition (5.3.1) as

$$P(n) = |\langle n | \alpha \rangle|^2 = \exp\left(-|\alpha|^2\right) \frac{|\alpha|^{2n}}{n!} = e^{-\langle n \rangle} \frac{\langle n \rangle^n}{n!}. \tag{5.3.21}$$

This is a Poisson probability distribution, similar to that in eqn (3.9.14), and eqn (5.3.19) is the usual expression for the variance of this distribution. The form of the Poisson distribution is illustrated in Fig. 5.3 for three values of the mean photon number $\langle n \rangle$. It has a peaked form in contrast to the monotonically decreasing photon distribution for chaotic light shown in Fig. 1.7. The Poisson distribution approaches a Gaussian distribution for large values of $\langle n \rangle$.

Problem 5.6   Use the property $\langle n \rangle^n = \exp\{n \ln\langle n \rangle\}$ and Stirling's approximation to $n!$ to show that eqn (5.3.21) can be written in the Gaussian form

$$P(n) \approx \frac{1}{\sqrt{2\pi\langle n \rangle}} \exp\left\{-\frac{(n - \langle n \rangle)^2}{2\langle n \rangle}\right\} \tag{5.3.22}$$

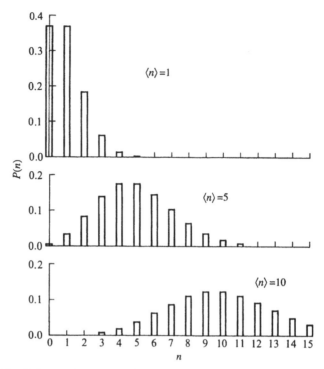

Fig. 5.3. The Poisson photon-number probability distributions of coherent states with the mean photon numbers shown.

for $\langle n \rangle \gg 1$. By taking n as a continuous variable, verify the normalization of this distribution and derive the result (5.3.19) for the variance.

It is easily shown that the Poisson distribution of eqn (5.3.21) has an rth factorial moment of the simple form

$$\langle n(n-1)(n-2)...(n-r+1) \rangle = \langle n \rangle^r. \tag{5.3.23}$$

The degree of second-order coherence for the coherent state $|\alpha\rangle$ thus follows from eqn (5.1.15) as

$$g^{(2)}(\tau) = 1. \tag{5.3.24}$$

The state is therefore second-order coherent according to the prescription of eqn (3.7.23), similar to the classical stable wave considered in eqn (3.7.21), and it is this property that leads to the nomenclature of the state $|\alpha\rangle$.

The coherent-state expectation values of the quadrature operators defined in eqn (4.3.36) are easily found with use of the eigenvalue properties (5.3.5) and (5.3.6),

$$\langle\alpha|\hat{X}|\alpha\rangle = \tfrac{1}{2}\langle\alpha|\hat{a}^\dagger + \hat{a}|\alpha\rangle = \tfrac{1}{2}\left(\alpha^* + \alpha\right) = \text{Re}\,\alpha = |\alpha|\cos\theta, \qquad (5.3.25)$$

where we have put

$$\alpha = |\alpha|e^{i\theta}. \qquad (5.3.26)$$

Similarly

$$\langle\alpha|\hat{Y}|\alpha\rangle = \text{Im}\,\alpha = |\alpha|\sin\theta. \qquad (5.3.27)$$

The expectation values of the squares of the quadrature operators are found with the help of the normal-ordering procedure. Thus

$$\hat{X}^2 = \tfrac{1}{4}\left(\hat{a}^\dagger\hat{a}^\dagger + 2\hat{a}^\dagger\hat{a} + \hat{a}\hat{a} + 1\right), \qquad (5.3.28)$$

$$\hat{Y}^2 = \tfrac{1}{4}\left(-\hat{a}^\dagger\hat{a}^\dagger + 2\hat{a}^\dagger\hat{a} - \hat{a}\hat{a} + 1\right), \qquad (5.3.29)$$

and it follows that the quadrature variances are

$$(\Delta X)^2 = (\Delta Y)^2 = \tfrac{1}{4}. \qquad (5.3.30)$$

Thus, in contrast to the variances for the number-state given by eqn (5.2.7), the coherent state is a quadrature minimum-uncertainty state for all mean photon numbers $|\alpha|^2$.

The expectation values of the field operator (5.1.3) are easily determined by very similar calculations. The coherent-state superposition of eqn (5.3.1) contains adjacent number states and thus the mean field does not vanish. The coherent signal is

$$S = \langle\alpha|\hat{E}(\chi)|\alpha\rangle = |\alpha|\cos(\chi - \theta) \qquad (5.3.31)$$

and the field variance, or noise, is

$$\mathcal{N} = \left(\Delta E(\chi)\right)^2 = \tfrac{1}{4}. \qquad (5.3.32)$$

The noise is thus phase-independent and it has the minimum value allowed by eqn (5.1.9), the same as for the vacuum state in eqn (5.1.8), for all values of $|\alpha|$. The signal-to-noise ratio, as defined in eqn (5.1.12), is

$$\text{SNR} = 4|\alpha|^2\cos^2(\chi - \theta) = 4\langle n\rangle\cos^2(\chi - \theta), \qquad (5.3.33)$$

with a maximum value of $4\langle n\rangle$ for $\chi = \theta$.

Figure 5.4 shows a representation of the field expectation values. The mean amplitude associated with the coherent state is shown by the heavy arrow of length $|\alpha| = \langle n\rangle^{1/2}$ inclined at angle $\chi - \theta$ to the real field axis. The projection of

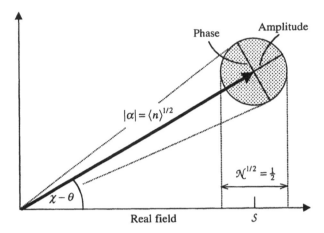

Fig. 5.4. Representation of the electric-field properties of the coherent state with mean photon number $\langle n \rangle = 4$, showing the mean and the uncertainty in the field.

the arrow on to the real field axis gives the mean value of eqn (5.3.31). The two angles that occur in this expression are physically distinct, as χ defined by eqn (5.1.2) is determined by the position and time of evaluation of the field averages, while θ, defined by eqn (5.3.26), is the mean phase of the coherent-state excitation of the mode. Thus χ is a property of the measurement and it is in principle under the control of the experimentalist. By contrast, θ is a property of the field excitation on which the measurement is made and it is in principle outside the control of the experimentalist.

The field uncertainty is shown in Fig. 5.4 by the circular disc of diameter 1/2 and the projection of the disc on to the real field axis gives the square-root of the phase-independent variance (5.3.32). The uncertainty disc is resolved into amplitude and phase contributions, shown by the orthogonal diameters. The amplitude contribution reproduces the exact result of eqn (5.3.19) for the photon-number uncertainty according to

$$\Delta n = \left(\langle n \rangle^{1/2} + \tfrac{1}{4} \right)^2 - \left(\langle n \rangle^{1/2} - \tfrac{1}{4} \right)^2 = \langle n \rangle^{1/2}. \tag{5.3.34}$$

The phase contribution represents an uncertainty in the coherent-state phase that is obtained as the angle $\Delta \varphi$ subtended by the diameter of the disc at the origin. Provided that $|\alpha| \gg 1$, the arc rule gives an approximate expression for the phase uncertainty in the form

$$\Delta \varphi = \frac{1}{2|\alpha|} = \frac{1}{2\langle n \rangle^{1/2}}. \tag{5.3.35}$$

The product of photon-number and phase uncertainties is therefore

$$\Delta n \Delta \varphi = \tfrac{1}{2}. \tag{5.3.36}$$

This result has the form of an uncertainty relation but its qualitative derivation is based on geometrical arguments, and it is not the consequence of a commutation relation between number and phase operators. Nevertheless, it correctly represents a trade-off between the values of the amplitude and phase uncertainties of the electric field associated with the coherent state. It is seen from eqns (5.3.20) and (5.3.35) that the fractional uncertainty in photon number and the phase uncertainty both vary like $1/|\alpha|$; as the mean photon number is increased, the electromagnetic wave becomes better defined both in amplitude and phase angle. The coherent states have some analogy to wavepackets in particle mechanics, which have uncertainty spreads in both position and momentum, but which reproduce the classical behaviour in the large-mass limit.

The phase dependence of the electric field is illustrated in Fig. 5.5, where the continuous cosine wave represents the mean field of eqn (5.3.31). The dashed lines, with a vertical separation equal to the field uncertainty of 1/2 obtained from the square root of eqn (5.3.32), are the envelope of the noise band associated with the mean field. It is clear from the figure that the electric-field variation resembles that of a classical stable wave more and more closely as the mean number of photons in the coherent state is increased and the constant uncertainty spread of the cosine wave becomes less significant. In the opposite limit of the vacuum state $|0\rangle$, where $\alpha = 0$, the mean field vanishes but the same noise band remains.

Problem 5.7   Calculate the range of phase angles over which the noise band crosses the horizontal axis in Fig. 5.5 when $|\alpha| \gg 1$ and show that it agrees with the phase uncertainty calculated in eqn (5.3.35).

The phase properties of the coherent state are calculated more formally with the use of the distribution defined in eqn (5.2.21). Thus, with the forms of the pure state coefficients $c(n)$ obtained from the number-state linear superposition in eqn (5.3.1), the phase probability distribution is

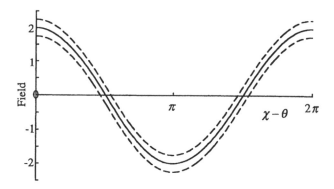

Fig. 5.5. Phase dependence of the electric field of a single-mode coherent state with $\langle n \rangle = 4$, showing the cosine oscillation of the mean, or signal, and the uncertainty, or noise, band with constant vertical width $\mathcal{N}^{1/2} = 1/2$.

$$P(\varphi) = \frac{1}{2\pi} \left| \sum_n \exp\left(-\tfrac{1}{2}|\alpha|^2\right) \frac{|\alpha|^n}{(n!)^{1/2}} e^{in(\theta-\varphi)} \right|^2,$$ (5.3.37)

where α is expressed in terms of its amplitude and phase in accordance with eqn (5.3.26). The phases and their differences are again defined modulo 2π, and it is convenient in this example to choose a range of phase angles that extends from $\theta - \pi$ to $\theta + \pi$. It is then seen that $P(\varphi)$ is an even function of $\theta - \varphi$, and it follows that the mean phase of the coherent state equals θ,

$$\langle \varphi \rangle = \theta.$$ (5.3.38)

The variance of the phase is more difficult to determine and the continuous curve in Fig. 5.6 shows the results of a numerical evaluation of $(\Delta\varphi)^2$ as a function of $\langle n \rangle$, in good agreement with experimental points [6]. The variance equals $\pi^2/3$ for the vacuum state, in agreement with eqn (5.2.23), as $|0\rangle$ is an example of a state with a uniform phase distribution. For large values of the mean photon number, the variance falls off as $1/4\langle n \rangle$ in agreement with eqn (5.3.35). This latter approximation to the phase variance of the coherent state can be proved analytically in the appropriate limit of the exact distribution of eqn (5.3.37).

Problem 5.8   Prove, with use of the Gaussian form of the photon number distribution for the coherent state from eqn (5.3.22), that the phase distribution from eqn (5.3.37) is approximately

$$P(\varphi) = \left(\frac{2\langle n \rangle}{\pi}\right)^{1/2} \exp\left[-2\langle n \rangle (\theta - \varphi)^2\right]$$ (5.3.39)

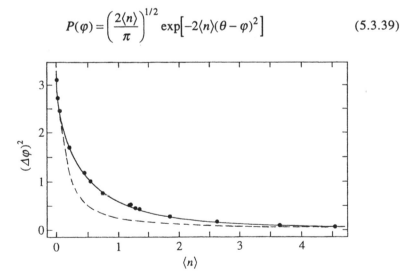

Fig. 5.6. Dependence of the phase variance $(\Delta\varphi)^2$ on the mean photon number $\langle n \rangle$ for a coherent state. The continuous curve is the result of a numerical evaluation based on eqn (5.3.37), in good agreement with measured values. The dashed curve shows the large $\langle n \rangle$ approximation of eqn (5.3.35). (After [6])

when $\langle n \rangle \gg 1$. Verify that this distribution reproduces the phase uncertainty given in eqn (5.3.35).

5.4 Chaotic light

Consider a chaotic light beam in an optical cavity and suppose that the contributions of all except one of the modes are removed by a filter, as in Problems 3.1 and 3.7. The properties of single-mode chaotic light are extensively treated in §§1.4 and 4.6, where it is shown that the photon number probabilities follow the geometric distribution

$$P(n) = \frac{\langle n \rangle^n}{\left(1 + \langle n \rangle\right)^{1+n}}, \tag{5.4.1}$$

illustrated in Fig. 1.7. It is shown in eqn (1.4.7) that the photon-number variance is related to the mean by

$$(\Delta n)^2 = \langle n \rangle^2 + \langle n \rangle. \tag{5.4.2}$$

The first term on the right-hand side represents fluctuations additional to those of coherent light, whose variance is given by the second term on the right, equal to the mean photon number. Light whose photon-number variance exceeds $\langle n \rangle$, as in chaotic light, is said to exhibit *super-Poissonian* fluctuations. By contrast, light whose photon-number variance falls below the value $\langle n \rangle$ for coherent light is said to exhibit *sub-Poissonian* fluctuations; number-state light, with the zero variance given in eqn (5.2.1), belongs to the sub-Poissonian category.

The photon-number variance of chaotic light can be compared with its intensity variance in the classical theory, calculated in eqn (3.6.6) as equal to the square of the mean intensity. The first term on the right-hand side of eqn (5.4.2) is the analogous result in the quantum theory, and this is sometimes called the *wave contribution*. The second term on the right is the *particle contribution* $\langle n \rangle$ that occurs as the photon-number variance of coherent light in eqn (5.3.19), and it has no analogue in the classical theory. There is, however, an analogous shot noise contribution in the semiclassical photocount variance of eqn (3.9.19). The particle term is negligible in comparison to the wave term when $\langle n \rangle \gg 1$, and in this case the quantum photon-number variance and the classical intensity variance have the same structures. More generally, the geometric distribution of eqn (5.4.1) is approximated for large n and $\langle n \rangle$ as

$$P(n) = \frac{1}{1 + \langle n \rangle} \left(\frac{\langle n \rangle}{1 + \langle n \rangle} \right)^n \approx \frac{1}{\langle n \rangle} \left(1 - \frac{1}{\langle n \rangle} \right)^n \approx \frac{1}{\langle n \rangle} \exp(-n/\langle n \rangle) \tag{5.4.3}$$

where the usual exponential limit is used. This is clearly the photon-number analogue of the classical intensity distribution for chaotic light given in eqn (3.6.12).

The geometric distribution of eqn (5.4.1) has an rth photon-number factorial moment given in eqn (1.4.4) as

$$\langle n(n-1)(n-2)...(n-r+1) \rangle = r!\langle n \rangle^r. \tag{5.4.4}$$

The degree of second-order coherence of chaotic light thus follows from eqn (5.1.15) as

$$g^{(2)}(\tau) = 2, \tag{5.4.5}$$

which is the same value as found from the classical theory in eqn (3.7.25).

The quadrature and electric-field properties of chaotic light resemble those of the photon number states. Thus with the density operator

$$\hat{\rho} = \sum_n P(n)|n\rangle\langle n| \tag{5.4.6}$$

from eqn (4.6.16), the expectation value of quadrature operator $\hat{X}$ is

$$\langle X \rangle = \sum_n P(n)\langle n|\hat{X}|n\rangle = 0. \tag{5.4.7}$$

The other expectation values are similarly given by averages of the corresponding expectation values for the number states and the results are

$$\langle Y \rangle = 0, \tag{5.4.8}$$

$$(\Delta X)^2 = (\Delta Y)^2 = \tfrac{1}{2}\left(\langle n \rangle + \tfrac{1}{2}\right), \tag{5.4.9}$$

$$\langle E(\chi) \rangle = 0 \tag{5.4.10}$$

and

$$(\Delta E(\chi))^2 = \tfrac{1}{2}\left(\langle n \rangle + \tfrac{1}{2}\right). \tag{5.4.11}$$

The quadrature and field averages of chaotic light are thus the same as found for the photon number state in §5.2 but with n replaced by $\langle n \rangle$. Note however that the eigenvalue property in eqn (5.2.5) does not hold for chaotic light and its noise properties are more accurately represented by a shaded disc, similar to that shown in Fig. 3.11, than by the representation in Fig. 5.1. Chaotic light cannot carry a coherent signal and its phase is completely uncertain. The more formal phase averages of chaotic light are also the same as for the number state, with the mean and variance as given in eqn (5.2.23).

5.5 The squeezed vacuum

All of the single-mode states treated so far have phase-independent electric-field uncertainty, or noise, and the inequality of eqn (5.1.9) applies to their electric-field variances. A field excitation is said to be *quadrature-squeezed* when its electric-field variance lies in the range specified by

$$0 \le (\Delta E(\chi))^2 < \tfrac{1}{4} \tag{5.5.1}$$

for some values of the measurement phase angle χ. It follows from the inequality (5.1.5) that the occurrence of a variance smaller than 1/4 at phase angle χ must be compensated by a variance larger than 1/4 at the phase angle perpendicular to χ. We treat a simple example of a squeezed state and show that its phase-dependent noise satisfies the squeezing condition (5.5.1).

Consider the single-mode *quadrature-squeezed vacuum state* defined by

$$|\zeta\rangle = \hat{S}(\zeta)|0\rangle, \tag{5.5.2}$$

where the *squeeze operator* is

$$\hat{S}(\zeta) = \exp\!\left(\tfrac{1}{2}\zeta^* \hat{a}^2 - \tfrac{1}{2}\zeta(\hat{a}^\dagger)^2\right) \tag{5.5.3}$$

and ζ is the *complex squeeze parameter* with amplitude and phase defined by

$$\zeta = s e^{i\vartheta}. \tag{5.5.4}$$

It is seen by inspection that $\hat{S}(\zeta)$ satisfies the conditions for a unitary operator [1],

$$\hat{S}^\dagger(\zeta)\hat{S}(\zeta) = \hat{S}(\zeta)\hat{S}^\dagger(\zeta) = 1. \tag{5.5.5}$$

The definition in eqn (5.5.2) resembles that of the coherent state in eqn (5.3.11) but with the linear displacement operator of eqn (5.3.12) replaced by the squeeze operator of eqn (5.5.3), whose exponent is quadratic in the mode creation and destruction operators. We show in §9.4 that the squeezed vacuum state is generated by the nonlinear optical process of degenerate parametric down-conversion.

It is clear from the form of the exponent in eqn (5.5.3) that the squeezed vacuum state must consist of a superposition only of the number states with even n. However, the procedure of Problem 5.5 does not apply to quadratic terms in an exponent and it is necessary to use more general operator-ordering theorems. It can be shown in this way that the number-state expansion of the squeezed vacuum state, analogous to the coherent-state superposition of number states given in eqn (5.3.1), is [3]

$$|\zeta\rangle = (\mathrm{sech}\,s)^{1/2} \sum_{n=0}^{\infty} \frac{[(2n)!]^{1/2}}{n!} \left[-\tfrac{1}{2}\exp(i\vartheta)\tanh s\right]^n |2n\rangle. \tag{5.5.6}$$

The number-state expansion can be used to evaluate the expectation values of the various combinations of creation and destruction operators that occur in the photon-number and electric-field means and variances.

These expectation values are, however, sometimes more easily evaluated by an alternative method. Consider the photon-number expectation value,

$$\langle n \rangle = \langle \zeta | \hat{a}^\dagger \hat{a} | \zeta \rangle = \langle 0 | \hat{S}^\dagger(\zeta) \hat{a}^\dagger \hat{S}(\zeta) \hat{S}^\dagger(\zeta) \hat{a} \hat{S}(\zeta) | 0 \rangle, \tag{5.5.7}$$

where the unitarity condition of eqn (5.5.5) is used to insert the unit quantity $\hat{S}\hat{S}^\dagger$ in the centre.

Problem 5.9   Prove the general relation

$$e^{-\hat{O}} \hat{a} e^{\hat{O}} = \hat{a} + \left[\hat{a}, \hat{O} \right] + \tfrac{1}{2!} \left[\left[\hat{a}, \hat{O} \right], \hat{O} \right] + \ldots \tag{5.5.8}$$

by expansion of the exponentials, where $\hat{O}$ is any operator. Hence derive the transformation property

$$\hat{S}^\dagger(\zeta) \hat{a} \hat{S}(\zeta) = \hat{a} \cosh s - \hat{a}^\dagger e^{i\vartheta} \sinh s. \tag{5.5.9}$$

Equation (5.5.9) and the Hermitian conjugate expression

$$\hat{S}^\dagger(\zeta) \hat{a}^\dagger \hat{S}(\zeta) = \hat{a}^\dagger \cosh s - \hat{a} e^{-i\vartheta} \sinh s \tag{5.5.10}$$

are the basic transformations needed to evaluate expectation values. The calculation of the mean photon number from eqn (5.5.7) is thus completed to yield

$$\langle n \rangle = \sinh^2 s, \tag{5.5.11}$$

and the same result is found by use of the number-state expansion of eqn (5.5.6) [3]. The mean photon number vanishes in the absence of any squeezing, when $s = 0$, and the squeezed vacuum state then reduces to the ordinary vacuum state, but $\langle n \rangle$ increases sharply as the magnitude of the squeeze parameter is increased.

Higher-order moments of the photon number are straightforwardly evaluated with use of the basic relations in eqns (5.5.9) and (5.5.10).

Problem 5.10   Prove that the mean-square photon number for the squeezed vacuum state is given by

$$\langle n^2 \rangle = 3 \sinh^4 s + 2 \sinh^2 s = 3 \langle n \rangle^2 + 2 \langle n \rangle. \tag{5.5.12}$$

The photon-number variance is accordingly

$$(\Delta n)^2 = 2 \langle n \rangle (\langle n \rangle + 1) \tag{5.5.13}$$

and the degree of second-order coherence is obtained from eqn (5.1.15) as

$$g^{(2)}(\tau) = 3 + \frac{1}{\langle n \rangle}. \tag{5.5.14}$$

The photon-number fluctuations in squeezed-vacuum-state light are thus super-Poissonian, in accordance with the categorization given after eqn (5.4.2), and indeed they exceed the fluctuations in chaotic light with the same mean photon number. Note that the expression (5.5.14) does not reduce to the coherent state value of unity for $\langle n \rangle = 0$, when the numerator and denominator of the general expression in eqn (5.1.15) both vanish. The value of the degree of second-order coherence thus depends on the way in which the vacuum limit is approached.

The basic transformations (5.5.9) and (5.5.10) are used to establish an eigenvalue equation for the squeezed vacuum state.

Problem 5.11 Prove the eigenvalue relation

$$\left(\hat{a} \cosh s + \hat{a}^\dagger e^{i\vartheta} \sinh s \right) |\zeta\rangle = 0. \tag{5.5.15}$$

This result is similar to the vacuum-state condition of eqn (4.3.23), to which it reduces in the absence of squeezing with $s = 0$ and $\zeta = 0$.

The particular interest in the squeezed vacuum state lies not in its photon-number distribution but in its quadrature-operator properties. It is clear from eqns (5.5.9) and (5.5.10) that

$$\langle \zeta | \hat{a} | \zeta \rangle = \langle \zeta | \hat{a}^\dagger | \zeta \rangle = 0, \tag{5.5.16}$$

and indeed these results are immediate consequences of the composition of state $|\zeta\rangle$ as a superposition only of the number states with even n. The expectation values of the quadrature operators defined in eqn (4.3.36) thus also vanish

$$\langle \zeta | \hat{X} | \zeta \rangle = \langle \zeta | \hat{Y} | \zeta \rangle = 0. \tag{5.5.17}$$

Consider however the variances of the quadrature operators, which are determined by the basic expectation values

$$\langle \zeta | \hat{a}\hat{a} | \zeta \rangle = -e^{i\vartheta} \sinh s \cosh s \tag{5.5.18}$$

and

$$\langle \zeta | \hat{a}^\dagger \hat{a}^\dagger | \zeta \rangle = -e^{-i\vartheta} \sinh s \cosh s. \tag{5.5.19}$$

Problem 5.12 Prove with the use of eqns (5.3.28) and (5.3.29) that the quadrature operator variances can be expressed in the forms

$$(\Delta X)^2 = \tfrac{1}{4} \left\{ e^{2s} \sin^2 \left(\tfrac{1}{2} \vartheta \right) + e^{-2s} \cos^2 \left(\tfrac{1}{2} \vartheta \right) \right\} \tag{5.5.20}$$

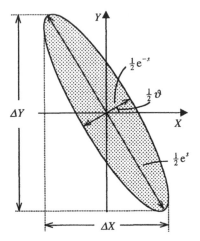

Fig. 5.7. Representation of the quadrature operator means and uncertainties for the squeezed vacuum state with s parameter given by $\exp(s) = 2$.

and

$$(\Delta Y)^2 = \tfrac{1}{4}\left\{ e^{2s}\cos^2\!\left(\tfrac{1}{2}\vartheta\right) + e^{-2s}\sin^2\!\left(\tfrac{1}{2}\vartheta\right) \right\}. \qquad (5.5.21)$$

Figure 5.7 shows a representation of the quadrature expectation values. Their zero mean values are represented by the centering of an elliptical uncertainty disc on the origin of the X and Y axes. The ellipse has major and minor axes of lengths $\exp(s)/2$ and $\exp(-s)/2$ respectively, inclined at angles $\vartheta/2$ to the axes. The square roots of the quadrature variances are represented by the projections of the disc on to these two axes and it is verified from standard properties of the ellipse [7] that the projections reproduce the expressions in eqns (5.5.20) and (5.5.21). The variances reproduce the coherent state values given in eqn (5.3.30) in the absence of squeezing with $s = 0$, but they otherwise take values that may be either greater or smaller than $1/4$.

The expectation values of the field operator (5.1.3) are easily obtained by very similar calculations. The mean field, or coherent signal, vanishes,

$$S = \langle \zeta | \hat{E}(\chi) | \zeta \rangle = 0, \qquad (5.5.22)$$

and the field variance, or noise, is

$$\mathcal{N} = (\Delta E(\chi))^2 = \tfrac{1}{4}\left\{ e^{2s}\sin^2\!\left(\chi - \tfrac{1}{2}\vartheta\right) + e^{-2s}\cos^2\!\left(\chi - \tfrac{1}{2}\vartheta\right) \right\}. \qquad (5.5.23)$$

The noise is clearly phase dependent and Fig. 5.8 illustrates the variation of the electric field with $\chi - (\vartheta/2)$, where the zero mean value lies along the horizontal axis and the separation between long-dashed curves represents the field uncertainty obtained from the square root of eqn (5.5.23). The short-dashed lines show the

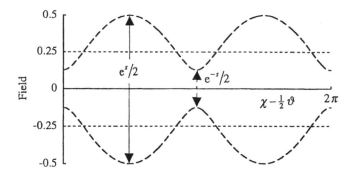

Fig. 5.8. The envelope of the noise band of the squeezed vacuum state for $\exp(s) = 2$ is shown by the long-dashed lines. The short-dashed lines show the noise band for the coherent vacuum state with $s = 0$.

corresponding field uncertainty of the coherent vacuum state obtained when $s = 0$. It is seen from eqn (5.5.23) that the minimum value of the uncertainty is

$$\Delta E_{\min} = \Delta E\left(\tfrac{1}{2}\vartheta + m\pi\right) = \tfrac{1}{2}e^{-s}, \tag{5.5.24}$$

where m is zero or an integer, and this value indicates a field that is quadrature-squeezed in accordance with the definition in eqn (5.5.1). The maximum value is

$$\Delta E_{\max} = \Delta E\left(\tfrac{1}{2}(\vartheta + \pi) + m\pi\right) = \tfrac{1}{2}e^{s}, \tag{5.5.25}$$

where m is again zero or an integer. The product of these extremal values,

$$\Delta E_{\min}\Delta E_{\max} = \tfrac{1}{4}, \tag{5.5.26}$$

satisfies the uncertainty relation (5.1.5) as an equality.

Problem 5.13 Prove that the squeezed vacuum state is quadrature-squeezed for all values of the state parameters s and ϑ and of the measurement phase angle χ that satisfy the condition

$$\cos(2\chi - \vartheta) > \tanh s. \tag{5.5.27}$$

Show that for strong squeezing with $s \gg 1$, the smallest phase differences that satisfy this condition are given approximately by

$$\left|\chi - \tfrac{1}{2}\vartheta\right| < e^{-s}. \tag{5.5.28}$$

The range of phase angles for which squeezing occurs thus diminishes with increasing magnitude of the squeeze parameter.

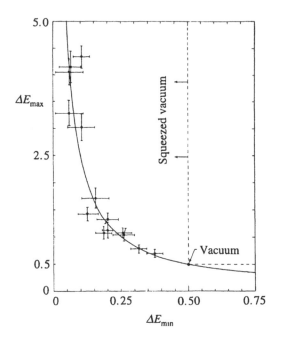

Fig. 5.9. The minimum and maximum field uncertainties for squeezed vacuum states generated by parametric down-conversion, determined by scans of the measurement phase angle χ. Different points refer to squeezed vacuum states with different magnitudes s of the squeeze parameter. The continuous curve shows the relation (5.5.26). (After [8])

Figure 5.9 shows values of the minimum and maximum field uncertainties inferred from measurements on a range of quadrature-squeezed vacuum states generated by parametric down-conversion [8]. The experimental points lie close to the continuous hyperbola, which is a plot of the theoretical result of eqn (5.5.26) for the product of minimum and maximum uncertainties. The parametric down-conversion process is treated in §9.4.

5.6 Squeezed coherent states

The reduced noise of the quadrature-squeezed vacuum state relative to the coherent state, for appropriate phase angles, is potentially valuable in practical applications. However, the absence of any coherent signal for the squeezed vacuum state, shown by eqn (5.5.22), limits its use as a carrier of information in systems that use homodyne detection. This feature is overcome in the single-mode *quadrature-squeezed coherent state*, defined by

$$|\alpha,\zeta\rangle = \hat{D}(\alpha)\hat{S}(\zeta)|0\rangle, \tag{5.6.1}$$

where $\hat{D}(\alpha)$ is the coherent-state displacement operator from eqn (5.3.12) and $\hat{S}(\zeta)$ is the squeeze operator from eqn (5.5.3). We show that the squeezed coherent state retains the reduced noise of the quadrature-squeezed vacuum state but it also acquires the nonzero signal of the ordinary coherent state. The form of the definition in eqn (5.6.1) parallels that of the *ideal squeezed state* [9]. An equivalent form of definition of the *two-photon coherent state* [10] reverses the order of the $\hat{D}(\alpha)$ and $\hat{S}(\zeta)$ operators. The account that follows uses the definitions and the notation of [9], but many of the basic properties of the squeezed coherent state are derived in [10]. The relations between the two formalisms are discussed in [11,12].

The number-state expansion of the squeezed coherent state is considerably more complicated than that of the squeezed vacuum state, given in eqn (5.5.6), and its form [11,12] is not reproduced here. Expectation values for the squeezed coherent state can be derived by simple generalizations of the basic transformations given by eqns (5.5.9) and (5.5.10) for the squeezed vacuum state. Use of the properties of the coherent-state displacement operator given in eqns (5.3.14) and (5.3.15) leads to the relations

$$\hat{S}^\dagger(\zeta)\hat{D}^\dagger(\alpha)\hat{a}\hat{D}(\alpha)\hat{S}(\zeta) = \hat{a}\cosh s - \hat{a}^\dagger e^{i\vartheta}\sinh s + \alpha \tag{5.6.2}$$

and

$$\hat{S}^\dagger(\zeta)\hat{D}^\dagger(\alpha)\hat{a}^\dagger\hat{D}(\alpha)\hat{S}(\zeta) = \hat{a}^\dagger\cosh s - \hat{a}e^{-i\vartheta}\sinh s + \alpha^*. \tag{5.6.3}$$

These transformations provide an eigenvalue relation for the squeezed coherent state by the same procedure as used in the derivation of eqn (5.5.15) for the squeezed vacuum state, and the result is

$$\left(\hat{a}\cosh s + \hat{a}^\dagger e^{i\vartheta}\sinh s\right)|\alpha,\zeta\rangle = \left(\alpha\cosh s + \alpha^* e^{i\vartheta}\sinh s\right)|\alpha,\zeta\rangle. \tag{5.6.4}$$

The eigenvalue equation reduces to that for the squeezed vacuum state in eqn (5.5.15) when $\alpha = 0$ and to that for the coherent state in eqn (5.3.5) when $s = 0$.

The calculation of the mean photon number follows the same procedure as in eqns (5.5.7) to (5.5.11) and the result is

$$\langle n \rangle = |\alpha|^2 + \sinh^2 s. \tag{5.6.5}$$

The mean photon number is thus the sum of contributions identical to those of the coherent state in eqn (5.3.16) and of the squeezed vacuum state in eqn (5.5.11). The photon-number variance is similarly obtained as

$$(\Delta n)^2 = |\alpha|^2\left\{e^{2s}\sin^2\left(\theta - \tfrac{1}{2}\vartheta\right) + e^{-2s}\cos^2\left(\theta - \tfrac{1}{2}\vartheta\right)\right\} + 2\sinh^2 s\left(\sinh^2 s + 1\right), \tag{5.6.6}$$

and the expression reduces to the results given in eqn (5.3.19) for the coherent state and in eqn (5.5.13) for the squeezed vacuum state in the appropriate limits. An expression for the degree of second-order coherence of the quadrature-squeezed coherent state is easily written down by substitution of eqns (5.6.5) and (5.6.6) into eqn (5.1.15).

The main interest of the squeezed coherent state again lies in its quadrature operator properties rather than in its photon-number statistics. It follows from eqns (5.6.1) to (5.6.3) that

$$\langle \alpha, \zeta | \hat{X} | \alpha, \zeta \rangle = \operatorname{Re} \alpha = |\alpha| \cos \theta \tag{5.6.7}$$

and

$$\langle \alpha, \zeta | \hat{Y} | \alpha, \zeta \rangle = \operatorname{Im} \alpha = |\alpha| \sin \theta, \tag{5.6.8}$$

identical to the results of eqns (5.3.25) and (5.3.27) for the ordinary coherent state. Similarly, it follows with use of eqns (5.3.28) and (5.3.29) that the quadrature variances are

$$(\Delta X)^2 = \tfrac{1}{4} \left\{ e^{2s} \sin^2\left(\tfrac{1}{2}\vartheta\right) + e^{-2s} \cos^2\left(\tfrac{1}{2}\vartheta\right) \right\} \tag{5.6.9}$$

and

$$(\Delta Y)^2 = \tfrac{1}{4} \left\{ e^{2s} \cos^2\left(\tfrac{1}{2}\vartheta\right) + e^{-2s} \sin^2\left(\tfrac{1}{2}\vartheta\right) \right\}, \tag{5.6.10}$$

identical to the results of eqns (5.5.20) and (5.5.21) for the squeezed vacuum state. Figure 5.10 illustrates the development of the quadrature properties from the factors on the right-hand side of eqn (5.6.1), beginning with the circular noise

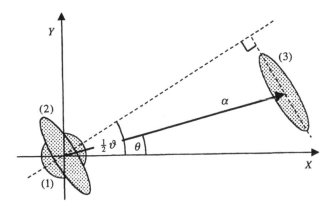

Fig. 5.10. Representations of the quadrature operator means and variances for (1) the ordinary vacuum state, (2) the squeezed vacuum state, and (3) the squeezed coherent state.

disc of the ordinary vacuum state centered on the origin in the first step, converted into the elliptical noise disc of the squeezed vacuum state by application of the operator $\hat{S}(\zeta)$ in the second step, and transformed into the shifted elliptical noise disc of the squeezed coherent state by application of the operator $\hat{D}(\alpha)$ in the third and final step. The figure shows how the mean values of the quadrature operators come to be determined solely by the coherent parameter α, while the variances of the operators are determined solely by the squeeze parameter ζ. The properties of the noise ellipse are identical to those for the squeezed vacuum state illustrated in Fig. 5.7.

The expectation values of the electric field operator of eqn (5.1.3) are easily obtained from the above results. Thus the mean field, or coherent signal, is

$$S = \langle \alpha | \hat{E}(\chi) | \alpha \rangle = |\alpha| \cos(\chi - \theta), \tag{5.6.11}$$

identical to the coherent-state signal of eqn (5.3.31), and the field variance, or noise, is

$$\mathcal{N} = (\Delta E(\chi))^2 = \tfrac{1}{4} \left\{ e^{2s} \sin^2\!\left(\chi - \tfrac{1}{2}\vartheta\right) + e^{-2s} \cos^2\!\left(\chi - \tfrac{1}{2}\vartheta\right) \right\}, \tag{5.6.12}$$

identical to the phase-dependent squeezed-vacuum noise of eqn (5.5.23). The signal-to-noise ratio, as defined in eqn (5.1.12), is

$$\mathrm{SNR} = \frac{4|\alpha|^2 \cos^2(\chi - \theta)}{e^{2s} \sin^2\!\left(\chi - \tfrac{1}{2}\vartheta\right) + e^{-2s} \cos^2\!\left(\chi - \tfrac{1}{2}\vartheta\right)}. \tag{5.6.13}$$

The signal and the noise are controlled by distinct phase angles θ and ϑ respectively, whose relative values depend on the method of generation of the squeezed coherent state. The source of the light can in principle be adjusted to optimize a desired feature of the beam. Thus the maximum signal-to-noise ratio is

$$\mathrm{SNR}_{\max} = 4e^{2s}|\alpha|^2 \quad \text{for} \quad \chi = \theta = \tfrac{1}{2}\vartheta, \tag{5.6.14}$$

which represents an enhancement of the ordinary coherent-state value by the exponential factor. The quadrature-squeezed coherent state benefits from the best of both worlds, in that its coherent signal component is the same as that of the coherent state but its noise can in principle be reduced significantly below the coherent-state value, matching the noise of the quadrature-squeezed vacuum state.

Figure 5.11 shows a representation of the field expectation values for phase angles chosen so that $\theta = \vartheta/2$, when the minor axis of the noise ellipse is parallel to the coherent amplitude and the maximum value of signal-to-noise ratio is achieved, as in eqn (5.6.14). The conventions in the figure are the same as those in Figs. 5.4 for the coherent state and 5.7 for the squeezed vacuum state. In particular, the minor and major axes of the elliptical noise disc show its resolution into amplitude and phase contributions. The light in this case is said to be *amplitude-squeezed*. Simple expressions for the photon-number and phase

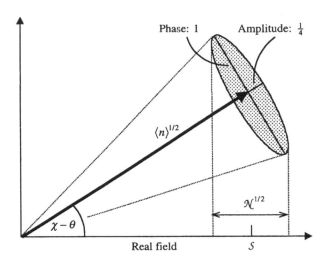

Fig. 5.11. Representation of a single-mode amplitude-squeezed coherent state with $\langle n \rangle = 4$, $\theta = \vartheta/2$ and $\exp(s) = 2$, showing the mean and the uncertainty in the electric field.

uncertainties occur in the limit where the coherent contribution to the mean photon number in eqn (5.6.5) greatly exceeds that of the squeezing, that is

$$|\alpha| \gg e^s, \quad \langle n \rangle \approx |\alpha|^2. \tag{5.6.15}$$

Then, analogous to eqn (5.3.34) for the coherent state,

$$\Delta n = \left(\langle n \rangle^{1/2} + \tfrac{1}{4}e^{-s} \right)^2 - \left(\langle n \rangle^{1/2} - \tfrac{1}{4}e^{-s} \right)^2 = \langle n \rangle^{1/2} e^{-s}, \tag{5.6.16}$$

in agreement with the result of eqn (5.6.6) for the same choice of phase angles and in the same limit given by eqn (5.6.15). The uncertainty in phase angle, analogous to eqn (5.3.35), is correspondingly

$$\Delta\varphi = \frac{e^s}{2|\alpha|} = \frac{e^s}{2\langle n \rangle^{1/2}}, \tag{5.6.17}$$

and the product of photon number and phase uncertainties is again

$$\Delta n \Delta\varphi = \tfrac{1}{2}. \tag{5.6.18}$$

It is seen that, in comparison with the coherent state, the amplitude-squeezed coherent state for $\theta = \vartheta/2$ has a reduced photon-number uncertainty and an enhanced phase uncertainty. The photon-number fluctuations in this case are sub-Poissonian. In the limit of large $|\alpha|$, the phase uncertainty in eqn (5.6.17) agrees with the value calculated more formally from the phase probability distribution of

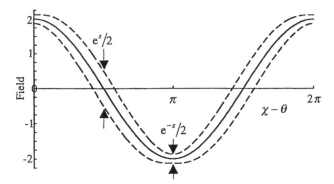

Fig. 5.12. Phase dependence of the electric field of the same amplitude-squeezed coherent state as in Fig. 5.11, showing the cosine variation of the mean field and the phase-dependent noise band.

the squeezed coherent state by a procedure parallel to that used for the coherent state in eqns (5.3.37) to (5.3.39), but we do not reproduce the details here.

Problem 5.14 Show that sub-Poissonian photon-number fluctuations occur in the amplitude-squeezed coherent state when the condition

$$|\alpha|^2 > \tfrac{1}{4}\left(e^{2s} - 1\right)\cosh(2s). \tag{5.6.19}$$

is satisfied.

The phase dependence of the electric field of the same amplitude-squeezed coherent state as in Fig. 5.11 is illustrated in Fig. 5.12. The continuous cosine wave represents the mean field of eqn (5.6.11) and the dashed lines show the envelope of the noise band obtained from the square root of eqn (5.6.12). The reduction in amplitude uncertainty and the increase in phase uncertainty are clearly evident in the comparison of this figure with Fig. 5.5. If the coherent amplitude is set equal to zero, the mean field vanishes but the noise band of the squeezed vacuum state, illustrated in Fig. 5.8, remains.

Figure 5.13 shows a representation of the electric field expectation values for an alternative choice of phase angles with $\theta = (\vartheta + \pi)/2$. The light in this case is said to be *phase-squeezed*. The effects of the squeezing on the amplitude and phase uncertainties are now interchanged, with the approximate magnitudes

$$\Delta n = \langle n \rangle^{1/2} e^{s} \tag{5.6.20}$$

and

$$\Delta \varphi = e^{-s}/2\langle n \rangle^{1/2}, \tag{5.6.21}$$

valid in the limit given in eqn (5.6.15). The amplitude uncertainty, represented by

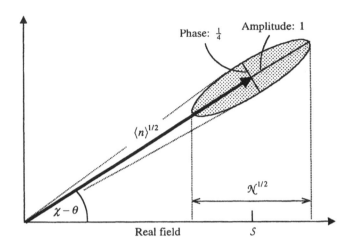

Fig. 5.13. Representation of a single-mode phase-squeezed coherent state similar to Fig. 5.11 but with $\theta = (\vartheta + \pi)/2$.

the major axis of the noise ellipse, is enhanced and the phase uncertainty, represented by the minor axis, is reduced relative to the coherent state.

These features of the phase-squeezed coherent state are evident in trace (d) of Fig. 5.14, which shows experimental results [13] for the noise. The measurements are made by homodyne detection with a local-oscillator phase angle that varies linearly with the time. The first three traces show respectively the coherent state, the squeezed vacuum state and the amplitude-squeezed coherent state. Their forms resemble the theoretical variations shown in Figs. 5.5, 5.8 and 5.12 respectively. Note, however, the enhanced 'pinching' of the measured noise band at the maxima and minima of the cosine wave in trace (c) compared to the calculated form shown in Fig. 5.12; the enhancement results from the larger ratio of squeezing s to coherent amplitude $|\alpha|$ in the experiment. More complicated graphs occur when the coherent phase θ and the squeezing phase ϑ are not related in any simple way, and an example is shown in trace (e) of Fig. 5.14. The distinctions between amplitude, phase and intermediate varieties of quadrature squeezing disappear as the coherent amplitude $|\alpha|$ is reduced below the squeezing exponential. The squeezed coherent states all tend to squeezed vacuum states for $|\alpha| \to 0$.

5.7 Beam-splitter input–output relations

The optical beam splitter is an important component in many of the experiments that study the quantum nature of light, as displayed, for example, by the different varieties of single-mode state treated in the present chapter. Thus the Mach–Zehnder and Brown–Twiss interference experiments, originally explained in the classical theory, take on additional significances when they are interpreted in the quantum theory. The latter experiment in particular plays a key role in the observation of nonclassical effects. Again, beam splitters are central to the measurement

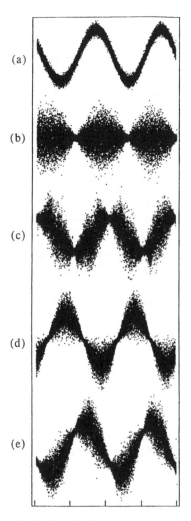

Fig. 5.14. Measured electric fields of (a) a coherent state, (b) a squeezed vacuum state, (c) an amplitude-squeezed state, (d) a phase-squeezed state and (e) a squeezed state with 48° between the coherent vector and the axis of the noise ellipse. The scale on the horizontal axis is proportional to the local oscillator phase. (After [13])

of the two-photon interference phenomena that provide illustrations of some of the most fundamental facets of quantum mechanics (see §6.8). The essential property of the beam splitter is its ability to convert an incoming photon state into a linear superposition of output states, a basic quantum-mechanical manipulation that is less easily achieved and studied in other physical systems.

Figure 5.15 shows a representation of the beam splitter with its input and output arms labelled by the destruction operators that enter the respective quan-

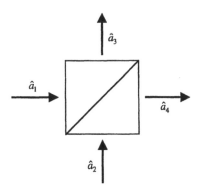

Fig. 5.15. Representation of a lossless beam-splitter showing the notation for the destruction operators associated with the input and output fields.

tized field operators. The relations between the classical input and output fields at a beam splitter are discussed in §3.2 on the basis of the conservation of the classical energy flow. These relations are ultimately determined by the boundary conditions for the electromagnetic fields at the partially reflecting and transmitting interface within the beam splitter. The boundary conditions are the same for the classical fields and for the quantum-mechanical field operators. It follows that the basic relations (3.2.12) satisfied by the reflection and transmission coefficients of a symmetric beam splitter for monochromatic incident fields, namely

$$|\mathcal{R}|^2 + |\mathcal{T}|^2 = 1 \quad \text{and} \quad \mathcal{R}\mathcal{T}^* + \mathcal{T}\mathcal{R}^* = 0, \tag{5.7.1}$$

remain the same. Similarly, the relations (3.2.1) between the classical input and output fields convert into analogous relations between the quantized field operators. For a symmetric beam splitter, these are expressed in the relations

$$\hat{a}_3 = \mathcal{R}\hat{a}_1 + \mathcal{T}\hat{a}_2 \quad \text{and} \quad \hat{a}_4 = \mathcal{T}\hat{a}_1 + \mathcal{R}\hat{a}_2 \tag{5.7.2}$$

between the destruction operators for the input and output modes. The inverse relations

$$\hat{a}_1 = \mathcal{R}^*\hat{a}_3 + \mathcal{T}^*\hat{a}_4 \quad \text{and} \quad \hat{a}_2 = \mathcal{T}^*\hat{a}_3 + \mathcal{R}^*\hat{a}_4 \tag{5.7.3}$$

are sometimes useful, and these are readily obtained from eqn (5.7.2) with the use of eqn (5.7.1). The corresponding relations between the input and output creation operators are given by the Hermitian conjugates of eqns (5.7.2) and (5.7.3).

We assume that the input fields in arms 1 and 2 are independent, with creation and destruction operators that satisfy the boson commutation relations

$$\left[\hat{a}_1, \hat{a}_1^\dagger\right] = \left[\hat{a}_2, \hat{a}_2^\dagger\right] = 1 \tag{5.7.4}$$

and

$$\left[\hat{a}_1, \hat{a}_2^\dagger\right] = \left[\hat{a}_2, \hat{a}_1^\dagger\right] = 0. \tag{5.7.5}$$

Then with the use of eqns (5.7.1) and (5.7.2),

$$\left[\hat{a}_3, \hat{a}_3^\dagger\right] = \left[\mathcal{R}\hat{a}_1 + \mathcal{T}\hat{a}_2, \mathcal{R}^*\hat{a}_1^\dagger + \mathcal{T}^*\hat{a}_2^\dagger\right] = |\mathcal{R}|^2 + |\mathcal{T}|^2 = 1, \tag{5.7.6}$$

$$\left[\hat{a}_3, \hat{a}_4^\dagger\right] = \left[\mathcal{R}\hat{a}_1 + \mathcal{T}\hat{a}_2, \mathcal{T}^*\hat{a}_1^\dagger + \mathcal{R}^*\hat{a}_2^\dagger\right] = \mathcal{R}\mathcal{T}^* + \mathcal{T}\mathcal{R}^* = 0 \tag{5.7.7}$$

and similarly

$$\left[\hat{a}_4, \hat{a}_4^\dagger\right] = 1. \tag{5.7.8}$$

Thus the output mode operators also have independent boson commutation relations. An alternative, and more fundamental, approach to the beam-splitter theory begins with the basic requirement that the output mode operators should have independent boson commutators, when the relations (5.7.1) between the reflection and transmission coefficients follow as consequences.

The input–output relations (5.7.2) provide immediate connections between the single-mode electric-field operators, as defined by eqn (5.1.3), for the four arms of the beam splitter. Thus with the arm labels denoted by subscripts on the fields, it follows that

$$\begin{cases} \hat{E}_3(\chi) = |\mathcal{R}|\hat{E}_1(\chi - \phi_\mathcal{R}) + |\mathcal{T}|\hat{E}_2(\chi - \phi_\mathcal{T}) \\ \hat{E}_4(\chi) = |\mathcal{T}|\hat{E}_1(\chi - \phi_\mathcal{T}) + |\mathcal{R}|\hat{E}_2(\chi - \phi_\mathcal{R}), \end{cases} \tag{5.7.9}$$

where $\phi_\mathcal{R}$ and $\phi_\mathcal{T}$ are the phases of the beam-splitter reflection and transmission coefficients as defined in eqn (3.2.11). The mean values of the input and output fields satisfy the same relations as in eqn (5.7.9). It is easily shown that their variances, defined as in eqn (5.1.6), are related by

$$\begin{cases} \left(\Delta E_3(\chi)\right)^2 = |\mathcal{R}|^2\left(\Delta E_1(\chi - \phi_\mathcal{R})\right)^2 + |\mathcal{T}|^2\left(\Delta E_2(\chi - \phi_\mathcal{T})\right)^2 \\ \left(\Delta E_4(\chi)\right)^2 = |\mathcal{T}|^2\left(\Delta E_1(\chi - \phi_\mathcal{T})\right)^2 + |\mathcal{R}|^2\left(\Delta E_2(\chi - \phi_\mathcal{R})\right)^2, \end{cases} \tag{5.7.10}$$

where the input fields are assumed to be uncorrelated. The beam splitter thus transmits the field fluctuations with appropriate coefficients $|\mathcal{R}|^2$ and $|\mathcal{T}|^2$ and changes of phase $\phi_\mathcal{R}$ and $\phi_\mathcal{T}$. These phase shifts in the noise are of course only significant for input states with phase-dependent noise, for example the squeezed states treated in §§5.5 and 5.6. The relations (5.7.10) represent a kind of conservation of quadrature noise, or fluctuations, between the beam-splitter input and output arms.

The photon number operators for the beam-splitter arms are defined as

$$\hat{n}_i = \hat{a}_i^\dagger \hat{a}_i \quad (i = 1,2,3,4) \tag{5.7.11}$$

and it follows from eqn (5.7.2) that

$$\hat{n}_3 = |\mathcal{R}|^2 \hat{a}_1^\dagger \hat{a}_1 + \mathcal{R}^* \mathcal{T} \hat{a}_1^\dagger \hat{a}_2 + \mathcal{T}^* \mathcal{R} \hat{a}_2^\dagger \hat{a}_1 + |\mathcal{T}|^2 \hat{a}_2^\dagger \hat{a}_2 \tag{5.7.12}$$

and

$$\hat{n}_4 = |\mathcal{T}|^2 \hat{a}_1^\dagger \hat{a}_1 + \mathcal{T}^* \mathcal{R} \hat{a}_1^\dagger \hat{a}_2 + \mathcal{R}^* \mathcal{T} \hat{a}_2^\dagger \hat{a}_1 + |\mathcal{R}|^2 \hat{a}_2^\dagger \hat{a}_2. \tag{5.7.13}$$

Addition of eqn (5.7.12) to eqn (5.7.13) with the use of eqn (5.7.1) gives

$$\hat{n}_3 + \hat{n}_4 = \hat{n}_1 + \hat{n}_2, \tag{5.7.14}$$

which represents photon-number conservation between the input and output arms of the beam splitter. Again, the basic relations (5.7.1) for the beam splitter reflection and transmission coefficients could be derived by requiring the validity of the conservation law (5.7.14) and indeed this is the quantum analogue of the classical energy-conservation method used to derive the relations in §3.2.

The effects of transmission through the beam splitter on the photon number fluctuations are calculated with the use of eqns (5.7.12) and (5.7.13). The expressions for the photon-number variances of the two output arms are quite complicated for general input states and we restrict attention to the common experimental arrangement where only one of the inputs is illuminated. Thus with the arm 2 input in its vacuum state, the output variances are

$$\begin{cases} (\Delta n_3)^2 = |\mathcal{R}|^4 (\Delta n_1)^2 + |\mathcal{R}|^2 |\mathcal{T}|^2 \langle n_1 \rangle \\ (\Delta n_4)^2 = |\mathcal{T}|^4 (\Delta n_1)^2 + |\mathcal{T}|^2 |\mathcal{R}|^2 \langle n_1 \rangle. \end{cases} \tag{5.7.15}$$

Each output photon-number variance has a contribution from the input variance, with appropriate scaling, plus an additional contribution proportional to the mean input photon number. This second contribution is regarded as generated by a beating of the input field in arm 1 with the vacuum field fluctuation in arm 2, or as a *partition noise* caused by the random division of the input photon stream at the beam splitter with probabilities $|\mathcal{R}|^2$ and $|\mathcal{T}|^2$ for the two output arms.

5.8 Single-photon input

The simplest application of the beam-splitter input–output relations occurs when a single photon is incident in arm 1, with arm 2 in its vacuum state. The input state is denoted

$$|1\rangle_1|0\rangle_2 = \hat{a}_1^\dagger|0\rangle, \tag{5.8.1}$$

where we have made use of the standard ladder property (4.3.33) of the harmonic oscillator states and $|0\rangle$ denotes the joint vacuum states of the beam-splitter arms. The input state is converted to the corresponding output state by use of the Hermitian conjugate of eqn (5.7.3)

$$|1\rangle_1|0\rangle_2 = \left(\mathcal{R}\hat{a}_3^\dagger + \mathcal{T}\hat{a}_4^\dagger\right)|0\rangle = \mathcal{R}|1\rangle_3|0\rangle_4 + \mathcal{T}|0\rangle_3|1\rangle_4. \tag{5.8.2}$$

This conversion of the input state to a linear superposition of the two possible output states is the basic quantum-mechanical process performed by the beam splitter. A state of the form shown on the right-hand side of eqn (5.8.2), with the property that each contribution to the superposition is a product of states for different systems (output arms), is said to be *entangled*. Note that although the state as a whole is pure, the individual output arms are not in pure states. Both product states in the superposition represent a photon in one arm and none in the other. Such superpositions occur only in quantum mechanics and they have no analogues in the classical theory. The entanglement is responsible for the importance of the beam splitter in an extensive range of experiments in quantum optics.

Consider first the effects on the beam-splitter output state of observations made on one of the output arms; for example, arm 3. According to von Neumann measurement theory [14], the state of the system *after* the observation is given by the projection of the state *before* the observation on to the state determined by the measurement. Suppose that the observation finds the output in arm 3 to be in its vacuum state $|0\rangle_3$. The state of the system before the observation is given by eqn (5.8.2) and the state after the observation is

$$N_3\langle 0|\{\mathcal{R}|1\rangle_3|0\rangle_4 + \mathcal{T}|0\rangle_3|1\rangle_4\} = N\mathcal{T}|1\rangle_4 = |1\rangle_4, \tag{5.8.3}$$

where N is a normalization constant, here equal to $1/\mathcal{T}$. In words, the state of the beam-splitter output *conditioned* on the observation of the vacuum state in arm 3 is a single photon in arm 4. This description of the effects of a measurement extends to arbitrary beam-splitter input states [15].

Now consider the electric-field expectation values for the output state on the right-hand side of eqn (5.8.2). It is easily shown, with use of the definition of the single-mode field operator in eqn (5.1.3), that the mean fields vanish in all arms of the beam splitter. The field variances are also calculated without difficulty.

Problem 5.15 Prove that

$$\begin{cases} \left(\Delta E_3(\chi_3)\right)^2 = \tfrac{1}{2}|\mathcal{R}|^2 + \tfrac{1}{4} \\ \left(\Delta E_4(\chi_4)\right)^2 = \tfrac{1}{2}|\mathcal{T}|^2 + \tfrac{1}{4}, \end{cases} \tag{5.8.4}$$

in agreement with eqn (5.7.10) when the input variances are

inserted from eqns (5.1.8) and (5.2.9). Show also that the correlation between output fields is

$$\left\langle \hat{E}_3(\chi_3)\hat{E}_4(\chi_4)\right\rangle = \tfrac{1}{2}|\mathcal{R}||T|\cos\left(\phi_{\mathcal{R}} - \phi_{T} - \chi_3 + \chi_4\right). \quad (5.8.5)$$

This nonzero correlation is important for the formation of fringes in the Mach–Zehnder interferometer.

The mean photon numbers in the two output arms are determined by replacement of output operators by input operators with the use of eqn (5.7.3). Thus,

$$\left\langle n_3\right\rangle = {}_2\langle 0|_1\langle 1|\hat{n}_3|1\rangle_1|0\rangle_2 = {}_2\langle 0|_1\langle 1|\left(\mathcal{R}^*\hat{a}_1^\dagger + T^*\hat{a}_2^\dagger\right)\left(\mathcal{R}\hat{a}_1 + T\hat{a}_2\right)|1\rangle_1|0\rangle_2 = |\mathcal{R}|^2,$$

$$(5.8.6)$$

where the ground state property of eqn (4.3.23) is used for arm 2, and similarly

$$\left\langle n_4\right\rangle = |T|^2. \quad (5.8.7)$$

Alternatively, the same results are obtained by evaluating the number operators expressed in terms of creation and destruction operators for arms 3 and 4 for the output state on the right of eqn (5.8.2).

These expressions for the mean output photon numbers resemble the corresponding classical results for the division of the input electromagnetic energy in accordance with the intensity reflection and transmission coefficients $|\mathcal{R}|^2$ and $|T|^2$ respectively. However, a result very different from the classical theory is found for the correlation between output photon numbers, as measured in the quantum interpretation of the Brown–Twiss interferometer. Consider a series of identical experiments for the single input photon, in which the numbers of photons observed in arms 3 and 4 are multiplied together. The quantum-mechanical average for this product is

$$\begin{aligned}
\left\langle n_3 n_4\right\rangle &= {}_2\langle 0|_1\langle 1|\hat{n}_3\hat{n}_4|1\rangle_1|0\rangle_2 \\
&= {}_2\langle 0|_1\langle 1|\left(\mathcal{R}^*\hat{a}_1^\dagger + T^*\hat{a}_2^\dagger\right)\left(\mathcal{R}\hat{a}_1 + T\hat{a}_2\right)\left(T^*\hat{a}_1^\dagger + \mathcal{R}^*\hat{a}_2^\dagger\right)\left(T\hat{a}_1 + \mathcal{R}\hat{a}_2\right)|1\rangle_1|0\rangle_2 \\
&= \mathcal{R}^*\mathcal{R}T^*T\,{}_1\langle 1|\hat{a}_1^\dagger\hat{a}_1\hat{a}_1^\dagger\hat{a}_1|1\rangle_1 + \mathcal{R}^*T\mathcal{R}^*T\,{}_2\langle 0|\hat{a}_2\hat{a}_2^\dagger|0\rangle_2\,{}_1\langle 1|\hat{a}_1^\dagger\hat{a}_1|1\rangle_1 \\
&= \mathcal{R}^*\left(\mathcal{R}T^* + T\mathcal{R}^*\right)T = 0,
\end{aligned}$$

$$(5.8.8)$$

where eqn (5.7.1) is used. The zero average is a clear consequence of the particle-like aspect of the single photon, in that its presence in one output arm requires its absence in the other and each experimental run produces a correlation 1×0 or 0×1. By contrast, no input field that can be described by the classical theory is capable of producing a zero correlation of output intensities.

Now consider the more complicated example of a Mach–Zehnder interfero-

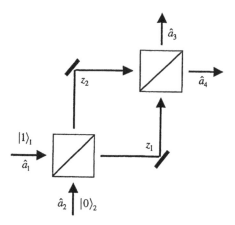

Fig. 5.16. Representation of a Mach–Zehnder interferometer showing the notation for input and output field operators and for the internal path lengths.

meter, as represented in Fig. 5.16. The two beam splitters are assumed to be symmetrical and identical, with the properties given in eqn (5.7.1). The complete interferometer can be regarded as a composite beam splitter, whose two output operators are related to the two input operators by

$$\hat{a}_3 = \mathcal{R}_{MZ}\hat{a}_1 + \mathcal{T}_{MZ}\hat{a}_2 \quad \text{and} \quad \hat{a}_4 = \mathcal{T}_{MZ}\hat{a}_1 + \mathcal{R}'_{MZ}\hat{a}_2. \tag{5.8.9}$$

The composite reflection and transmission coefficients are given by

$$\begin{cases} \mathcal{R}_{MZ} = \mathcal{R}^2 e^{ikz_1} + \mathcal{T}^2 e^{ikz_2}, \quad \mathcal{R}'_{MZ} = \mathcal{T}^2 e^{ikz_1} + \mathcal{R}^2 e^{ikz_2} \\ \mathcal{T}_{MZ} = \mathcal{R}\mathcal{T}\left(e^{ikz_1} + e^{ikz_2}\right), \end{cases} \tag{5.8.10}$$

where $k = \omega/c$ is the optical wavevector and z_1 and z_2 are the lengths of the two internal paths of the interferometer. It is readily verified that these composite coefficients satisfy the relations

$$|\mathcal{R}_{MZ}|^2 + |\mathcal{T}_{MZ}|^2 = |\mathcal{R}'_{MZ}|^2 + |\mathcal{T}_{MZ}|^2 = 1 \quad \text{and} \quad \mathcal{R}'_{MZ}\mathcal{T}^*_{MZ} + \mathcal{T}_{MZ}\mathcal{R}^*_{MZ} = 0, \tag{5.8.11}$$

in accordance with the general beam-splitter properties given by eqns (3.2.6) and (3.2.8). The Mach–Zehnder interferometer as a whole is *not* equivalent to a symmetric beam splitter. The amplitudes of the two composite reflection coefficients in eqn (5.8.10) are the same but their phases are different. The mean photon number in output arm 4 is now given by eqn (5.8.7) but with substitution of the composite transmission coefficient,

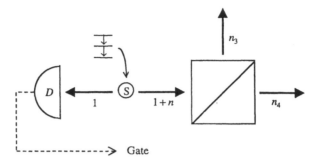

Fig. 5.17. Brown–Twiss interferometer using a single-photon input obtained from cascade emission with an electronic gate.

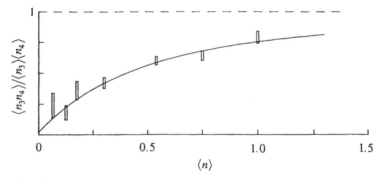

Fig. 5.18. Normalized output correlation, or degree of second-order coherence, as a function of the additional mean photon number $\langle n \rangle$, as measured in the experiment represented in Fig. 5.17. The continuous curve shows the function derived in eqn (5.9.6). (After [4])

$$\langle n_4 \rangle = |\mathcal{T}_{MZ}|^2 = 4|\mathcal{R}|^2|\mathcal{T}|^2 \cos^2\left[\tfrac{1}{2}k(z_1 - z_2)\right]. \tag{5.8.12}$$

The fringe pattern has the same dependence on wavevector and path difference as found in the classical theory, given by eqn (3.5.3) in conditions where the light that travels through the interferometer by the two internal paths has first-order coherence.

Both of the single-photon experiments outlined above have been performed [4] and some details are shown in Figs. 5.17 to 5.19. The single-photon inputs are obtained from cascade emission by atomic Ca, in which two photons are emitted in succession, as described in detail in §8.6. As is represented in Fig. 5.17, the detection of one of the photons on the left-hand side is used to operate an electronic gate that activates a measurement on the other photon of the pair on the right-hand side. The experiment measures the normalized correlation of the photon numbers n_3 and n_4 in the two output arms of the beam splitter, which is expected to vanish in accordance with eqn (5.8.8). However, in addition to the twin of the photon that opens the gate, a further n photons may enter the apparatus during the time for which the gate remains open. Figure 5.18 shows the normalized correlation as a function of the mean number $\langle n \rangle$ of additional photons received by the gate detector during the detection period, or integration time. It is clear that the output correlation tends towards zero as the integration time is made sufficiently short that only a single photon enters the apparatus. The detailed form of variation of the correlation with $\langle n \rangle$ is derived in eqn (5.9.6). The experiments essentially confirm the expected vanishing of the quantum Brown–Twiss correlation (5.8.8).

Figure 5.19 shows the results of a second experiment with the same gated cascade-emission source but with the Brown–Twiss interferometer replaced by the Mach–Zehnder. The results are built up from series of single-photon measurements with increasing path difference $z_1 - z_2$. The fringes have the Mach–Zehnder form of eqn (5.8.12) as expected, with a high 98% visibility, and there is no detectable difference from the classical fringe pattern of eqn (3.5.3).

5.9 Arbitrary single-arm input

The calculations of the preceding section are here generalized to allow for an arbitrary input state in arm 1, with arm 2 remaining in its vacuum state. The combined input state is denoted $|arb\rangle_1 |0\rangle_2$. It is straightforward to repeat the previous calculations, and the simplifications that result from the assumption of a vacuum state in arm 2 remain. The means and the variances of the electric fields in the output arms of the beam splitter are determined by eqns (5.7.9) and (5.7.10) with the

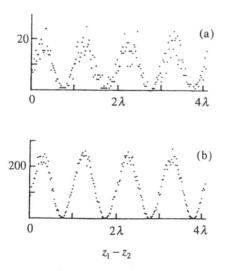

Fig. 5.19. Mach–Zehnder fringes observed with a single-photon input as a function of the path difference $z_1 - z_2$ expressed in terms of the wavelength λ. The vertical axis shows the numbers of photodetections in arm 4 for (a) a 1 s integration time and (b) a compilation of 15 such scans. (After [4])

vacuum input values taken from eqns (5.1.7) and (5.1.8). The mean output photon numbers of the single beam splitter from eqns (5.8.6) and (5.8.7) are generalized to

$$\langle n_3 \rangle = {}_2\langle 0|_1\langle \text{arb}\| \mathcal{R}|^2\, \hat{a}_1^\dagger \hat{a}_1 | \text{arb} \rangle_1 | 0 \rangle_2 = |\mathcal{R}|^2 \langle n_1 \rangle \qquad (5.9.1)$$

and

$$\langle n_4 \rangle = |\mathcal{T}|^2 \langle n_1 \rangle, \qquad (5.9.2)$$

where

$$\langle n_1 \rangle = {}_1\langle \text{arb}|\hat{a}_1^\dagger \hat{a}_1 | \text{arb} \rangle_1 \qquad (5.9.3)$$

is the mean number of photons in the arbitrary input state.

The photon-number correlation between the two output arms, as measured in the Brown–Twiss interferometer, is generalized from the single-photon result of eqn (5.8.8) to an arbitrary input as

$$
\begin{aligned}
\langle n_3 n_4 \rangle &= \mathcal{R}^* \mathcal{R} \mathcal{T}^* \mathcal{T} \,{}_1\langle \text{arb}|\hat{a}_1^\dagger \hat{a}_1 \hat{a}_1^\dagger \hat{a}_1 | \text{arb} \rangle_1 + \mathcal{R}^* \mathcal{T} \mathcal{R}^* \mathcal{T} \,{}_1\langle \text{arb}|\hat{a}_1^\dagger \hat{a}_1 | \text{arb} \rangle_1 \\
&= |\mathcal{R}|^2 |\mathcal{T}|^2 \,{}_1\langle \text{arb}|\hat{a}_1^\dagger \hat{a}_1 \hat{a}_1^\dagger \hat{a}_1 - \hat{a}_1^\dagger \hat{a}_1 | \text{arb} \rangle_1 \\
&= |\mathcal{R}|^2 |\mathcal{T}|^2 \,{}_1\langle \text{arb}|\hat{a}_1^\dagger \hat{a}_1^\dagger \hat{a}_1 \hat{a}_1 | \text{arb} \rangle_1 \\
&= |\mathcal{R}|^2 |\mathcal{T}|^2 \langle n_1(n_1 - 1) \rangle,
\end{aligned}
\qquad (5.9.4)
$$

where eqns (5.7.1) and (5.7.4) are used. The above results reduce to eqns (5.8.6) to (5.8.8) when the input in arm 1 is taken as the single-photon number state $|1\rangle_1$. It is seen from eqn (5.9.4) that the correlation $\langle n_3 n_4 \rangle$ vanishes only when the arm 1 input is the vacuum state $|0\rangle_1$ or the number state $|1\rangle_1$, or a superposition of the two.

The above analysis of the Brown–Twiss interferometer is expressed in terms of the quantum degree of second-order coherence, similar to the classical treatment in §3.8. Thus with the same notation for the normalized correlation of output beams as used in eqn (3.8.3), eqns (5.9.1), (5.9.2) and (5.9.4) give

$$g_{3,4}^{(2)}(\tau) = \frac{\langle n_3 n_4 \rangle}{\langle n_3 \rangle \langle n_4 \rangle} = \frac{\langle n_1(n_1 - 1) \rangle}{\langle n_1 \rangle^2} = g_{1,1}^{(2)}(\tau), \qquad (5.9.5)$$

where the subscripts denote the arms of the beam splitter, as before, and the expression from eqn (5.1.15) for the quantum degree of second-order coherence of the single-mode input is used. The quantum theory of the Brown–Twiss interferometer thus produces results that are the same as in the classical theory of §3.8 but with the classical degree of second-order coherence replaced by its quantum counterpart. In particular, the quantum Brown–Twiss correlation is obtained from eqn (3.8.5) by substitution of the quantum degree of second-order coherence.

The measurement of the Brown–Twiss correlation described at the end of §5.8 was not strictly performed with single-photon inputs, as the gate detector generally received more than one photon during the detection period and a correspondingly larger number of photons also entered the interferometer. The additional photons are produced by uncorrelated random emissions by different atoms in the source.

Problem 5.16   Consider the Brown–Twiss interferometer with an input photon number $1 + n$, where n has a Poisson distribution of mean $\langle n \rangle$. Show that the measured degree of second-order coherence is

$$g^{(2)}(\tau) = 1 - \frac{1}{\left(1 + \langle n \rangle\right)^2}. \tag{5.9.6}$$

This form of variation of the degree of second-order coherence with $\langle n \rangle$ provides a good match to the experimental points shown in Fig. 5.18. The single-mode theory used here is not strictly applicable to the $1 + n$ input photons, which were not all in the same mode in the conditions of the experiment [4]. However, the same expression in eqn (5.9.6) is found in a more realistic calculation.

The Mach–Zehnder output of eqn (5.8.12) is similarly converted to

$$\langle n_4 \rangle = |\mathcal{T}_{\mathrm{MZ}}|^2 \langle n_1 \rangle = 4|\mathcal{R}|^2 |\mathcal{T}|^2 \langle n_1 \rangle \cos^2\left[\tfrac{1}{2} k(z_1 - z_2)\right], \tag{5.9.7}$$

and the fringe pattern is independent of any statistical properties of the input state other than its mean photon number. Thus the same intensity distribution in the fringes as observed in an interference experiment with $\langle n_1 \rangle$ incident photons is built up by a series of $\langle n_1 \rangle$ identical experiments each sending a single photon through the interferometer [16]. This principle is verified experimentally by the interference experiments [4] that isolate single-photon inputs by means of the cascade-emission light source with gating of the interferometer. Much earlier Young's slit interference experiments [17], with a classical chaotic light source, had also failed to detect any change in the fringes when the light was attenuated to a level where only one photon at a time was present in the apparatus. It was found that the interference pattern under single-photon conditions was exactly the same as that for a normally strong light source, although the noisy statistics of chaotic light, discussed in §5.4, prevented the convincing demonstration of the nature of the interference achieved in the more recent experiments [4], shown in Fig. 5.19. The latter experiments show clearly that the interference is a single-photon effect, which does not depend in any way on the interaction of photons with each other.

For the Mach–Zehnder interferometer, each incident photon must propagate through the apparatus in such a way that the probability of its leaving the interferometer by arm 4 is proportional to the calculated mean photon number in that arm. This is achieved only if each photon excites both internal paths of the interferometer, so that the input state at the second beam splitter is determined by the complete interferometer geometry. This geometry is inherent in the superposition of output states from the first beam splitter in eqn (5.8.2) and in the propagation phase factors in eqn (5.8.10), with appropriate probability amplitudes for the two internal paths. The 'photon' in the Mach–Zehnder interferometer should thus be viewed as a composite excitation of input arm, internal paths and output arms, equivalent to the spatial field distribution produced by illumination of the input by a classical light beam.

There is indeed no way in which a photon can simultaneously be assigned to a particular internal path *and* contribute to the interference pattern. If a phototube is placed in one of the output arms of the first beam splitter to detect photons in the corresponding path, known as a 'which way' or 'welcher Weg' experiment, then it is not possible to avoid obscuring that path, with consequent destruction of the interference pattern. A succession of suggestions for more and more ingeneous 'which way' experiments has failed to provide any method for simultaneous path and fringe observations; a complete determination of the one leads to a total loss of resolution of the other, while a partial determination of the one leads to an accompanying partial loss of resolution of the other [18].

The interference experiments discussed above need only expressions for the mean and the correlation of the output photon numbers from a beam splitter, and the general results in eqns (5.9.4) and (5.9.7) apply for any variety of input light. The properties of the output photon distributions can be calculated in greater detail for specific input states in arm 1. Consider the effects of an input number state $|n_1\rangle_1$. Then the output state is obtained by generalization of eqn (5.8.2) as

$$
|n_1\rangle_1|0\rangle_2 = \frac{1}{(n_1!)^{1/2}}\left(\hat{a}_1^\dagger\right)^{n_1}|0\rangle = \frac{1}{(n_1!)^{1/2}}\left(\mathcal{R}\hat{a}_3^\dagger + \mathcal{T}\hat{a}_4^\dagger\right)^{n_1}|0\rangle
$$

$$
= \sum_{m=0}^{n_1}\left(\frac{n_1!}{m!(n_1-m)!}\right)^{1/2}\mathcal{R}^m\mathcal{T}^{n_1-m}|m\rangle_3|n_1-m\rangle_4,
$$

(5.9.8)

where eqns (4.3.45) and (4.3.46) are used and $|0\rangle$ is again the joint vacuum state of the beam-splitter arms. The summation in eqn (5.9.8) expresses the entangled superposition of number states in the output arms 3 and 4 produced by the input $|n_1\rangle_1$ in arm 1.

Let $P_1(n_1)$ denote the photon-number distribution for input arm 1. The probability distribution for the numbers of photons in the two output arms is obtained from the product of the appropriate element of the input distribution with the square modulus of the corresponding coefficient in eqn (5.9.8) as

$$
P_{3,4}(n_3,n_4) = P_1(n_3+n_4)\frac{(n_3+n_4)!}{n_3!n_4!}|\mathcal{R}|^{2n_3}|\mathcal{T}|^{2n_4}.
$$

(5.9.9)

This is a *binomial distribution* of the kind that governs the random partition of n identical classical objects into two distinct categories with respective probabilities $|\mathcal{R}|^2$ and $|\mathcal{T}|^2$. The partition process itself is known as *Bernoulli sampling* of the input photons. Measurements that involve only output arm 4 are controlled by the reduced distribution

$$P_4(n_4) = \sum_{n_3=0}^{\infty} P_{3,4}(n_3,n_4) = \sum_{n_1=n_4}^{\infty} P_1(n_1) \frac{n_1!}{(n_1-n_4)!n_4!} \left(1-|\mathcal{T}|^2\right)^{n_1-n_4} |\mathcal{T}|^{2n_4}.$$

$$(5.9.10)$$

and $P_3(n_3)$ is given by a similar expression.

These distributions are used to evaluate various averages for the output photon numbers.

Problem 5.17 Prove the averages

$$\begin{cases} \langle n_3(n_3-1)\rangle = |\mathcal{R}|^4 \langle n_1(n_1-1)\rangle \\ \langle n_4(n_4-1)\rangle = |\mathcal{T}|^4 \langle n_1(n_1-1)\rangle \end{cases}$$

$$(5.9.11)$$

and verify eqn (5.9.4). Hence show that

$$\langle (n_3-n_4)^2 \rangle = \left(|\mathcal{R}|^2 - |\mathcal{T}|^2\right)^2 \langle n_1(n_1-1)\rangle + \langle n_1\rangle.$$

$$(5.9.12)$$

The results (5.9.1), (5.9.2) and (5.9.11) show that

$$g_{3,3}^{(2)}(\tau) = g_{4,4}^{(2)}(\tau) = g_{1,1}^{(2)}(\tau),$$

$$(5.9.13)$$

and the degree of second-order coherence is unchanged from its input value by reflection from, or transmission through, a beam splitter. In the special case of a 50:50 beam splitter, eqn (5.9.12) shows that the mean-square difference in output photon number equals the mean input-photon number.

Consider single-mode chaotic light as a specific example of beam-splitter input state. The number of photons in input arm 1 is now distributed in accordance with the probability $P_1(n_1)$ obtained from eqn (5.4.1). The output distribution derived in accordance with eqn (5.9.9) is

$$P_{3,4}(n_3,n_4) = \frac{1}{1+\langle n_1\rangle} \frac{(n_3+n_4)!}{n_3!n_4!} \left(\frac{|\mathcal{R}|^2 \langle n_1\rangle}{1+\langle n_1\rangle}\right)^{n_3} \left(\frac{|\mathcal{T}|^2 \langle n_1\rangle}{1+\langle n_1\rangle}\right)^{n_4}.$$

$$(5.9.14)$$

The reduced distribution for measurements on arm 4 alone, obtained in accordance with eqn (5.9.10) is

$$P_4(n_4) = \frac{\left(|\mathcal{T}|^2 \langle n_1\rangle\right)^{n_4}}{\left(1+|\mathcal{T}|^2 \langle n_1\rangle\right)^{1+n_4}},$$

$$(5.9.15)$$

which describes a single-mode chaotic distribution with mean photon number $|\mathcal{T}|^2\langle n_1 \rangle$. A similar distribution holds for measurements on arm 3 alone, with the mean photon number changed to $|\mathcal{R}|^2\langle n_1 \rangle$. However, the chaotic input does not produce independent chaotic beams in the two outputs.

Problem 5.18 Prove, with use of the output distribution in eqn (5.9.14), that the photon-number correlation between the two output arms of the beam splitter for a chaotic input in arm 1 is

$$\langle n_3 n_4 \rangle = 2|\mathcal{R}|^2|\mathcal{T}|^2\langle n_1 \rangle^2 = 2\langle n_3 \rangle\langle n_4 \rangle. \qquad (5.9.16)$$

This expression has an additional factor of 2 compared to the corresponding result for independent light beams, derived in eqn (5.9.21) below.

The output distribution for an input coherent state is most easily derived with use of the coherent-state creation operator for arm 1, as defined in eqn (5.3.12). The input state is thus

$$|\alpha\rangle_1|0\rangle_2 = \hat{D}_1(\alpha)|0\rangle. \qquad (5.9.17)$$

Then, with the use of eqn (5.7.3) and the Hermitian conjugates,

$$\hat{D}_1(\alpha) = \exp\left(\alpha \hat{a}_1^\dagger + \alpha^* \hat{a}_1\right)$$
$$= \exp\left(\alpha \mathcal{R} \hat{a}_3^\dagger + \alpha^* \mathcal{R}^* \hat{a}_3 + \alpha \mathcal{T} \hat{a}_4^\dagger + \alpha^* \mathcal{T}^* \hat{a}_4\right) = \hat{D}_3(\mathcal{R}\alpha)\hat{D}_4(\mathcal{T}\alpha), \qquad (5.9.18)$$

where the factorization in the final step is allowed because the operators for output arms 3 and 4 commute, in accordance with eqn (5.7.7). It follows that the output is a product of coherent states,

$$|\alpha\rangle_1|0\rangle_2 = |\mathcal{R}\alpha\rangle_3|\mathcal{T}\alpha\rangle_4, \qquad (5.9.19)$$

and no entanglement occurs. The output photon-number distributions thus have the Poisson forms of eqn (5.3.21) and, for example,

$$P_4(n_4) = \exp\left(-|\mathcal{T}|^2\langle n_1 \rangle\right)\frac{\left(|\mathcal{T}|^2\langle n_1 \rangle\right)^{n_4}}{n_4!}. \qquad (5.9.20)$$

The input coherent state has the unique property that its division at a beam splitter produces two outputs that are indistinguishable from coherent light beams produced by independent sources [19]. Thus all correlations between the outputs factorize and, for example, it is easily shown that

$$\langle n_3 n_4 \rangle = \langle n_3 \rangle\langle n_4 \rangle \qquad (5.9.21)$$

and

$$\left\langle \hat{E}_3(\chi_3)\hat{E}_4(\chi_4)\right\rangle = \left\langle \hat{E}_3(\chi_3)\right\rangle\left\langle \hat{E}_4(\chi_4)\right\rangle, \tag{5.9.22}$$

where the expectation values for the individual coherent states are as given in eqns (5.3.16) and (5.3.31) respectively, with substitution of the complex amplitudes $\mathcal{R}\alpha$ and $\mathcal{T}\alpha$ for the output arms.

The fringe pattern for a Mach–Zehnder interferometer with an input coherent state is given by eqn (5.9.7), with $|\alpha|^2$ substituted for the mean input photon number. The above identification of the output from the first beam splitter with independent coherent states suggests the similar fringes should occur in the superposition of light beams from independent sources. Such fringes are indeed observed in the field formed by superposition of light from two independent lasers [20,21]. The interference occurs between the probability amplitudes that a detected photon was emitted by one source or the other. The interpretation of the experiment is the same as that of the Mach–Zehnder interferometer with coherent-state excitation, in that there is no way in which a photon can simultaneously contribute to the interference effects and be assigned to a definite laser source. The 'photon' in this case excites a superposition state similar to eqn (5.8.2), where 3 and 4 now label the two laser output beams and $\mathcal{R}$ and $\mathcal{T}$ are proportional to the beam amplitudes.

The output photon-number distributions for chaotic and coherent input light given by eqns (5.9.15) and (5.9.20) have the same functional forms as the input distributions. This is, however, not a general property of transmission through a beam splitter and a counter-example is provided by the input number state $\left| n_1 \right\rangle_1$, where eqn (5.9.10) gives

$$P_4(n_4) = \frac{n_1!}{(n_1 - n_4)!n_4!}\left(1 - |\mathcal{T}|^2\right)^{n_1 - n_4}|\mathcal{T}|^{2n_4}. \tag{5.9.23}$$

This probability is generally nonzero for all values of n_4 from 0 to n_1, unlike the single nonzero element of the input distribution, and the example of $n_1 = 4$ is shown in Fig. 5.20. The equality of input and output degrees of second-order coherence, expressed by eqn (5.9.13), remains valid.

Beam splitters play important roles in experimental studies of the squeezed states described in §§5.5 and 5.6, but consideration of this topic is deferred to the treatment of homodyne detection in §6.11. The technique uses states that excite both input arms of the beam splitter and these are treated in §6.8 (see also [22]).

5.10 Nonclassical light

The quantum degrees of second-order coherence of the various single-mode field excitations derived in the present chapter are all independent of the time delay τ, but it is already possible to make some useful comparisons with the corresponding classical theory. The classical degree of second-order coherence defined and discussed in §3.7 has magnitudes that lie in the ranges given by eqns (3.7.7) and (3.7.8) as

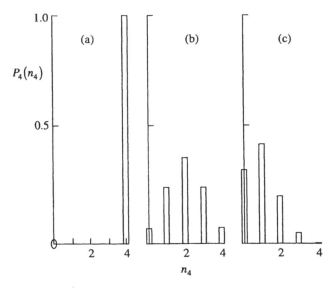

Fig. 5.20. Photon-number probability distributions for an input number state with $n_1 = 4$ and the transmitted output states of beam splitters with $|T|^2$ equal to (a) 1, (b) 1/2 and (c) 1/4. The degree of second-order coherence always equals 3/4.

$$1 \leq g^{(2)}(0) \leq \infty \quad \tau = 0 \tag{5.10.1}$$

and

$$0 \leq g^{(2)}(\tau) \leq \infty \quad \tau \neq 0. \tag{5.10.2}$$

The magnitude of the quantum degree of second-order coherence defined in eqn (5.1.14) has no restriction similar to than in eqn (5.10.1) and for single-mode excitations it can lie in the range obtained from eqn (5.1.17) as

$$1 - \frac{1}{\langle n \rangle} \leq g^{(2)}(\tau) \leq \infty \quad \text{for } \langle n \rangle \geq 1 \text{ and for all } \tau. \tag{5.10.3}$$

A lower limit of 0 applies for $\langle n \rangle < 1$. The lower limit is a specialization for single-mode states of the more general relation (4.12.10).

It is seen by comparison of the classical and quantum ranges in eqns (5.10.1) and (5.10.3) that the quantum degree of second-order coherence has access to an additional range of values

$$1 - \frac{1}{\langle n \rangle} \leq g^{(2)}(0) < 1 \quad \text{for } \langle n \rangle \geq 1, \tag{5.10.4}$$

and the range extends from 0 to 1 for $\langle n \rangle < 1$. Light whose degree of second-order coherence lies in the range of eqn (5.10.4) is an example of *nonclassical light.* Figure 5.21 shows the degrees of second-order coherence of the single-mode states derived in the present chapter as functions of $\langle n \rangle$. It is seen that the photon number state has a degree of second-order coherence that lies in the nonclassical range but the other varieties of light illustrated all have degrees of second-order coherence greater than or equal to unity. Note also how different values of the degree of second-order coherence for the vacuum state occur for the $\langle n \rangle \to 0$ limits of the different kinds of light.

The basic reason for the difference between the ranges of values of the degree of second-order coherence in the classical and quantum theories lies in the forms of their numerators, respectively $\langle I^2 \rangle$ and $\langle n(n-1) \rangle$. We have seen that these quantities describe experiments of the Brown–Twiss variety, which essentially measure the intensity or photon number of the light beam twice in succession. A classical measurement of the intensity leaves its value I unchanged for the second measurement and I^2 is the quantity to be averaged. However, the quantum average of $n(n-1)$ exemplifies the principle that a quantum measurement generally interferes with the measured system. Thus, in the present example, the measurement acts by determination of the rate of photon absorption and it reduces the photon number by 1, so that the second measurement finds only $n-1$ photons. This effect lies at the root of the differences between the degrees of second-order coherence in the classical and quantum theories.

The quantum measurement effect also accounts for the occurence of the same values of the degrees of second-order coherence for coherent and chaotic light in the classical and quantum theories. Thus, as discussed after eqn (5.3.19) for coherent light and after eqn (5.4.2) for chaotic light, it appears that the quantum theory produces particle contributions of magnitude $\langle n \rangle$ to the photon-number

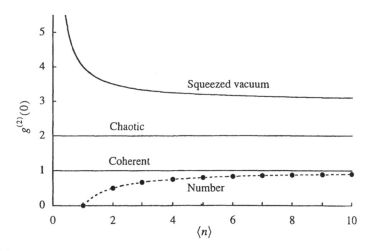

Fig. 5.21. Degrees of second-order coherence of the single-mode states indicated. The dashed curve shows the lower limit from eqn (5.10.4) and the points on this curve show the results for the photon number states.

variance, additional to the wave contributions of the classical theory. However, this additional contribution is exactly removed by the occurrence of $\langle n(n-1) \rangle$ rather than $\langle n^2 \rangle$ in the quantum degree of second-order coherence.

The degree of second-order coherence provides a single indication of the nature of the complete photon state that is often sufficient to characterize the different kinds of light. It has a distinct range of values that clearly identifies nonclassical light and it corresponds to a reasonably straightforward measurement in the laboratory.

An alternative measure of the second-order photon-statistical properties of a light beam is provided by *Mandel's Q parameter* [23], defined as

$$Q = \frac{(\Delta n)^2 - \langle n \rangle}{\langle n \rangle} = \frac{\langle n(n-1) \rangle - \langle n \rangle^2}{\langle n \rangle} = \langle n \rangle \left[g^{(2)}(0) - 1 \right], \tag{5.10.5}$$

with a nonclassical range of values obtained from eqn (5.10.4) as

$$-1 \leq Q < 0 \quad \text{or} \quad 0 \leq (\Delta n)^2 \leq \langle n \rangle. \tag{5.10.6}$$

The closely-related *Fano factor*, equal to $Q+1$, is sometimes used instead of Q. It follows from eqn (5.3.19) that

$$Q = 0 \tag{5.10.7}$$

for the coherent state, and this zero value of Q is a characteristic of the Poisson distribution. Sub-Poissonian fluctuations, defined after eqn (5.4.2), give negative values of Q and $g^{(2)}(0)$ less than unity. Examples of sub-Poissonian distributions are provided by number-state light from eqn (5.2.1), which gives the minimum allowed nonclassical value of

$$Q = -1, \tag{5.10.8}$$

and the amplitude-squeezed coherent state treated in eqn (5.6.16), with

$$Q = e^{-2s} - 1. \tag{5.10.9}$$

Light that has sub-Poissonian statistics in accordance with eqn (5.10.6) is sometimes described as *photon-number squeezed*, by analogy with the quadrature squeezing defined by the condition in eqn (5.5.1). Photon-number squeezing and amplitude squeezing often occur together, in accordance with the condition in eqn (5.6.19).

Super-Poissonian fluctuations give positive values of Q, as in the examples of single-mode chaotic light from eqn (5.4.2) with

$$Q = \langle n \rangle, \tag{5.10.10}$$

squeezed vacuum light from eqn (5.5.12) with

$$Q = 2\langle n \rangle + 1 \tag{5.10.11}$$

and the phase-squeezed light described by eqn (5.6.20) with

$$Q = e^{2s} - 1. \tag{5.10.12}$$

The Q parameter is thus particularly useful in the characterization of photon number fluctuations with respect to those for the Poisson distribution and in the identification of nonclassical light.

The existence of light whose degree of second-order coherence lies in the range given in eqn (5.10.4), or whose Q parameter lies in the range given in eqn (5.10.6), establishes the need for a quantum theory of light to treat the phenomena that can be measured with the appropriate light sources. The characteristics described in the present section are examples of a range of observable properties that are used to distinguish nonclassical light. The treatment is generalized in Chapter 6 to cover the continuous-mode excitations on which most measurements of the second-order statistical properties are made. Continuous-mode theory embraces light beams with time-dependent fluctuations, similar to the classical theory of Chapter 3.

The observation of other nonclassical aspects of the squeezed light treated in §§5.5 and 5.6 is discussed in §§6.9 and 6.11, the nonclassical correlations between pairs of photons are covered in §6.7, and the nonclassical phenomenon of photon antibunching is described in §§6.5, 7.9, 8.3 and 8.4.

References

[1]   Atkins, P.W. and Friedman, R.S., *Molecular Quantum Mechanics*, 3rd edn (Oxford University Press, Oxford, 1997); Merzbacher, E., *Quantum Mechanics*, 3rd edn (Wiley, New York, 1998).

[2]   Pegg, D.T. and Barnett, S.M., Tutorial review: Quantum optical phase, *J. Mod. Opt.* **44**, 225–64 (1997).

[3]   Barnett, S.M. and Radmore, P.M., *Methods in theoretical quantum optics* (Oxford University Press, Oxford, 1997).

[4]   Grangier, P., Roger, G. and Aspect, A., Experimental evidence for a photon anticorrelation effect on a beam splitter: a new light on single-photon interferences, *Europhys. Lett.* **1**, 173–9 (1986).

[5]   Glauber, R.J., Coherent and incoherent states of the radiation field, *Phys. Rev.* **131**, 2766–88 (1963).

[6]   Beck, M., Smithey, D.T., Cooper, J. and Raymer, M.G., Experimental determination of number–phase uncertainty relations, *Opt. Lett.* **18**, 1259–61 (1993).

[7]   Loudon, R., Graphical representation of squeezed-state variances, *Opt. Comm.* **70**, 109–14 (1989).

[8]   Wu, L.-A., Xiao, M. and Kimble, H.J., Squeezed states of light from an optical parametric oscillator, *J. Opt. Soc. Am. B* **4**, 1465–75 (1987).

[9]   Caves, C.M., Quantum-mechanical noise in an interferometer, *Phys. Rev. D* **23**, 1693–708 (1981).

[10]   Yuen, H.P., Two-photon coherent states of the radiation field, *Phys. Rev. A* **13**, 2226–43 (1976).

[11]   Loudon, R. and Knight, P.L., Squeezed light, *J. Mod. Opt.* **34**, 709–59 (1987).

[12]   Zaheer, K. and Zubairy, M.S., Squeezed states of the radiation field, *Adv. At. Mol. Opt. Phys.* **28**, 143–235 (1991).

[13]   Breitenbach, G., Schiller, S. and Mlynek, J., Measurement of the quantum states of squeezed light, *Nature* **387**, 471–5 (1997); Breitenbach, G. and Schiller, S., Homodyne tomography of classical and nonclassical light, *J. Mod. Opt.* **44**, 2207–25 (1997).

[14]   von Neumann, J., *Mathematical Foundations of Quantum Mechanics* (Princeton University Press, Princeton, 1955).

[15]   Clausen, J., Dakna, M., Knöll, L. and Welsch, D.G., Conditional quantum-state transformation at a beam splitter, *J. Opt. B* **1**, 332–8 (1999).

[16]   Walls, D.F., A simple field theoretic description of photon interference, *Am. J. Phys.* **45**, 952–6 (1977).

[17]   Taylor, G.I., Interference fringes with feeble light, *Proc. Camb. Phil. Soc.* **15**, 114–5 (1909).

[18]   Scully, M.O. and Zubairy, M.S., *Quantum Optics* (Cambridge University Press, Cambridge, 1997).

[19]   Aharonov, Y., Falkoff, D., Lerner, E. and Pendleton, H., A quantum characterization of classical radiation, *Ann. Phys.* **39**, 498–512 (1966); Titulaer, U.M. and Glauber, R.J., Density operators for coherent fields, *Phys. Rev.* **145**, 1041–50 (1966).

[20]   Pfleegor, R.L. and Mandel, L., Interference of independent photon beams, *Phys. Rev.* **159**, 1084–8 (1967).

[21]   Louradour, F., Reynaud, F., Colombeau, B. and Froehly, C., Interference fringes between two separate lasers, *Am. J. Phys.* **61**, 242–5 (1993).

[22]   Campos, R.A., Saleh, B.E.A. and Teich, M.C., Quantum-mechanical lossless beam splitter: SU(2) symmetry and photon statistics, *Phys. Rev. A* **40**, 1371–84 (1989).

[23]   Mandel, L., Sub-Poissonian photon statistics in resonance fluorescence, *Opt. Lett.* **4**, 205–7 (1979).

6 Multimode and continuous-mode quantum optics

The single-mode theory of Chapter 5 is inherently limited to the description of experiments that use time-independent light beams. The light is essentially non-ergodic, in that its statistical fluctuations may be described by an ensemble, but these are not apparent in a time series of experiments made on the single member of the ensemble corresponding to a practical realization of the state. Identical measurements always give the same results and there are no time-dependent correlations of the kind considered in the classical theory of Chapter 3. Many of the nonclassical properties of light can be understood in terms of single-mode theory but other properties involve excitations of two or more modes. Practical observations of nonclassical effects are usually made on time-dependent light beams.

All real light beams are in fact time-dependent, in the sense that they do not continue for ever, and almost all of them exhibit variations on time scales comparable to, or shorter than, the observation times in practical experiments. Thus most of the natural light in the universe is chaotic, as is the light from conventional sources, with characteristic time-dependent intensity correlations and fluctuations. The nonclassical light generated in many experiments comes in the form of short optical pulses whose spectra cover a correspondingly extended range of frequencies.

The representation of time dependence requires the use of two or more modes of the optical system. In general, it is necessary to consider the excitation of all of an infinite range of modes, even when the light beam is made up from contributions of modes whose wavevectors all point in a common direction. Some experiments use an optical cavity, where the mode wavevectors have the discrete values defined in eqn (1.10.11). The quantum-optical theory in this case is described as *multimode*, and a brief treatment is given in §6.1. However, the majority of optical experiments have no identifiable cavity, but rather the intensity in the light beam flows from source to detector through some kind of interaction region. The positions of source and detector do not themselves define an optical cavity as they are parts of a continuous unidirectional flow of optical energy with no significant reflection or recycling. For experiments of this nature, it is preferable to quantize the electromagnetic field in free space with a set of modes characterized by a continuous wavevector. The quantum-optical theory in this case is described as *continuous-mode*, and the theory is presented and used in the remainder of the present chapter.

6.1 Multimode states

The formalisms for the quantization of the radiation field and its interaction with atoms given in Chapter 4 embrace all of the modes of the radiation field in a cavity. The modes are labelled by a discrete wavevector $\mathbf{k}$ and a polarization label λ. The multimode number states, or Fock states, $|\{n_{\mathbf{k}\lambda}\}\rangle$ defined in eqn (4.4.6) provide a complete set. Thus a general pure state is expressed in the form of eqn (4.6.1) and statistical mixture states are described by the multimode density operator, as in the example of a thermally-excited cavity treated in eqns (4.6.19) to (4.6.22). The quantum degrees of first and second-order coherence defined in §4.11 apply to multimode excitations when the complete electric field operators from eqns (4.4.14) and (4.4.15) are used.

The natures of the multimode states are most easily illustrated by means of some simple examples that provide quantum-optical versions of some of the light beams treated in the classical theory of Chapter 3. We assume parallel beams with plane wavefronts and a single polarization throughout the present section, so that the wavevectors involved all point in the z direction and the polarization index λ can be dropped. The commutation relation (4.4.5) thus reduces to

$$\left[\hat{a}_k, \hat{a}_{k'}^{\dagger}\right] = \delta_{k,k'}. \tag{6.1.1}$$

Consider first the multimode coherent state

$$|\{\alpha\}\rangle = |\alpha_{k_1}\rangle |\alpha_{k_2}\rangle |\alpha_{k_3}\rangle \cdots, \tag{6.1.2}$$

in which the modes of all wavevectors parallel to the z axis are individually excited to coherent states of the form defined and discussed in §5.3. Here $\{\alpha\}$ denotes the complete set of complex amplitudes that specify the coherent states in each excited mode of the cavity. It follows from the property (5.3.5) possessed by the state of each of these modes and from the form of the interaction-picture field operator in eqn (4.4.14) that

$$\hat{E}_{\mathrm{T}}^{-}(z,t)|\{\alpha\}\rangle = \sum_k \left(\hbar\omega_k/2\varepsilon_0 V\right)^{1/2} \alpha_k \exp\left[-i\chi_k(z,t)\right]|\{\alpha\}\rangle, \tag{6.1.3}$$

where the phase angle is defined as

$$\chi_k(z,t) = \omega_k t - kz - \tfrac{\pi}{2}, \tag{6.1.4}$$

similar to eqns (4.4.12) and (5.1.2). The multimode coherent state is thus an eigenstate of the positive-frequency part of the electric-field operator. The field eigenvalue in eqn (6.1.3) is generally a function of position and time determined by the dependence of the α_k on k. The multimode field may, for example, have a pulsed structure and the light does not in general have stationary properties.

It follows from the electric-field eigenvalue relation and from the expressions for the degrees of first- and second-order coherence in eqns (4.12.4) and (4.12.8) that

$$\left| g^{(1)}(z_1,t_1;z_2,t_2) \right| = g^{(2)}(z_1,t_1;z_2,t_2) = 1, \tag{6.1.5}$$

independent of the pairs of space–time points. The multimode coherent beam is thus second-order coherent at all pairs of space–time points in accordance with the prescription in eqn (3.7.23). This state is the quantum-optical analogue of an excitation of classical stable waves defined in eqn (3.4.20) in each mode of a parallel beam.

A second example of multimode light is the thermal excitation described by the density operator of eqns (4.6.19) and (4.6.20). With the observations again restricted to a single polarization and mode wavevectors parallel to the z axis, the degree of first-order coherence in eqn (4.12.5) is readily calculated.

Problem 6.1   Derive the degree of first-order coherence of a parallel beam of thermal light as

$$g^{(1)}(\tau) = \frac{\sum_k \omega_k \langle n_k \rangle \exp(-i\omega_k \tau)}{\sum_k \omega_k \langle n_k \rangle}. \tag{6.1.6}$$

This expression holds for all varieties of multimode light whose components have uncorrelated field amplitudes.

The mean photon numbers for thermal light are given by eqn (4.6.21) but the form of the density operator is valid for all kinds of chaotic light. The expression (6.1.6) is thus also valid for all forms of chaotic light. For a beam of light with the Lorentzian spectrum of eqn (2.5.20), the intensity at frequency ω_k is proportional to $\omega_k \langle n_k \rangle$, according to eqn (4.11.7), and we can write

$$\omega_k \langle n_k \rangle \propto \frac{\gamma/\pi}{(\omega_0 - \omega_k)^2 + \gamma^2}. \tag{6.1.7}$$

The summations over k in eqn (6.1.6) are converted to integrations by the one-dimensional analogue of eqn (1.1.11) but with travelling-mode wavevectors of the form given in eqn (4.2.5) for a cavity of length L,

$$\sum_k \to \int_0^\infty dk(L/2\pi) \to \int_0^\infty d\omega_k(L/2\pi c). \tag{6.1.8}$$

Thus, for chaotic light that excites a large number of modes, the degree of first-order coherence from eqn (6.1.6) becomes

$$g^{(1)}(\tau) = \int_0^\infty d\omega_k \frac{(\gamma/\pi)\exp(-i\omega_k\tau)}{(\omega_0 - \omega_k)^2 + \gamma^2}. \tag{6.1.9}$$

The lower limit on the integral can be replaced by $-\infty$ for a beam of narrow frequency width, $\gamma \ll \omega_0$, without significant change in its value. A simple contour integration then gives the result

$$g^{(1)}(\tau) = \exp(-i\omega_0\tau - \gamma|\tau|). \tag{6.1.10}$$

This has an identical form to the classical expression of eqn (3.4.7). The quantum calculation is extended without difficulty to include the effects of Doppler broadening and the results agree with the classical form given in eqn (3.4.25).

The degree of second-order coherence is likewise calculated straightforwardly from eqn (4.12.9).

Problem 6.2   Prove that the degree of second-order coherence of chaotic light is related to its degree of first-order coherence by

$$g^{(2)}(\tau) = 1 + \left|g^{(1)}(\tau)\right|^2, \tag{6.1.11}$$

where the light is assumed to excite a large number of modes.

The quantum degree of second-order coherence of a chaotic light beam is given by the same expression (3.7.16) as in the classical theory.

The above results show that the classical and quantum theories give the same degrees of first- and second-order coherence for the important examples of multimode coherent and chaotic light. Thus the two theories predict identical results in these cases for the Mach–Zehnder interference fringes and for the Brown–Twiss intensity interferometer. The quantum degrees of coherence for orders r higher than the second are defined in a similar way to the classical rth-order coherence considered in §3.7. The general expressions have the forms of eqns (4.12.3) and (4.12.7) but with the expectation values of products of r field operators $\hat{E}_T^-$ followed by r operators $\hat{E}_T^+$ in their numerators and products of square roots of $\hat{E}_T^-\hat{E}_T^+$ expectation values in their denominators. It is not difficult to prove that the multimode coherent state of eqn (6.1.2) satisfies the same relation (3.7.31) as the classical stable wave and the state is thus coherent in all orders. Likewise, the multimode plane-polarized parallel beam of chaotic light satisfies the same relation (3.7.32) for the degree of rth order coherence at a single space–time point as its classical counterpart. In physical terms, the degree of rth-order coherence determines the outcomes of experiments that measure r-photon correlations.

The results of the present section show that the quantum properties of multimode coherent and chaotic light are essentially the same as their classical properties. However, it is shown in §5.10 that single-mode quantum field excitations do not necessarily obey the classical limitation of eqn (3.7.7) on their degrees of second-order coherence, and we need to extend the treatment of non-classical light to multimode excitations. The discrete-mode formalism is somewhat awkward for these calculations and, similar to the derivation in eqns (6.1.6) to (6.1.10), it is simpler to convert to a continuous-mode variable. The treatment of nonclassical effects is resumed after an appropriate reformulation of the theory.

6.2 Continuous-mode field operators

A typical optical experiment uses light beams that travel in straight lines from sources to detectors through a region that is not usually contained within a real optical cavity. For such systems, it is advantageous to take the limit of a quantization axis of infinite extent parallel to the z axis but to retain a finite cross-sectional area A perpendicular to the axis [1]. The one-dimensional continuous-mode variable can be taken equivalently as the wavevector k or the frequency ω_k. The frequency variable is chosen here, with the k subscript removed. The one-dimensional mode spacing is thus

$$\Delta\omega = 2\pi c/L, \tag{6.2.1}$$

which tends to zero as the length L tends to infinity. The conversion from sum to integral as in eqn (6.1.8) is

$$\sum_k \to \frac{1}{\Delta\omega} \int d\omega. \tag{6.2.2}$$

The discrete Kronecker delta and the continuous Dirac delta-function are related by

$$\delta_{k,k'} \to \Delta\omega\,\delta(\omega - \omega'), \tag{6.2.3}$$

and it is readily verified that these conversions are consistent with the Dirac delta-function property of eqn (2.4.2).

The continuous-mode creation and destruction operators, denoted $\hat{a}(\omega)$ and $\hat{a}^\dagger(\omega)$, are related to their discrete-mode counterparts by

$$\hat{a}_k \to (\Delta\omega)^{1/2}\hat{a}(\omega) \quad \text{and} \quad \hat{a}_k^\dagger \to (\Delta\omega)^{1/2}\hat{a}^\dagger(\omega). \tag{6.2.4}$$

The continuous-mode commutation relation is thus obtained by conversion of the discrete-mode relation (6.1.1) as

$$\left[\hat{a}(\omega),\hat{a}^\dagger(\omega')\right] = \delta(\omega - \omega'). \tag{6.2.5}$$

Despite the vanishing of the mode spacing in the limit of infinite L, the expectation values of physical observables in the continuous-mode theory are well-behaved. For additional simplicity, the light is again assumed to have a fixed linear polarization and the λ subscript is accordingly dropped.

The continuous-mode quantized field operators in the interaction picture are obtained from their discrete-mode counterparts in eqns (4.4.12) to (4.4.18) by the above replacements and the main results for the positive-frequency parts are

$$\hat{E}_T^+(z,t) = i\int_0^\infty d\omega \left(\frac{\hbar\omega}{4\pi\varepsilon_0 cA}\right)^{1/2} \hat{a}(\omega)\exp\left[-i\omega\left(t-\frac{z}{c}\right)\right]$$

(6.2.6)

and

$$\hat{B}^+(z,t) = i\int_0^\infty d\omega \left(\frac{\hbar\omega}{4\pi\varepsilon_0 c^3 A}\right)^{1/2} \hat{a}(\omega)\exp\left[-i\omega\left(t-\frac{z}{c}\right)\right].$$

(6.2.7)

The complete field operators are obtained by addition of Hermitian conjugates as in eqns (4.4.13) and (4.4.16). With the frequency ω taken as positive, only the parts of the fields that correspond to propagation in the positive z direction are included in these expressions. The electric and magnetic fields are taken to be oriented in the x and y directions, respectively, and the field operators are written as scalars.

The electromagnetic field Hamiltonian is obtained from eqn (4.4.21), with the integration volume now taken to have a finite cross-section A but an infinite extent along the z axis. The derivation needs care, as the operator combination $\hat{a}(\omega)\hat{a}^\dagger(\omega)$ is undefined owing to the Dirac delta-function in the commutator of eqn (6.2.5) [2]. The problem is associated with an additional divergence in the zero-point or vacuum energy for an infinite length of z axis, this energy being already infinite even for a finite volume. The calculation is best performed by conversion from discrete to continuous-mode formalism after the operators have been put in normal order.

Problem 6.3   Show that the electromagnetic field Hamiltonian defined by suitable adaptation and manipulation of eqn (4.4.21) is obtained from the above field operators as

$$\hat{\mathcal{H}}_R = \int_0^\infty d\omega\hbar\omega\hat{a}^\dagger(\omega)\hat{a}(\omega) + \text{vacuum energy}.$$

(6.2.8)

The vacuum energy is ignored, as in §4.4, until later in the chapter. Equation (6.2.8) shows that the excitation energy of the electromagnetic field above its ground state is given by insertion of the photon energy $\hbar\omega$ in the integrand of a number operator

$$\hat{n} = \int d\omega\hat{a}^\dagger(\omega)\hat{a}(\omega).$$

(6.2.9)

This number operator is the continuous-mode analogue of the discrete-mode operator obtained by summation of eqn (4.4.3) over a set of one-dimensional wavevectors. The zero-point or vacuum energy is considered in §6.12.

The electromagnetic Hamiltonian and the total field energy are natural descriptions of the energy content for standing-wave fields confined inside optical cavities. However, the energy of a travelling-wave field is more conveniently expressed in terms of its intensity, or Poynting vector. The quantum-optical intensity operator defined in eqn (4.11.6) represents the quantity that is measured in most relevant experiments and it is free of any zero-point contribution. It is readily evaluated with use of the form of electric-field operator in eqn (6.2.6) as

$$\hat{I}(z,t) = \frac{\hbar}{2\pi A} \int d\omega \int d\omega' (\omega\omega')^{1/2} \hat{a}^\dagger(\omega)\hat{a}(\omega') \exp\left[i(\omega - \omega')\left(t - \frac{z}{c} \right) \right], \quad (6.2.10)$$

oriented parallel to the z axis. This somewhat complicated expression takes a very simple form upon integration over all time to give the total energy that flows through a plane of constant z as

$$A \int_{-\infty}^{\infty} dt \hat{I}(z,t) = \int d\omega \hbar\omega \hat{a}^\dagger(\omega)\hat{a}(\omega). \quad (6.2.11)$$

This result shows that all of the energy contained in the field must pass each point on the axis in the full course of time. Similarly, the total energy flow over the entire length of the z axis at a given instant of time is

$$A \int_{-\infty}^{\infty} dz \hat{I}(z,t) = c \int d\omega \hbar\omega \hat{a}^\dagger(\omega)\hat{a}(\omega), \quad (6.2.12)$$

with a similarly transparent physical interpretation. Practical light beams do not persist for ever nor do they extend over an infinite distance, but these last two expressions confirm the interpretation of the operator $\hat{n}$ in eqn (6.2.9) as representing the number of continuous-mode photons.

The range of integration over ω in the above expressions strictly extends only from 0 to ∞, as the frequency is defined to be positive. However, the range can be extended to cover the entire frequency axis from $-\infty$ to ∞ without significant error, when the bandwidth of the field excitation is much smaller than its central frequency. This condition is exemplified by the narrow-band Lorentzian assumed in eqn (6.1.9) and it applies to many light beams used in practice. With the range of integration extended in this way, it is useful to define Fourier-transformed operators by

$$\hat{a}(t) = (2\pi)^{-1/2} \int_{-\infty}^{\infty} d\omega \hat{a}(\omega) \exp(-i\omega t). \quad (6.2.13)$$

The commutation relation for these operators is obtained with the use of eqn (6.2.5) as

$$\left[\hat{a}(t), \hat{a}^\dagger(t')\right] = \delta(t - t').$$

(6.2.14)

The inverse Fourier transform is

$$\hat{a}(\omega) = (2\pi)^{-1/2} \int_{-\infty}^{\infty} dt\,\hat{a}(t)\exp(i\omega t),$$

(6.2.15)

and the number operator from eqn (6.2.9) is expressed in the equivalent form

$$\hat{n} = \int dt\,\hat{a}^\dagger(t)\hat{a}(t).$$

(6.2.16)

The continuous-mode vacuum state $|0\rangle$ satisfies the conditions

$$\hat{a}(\omega)|0\rangle = \hat{a}(t)|0\rangle = 0,$$

(6.2.17)

analogous to the discrete-mode ground-state condition of eqn (4.3.23).

The narrow-bandwidth approximation can be taken further by putting the square-root frequency factor in the integrand of the electric-field operator in eqn (6.2.6) equal to its value at the central frequency ω_0. The square-root factor is then taken outside the integral and use of eqn (6.2.13) gives

$$\hat{E}_T^+(z,t) = i\left(\frac{\hbar\omega_0}{2\varepsilon_0 cA}\right)^{1/2}\hat{a}\left(t - \frac{z}{c}\right).$$

(6.2.18)

It is emphasized that this expression is an approximation valid only for calculations in which expectation values of the field operator are to be evaluated for narrow-bandwidth excitations. A similar treatment of the Poynting operator from eqn (6.2.10) gives

$$A\hat{I}(z,t) = \hbar\omega_0\hat{a}^\dagger\left(t - \frac{z}{c}\right)\hat{a}\left(t - \frac{z}{c}\right).$$

(6.2.19)

The operator on the right-hand side of this equation represents the flux of the light beam in units of photons per second and it is convenient to define a flux operator

$$\hat{f}(t) = \hat{a}^\dagger(t)\hat{a}(t).$$

(6.2.20)

The mean photon flux is obtained with the use of eqn (6.2.13) as

$$f(t) = \langle\hat{f}(t)\rangle = (2\pi)^{-1}\int d\omega\int d\omega'\langle\hat{a}^\dagger(\omega)\hat{a}(\omega')\rangle\exp[i(\omega - \omega')t],$$

(6.2.21)

and the mean photon number follows from eqn (6.2.16) as

$$\langle n \rangle = \int dt\, f(t). \tag{6.2.22}$$

A special case of continuous-mode excitation is the stationary light beam defined in §3.3 as one produced by a source whose fluctuation properties do not change with the time. Such an excitation must persist for ever and, although this is not strictly possible in practice, stationary light is often approximately realized in experiments where the total duration of the steady beam far exceeds all other characteristic time scales of the system. The frequency correlation for a stationary beam can always be written as

$$\left\langle \hat{a}^{\dagger}(\omega)\hat{a}(\omega')\right\rangle = 2\pi f(\omega)\delta(\omega - \omega'), \tag{6.2.23}$$

and substitution into eqn (6.2.21) produces a mean photon flux of the required time-independent form

$$f(t) = \int d\omega\, f(\omega) \equiv F. \tag{6.2.24}$$

Thus $f(\omega)$ is the dimensionless mean photon flux per unit angular-frequency bandwidth at angular frequency ω. Note that the expectation value of the number operator $\hat{n}$ in eqns (6.2.9) or (6.2.16) is infinite for all stationary light beams, as is expected from eqn (6.2.22) for a state that has nonzero photon flux for all time.

The degrees of coherence defined in eqns (4.12.4) and (4.12.8) are straightforwardly expressed in terms of continuous-mode operators with the use of eqn (6.2.18), when the narrow-bandwidth approximation is valid. When the light is also stationary, the expressions reduce to

$$g^{(1)}(\tau) = \frac{\left\langle \hat{a}^{\dagger}(t)\hat{a}(t+\tau)\right\rangle}{\left\langle \hat{a}^{\dagger}(t)\hat{a}(t)\right\rangle} \tag{6.2.25}$$

and

$$g^{(2)}(\tau) = \frac{\left\langle \hat{a}^{\dagger}(t)\hat{a}^{\dagger}(t+\tau)\hat{a}(t+\tau)\hat{a}(t)\right\rangle}{\left\langle \hat{a}^{\dagger}(t)\hat{a}(t)\right\rangle^2} \tag{6.2.26}$$

where

$$\tau = t_2 - t_1 - \frac{z_2 - z_1}{c} \tag{6.2.27}$$

as in eqn (3.4.16). These are the continuous-mode forms of the degrees of coherence for stationary light given in eqns (4.12.5) and (4.12.9).

The states defined by

$$|1_\omega\rangle = \hat{a}^\dagger(\omega)|0\rangle \tag{6.2.28}$$

are continuous-mode analogues of the single-photon number states defined in the discrete-mode formalism of §§4.4 and 6.1. It follows by use of the commutator in eqn (6.2.5) and the ground-state condition in eqn (6.2.17) that the states satisfy the orthonormality relation

$$\langle 1_\omega | 1_{\omega'} \rangle = \delta(\omega - \omega'). \tag{6.2.29}$$

The normalization in terms of an infinite Dirac delta-function contrasts with the Kronecker delta-function normalization of the discrete-mode functions. The states defined in eqn (6.2.28) are an example of the continuous final states assumed in the integral form of Fermi's golden rule in eqn (2.4.16). However, these single-frequency states do not in fact provide a realistic representation of the photon-number states generated in experiments, which always have non-zero bandwidths and satisfy the usual normalization conditions. The physical photon-number states are best described by wavepackets covering a range of frequencies, and these are treated in the following section.

6.3 Number states

Any continuous-mode state that contains a finite number of photons must have the form of a pulse, or succession of pulses, for the energy and intensity in the beam are otherwise spread over the infinite z axis, with zero magnitudes at all points. Thus, for example, the single-photon states described in §5.8 are generated by cascade emission processes in which the detection of one of each pair of emitted photons is used to open an electronic gate that allows the isolation of the other photon of the pair. A similar technique is used on the pair states generated in parametric down-conversion [3]. The surviving photon in each case is described by a wavepacket, whose spectrum and time-dependent amplitude are determined by the parameters of the source and the detection process of its partner photon. The method of calculation is considered in §6.7.

It is convenient, for examples of single and multiple photon states, to use the Gaussian pulse shape, with a spectral amplitude represented by the function

$$\xi(\omega) = \left(2\pi\Delta^2\right)^{-1/4} \exp\left\{-i(\omega_0 - \omega)t_0 - \frac{(\omega_0 - \omega)^2}{4\Delta^2}\right\}. \tag{6.3.1}$$

Here ω_0 is the central frequency of the pulse spectrum and t_0 is the time at which the peak of the pulse passes the coordinate origin at $z = 0$. The intensity spectrum, given by the square modulus of $\xi(\omega)$, has a variance equal to Δ^2. The bandwidth of the spectrum is assumed throughout to be small, with $\Delta \ll \omega_0$, so that the narrow-bandwidth approximations in the previous section are valid. The Fourier transform of eqn (6.3.1), defined by the same convention as eqn (6.2.13), is the wavepacket amplitude

$$\xi(t) = \left(2\Delta^2/\pi\right)^{1/4} \exp\left\{-i\omega_0 t - \Delta^2(t_0 - t)^2\right\}. \tag{6.3.2}$$

It is seen by comparison with eqns (2.6.11) and (2.6.12) that the two amplitudes satisfy the normalization conditions

$$\int d\omega |\xi(\omega)|^2 = \int dt |\xi(t)|^2 = 1. \tag{6.3.3}$$

The formalism that follows is valid for any normalized functions $\xi(\omega)$ and $\xi(t)$, but the Gaussian form is used to illustrate the results.

Consider the *photon-wavepacket creation operator*

$$\hat{a}_\xi^\dagger = \int d\omega \xi(\omega)\hat{a}^\dagger(\omega) = \int dt \xi(t)\hat{a}^\dagger(t), \tag{6.3.4}$$

where the equivalence of the frequency and time-dependent forms is easily established by use of the Fourier transforms of the functions in the integrands. The commutator of this operator with its Hermitian-conjugate destruction operator is obtained with the use of eqns (6.2.5) or (6.2.14) and (6.3.3) as

$$\left[\hat{a}_\xi, \hat{a}_\xi^\dagger\right] = 1, \tag{6.3.5}$$

similar to the discrete-mode commutator of eqn (4.3.9). The wavepacket operator is used to construct continuous-mode photon number states by a prescription analogous to that for the discrete-mode number states in eqns (4.3.45) and (4.3.46),

$$\left|n_\xi\right\rangle = (n!)^{-1/2}\left(\hat{a}_\xi^\dagger\right)^n |0\rangle, \tag{6.3.6}$$

where $|0\rangle$ is again the continuous-mode vacuum state. The number states and the wavepacket creation and destruction operators are subscripted with ξ to identify the pulse function that governs their spectrum and time dependence.

The continuous-mode field operators treated in §6.2 are expressed in terms of the $\hat{a}(\omega)$ and $\hat{a}(t)$ operators. The evaluation of the various properties of the number states requires the commutators of these operators with the wavepacket operators defined in eqn (6.3.4).

Problem 6.4   Prove the commutation relation

$$\left[\hat{a}(\omega), \left(\hat{a}_\xi^\dagger\right)^n\right] = n\xi(\omega)\left(\hat{a}_\xi^\dagger\right)^{n-1}. \tag{6.3.7}$$

The same relation holds with ω replaced by t. Hence show that the continuous-mode number state defined in eqn (6.3.6) is an eigenfunction of the number operator $\hat{n}$ of eqn (6.2.9) or (6.2.16) with

$$\hat{n}|n_\xi\rangle = n|n_\xi\rangle. \tag{6.3.8}$$

Expectation values of other field operators are evaluated straightforwardly with use of the commutation relation (6.3.7) and the ground-state conditions in eqn (6.2.17). Thus for the electric-field operator of eqn (6.2.18),

$$\langle n_\xi|\hat{E}_T(z,t)|n_\xi\rangle = 0, \tag{6.3.9}$$

identical to the property of the single-mode number state in eqn (5.2.8). The expectation value of the photon flux operator defined by eqn (6.2.20) is

$$\langle n_\xi|\hat{f}(t)|n_\xi\rangle = n|\xi(t)|^2 \tag{6.3.10}$$

and, with insertion of the Gaussian pulse shape from eqn (6.3.2), the expectation value of the Poynting intensity operator from eqn (6.2.19) is

$$\langle n_\xi|\hat{I}(z,t)|n_\xi\rangle = \left(\frac{2}{\pi}\right)^{1/2}\frac{n\hbar\omega_0\Delta}{A}\exp\left\{-2\Delta^2\left(t-t_0-\frac{z}{c}\right)^2\right\}. \tag{6.3.11}$$

This expression shows an explicit form of a number-state pulse or wavepacket in the continuous-mode theory. The detailed shape of the pulse varies according to the nature of the experiment that produces it, but its limited spatial extent, here of order c/Δ, and its propagating nature are general characteristic features.

The degrees of coherence of the number-state pulse are calculated by evaluation of the expressions given in §4.12, with substitution of the electric-field operator from eqn (6.2.18). The wavepacket excitations assumed here are not, of course, stationary and the degree of first-order coherence must be evaluated from the general one-dimensional form in eqn (4.12.4), with the result

$$g^{(1)}(z_1,t_1;z_2,t_2) = \frac{\xi^*\left(t_1-\frac{z_1}{c}\right)\xi\left(t_2-\frac{z_2}{c}\right)}{\left|\xi\left(t_1-\frac{z_1}{c}\right)\xi\left(t_2-\frac{z_2}{c}\right)\right|}. \tag{6.3.12}$$

This expression is valid for any wavepacket amplitude, and for the Gaussian of eqn (6.3.2) it reduces to the simple form

$$g^{(1)}(z_1,t_1;z_2,t_2) = \exp(-i\omega_0\tau), \tag{6.3.13}$$

where τ is defined in eqn (6.2.27). This expression is identical to the degree of first-order coherence of an arbitrary discrete single-mode state, as derived in eqn (5.1.13), and the same result occurs for other forms of wavepacket amplitude. The continuous-mode number states are thus first-order coherent in accordance with the definition in eqn (3.4.15).

The degree of second-order coherence of the continuous-mode number state is similarly calculated from the expression in eqn (4.12.8) as

$$g^{(2)}(z_1,t_1;z_2,t_2) = 1 - \frac{1}{n}. \tag{6.3.14}$$

This is again the same as the corresponding result for the single-mode number states derived in eqn (5.2.3) and it violates the classical inequality (3.7.6). The result is strikingly independent of the positions and times in the field operators, despite the travelling-wave pulsed form of the field excitation. It thus transpires that the single-discrete-mode number state, with its field amplitude uniformly distributed along a finite quantization axis, and the continuous-mode photon-number state, with its field amplitude localized in a narrow-band wavepacket, have identical degrees of first and second-order coherence. The remarks in §5.10 about the nonclassical nature of number-state light apply to both varieties of number state.

The main application of the wavepacket number states described above is to single-photon excitations, as it is not easily possible in practice to generate states of two or more photons with identical wavepackets. There are several kinds of experiment that do generate simultaneous pairs of photons but these require a different description, which is provided in §6.7.

6.4 Coherent states

Continuous-mode coherent states are generated from the vacuum state by a generalization of the single-mode displacement operator of eqn (5.3.12),

$$|\{\alpha\}\rangle = \exp(\hat{a}_\alpha^\dagger - \hat{a}_\alpha)|0\rangle, \tag{6.4.1}$$

where the wavepacket operators are defined by eqn (6.3.4) and its Hermitian conjugate but with the ξ functions replaced by $\alpha(\omega)$ and $\alpha(t)$. Use of the same notation $|\{\alpha\}\rangle$ for the continuous-mode coherent states as for the multimode coherent state of (6.1.2) should not cause any confusion. The continuous-mode coherent states are automatically normalized for any amplitudes $\alpha(\omega)$ and $\alpha(t)$, and, unlike the number-state-wavepacket functions $\xi(\omega)$ and $\xi(t)$, these are not required to satisfy the conditions in eqn (6.3.3), which are replaced by

$$\int d\omega |\alpha(\omega)|^2 = \int dt |\alpha(t)|^2 = \langle n \rangle. \tag{6.4.2}$$

It is shown below that $\langle n \rangle$ is indeed the mean number of photons in the coherent state.

The main properties of the continuous-mode coherent states parallel those of their single-mode counterparts. Thus, the reverse procedure of that given in eqns (5.3.7) to (5.3.10) shows that the definition (6.4.1) can be rearranged as

$$|\{\alpha\}\rangle = \exp\!\left(\hat{a}^\dagger_\alpha - \tfrac{1}{2}\langle n \rangle\right)|0\rangle. \tag{6.4.3}$$

The commutation relation of eqn (6.3.7) remains valid with ξ replaced by α, and its use leads to

$$\hat{a}(\omega)|\{\alpha\}\rangle = \alpha(\omega)|\{\alpha\}\rangle. \tag{6.4.4}$$

The relation

$$\hat{a}(t)|\{\alpha\}\rangle = \alpha(t)|\{\alpha\}\rangle \tag{6.4.5}$$

is proved in a similar fashion, where the use of time-dependent operators is again valid only for narrow-band excitations. The coherent state is therefore an eigenstate of the continuous-mode destruction operators. It follows from eqns (6.4.4) and (6.4.5) that the expectation value of the continuous-mode number operator defined in eqn (6.2.9) or (6.2.16) equals the quantity $\langle n \rangle$ introduced above.

It also follows from the definition of the coherent-state-wavepacket destruction operator that

$$\hat{a}_\alpha|\{\alpha\}\rangle = \langle n \rangle|\{\alpha\}\rangle. \tag{6.4.6}$$

The coherent-state wavepacket operators do not satisfy the same standard boson commutation relation as the number-state operators, given in eqn (6.3.5), which is replaced by

$$\left[\hat{a}_\alpha, \hat{a}^\dagger_\alpha\right] = \langle n \rangle. \tag{6.4.7}$$

The mean photon flux is obtained from the expectation value of the flux operator, defined in eqn (6.2.20), as

$$f(t) = |\alpha(t)|^2 \tag{6.4.8}$$

and the mean intensity is obtained from the expectation value of the Poynting operator, defined in eqn (6.2.19), as

$$\langle\{\alpha\}|\hat{I}(z,t)|\{\alpha\}\rangle = \frac{\hbar\omega_0}{A}\left|\alpha\!\left(t - \frac{z}{c}\right)\right|^2. \tag{6.4.9}$$

The wavepacket amplitude $\alpha(t)$ may, for example, have the Gaussian form $\xi(t)$

of eqn (6.3.2) but with an additional factor of the square root of the mean photon number, when the mean intensity is the same as in eqn (6.3.11) for the photon-number state but with n replaced by $\langle n \rangle$.

The continuous-mode coherent state of eqn (6.4.1) is an eigenstate of the positive-frequency part of the electric-field operator defined in eqn (6.2.6), with an eigenvalue obtained by the simple replacement of $\hat{a}(\omega)$ by $\alpha(\omega)$ in the frequency integrand. It follows that the degrees of first and second-order coherence have the same values as given in eqn (6.1.5) for the multimode coherent state, and $|\{\alpha\}\rangle$ is accordingly second-order coherent at all pairs of space–time points. The coherent states can be expressed as linear superpositions of photon-number states, obtained by expansion of the exponent in eqn (6.4.3) and with suitable scaling of the coefficients in terms of the mean photon number $\langle n \rangle$. The superposition removes any nonclassical features from the coherent states, whose degree of second-order coherence has the same unit value as that of the classical stable wave derived in eqn (3.7.21). The coherence properties of the states thus resemble those of a superposition of classical stable waves.

Experiments sometimes use the output beam from a laser in which only a single internal mode is excited. The external beam often has a bandwidth that is much smaller than that of any other components in the experiment. The light in this case can be represented by a continuous-mode coherent state with spectral and wavepacket amplitudes given by

$$\alpha(\omega) = (2\pi F)^{1/2} e^{i\theta} \delta(\omega - \omega_0) \tag{6.4.10}$$

and

$$\alpha(t) = F^{1/2} \exp(-i\omega_0 t + i\theta), \tag{6.4.11}$$

where F is the time-independent mean photon flux, in agreement with eqn (6.4.8), θ is the phase and ω_0 is the frequency. The wavepacket creation operator is obtained from eqn (6.3.4), with $\xi(\omega)$ replaced by $\alpha(\omega)$ from eqn (6.4.10), as

$$\hat{a}_\alpha^\dagger = (2\pi F)^{1/2} e^{i\theta} \hat{a}^\dagger(\omega_0) . \tag{6.4.12}$$

It should be noted that, in contrast to eqn (6.2.28), this operator is exponentiated before application to the vacuum state in the definition (6.4.1) of the continuous-mode coherent state.

The state represented by these amplitudes and creation operator is stationary in accordance with eqn (6.2.23). Its mean photon flux per unit angular frequency bandwidth is

$$f(\omega) = F\delta(\omega - \omega_0). \tag{6.4.13}$$

The correlation function that enters the degree of second-order coherence in eqn

(6.2.26) is obtained from the eigenvalue property in eqn (6.4.5) as

$$\left\langle \hat{a}^\dagger(t)\hat{a}^\dagger(t+\tau)\hat{a}(t+\tau)\hat{a}(t)\right\rangle = F^2, \tag{6.4.14}$$

giving again a degree of second-order coherence equal to unity for all t and τ. The state has the nature of a 'single-mode' excitation in a continuous-mode theory. The mean number of photons $\langle n \rangle$ determined by eqn (6.4.2) is infinite for this stationary coherent state and its spectral amplitude cannot be normalized. However, with sufficient care, it provides a useful representation of the light used in many experiments.

6.5 Chaotic light: photon bunching and antibunching

A stationary beam of chaotic light is essentially a statistical mixture state of the radiation field, described by an appropriate density operator in the formalism of §4.6. The density operator in the continuous-mode theory has the form of an integral over a continuous mode variable, instead of a discrete summation of the kind shown in eqn (4.6.19). However, the properties of the light can be specified by a frequency correlation function of the kind shown in eqn (6.2.23) and it is not necessary to work directly with the density operator itself.

Consider a beam of chaotic light with a Lorentzian spectrum, where the mean photon flux per angular frequency bandwidth is

$$f(\omega) = \frac{\gamma/\pi}{(\omega_0 - \omega)^2 + \gamma^2} F, \tag{6.5.1}$$

and F is the total mean flux defined by eqn (6.2.24). The frequency correlation is given by eqn (6.2.23) and its Fourier transform gives the time-dependent correlation function

$$\left\langle \hat{a}^\dagger(t)\hat{a}(t+\tau)\right\rangle = F\exp(-i\omega_0\tau - \gamma|\tau|). \tag{6.5.2}$$

The degree of first-order coherence of the light is readily obtained from eqn (6.2.25) and the result is the same as that found from the multimode theory in eqn (6.1.10).

The corresponding calculation of the degree of second-order coherence from eqn (6.2.26) requires the correlation function of two creation with two destruction operators. This is obtained by a factorization rule [2], valid for chaotic light,

$$\left\langle \hat{a}^\dagger(\omega)\hat{a}^\dagger(\omega')\hat{a}(\omega'')\hat{a}(\omega''')\right\rangle$$
$$= \left\langle \hat{a}^\dagger(\omega)\hat{a}(\omega''')\right\rangle\left\langle \hat{a}^\dagger(\omega')\hat{a}(\omega'')\right\rangle + \left\langle \hat{a}^\dagger(\omega)\hat{a}(\omega'')\right\rangle\left\langle \hat{a}^\dagger(\omega')\hat{a}(\omega''')\right\rangle, \tag{6.5.3}$$

whose Fourier transform gives

$$\left\langle \hat{a}^\dagger(t)\hat{a}^\dagger(t+\tau)\hat{a}(t+\tau)\hat{a}(t)\right\rangle$$

$$= \left\langle \hat{a}^\dagger(t)\hat{a}(t)\right\rangle\left\langle \hat{a}^\dagger(t+\tau)\hat{a}(t+\tau)\right\rangle + \left\langle \hat{a}^\dagger(t)\hat{a}(t+\tau)\right\rangle\left\langle \hat{a}^\dagger(t+\tau)\hat{a}(t)\right\rangle. \tag{6.5.4}$$

Substitution of the binary correlation functions from eqn (6.5.2) gives

$$\left\langle \hat{a}^\dagger(t)\hat{a}^\dagger(t+\tau)\hat{a}(t+\tau)\hat{a}(t)\right\rangle = F^2\left[1 + \exp(-2\gamma|\tau|)\right], \tag{6.5.5}$$

and the degree of second-order coherence from eqn (6.2.26) is identical with the multimode expression in eqn (6.1.11). The degrees of coherence of stationary chaotic light in the continuous-mode theory are thus the same as in the multi-mode theory, and both agree with the classical expressions in eqns (3.4.7) and (3.7.16).

The excess positive correlation represented by the second term in the large bracket of eqn (6.5.5), compared to the corresponding expression for coherent light in eqn (6.4.14), is described as the *photon-bunching effect*. This is just the translation into quantum language of the kind of intensity fluctuations that occur in the classical description of chaotic light, shown in Figs. 3.4 and 3.5. The excess correlation in eqn (6.5.5) ensures that the classical inequality,

$$g^{(2)}(\tau) \le g^{(2)}(0), \tag{6.5.6}$$

of eqn (3.7.11) is satisfied. Similar excess fluctuations occur in the semiclassical photocount distribution of chaotic light evaluated in §3.9 and in the identical quantum distribution evaluated below in §6.10. In terms of the detection of photons at a photodetector, the arrival of a beam of light generates a succession of photocounts whose statistics mimic those of the electromagnetic field excitation. Thus the arrival of a high-intensity fluctuation, or 'photon bunch', generates closely spaced photocounts, while the arrival of an intensity trough produces very few photocounts. Figure 6.1(a) shows a schematic representation of the occurrence of photocounts as a function of the time for chaotic light. The clusters of counts, associated with the photon bunches in the incident light beam, are clearly visible. The extents in time of the bunches are of the order of the coherence time $\tau_c = 1/\gamma$.

Coherent light has no intensity fluctuations in the classical picture and it has no photon bunches in the quantum picture. The degree of second-order coherence is constant,

$$g^{(2)}(\tau) = 1, \tag{6.5.7}$$

as in eqn (3.7.21). The time-independent pair correlation function given by eqn (6.4.14) corresponds to a random stream of photocounts. The Poisson photocount distribution for coherent light is found both in the semiclassical theory of §3.9 and in the quantum theory of §6.10. Figure 6.1(b) shows the random succession of photocounts produced by coherent light.

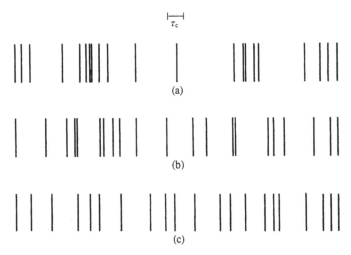

Fig. 6.1. Schematic representations of time series of photocounts for light beams that are (a) bunched, (b) random, and (c) antibunched.

Figure 6.1(c) shows a succession of photocounts in which the frequency of occurrence of closely-spaced events is depressed below the frequency of events with larger spacings. A photocount pattern of this form is associated with so-called *antibunched light*, whose degree of second-order coherence satisfies the inequality

$$g^{(2)}(\tau) > g^{(2)}(0), \tag{6.5.8}$$

in violation of eqn (3.7.11). Such light is generated by sources in which the emissions of successive photons are more likely to occur with longer time separations than at shorter time intervals. Examples are provided by light-emitting devices driven by electron beams or currents, whose mutually repulsive charges acquire characteristic spatial separations that are partly transferred to the emitted photons [4]. Another example, treated in §§7.9, and 8.2 to 8.4, is the resonance fluorescence by a single atom, where the atomic recovery between one emission and the next introduces time delays between successive photons. Photon anti-bunching is significant for the quantum theory of light, as it cannot occur in the semiclassical theory of §3.9 and its treatment requires the quantum photocount theory of §6.10. This nonclassical effect is distinct from the sub-Poissonian statistics, or photon-number squeezing, identified by the condition

$$g^{(2)}(0) < 1, \tag{6.5.9}$$

from eqn (5.10.4). The two nonclassical effects often occur together but each can occur in the absence of the other [5].

6.6 The Mach–Zehnder interferometer

Some of the time-dependent properties associated with continuous-mode states are illustrated by a reconsideration of the Mach–Zehnder interferometer. We discuss the continuous-mode formalism of the device and its performance with an incident single-photon wavepacket.

Beam splitters in general have reflection and transmission coefficients that vary with the frequency, but these are often constant over the range of frequencies in a narrow-band field excitation. With this assumption and for a symmetric lossless beam splitter, the continuous-mode input–output relations are the same as those for the discrete modes of §5.7, given by

$$\hat{a}_3(\omega) = \mathcal{R}\hat{a}_1(\omega) + \mathcal{T}\hat{a}_2(\omega) \quad \text{and} \quad \hat{a}_4(\omega) = \mathcal{T}\hat{a}_1(\omega) + \mathcal{R}\hat{a}_2(\omega), \tag{6.6.1}$$

where the notation for the arms of the beam splitter is that shown in Fig. 5.15. Again with the assumption of narrow-band spectra, the continuous-mode time-dependent destruction operators may be used and their input–output relations have the same forms,

$$\hat{a}_3(t) = \mathcal{R}\hat{a}_1(t) + \mathcal{T}\hat{a}_2(t) \quad \text{and} \quad \hat{a}_4(t) = \mathcal{T}\hat{a}_1(t) + \mathcal{R}\hat{a}_2(t). \tag{6.6.2}$$

The reflection and transmission coefficients continue to satisfy the relations in eqn (5.7.1), and the commutation relations in eqns (5.7.4) to (5.7.8) transfer to the continuous-mode formalism with frequency and time-dependent versions similar to those in eqns (6.2.5) and (6.2.14).

The continuous-mode theory for the Mach–Zehnder interferometer, represented in Fig. 5.16, parallels the discrete-mode treatment in §5.8. The complete device is equivalent to a composite beam splitter with input–output relations of the form given in eqn (5.8.9). We consider, as before, only the output in arm 4 with the required reflection and transmission coefficients obtained from eqn (5.8.10) as

$$\mathcal{R}'_{MZ}(\omega) = \mathcal{T}^2 \exp(i\omega z_1/c) + \mathcal{R}^2 \exp(i\omega z_2/c) \tag{6.6.3}$$

and

$$\mathcal{T}_{MZ}(\omega) = \mathcal{R}\mathcal{T}\left[\exp(i\omega z_1/c) + \exp(i\omega z_2/c)\right]. \tag{6.6.4}$$

The variations of the composite coefficients with ω must be inserted into the frequency-dependent input–output relations (6.6.1) before Fourier transformation into time-dependent relations analogous to eqn (6.6.2). The time-dependent destruction operator for arm 4 is thus obtained with the use of eqn (6.2.13) as

$$\hat{a}_4(t) = \mathcal{R}\mathcal{T}\hat{a}_1\left(t - \frac{z_1}{c}\right) + \mathcal{R}\mathcal{T}\hat{a}_1\left(t - \frac{z_2}{c}\right) + \mathcal{T}^2\hat{a}_2\left(t - \frac{z_1}{c}\right) + \mathcal{R}^2\hat{a}_2\left(t - \frac{z_2}{c}\right). \tag{6.6.5}$$

Thus, with the arm 2 input in its vacuum state, the expectation value of the photon flux operator in arm 4 is

$$f_4(t) = |\mathcal{R}|^2 |\mathcal{T}|^2 \left\langle \left[\hat{a}_1^\dagger \left(t - \frac{z_1}{c} \right) + \hat{a}_1^\dagger \left(t - \frac{z_2}{c} \right) \right] \left[\hat{a}_1 \left(t - \frac{z_1}{c} \right) + \hat{a}_1 \left(t - \frac{z_2}{c} \right) \right] \right\rangle. \quad (6.6.6)$$

The treatment so far applies generally to any form of continuous-mode excitation in arm 1. It is instructive to evaluate the output flux produced by an input single-photon number state defined in accordance with eqn (6.3.6). Then with use of the commutation relation (6.2.14) and the vacuum-state condition of eqn (6.2.17), the output flux is

$$f_4(t) = |\mathcal{R}|^2 |\mathcal{T}|^2 \left| \xi \left(t - \frac{z_1}{c} \right) + \xi \left(t - \frac{z_2}{c} \right) \right|^2. \quad (6.6.7)$$

A more explicit expression is obtained by substitution for $\xi(t)$ of the Gaussian wavepacket amplitude given in eqn (6.3.2). In practice, experiments on single-photon interference [6] measure the total photon number in output arm 4.

Problem 6.5   Show with the use of eqn (6.2.22) that the mean output photon number in arm 4 of the Mach–Zehnder interferometer for the input wavepacket amplitude in eqn (6.3.2) is

$$\langle n_4 \rangle = 2 |\mathcal{R}|^2 |\mathcal{T}|^2 \left\{ 1 + \exp \left[-\frac{\Delta^2 (z_1 - z_2)^2}{2c^2} \right] \cos \left[\frac{\omega_0 (z_1 - z_2)}{c} \right] \right\},$$

$$(6.6.8)$$

where t_0 is set equal to zero.

The form of the output differs significantly from the discrete single-mode expression in eqn (5.8.12). The strength of the interference fringes represented by the cosine is reduced by its Gaussian prefactor. In physical terms, the fringes survive at full strength if the path difference $z_1 - z_2$ is much smaller than the pulse length c/Δ, so that the pulse contributions from the two internal paths of the interferometer coincide in time at the second beam splitter. This condition is satisfied in the experimental results shown in Fig. 5.19. However, if the internal path difference greatly exceeds the pulse length, the two pulse contributions arrive separately at the second beam splitter and the interference effect is removed. Very similar reductions in Mach–Zehnder fringe strength with increasing internal path difference occur for input pulses of coherent light or indeed for classical input pulses.

It therefore remains the case that nonclassical effects must be sought in higher-order experiments, for example the Brown–Twiss interferometer.

Problem 6.6   Prove, with the use of eqns (6.2.14), (6.2.16) and the input–output relations in eqn (6.6.2), that the means and the correlation of the total photon numbers in the two output arms of the beam splitter shown in Fig. 5.15 are

$$\langle n_3 \rangle = |\mathcal{R}|^2 \langle n_1 \rangle, \quad \langle n_4 \rangle = |\mathcal{T}|^2 \langle n_1 \rangle \tag{6.6.9}$$

and

$$\langle n_3 n_4 \rangle = |\mathcal{R}|^2 |\mathcal{T}|^2 \int dt \int dt' \langle \hat{a}_1^\dagger(t) \hat{a}_1^\dagger(t') \hat{a}_1(t') \hat{a}_1(t) \rangle. \tag{6.6.10}$$

The Brown–Twiss correlation thus continues to vanish for a single-photon wave-packet, in agreement with the experimental results shown in Fig. 5.18. A different variety of nonclassical effect at a beam splitter occurs for the two-photon interference experiments discussed in §6.8.

6.7 Photon pair states

Several experiments, for example cascade emission (see §8.6) and parametric fluorescence or down-conversion (see §9.4), generate photons in pairs. Practical sources often generate such light with stationary statistical properties, corresponding to a stream of photon pairs that may excite a single beam or may excite two distinct beams with one photon in each. It is convenient to begin the treat-ment with the theory of a single pair of photons. The more general excitations of multiple photon pairs, stationary or nonstationary, are considered in §6.9. The discussion in the present section covers firstly a pair of photons in the same continuous-mode field and secondly a pair of photons in each of two different continuous-mode fields.

The number state defined by eqn (6.3.6) for $n = 2$ has the form

$$|2_\xi\rangle = 2^{-1/2} \left(\hat{a}_\xi^\dagger \right)^2 |0\rangle, \tag{6.7.1}$$

where the ξ subscript symbolizes the functions $\xi(\omega)$ and $\xi(t)$ that describe the spectrum and amplitude of the photon wavepacket. However, the state defined in this way represents two independent photons in the same wavepacket and it does not provide an adequate description of the light produced in many experiments. The two photons in a pair usually have different wavepacket functions and, more generally, they are often generated in some form of entangled state. For example, in the nonlinear optical process of parametric down-conversion, a photon in a pump beam of frequency ω_p splits into two photons of frequencies ω and ω' such that $\omega + \omega' = \omega_p$. It is not possible to include this restriction on the fre-quencies in the definition of the two-photon number state by eqn (6.7.1).

Consider instead the pair state created by a two-photon generalization of the single-photon wavepacket creation operator in eqn (6.3.4),

$$\left|(2_a)_\beta\right\rangle = \hat{P}^\dagger_{\beta aa}|0\rangle,$$

(6.7.2)

where the *photon-pair creation operator* is

$$\hat{P}^\dagger_{\beta aa} = 2^{-1/2}\int d\omega \int d\omega' \beta(\omega,\omega')\hat{a}^\dagger(\omega)\hat{a}^\dagger(\omega').$$

(6.7.3)

The state notation specifies the creation of both photons in the *same* continuous-mode field, denoted by label a, in contrast to states treated later in the present section where the two photons are in *different* continuous-mode fields. The joint two-photon wavepacket spectrum is normalized according to

$$\int d\omega \int d\omega' |\beta(\omega,\omega')|^2 = 1.$$

(6.7.4)

The Fourier transform of the two-photon spectrum for narrow-band excitations is defined as

$$\beta(t,t') = (2\pi)^{-1}\int_{-\infty}^{\infty} d\omega \int_{-\infty}^{\infty} d\omega' \beta(\omega,\omega')\exp(-i\omega t - i\omega' t'),$$

(6.7.5)

and the photon-pair creation operator from eqn (6.7.3) can be written in the equivalent form

$$\hat{P}^\dagger_{\beta aa} = 2^{-1/2}\int dt \int dt' \beta(t,t')\hat{a}^\dagger(t)\hat{a}^\dagger(t'),$$

(6.7.6)

with the normalization condition

$$\int dt \int dt' |\beta(t,t')|^2 = 1.$$

(6.7.7)

The orders of frequencies or times in the β functions are not significant, because the two creation operators in eqns (6.7.3) and (6.7.6) refer to the same field, so that these functions satisfy the symmetry relations

$$\beta(\omega',\omega) = \beta(\omega,\omega') \quad \text{and} \quad \beta(t',t) = \beta(t,t').$$

(6.7.8)

The basic formal properties of the photon pair states are derived by use of the commutation relations (6.2.5) and (6.2.14) together with the ground-state conditions of eqn (6.2.17). Thus, normalization of the state defined in eqns (6.7.2) and (6.7.3) requires the evaluation of the expectation value

$$\langle 0|\hat{a}(\omega)\hat{a}(\omega')\hat{a}^\dagger(\omega'')\hat{a}^\dagger(\omega''')|0\rangle$$

$$= \langle 0|\hat{a}(\omega)\left[\hat{a}^\dagger(\omega'')\hat{a}(\omega') + \delta(\omega'-\omega'')\right]\hat{a}^\dagger(\omega''')|0\rangle$$

(6.7.9)

$$= \delta(\omega-\omega'')\delta(\omega'-\omega''') + \delta(\omega'-\omega'')\delta(\omega-\omega''').$$

It therefore follows with the use of eqn (6.7.8) that

$$\left\langle (2_a)_\beta \middle| (2_a)_\beta \right\rangle = 1 \tag{6.7.10}$$

and the pair state is indeed normalized. Similar techniques are used to prove that the state is an eigenfunction of the photon-number operator, defined by the frequency integral in eqn (6.2.9) or the time integral in eqn (6.2.16),

$$\hat{n} \left| (2_a)_\beta \right\rangle = 2 \left| (2_a)_\beta \right\rangle \tag{6.7.11}$$

with eigenvalue 2, as expected for a pair state.

Problem 6.7   Prove that the mean photon flux of the pair state, defined as in eqns (6.2.20) and (6.2.21), is

$$f(t) = \tfrac{1}{2} \int dt' |\beta(t,t') + \beta(t',t)|^2 = 2 \int dt' |\beta(t,t')|^2, \tag{6.7.12}$$

with use of the symmetry property in eqn (6.7.8).

The mean photon number obtained by integration of this expression over t in accordance with eqn (6.2.22) is

$$\langle n \rangle = \int dt\, f(t) = 2, \tag{6.7.13}$$

again as expected for a state that contains 2 photons. Similar to the photon number states, there is no possibility of a stationary state for a single photon pair, as the corresponding joint wavepacket amplitude could not be normalized.

The degree of first-order coherence of the photon pair state is obtained from the general one-dimensional form in eqn (4.12.4) as

$$g^{(1)}(z_1, t_1; z_2, t_2) = \frac{2 \int dt\, \beta^*\left(t_1 - \dfrac{z_1}{c}, t \right) \beta\left(t, t_2 - \dfrac{z_2}{c} \right)}{\left[f\left(t_1 - \dfrac{z_1}{c} \right) f\left(t_2 - \dfrac{z_2}{c} \right) \right]^{1/2}}, \tag{6.7.14}$$

where the pair-state wavepacket spectrum is assumed to be sufficiently narrow for the form of electric-field operator in eqn (6.2.18) to be valid. This expression reduces to that found for the two-photon number state in eqn (6.3.12) in special cases where the two photons are not entangled and the pair-state wavepacket factorizes,

$$\beta(t,t') = \xi(t)\xi(t'), \tag{6.7.15}$$

but the degree of first-order coherence in eqn (6.7.14) applies more generally.

The degree of second-order coherence is likewise obtained from the expression in eqn (4.12.8) as

$$g^{(2)}(z_1,t_1;z_2,t_2) = \frac{2\left|\beta\left(t_1 - \dfrac{z_1}{c}, t_2 - \dfrac{z_2}{c}\right)\right|^2}{f\left(t_1 - \dfrac{z_1}{c}\right)f\left(t_2 - \dfrac{z_2}{c}\right)}. \tag{6.7.16}$$

This is the degree of coherence obtained by instantaneous measurements at the space–time points (z_1,t_1) and (z_2,t_2). However, the photon-number correlation in the numerator and the photon fluxes in the denominator are integrated over the time in many experiments as the photon pair state propagates through the apparatus. This is the case, for example, in the Brown–Twiss interferometer, discussed in §3.8, when the detector integration time T exceeds the wavepacket duration. The required time integrals given by eqns (6.7.7) and (6.7.13) then produce a degree of second-order coherence equal to 1/2, the same value as found for the single-mode $n = 2$ number state in eqn (5.2.3). This value also applies for the wavepacket factorization in eqn (6.7.15), which reproduces the two-photon number-state result from eqn (6.3.14).

The pair states treated above have both photons in the same continuous-mode field, corresponding to a single light beam. However, many experiments generate pairs of photons in distinguishable states, corresponding to two different continuous-mode fields. For example, they may be created in light beams that travel in different directions or they may have different polarizations. Parametric down-conversion can generate pairs of photons with the same or different properties depending on the composition and orientation of the nonlinear material. Consider the photon pair state defined by [7]

$$\left|(1_a,1_b)_\beta\right\rangle = \hat{P}^\dagger_{\beta ab}|0\rangle, \tag{6.7.17}$$

where the photon-pair creation operator is now

$$\hat{P}^\dagger_{\beta ab} = \int d\omega \int d\omega' \beta(\omega,\omega') \hat{a}^\dagger(\omega)\hat{b}^\dagger(\omega'). \tag{6.7.18}$$

The state notation specifies the creation of one photon in each of two different continuous-mode fields, and the a operators commute with the b operators. The two beams are distinguished by labels a and b in the following discussion. The joint two-photon wavepacket spectrum $\beta(\omega,\omega')$ is still normalized according to eqn (6.7.4) and its Fourier transform is defined by eqn (6.7.5). The photon-pair creation operator can be written in the equivalent form

$$\hat{P}^\dagger_{\beta ab} = \int dt \int dt' \beta(t,t') \hat{a}^\dagger(t)\hat{b}^\dagger(t'), \tag{6.7.19}$$

and the normalization condition in eqn (6.7.7) applies. The two frequencies in

$\beta(\omega,\omega')$ and the two times in $\beta(t,t')$ now refer respectively to the fields a and b. Their orders are significant and the symmetry properties in eqn (6.7.8) are no longer required to hold. The pair state itself is again normalized,

$$\left\langle (1_a,1_b)_\beta \middle| (1_a,1_b)_\beta \right\rangle = 1, \tag{6.7.20}$$

and the proof is similar to, but easier than, that of eqn (6.7.10) on account of the commuting of the a with the b operators.

The number operators for the photons of kinds a and b are

$$\hat{n}_a = \int dt\, \hat{a}^\dagger(t)\hat{a}(t) \quad \text{and} \quad \hat{n}_b = \int dt\, \hat{b}^\dagger(t)\hat{b}(t), \tag{6.7.21}$$

similar to eqn (6.2.16), and there are equivalent frequency-dependent expressions, similar to eqn (6.2.9). It follows from the definition in eqns (6.7.17) and (6.7.19) that the pair state satisfies the two eigenvalue relations

$$\hat{n}_a \left| (1_a,1_b)_\beta \right\rangle = \hat{n}_b \left| (1_a,1_b)_\beta \right\rangle = \left| (1_a,1_b)_\beta \right\rangle \tag{6.7.22}$$

and it thus contains one photon of each kind, as expected from its definition. The mean photon fluxes for the two beams,

$$f_a(t) = \int dt' |\beta(t,t')|^2 \quad \text{and} \quad f_b(t) = \int dt' |\beta(t',t)|^2, \tag{6.7.23}$$

generally have different time dependences but their mean photon numbers, obtained by integration over t, both equal unity. There is again no possibility of a stationary pair state.

The two-beam photon pair state has no analogue in the classical theory of light and its nonclassical features are illustrated by evaluation of its degrees of second-order coherence. Consider first the *intrabeam* degrees of coherence, where measurements are made on only one of the beams, a or b. We assume for simplicity that the field operators in the degree of second-order coherence for a nonstationary light beam, given by eqn (4.12.8), are evaluated at the same position in the beam. We also assume that the excitation spectra have narrow bandwidths so that the form of field operator in eqn (6.2.18) is again valid. It follows from the definition of the pair state in eqns (6.7.17) and (6.7.19) that

$$\left\langle (1_a,1_b)_\beta \middle| \hat{a}^\dagger(t_1)\hat{a}^\dagger(t_2)\hat{a}(t_2)\hat{a}(t_1) \middle| (1_a,1_b)_\beta \right\rangle$$
$$= \left\langle (1_a,1_b)_\beta \middle| \hat{b}^\dagger(t_1)\hat{b}^\dagger(t_2)\hat{b}(t_2)\hat{b}(t_1) \middle| (1_a,1_b)_\beta \right\rangle = 0 \tag{6.7.24}$$

and hence

$$g_{a,a}^{(2)}(t_1,t_2) = g_{b,b}^{(2)}(t_1,t_2) = 0. \tag{6.7.25}$$

These results violate the classical inequality of eqn (3.7.6). In physical terms, they express the impossibility of detecting two photons in beams that each contain only one photon.

For the *interbeam* degree of second-order coherence we need the expectation value

$$\left\langle (1_a,1_b)_\beta \left| \hat{a}^\dagger(t_1)\hat{b}^\dagger(t_2)\hat{b}(t_2)\hat{a}(t_1) \right| (1_a,1_b)_\beta \right\rangle = \left| \beta(t_1,t_2) \right|^2. \tag{6.7.26}$$

In physical terms, this quantity represents the correlation of the photon numbers between the two beams and this is, of course, nonzero for the two-beam pair state. The integral of the expectation value over t_1 and t_2, equal to unity in accordance with eqn (6.7.7), gives the total probability for the existence of a pair of photons, one in each beam. It follows that

$$g_{a,b}^{(2)}(t_1,t_2) = \frac{\left| \beta(t_1,t_2) \right|^2}{f_a(t_1)f_b(t_2)}. \tag{6.7.27}$$

It is again the case that many experiments measure time integrals that pick up the total contributions of the two-photon wavepacket to the quantities in the numerator and denominator of eqn (6.7.27). This time-integrated degree of second-order coherence equals unity. More generally, the interbeam degree of second-order coherence does not vanish and the classical inequality in eqn (3.7.27) is accordingly violated. The violated inequality reflects the inability of classical theory to describe the excitation of a single photon in each beam and the correlation of the single-photon excitations between the two beams.

A simple example of a two-photon wavepacket amplitude is the function

$$\beta(t,t') = \left[\frac{4\Delta^2(\Delta^2 + 2\delta^2)}{\pi^2} \right]^{1/4} \exp\left\{ -i\omega_0(t+t') - \Delta^2(t^2 + t'^2) - \delta^2(t-t')^2 \right\},$$

$$\tag{6.7.28}$$

where the first two terms in the exponent resemble the single-photon wavepacket amplitude in eqn (6.3.2), now centred on the origins of the time scales, and the final term represents the correlation between the photons. The durations of the wavepackets of the individual photons are of order $1/\Delta$, while the correlation limits their separation to a time of order $1/\delta$. This amplitude function is symmetrical in the two time variables and it is applicable to both single-mode and two-mode pair states.

Problem 6.8   Derive an expression for the mean photon fluxes of a two-beam pair state whose wavepacket amplitude is given by eqn (6.7.28). Show that the fluxes simplify to

$$f_a(t) = f_b(t) = \frac{2\Delta}{\pi^{1/2}} \exp\left(-4\Delta^2 t^2\right) \qquad (6.7.29)$$

when the photons are in close proximity compared to their individual wavepacket durations, that is $\delta \gg \Delta$.

An effective technique for the generation of a single-photon state relies on the detection of a one photon of a pair to isolate the second photon of the pair for a subsequent measurement [3]. This method is used, for example, in the single-photon experiments on Brown–Twiss and Mach–Zehnder interferometers [6] described in §5.8. The single-photon state obtained in this way is determined by a calculation similar to that made in eqn (5.8.3). Thus, suppose that the observation of the first photon finds its state in the continuous-mode field b to be

$$\left|1_{b\gamma}\right\rangle = \int dt \gamma(t) \hat{b}^\dagger(t) |0\rangle_b, \qquad (6.7.30)$$

where the form of the wavepacket amplitude $\gamma(t)$ is determined by the nature of the measurement. The vacuum states of fields a and b are distinguished by subscripts. Then, similar to eqn (5.8.3), the state of the system after the observation of one of the photons in the pair state defined in eqns (6.7.17) and (6.7.19) is

$$N\left\langle 1_{b\gamma} \left| (1_a, 1_b)_\beta \right\rangle = N \int dt \int dt' \beta(t, t') \gamma^*(t') \hat{a}^\dagger(t) |0\rangle_a. \qquad (6.7.31)$$

The state of field a, conditioned on the observation of a photon in field b, is thus a single-photon number state, as defined by eqns (6.3.4) and (6.3.6). The normalized wavepacket amplitude for field a is

$$\xi(t) = \frac{\int dt' \beta(t, t') \gamma^*(t')}{\left\{\int dt \left| \int dt' \beta(t, t') \gamma^*(t') \right|^2\right\}^{1/2}}. \qquad (6.7.32)$$

The single-photon state that is isolated by the observation of its partner is evaluated straightforwardly when the wavepacket amplitudes of the pair-state and of the measurement are known functions. Consider, for example, a measurement of the field b described by an amplitude function

$$\gamma(t) = \left(2D^2/\pi\right)^{1/4} \exp\left(-i\omega_0 t - D^2 t^2\right), \qquad (6.7.33)$$

which represents a measurement duration of order $1/D$ centred on $t = 0$.

Problem 6.9   Derive the form of the single-photon amplitude function $\xi(t)$ for field a produced by a field b measurement, with amplitude given by eqn (6.7.33), for a pair state whose amplitude is given by eqn (6.7.28). Show that the single-photon wavepacket function simplifies to

$$\xi(t) = \left[\frac{2(2\Delta^2 + D^2)}{\pi} \right]^{1/4} \exp\{-i\omega_0 t - (2\Delta^2 + D^2)t^2\} \quad (6.7.34)$$

when the photons are in close proximity compared to their individual wavepacket durations, that is $\delta \gg \Delta$.

The condition $\delta \gg \Delta$ corresponds closely to the pair states generated in parametric down-conversion (§9.4) but not to those produced by two-photon cascade emission (§8.6). When the condition is satisfied, the duration of the single-photon wavepacket is of the order of the inverse square-root of $2\Delta^2 + D^2$.

6.8 Two-photon interference

Some of the nonclassical features of the photon pair states are nicely illustrated by experiments on two-photon interference, where the photons are fed into the input arms of a beam splitter. The transmission of the photons through the beam splitter exhibits behaviour similar to that of classical particles when the pair state excites a single continuous-mode field in one input arm. However, striking nonclassical effects occur when the pair state excites both continuous-mode fields in the two input arms of the beam splitter. Although the relevant experiments generally use stationary beams of two-photon pairs, discussed in §6.9, the theory is adequately presented in terms of the single pair states of the previous section.

Consider first the incidence of a pair state in the *single* continuous-mode field of the input arm labelled 1 in Fig. 5.15. The state is defined by eqns (6.7.2) and (6.7.6), but with substitution of the arm 1 creation operators to give

$$\left| (2_1)_\beta \right\rangle = 2^{-1/2} \int dt \int dt' \beta(t,t') \hat{a}_1^\dagger(t) \hat{a}_1^\dagger(t') |0\rangle, \quad (6.8.1)$$

where the wavepacket amplitude $\beta(t,t')$ satisfies the symmetry relation in eqn (6.7.8). The input arm 2 is in its vacuum state. The relations between input and output creation operators are obtained from eqn (6.6.2), with use of the properties of the beam-splitter coefficients from eqn (5.7.1), as

$$\hat{a}_1^\dagger(t) = \mathcal{R}\hat{a}_3^\dagger(t) + \mathcal{T}\hat{a}_4^\dagger(t) \quad \text{and} \quad \hat{a}_2^\dagger(t) = \mathcal{T}\hat{a}_3^\dagger(t) + \mathcal{R}\hat{a}_4^\dagger(t). \quad (6.8.2)$$

Thus substitution for the $\hat{a}_1^\dagger$ operators in eqn (6.8.1) gives

$$\left| (2_1)_\beta \right\rangle = 2^{-1/2} \int dt \int dt' \beta(t,t')$$
$$\times \left[\mathcal{R}^2 \hat{a}_3^\dagger(t) \hat{a}_3^\dagger(t') + 2\mathcal{R}\mathcal{T}\hat{a}_3^\dagger(t) \hat{a}_4^\dagger(t') + \mathcal{T}^2 \hat{a}_4^\dagger(t) \hat{a}_4^\dagger(t') \right] |0\rangle, \quad (6.8.3)$$

with use of the symmetry property in eqn (6.7.8). The right-hand side can be expressed in terms of the single continuous-mode states of eqns (6.7.2) and (6.7.6) and the two continuous-mode states of eqns (6.7.17) and (6.7.19) as

$$\left|(2_1)_\beta\right\rangle = \mathcal{R}^2\left|(2_3)_\beta\right\rangle + 2^{1/2}\mathcal{R}\mathcal{T}\left|(1_3,1_4)_\beta\right\rangle + \mathcal{T}^2\left|(2_4)_\beta\right\rangle, \tag{6.8.4}$$

where the numerical subscripts indicate the beam-splitter output arms. This entangled output state is an extension to a pair-state input of the single-photon output state derived in eqn (5.8.2). The different notation for the output states used in eqn (6.8.4) reflects the entangled nature of the pair states themselves.

The result of the calculation is conveniently expressed in terms of the probability distribution for the two photons between the two output arms. In an obvious notation

$$P(2_3,0_4) = |\mathcal{R}|^4, \quad P(1_3,1_4) = 2|\mathcal{R}|^2|\mathcal{T}|^2, \quad P(0_3,2_4) = |\mathcal{T}|^4. \tag{6.8.5}$$

The probability distribution is thus independent of the form of the pair-state wavepacket. It reproduces the binomial form found for the distribution of classical particles between two channels with probabilities for each particle of $|\mathcal{R}|^2$ and $|\mathcal{T}|^2$. The two-photon output distribution has been measured in a series of experiments [8] with input photon pair states generated by parametric down-conversion incident on a series of beam splitters with different $|\mathcal{R}|^2$ and $|\mathcal{T}|^2$.

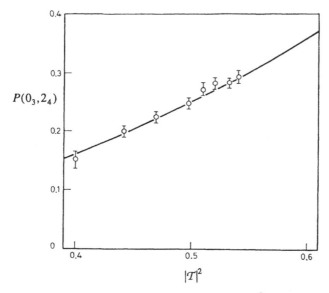

Fig. 6.2. Variation with beam-splitter coefficient $|\mathcal{T}|^2$ of the probability $P(0_3,2_4)$ for transmission to arm 4 of both photons in an input pair state in arm 1. The continuous curve shows the calculated probability of $|\mathcal{T}|^4$ from eqn (6.8.5). (After [8])

Figure 6.2 shows that the results for $P(0_3,2_4)$ are in close agreement with the expression from eqn (6.8.5).

Now consider the incidence of a photon pair state that excites two *different* continuous-mode fields. The corresponding light beams are arranged to arrive at the beam splitter in the input arms labelled 1 and 2 in Fig. 5.15. The state is defined by eqns (6.7.17) and (6.7.19), but with substitution of arm 1 and 2 creation operators to give

$$\left|(1_1,1_2)_\beta\right\rangle = \int dt \int dt' \beta(t,t') \hat{a}_1^\dagger(t) \hat{a}_2^\dagger(t') |0\rangle. \tag{6.8.6}$$

The times t and t' are now associated with arms 1 and 2 respectively, and the wavepacket amplitude $\beta(t,t')$ no longer satisfies the symmetry property in eqn (6.7.8). Substitution of the input operators by output operators with the use of eqn (6.8.2) gives

$$\left|(1_1,1_2)_\beta\right\rangle = \int dt \int dt' \beta(t,t') \Big[\mathcal{R}\mathcal{T}\hat{a}_3^\dagger(t)\hat{a}_3^\dagger(t') + \mathcal{R}^2 \hat{a}_3^\dagger(t)\hat{a}_4^\dagger(t')$$
$$+ \mathcal{T}^2 \hat{a}_4^\dagger(t)\hat{a}_3^\dagger(t') + \mathcal{T}\mathcal{R}\hat{a}_4^\dagger(t)\hat{a}_4^\dagger(t')\Big]|0\rangle. \tag{6.8.7}$$

The right-hand side is rearranged, with suitable interchanges of the time variables, as

$$\left|(1_1,1_2)_\beta\right\rangle = \int dt \int dt' \Big\{ \tfrac{1}{2}\mathcal{R}\mathcal{T}[\beta(t,t') + \beta(t',t)]\big[\hat{a}_3^\dagger(t)\hat{a}_3^\dagger(t') + \hat{a}_4^\dagger(t)\hat{a}_4^\dagger(t')\big]$$
$$+ \big[\mathcal{R}^2\beta(t,t') + \mathcal{T}^2\beta(t',t)\big]\hat{a}_3^\dagger(t)\hat{a}_4^\dagger(t')\Big\}|0\rangle. \tag{6.8.8}$$

The sum of β functions in the first term of the integrand is invariant under interchange of t and t'. Comparison of the two contributions from this term with the form of the single continuous-mode pair state in eqns (6.7.2) and (6.7.6) gives the probabilities for both photons to emerge in the same output arm as

$$P(2_3,0_4) = P(0_3,2_4) = \tfrac{1}{2}|\mathcal{R}|^2|\mathcal{T}|^2 \int dt \int dt' |\beta(t,t') + \beta(t',t)|^2. \tag{6.8.9}$$

Use of the normalization property in eqn (6.7.7) and further interchange of the time variables gives

$$P(2_3,0_4) = P(0_3,2_4) = |\mathcal{R}|^2|\mathcal{T}|^2\big(1 + |J|^2\big), \tag{6.8.10}$$

where the *overlap integral* of the two input states is defined by

$$|J|^2 = \int dt \int dt' \beta^*(t,t')\beta(t',t). \tag{6.8.11}$$

The double integral is easily shown to be a real quantity as it must be invariant under interchange of the integration variables t and t'. A similar treatment of the second term in the integrand of eqn (6.8.8) gives the probability for one photon to emerge in each output arm as

$$P(1_3, 1_4) = \int dt \int dt' \left| \mathcal{R}^2 \beta(t, t') + \mathcal{T}^2 \beta(t', t) \right|^2. \tag{6.8.12}$$

Further use of the normalization in eqn (6.7.7) and the standard beam-splitter properties in eqn (5.7.1) leads to

$$P(1_3, 1_4) = 1 - 2|\mathcal{R}|^2 |\mathcal{T}|^2 \left(1 + |J|^2 \right). \tag{6.8.13}$$

The output probability distribution for an input state with one photon each in arms 1 and 2 is thus given in general by eqns (6.8.10) and (6.8.13). In contrast to the distribution in eqn (6.8.5), the output does now depend on the form of the pair-state wavepacket. The distribution may also be compared with that for classical particles that reflect and transmit independently with probabilities $|\mathcal{R}|^2$ and $|\mathcal{T}|^2$, given by

$$P(2_3, 0_4) = P(0_3, 2_4) = |\mathcal{R}|^2 |\mathcal{T}|^2 \quad \text{and} \quad P(1_3, 1_4) = |\mathcal{R}|^4 + |\mathcal{T}|^4. \tag{6.8.14}$$

It is seen by comparison with eqn (6.8.10) that photons are more likely than classical particles to emerge in the same output arm, and correspondingly less likely to emerge in different arms.

The joint wavepacket amplitude factorizes as

$$\beta(t, t') = \xi_1(t) \xi_2(t') \tag{6.8.15}$$

for input states that are not entangled and the overlap integral from eqn (6.8.11) becomes

$$|J|^2 = \left| \int dt \, \xi_1^*(t) \xi_2(t) \right|^2. \tag{6.8.16}$$

The integral has a simple form for individual wavepacket amplitudes $\xi_1(t)$ and $\xi_2(t)$ of the Gaussian shape shown in eqn (6.3.2) but with different arrival times at the beam splitter, denoted respectively t_{01} and t_{02}, when

$$J = \exp\left\{ -\tfrac{1}{2} \Delta^2 (t_{01} - t_{02})^2 \right\}. \tag{6.8.17}$$

Thus for simultaneous arrival of the peaks of the two input-photon pulses at the beam splitter, with $t_{01} = t_{02}$, the unit overlap integral produces a probability distribution

$$P(2_3,0_4) = P(0_3,2_4) = 2|\mathcal{R}|^2|\mathcal{T}|^2 \qquad (6.8.18)$$

and

$$P(1_3,1_4) = \left(|\mathcal{R}|^2 - |\mathcal{T}|^2\right)^2 = \left|\mathcal{R}^2 + \mathcal{T}^2\right|^2. \qquad (6.8.19)$$

The probability for both photons to emerge in the same arm is now twice the classical value given in eqn (6.8.14). This property holds generally for scattering processes with identical incident boson particles [9] and its origin lies in the nature of the operator commutation relations imposed by the Bose–Einstein statistics. Quite different behaviour occurs for corresponding experiments with identical incident fermion particles, where the probability for the particles to emerge in the same output arm vanishes in accordance with the Pauli exclusion principle.

Another interesting special case is that of a 50:50 beam splitter with $|\mathcal{R}|^2$ and $|\mathcal{T}|^2$ both equal to 1/2, where the distribution in eqns (6.8.10) and (6.8.13) reduces to

$$P(2_3,0_4) = P(0_3,2_4) = \tfrac{1}{4}\left(1+|J|^2\right) \quad \text{and} \quad P(1_3,1_4) = \tfrac{1}{2}\left(1-|J|^2\right). \qquad (6.8.20)$$

Figure 6.3 shows results for the number of measured coincidences between photons in the two output arms of a 50:50 beam splitter as a function of the relative times of arrival of the incident photons [10]. The latter were adjusted by

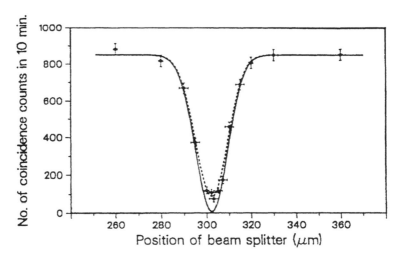

Fig. 6.3. Measured numbers of coincidences in the two output arms of a beam splitter produced by photon pair states that excite both input arms. The horizontal axis shows the relative times of arrival of the photon wavepackets, as varied by displacement of the beam splitter. The points, fitted by the dashed curve, show the measurements and the continuous curve shows the theoretical expression. (After [10])

shifts in the position of the beam splitter towards one input beam or the other. The coincidence probability is given by the final element of the probability distribution in eqn (6.8.20) and the experimental results are well described by this expression when the Gaussian form from eqn (6.8.17) is inserted. The output photon coincidences should be completely removed for simultaneous arrival of the photons, and this effect is observed experimentally, apart from a small residual coincidence rate caused by a less-than-perfect overlap of the input photon wavepackets. The pairs of incident photons were in entangled states generated by parametric down-conversion but the dependence of the overlap integral on relative times of arrival retains the features discussed above in the context of non-entangled states.

The totally destructive two-photon interference for simultaneous photon arrivals is a quantum-mechanical effect that does not occur in any classical description of the experiment. The interference occurs because there are two routes by which one photon can appear in each output arm. Thus each incident photon can be reflected, with probability amplitude $\mathcal{R}^2$, or each photon can be transmitted, with probability amplitude $\mathcal{T}^2$. These amplitudes are added in the second form of $P(1_3, 1_4)$ in eqn (6.8.19) and the standard properties of the coefficients then give the first form of the probability, with the clear total cancellation of the amplitudes for a 50:50 beam splitter. It is emphasized that the interference results from summation of the two probability amplitudes for a common outcome, and it is not a consequence of some kind of interference between the photons within the beam splitter.

Problem 6.10 Draw a sketch of the vectors that represent the beam-splitter reflection and transmission coefficients $\mathcal{R}$ and $\mathcal{T}$ in the complex plane, such that the standard relations (5.7.1) are satisfied. Show the vectors that represent $\mathcal{R}^2$ and $\mathcal{T}^2$, and verify the vanishing of $\mathcal{R}^2 + \mathcal{T}^2$ for a 50:50 beam splitter.

6.9 Squeezed light

The photon-pair creation operators defined in §6.7 are used to construct coherent pair-superposition states in much the same way as the single-photon wavepacket creation operator is used to form coherent states by the procedure outlined in §6.4. The pairs may be excitations in a single beam or in two distinct beams and both kinds of coherent superposition state have interesting properties. They represent different varieties of continuous-mode quadrature-squeezed vacuum state.

Consider first the single-beam states, where the relevant photon-pair creation operator is defined in eqns (6.7.2) and (6.7.3) or (6.7.6). This operator and its Hermitian conjugate are used to define *continuous-mode two-photon coherent states* as

$$|\{\{\beta aa\}\}\rangle = \exp\left(\hat{P}_{\beta aa} - \hat{P}^{\dagger}_{\beta aa}\right)|0\rangle, \tag{6.9.1}$$

similar to the definition of discrete-mode two-photon coherent states in §5.6 but with the coherent amplitude α set equal to zero. The definition is analogous to that for the single-photon coherent states in eqn (6.4.1) but with a conventional change of sign in the exponent. The two-photon coherent states are automatically normalized for any wavepacket amplitudes $\beta(\omega,\omega')$ or $\beta(t,t')$, and these functions are not required to satisfy the normalization conditions in eqns (6.7.4) and (6.7.7). The states are also known as *single-continuous-mode squeezed vacuum states*, and the continuous-mode exponential operator in eqn (6.9.1) resembles that in the single-discrete-mode squeeze operator in eqn (5.5.3).

An important special case of the two-photon coherent states occurs for a process in which each pair of photons is produced by the splitting of a single photon from a monochromatic pump beam of frequency ω_p. This nonlinear-optical process of parametric down-conversion is treated in detail in §9.4. The resulting wavepacket amplitude has the form

$$\beta(\omega,\omega') = 2^{-1/2}\zeta(\omega)\delta\big(\omega+\omega'-\omega_p\big) \quad 0 \le \omega, \omega' \le \omega_p, \tag{6.9.2}$$

where the delta-function ensures energy conservation in the splitting process. The complex amplitude function $\zeta(\omega)$ must satisfy the relation

$$\zeta\big(\omega_p - \omega\big) = \zeta(\omega) \tag{6.9.3}$$

to comply with the symmetry requirement in eqn (6.7.8). The photon-pair creation operator of eqn (6.7.3) reduces to

$$\hat{P}^\dagger_{\beta aa} = \tfrac{1}{2} \int_0^{\omega_p} d\omega \zeta(\omega) \hat{a}^\dagger(\omega) \hat{a}^\dagger\big(\omega_p - \omega\big). \tag{6.9.4}$$

The state defined by eqn (6.9.1) for the form of wavepacket amplitude in eqn (6.9.2) represents an entangled continuous-mode field in which there are correlations between the contributions of pairs of frequencies symmetrically displaced from a central frequency $\omega_p/2$. The wavepacket amplitude in eqn (6.9.2) cannot be normalized but the resulting squeezed states are themselves normalized and they have well-behaved observable properties. The Fourier-transformed amplitude obtained with the use of eqn (6.7.5) is

$$\beta(t,t') = \tfrac{1}{2}\pi^{-1/2}\zeta(t-t')\exp\big(-i\omega_p t'\big), \tag{6.9.5}$$

where $\zeta(t)$ is the Fourier transform of $\zeta(\omega)$. The treatment that follows is restricted to single-continuous-mode squeezed states with the special form of wavepacket amplitude defined by eqns (6.9.2) and (6.9.5).

The complex function $\zeta(\omega)$ is conveniently separated into a real amplitude and a phase factor as

$$\zeta(\omega) = s(\omega)\exp[i\vartheta(\omega)], \tag{6.9.6}$$

analogous to the separation of the discrete-mode operator in eqn (5.5.4). Properties analogous to the discrete-mode result in eqn (5.5.9) can also be proved.

Problem 6.11 Prove the transformation property

$$\exp\left(\hat{P}^{\dagger}_{\beta aa} - \hat{P}_{\beta aa}\right)\hat{a}(\omega)\exp\left(\hat{P}_{\beta aa} - \hat{P}^{\dagger}_{\beta aa}\right)$$
$$= \hat{a}(\omega)\cosh[s(\omega)] - \hat{a}^{\dagger}\left(\omega_{p} - \omega\right)\exp[i\vartheta(\omega)]\sinh[s(\omega)].$$

$$(6.9.7)$$

It also follows with the use of eqn (6.9.3) that

$$\exp\left(\hat{P}^{\dagger}_{\beta aa} - \hat{P}_{\beta aa}\right)\hat{a}\left(\omega_{p} - \omega\right)\exp\left(\hat{P}_{\beta aa} - \hat{P}^{\dagger}_{\beta aa}\right)$$
$$= \hat{a}\left(\omega_{p} - \omega\right)\cosh[s(\omega)] - \hat{a}^{\dagger}(\omega)\exp[i\vartheta(\omega)]\sinh[s(\omega)].$$

$$(6.9.8)$$

The frequency correlation function is readily calculated with the use of these transformation properties and their Hermitian conjugates as

$$\left\langle\hat{a}^{\dagger}(\omega)\hat{a}(\omega')\right\rangle = \sinh^{2}[s(\omega)]\delta(\omega - \omega').$$

$$(6.9.9)$$

The light beam is therefore stationary in accordance with eqn (6.2.23) and its time-independent mean photon flux is obtained from eqn (6.2.24) as

$$f(t) = (2\pi)^{-1}\int d\omega\sinh^{2}[s(\omega)] = F.$$

$$(6.9.10)$$

The correlation of two destruction operators is similarly obtained as

$$\left\langle\hat{a}(\omega)\hat{a}(\omega')\right\rangle = -\tfrac{1}{2}\exp[i\vartheta(\omega)]\sinh[2s(\omega)]\delta\left(\omega + \omega' - \omega_{p}\right),$$

$$(6.9.11)$$

and the corresponding correlation of two creation operators is given by the conjugate expression.

The squeezed vacuum state defined by eqn (6.9.1) is generally nonstationary and pulsed squeezed light is generated by experiments in which the pump beam is itself pulsed with frequencies ω_{p} that spread over a continuous range of values. The assumption of a monochromatic pump embodied in the two-photon wavepacket amplitude of eqn (6.9.2) produces a stationary beam of quadrature squeezed light analogous to the way in which the form of single-photon wavepacket amplitude in eqn (6.4.10) produces a stationary beam of coherent light.

The relations (6.9.7) and (6.9.8) are used to calculate higher-order expectation values of the continuous-mode operators, analogous to the discrete-mode results in eqns (5.5.12) to (5.5.14). However, the main interest and applications of squeezed light lie in its quadrature-operator properties, and these are studied experimentally by means of the homodyne detection technique treated in §6.11.

Further consideration of the single-continuous-mode squeezed light is deferred to this section.

Now consider the two-photon coherent states defined by

$$|\{\{\beta ab\}\}\rangle = \exp(\hat{P}_{\beta ab} - \hat{P}_{\beta ab}^\dagger)|0\rangle,$$ (6.9.12)

in which the photons are excited in different continuous-mode fields. The pair-state creation operator is given by eqn (6.7.18) or (6.7.19), and the above state is again normalized for arbitrary wavepacket amplitudes, without any need for the normalization conditions in eqns (6.7.4) and (6.7.7). There is no requirement that the symmetry properties in eqn (6.7.8) should apply to these states. The states are known as *two-continuous-mode squeezed vacuum states*. They are generally nonstationary and they may, for example, be generated in the form of a pulse or succession of pulses. However, we here make the assumption that they are produced by parametric splitting, or down-conversion, of the photons in a mono-chromatic pump beam. The wavepacket amplitude is accordingly taken in the form

$$\beta(\omega, \omega') = \zeta(\omega)\delta(\omega + \omega' - \omega_p) \quad 0 \le \omega, \omega' \le \omega_p,$$ (6.9.13)

similar to eqn (6.9.2) but without the symmetry property in eqn (6.9.3). The pair creation operator from eqn (6.7.18) is thus

$$\hat{P}_{\beta ab}^\dagger = \int_0^{\omega_p} d\omega \, \zeta(\omega)\hat{a}^\dagger(\omega)\hat{b}^\dagger(\omega_p - \omega).$$ (6.9.14)

This special variety of squeezed vacuum state corresponds to twin stationary light beams. The complex function $\zeta(\omega)$ is again expressed in terms of an amplitude and phase, as in eqn (6.9.6).

The transformation properties of eqns (6.9.7) and (6.9.8) are readily recalculated for the two-mode states as

$$\exp(\hat{P}_{\beta ab}^\dagger - \hat{P}_{\beta ab})\hat{a}(\omega)\exp(\hat{P}_{\beta ab} - \hat{P}_{\beta ab}^\dagger)$$
$$= \hat{a}(\omega)\cosh[s(\omega)] - \hat{b}^\dagger(\omega_p - \omega)\exp[i\vartheta(\omega)]\sinh[s(\omega)]$$ (6.9.15)

and

$$\exp(\hat{P}_{\beta ab}^\dagger - \hat{P}_{\beta ab})\hat{b}(\omega_p - \omega)\exp(\hat{P}_{\beta ab} - \hat{P}_{\beta ab}^\dagger)$$
$$= \hat{b}(\omega_p - \omega)\cosh[s(\omega)] - \hat{a}^\dagger(\omega)\exp[i\vartheta(\omega)]\sinh[s(\omega)].$$ (6.9.16)

These expressions and their Hermitian conjugates enable the calculation of expectation values of products of the creation and destruction operators. Thus the time-independent mean photon fluxes in the two beams are obtained from eqn (6.2.24) as

$$f_a(t) = f_b(t) = (2\pi)^{-1}\int d\omega \sinh^2[s(\omega)] = F. \tag{6.9.17}$$

Problem 6.12 Prove that the degrees of first-order coherence of the beams are

$$g_a^{(1)}(\tau) = g_b^{(1)}(\tau) = \frac{\displaystyle\int_0^{\omega_p} d\omega \exp(-i\omega\tau)\sinh^2[s(\omega)]}{\displaystyle\int_0^{\omega_p} d\omega \sinh^2[s(\omega)]}, \tag{6.9.18}$$

when the wavepacket spectrum has a narrow bandwidth.

The nonclassical properties of the single pair states derived in §6.7 are lost to some extent by the formation of the superposition of multiple pair states given by eqns (6.9.12) and (6.9.14). Thus the intrabeam expectation values analogous to eqn (6.7.24) no longer vanish, but are replaced by

$$\langle\{\{\beta ab\}\}|\hat{a}^\dagger(t_1)\hat{a}^\dagger(t_2)\hat{a}(t_2)\hat{a}(t_1)|\{\{\beta ab\}\}\rangle$$
$$= \langle\{\{\beta ab\}\}|\hat{b}^\dagger(t_1)\hat{b}^\dagger(t_2)\hat{b}(t_2)\hat{b}(t_1)|\{\{\beta ab\}\}\rangle \tag{6.9.19}$$
$$= F^2 + \left|\frac{1}{2\pi}\int_0^{\omega_p} d\omega \exp[-i\omega(t_2 - t_1)]\sinh^2[s(\omega)]\right|^2,$$

where F is the mean photon flux from eqn (6.9.17). The intrabeam degrees of second-order coherence defined by eqn (4.12.9) can thus be written in the forms

$$g_{a,a}^{(2)}(\tau) = g_{b,b}^{(2)}(\tau) = 1 + |g_a^{(1)}(\tau)|^2 = 1 + |g_b^{(1)}(\tau)|^2, \tag{6.9.20}$$

where the degrees of first-order coherence are given by eqn (6.9.18). The intrabeam degrees of first and second-order coherence are thus related by the same expression as eqn (6.1.11), characteristic of chaotic light. In particular, the degrees of second-order coherence equal 2 for zero time delay. The nonclassical properties of the single pair states expressed by eqn (6.7.25) do not therefore survive the formation of two-photon coherent states, or squeezed states. Indeed, the two beams individually display all the properties of chaotic, or thermal, light [2,11].

Consider, however, the interbeam expectation value analogous to the pair-state result in eqn (6.7.26), given by

$$\langle\{\{\beta ab\}\}|\hat{a}^{\dagger}(t_1)\hat{b}^{\dagger}(t_2)\hat{b}(t_2)\hat{a}(t_1)|\{\{\beta ab\}\}\rangle$$

$$= F^2 + \left|\frac{1}{2\pi}\int_0^{\omega_p}\!\!d\omega\exp\!\left[-i\omega(t_2-t_1)-i\vartheta(\omega)\right]\sinh[s(\omega)]\cosh[s(\omega)]\right|^2. \qquad (6.9.21)$$

The interbeam degree of second-order coherence is therefore

$$g_{a,b}^{(2)}(\tau) = 1 + \frac{\left|\displaystyle\int_0^{\omega_p}\!\!d\omega\exp\!\left[-i\omega\tau-i\vartheta(\omega)\right]\sinh[s(\omega)]\cosh[s(\omega)]\right|^2}{\left|\displaystyle\int_0^{\omega_p}\!\!d\omega\sinh^2[s(\omega)]\right|^2}. \qquad (6.9.22)$$

For a sufficiently slowly-varying phase function $\vartheta(\omega)$, it follows that, as the hyperbolic sine is smaller than the cosine,

$$\left[g_{a,b}^{(2)}(\tau)\right]^2 > g_{a,a}^{(2)}(\tau)g_{b,b}^{(2)}(\tau), \qquad (6.9.23)$$

and the classical inequality of eqn (3.7.27) is violated. The violation is particularly strong for $s(\omega) \ll 1$, when the photon fluxes of the beams are small. In this case, the correlation (6.9.19) between pairs of photons in the same beam is small but the construction of the state from pairs of photons in the different beams leads to the significant interbeam correlations described by eqn (6.9.21). These properties are best illustrated by a simple example.

Problem 6.13 Consider a two-continuous-mode squeezed state whose amplitude function in eqn (6.9.13) has the Gaussian form

$$\zeta(\omega) = \left(\frac{2\pi F^2}{\Delta^2}\right)^{1/4}\exp\!\left\{-\frac{\left[(\omega_p/2)-\omega\right]^2}{4\Delta^2}\right\}, \qquad (6.9.24)$$

where the parameters satisfy $F \ll \Delta \ll \omega_p$. Evaluate the beam fluxes and show that the intrabeam degrees of second-order coherence are given by the classical expression for Gaussian–Gaussian light in eqn (3.7.20). Derive the interbeam degree of second-order coherence as

$$g_{a,b}^{(2)}(\tau) \approx 1 + \left(\frac{2}{\pi}\right)^{1/2}\frac{\Delta}{F}\exp\!\left(-2\Delta^2\tau^2\right). \qquad (6.9.25)$$

The results for this example show the twin nature of the two light beams, as the average separation between adjacent photons in the same beam is of order $1/F$

but the average separation between the twin photons, one in each beam, is a much smaller time interval of order $1/\Delta$. The two-continuous-mode squeezed vacuum states thus combine the classical-like feature of chaotic fluctuations in the individual beams with the nonclassical feature of strong photon-pair correlations between the beams. The first measurements [12] of photon correlations in beams generated by parametric down-conversion gave results in which the left-hand side of the inequality (6.9.23) exceeded the right-hand side by a factor of 10^4. Similar beams are used for the two-photon interference experiments described in §6.8.

6.10 Quantum theory of direct detection

The results derived so far in the present chapter refer to excitations of the quantized electromagnetic field, without any consideration of the extents to which their properties can be measured. Most quantum-optical measurements make use of photoelectric detection and it is possible in principle to determine the statistical distributions of photocounts produced by specific kinds of light beam. The photocount distributions derived in §3.9 are based on the semiclassical expression (3.9.12). A more general theory is provided by a fully quantum-mechanical formulation of the photodetection statistics, based on the quantum theory of the phototube from §4.11. The quantum theory can handle field excitations that are not describable in classical terms and it provides the definitive results for all kinds of field excitation. The present section treats *direct detection*, where the photon flux of the light beam to be measured is directly incident on a phototube, as in the semiclassical theory of §3.9.

It is convenient to begin the account of direct detection with derivations of expressions for the measured mean and variance of the photon flux. These quantities are particularly important in the quantum theory of detection as they provide the indicators of the nonclassical natures of light beams discussed in §5.10. The complete photocount distribution is deferred to later in the present section. The photodetected light beams are assumed to be narrow-band excitations of the quantized field so that the time-dependent operators of eqn (6.2.13) are valid. The photodetection intensity operator derived in eqn (4.11.6) thus takes the form of the Poynting operator in eqn (6.2.19), which is proportional to the photon flux. The number of photons arriving at the detector during the integration time T is represented by the operator [1]

$$\hat{M}(t,T) = \int\limits_{t}^{t+T} \mathrm{d}t'\,\hat{f}(t') = \int\limits_{t}^{t+T} \mathrm{d}t'\,\hat{a}^\dagger(t')\hat{a}(t'), \qquad (6.10.1)$$

analogous to the classical integrated mean intensity defined in eqn (3.9.10). This operator is similar to the photon number operator defined in eqn (6.2.16), but the expectation value of $\hat{M}(t,T)$ remains finite even for stationary light beams because of the finite value of T. The operator represents the property of the light beam that the photodetection experiment seeks to measure. The incident flux depends on the position of the detector for a nonstationary light beam but this is left implicit in the notation of eqn (6.10.1).

The amplification of single photoionized electrons in a phototube to produce detectable current pulses, described in §3.9, ensures that the device is capable in principle of registering the arrival of individual photons. However, for a variety of reasons, practical detectors do not achieve complete conversion of incident photons to photocounts. For example, the spatial profile of the incident beam may not match the active area of the phototube, photons may fail to trigger an ionization event and the recovery time, or *dead time*, of the phototube after a successful detection may mask a subsequent photoionization. The fraction of photon arrivals that are registered as photocounts is the *quantum efficiency* of the detector, denoted by η. Typical values of η for phototubes lie in the range 0.1 to 0.4, although larger values can be achieved by other devices. It is accordingly necessary to include the quantum efficiency in any realistic model of photodetection.

Figure 6.4 shows how an inefficient photodetector can be represented as a hypothetical perfectly-efficient detector preceded by a beam splitter [13]. The beam splitter has reflection and transmission coefficients

$$\mathcal{R} = i(1 - \eta)^{1/2} \quad \text{and} \quad \mathcal{T} = \eta^{1/2} \tag{6.10.2}$$

that satisfy the standard requirements in eqn (5.7.1). The measured light beam is incident in arm 1 while the field in arm 2, with operator $\hat{v}(t)$, is in a continuous-mode vacuum state. The output in arm 3 is lost while the output in arm 4, with operator $\hat{d}(t)$, falls on to the perfect detector. The required beam-splitter input–output relation, similar to eqn (6.6.2), is

$$\hat{d}(t) = \eta^{1/2}\hat{a}(t) + i(1 - \eta)^{1/2}\hat{v}(t). \tag{6.10.3}$$

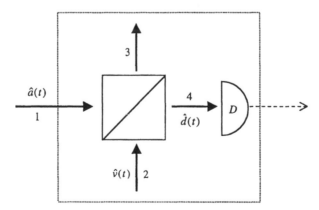

Fig. 6.4. Representation of an inefficient photodetector showing the input field $\hat{a}(t)$ entering on the left, the η beam splitter, the perfect detector D, and the output photocurrent leaving on the right.

The *photocount operator* is defined as a modified form of eqn (6.10.1) in which the $\hat{a}$ operators are replaced by $\hat{d}$ operators to give

$$\hat{M}_{\mathrm{D}}(t,T) = \int_{t}^{t+T} dt' \hat{d}^{\dagger}(t')\hat{d}(t'), \tag{6.10.4}$$

where the D subscript, here and below, indicates quantities measured by the inefficient detector. This expression returns to the form in eqn (6.10.1) for $\eta = 1$ but eqn (6.10.4) must be used for $\eta < 1$. The photocount operator represents the primary quantity measured by the inefficient photodetector.

The mean photocount is obtained from the average value of the photocount operator as

$$\langle m \rangle = \langle \hat{M}_{\mathrm{D}}(t,T) \rangle = \eta \langle \hat{M}(t,T) \rangle, \tag{6.10.5}$$

where m denotes the number of photocounts, as in the semiclassical theory of §3.9. The angle brackets on the right indicate quantum-mechanical expectation values and the standard vacuum-state condition is applied to the input operator $\hat{v}(t)$. The second factorial moment of the photocount numbers is

$$\langle m(m-1) \rangle = \langle \hat{M}_{\mathrm{D}}(t,T)\left[\hat{M}_{\mathrm{D}}(t,T)\right] - 1 \rangle = \langle : \left[\hat{M}_{\mathrm{D}}(t,T)\right]^{2} : \rangle$$
$$= \int_{t}^{t+T} dt' \int_{t}^{t+T} dt'' \langle \hat{d}^{\dagger}(t')\hat{d}^{\dagger}(t'')\hat{d}(t'')\hat{d}(t') \rangle. \tag{6.10.6}$$

The vacuum operators $\hat{v}(t)$ make no contribution to the normally-ordered expectation value when the d operators are substituted from eqn (6.10.3) and its Hermitian conjugate, and the result is

$$\langle m(m-1) \rangle = \eta^{2} \langle : \left[\hat{M}(t,T)\right]^{2} : \rangle = \eta^{2} \langle \hat{M}(t,T)\left[\hat{M}(t,T) - 1\right] \rangle. \tag{6.10.7}$$

As a consequence of the modelling of the inefficiency of the photodetector by an effective beam splitter, the relations (6.10.5) and (6.10.7) between the means and the second factorial moments of the photocount and the input photon number distributions are the same as the standard beam-splitter results of eqns (5.9.2) and (5.9.11).

The variance of the photocount distribution is obtained from eqns (6.10.5) and (6.10.7) as

$$(\Delta m)^{2} = \eta^{2} \langle \left[\Delta\hat{M}(t,T)\right]^{2} \rangle + \eta(1-\eta)\langle \hat{M}(t,T) \rangle. \tag{6.10.8}$$

The first contribution on the right is the variance of the integrated photon number of the light beam, represented by the operator defined by eqn (6.10.1), scaled by the square of the quantum efficiency. The second contribution is a *partition noise* that results from the random selection of a fraction η of the incident photons by the imperfect photodetector.

The measured degree of second-order coherence for zero time delay is defined in terms of the photocount averages as

$$g_D^{(2)}(0) = \frac{\langle m(m-1) \rangle}{\langle m \rangle^2} = \frac{\langle \hat{M}(t,T)[\hat{M}(t,T)-1] \rangle}{\langle \hat{M}(t,T) \rangle^2}, \tag{6.10.9}$$

where the expressions from eqns (6.10.5) and (6.10.7) are substituted. The quantum efficiency thus cancels and the measured degree of second-order coherence equals that of the incident light beam, with allowance for the finite integration time in the definition (6.10.1). This measured degree of coherence has a non-classical range of values

$$1 - \frac{1}{\langle \hat{M}(t,T) \rangle} \leq g_D^{(2)}(0) < 1, \tag{6.10.10}$$

derived in a way analogous to eqn (5.10.4). The measured form of the Mandel Q parameter, defined in eqn (5.10.5), is similarly

$$Q_D = \frac{(\Delta m)^2 - \langle m \rangle}{\langle m \rangle} = \eta \frac{\langle [\Delta \hat{M}(t,T)]^2 \rangle - \langle \hat{M}(t,T) \rangle}{\langle \hat{M}(t,T) \rangle}, \tag{6.10.11}$$

with a nonclassical range of values given by

$$-\eta \leq Q_D < 0, \tag{6.10.12}$$

similar to eqn (5.10.6) but reduced by the presence of the quantum efficiency.

As a first example of the quantum photocount averages, consider the photon number state treated in §6.3, with t_0 in the Gaussian wavepacket amplitude of eqn (6.3.2) now representing the time at which the peak of the pulse arrives at the photodetector. The mean photocount obtained from eqns (6.3.10), (6.10.1) and (6.10.5) is then

$$\langle m \rangle = \eta n \int_t^{t+T} dt' |\xi(t')|^2. \tag{6.10.13}$$

The second factorial moment is similarly obtained from eqn (6.10.7) as

$$\langle m(m-1) \rangle = \eta^2 n(n-1) \left[\int_t^{t+T} dt' |\xi(t')|^2 \right]^2, \tag{6.10.14}$$

and the photocount variance is therefore

$$(\Delta m)^2 = \langle m \rangle - \eta^2 n \left[\int\limits_{t}^{t+T} dt' |\xi(t')|^2 \right]^2 \tag{6.10.15}$$

The measured degree of second-order coherence and Mandel Q parameter are

$$g_D^{(2)}(0) = 1 - \frac{1}{n} \tag{6.10.16}$$

and

$$Q_D = -\eta \int\limits_{t}^{t+T} dt' |\xi(t')|^2, \tag{6.10.17}$$

with both values lying in their nonclassical ranges. The measured degree of second-order coherence is the same as that of the number state itself, given in eqn (6.3.14). The integrals in the above equations equal unity when the period from t to $t+T$ is timed so that essentially all of the pulse intensity falls within the integration time.

In contrast to the pulsed number state treated here, the main experiments on sub-Poissonian photocount statistics use stationary light beams. Important examples are the light emitted in single-atom resonance fluorescence, considered in §§7.9, 8.3 and 8.4, and the light-emitting diode, whose output generates a non-classical sub-Poissonian photocurrent observable in undergraduate experiments [14]. As is discussed in the above sections and in §6.5, these sources emit light that also has the nonclassical property of photon antibunching, where closely-spaced photocounts occur less often than more-widely-spaced counts, as illustrated in the time series of Fig. 6.1(c). Such light violates the classical inequality of eqn (3.7.11).

Further examples of the quantum photocount averages are provided by the coherent states treated in §6.4 and the chaotic light treated in §6.5.

Problem 6.14 For the coherent state defined in eqn (6.4.1), prove that

$$g_D^{(2)}(0) = 1 \tag{6.10.18}$$

Show that the photocount variance is given by

$$(\Delta m)^2 = \langle m \rangle \tag{6.10.19}$$

and hence that

$$Q_D = 0 \tag{6.10.20}$$

The measured properties (6.10.18) and (6.10.20) in this example mimic those of single-mode coherent light, treated in §5.3, while the photocount variance has the same shot-noise contribution as the semiclassical result of eqn (3.9.20). The photocount time series has the random Poissonian form illustrated in Fig. 6.1(b) in both the semiclassical and quantum theories. When the coherent light is stationary, with the properties given in eqns (6.4.10) to (6.4.13), the mean photocount given by eqn (6.10.5) is

$$\langle m \rangle = \eta F T, \tag{6.10.21}$$

and this expression is valid for all stationary light beams.

Problem 6.15 For the stationary beam of chaotic light described by the correlation function in eqn (6.5.2), prove that the photocount variance is given by the same expression as the semiclassical result in eqn (3.9.21), with $\gamma = 1/\tau_c$. It follows that the measured degree of second-order coherence is given by the same expression as eqn (3.8.7), while the Mandel Q parameter is

$$Q_D = \frac{\langle m \rangle}{2\gamma^2 T^2}[\exp(-2\gamma T) - 1 + 2\gamma T]. \tag{6.10.22}$$

The photocount distribution for chaotic light is super-Poissonian and the photocount time series has the bunched form illustrated in Fig. 6.1(a), in both the semiclassical and quantum theories.

The above discussion of the first and second moments of the photocount distribution provides the main information needed to distinguish the various kinds of light beam and to illustrate the differences between the semiclassical and quantum theories. The complete quantum-mechanical form of the photocount distribution can also be derived by an extension of the calculation that leads to the semiclassical form in eqn (3.9.12). The result of a somewhat lengthy derivation [15,16] is

$$P_m(t,T) = \left\langle : \frac{\left[\eta \hat{M}(t,T)\right]^m}{m!} \exp\left[-\eta \hat{M}(t,T)\right] : \right\rangle, \tag{6.10.23}$$

where the integrated photon-number operator is defined by eqn (6.10.1). The angle brackets now denote a quantum-mechanical expectation value, calculated with use of the appropriate density operator when the light beam is in a statistical mixture state. The colons again denote normal ordering of the creation and destruction operators, which results from the proper quantum-mechanical evaluation of the multiple photon absorption rates, as used in eqn (4.12.7) for two-photon absorption. The time t at which the measurement period begins is important for nonstationary light beams, for example optical pulses, but it becomes irrelevant for stationary light, when the photocount distribution is denoted $P_m(T)$.

The expression for the quantum photocount distribution resembles the semi-classical form in eqn (3.9.12) and the two formalisms lead to the same results for light beams that are adequately described by the classical theory. Thus the normal ordering in eqn (6.10.23) allows an immediate evaluation of the expectation value for stationary coherent light, with the result

$$P_m(T) = \frac{\langle m \rangle^m}{m!} e^{-\langle m \rangle}, \tag{6.10.24}$$

where the mean value $\langle m \rangle$ is given by eqn (6.10.21). This is the same Poisson photocount distribution as found from the semiclassical theory in eqn (3.9.14). The generation of a Poisson distribution in the inefficient photodetection is another consequence of the preservation of the statistics of quantum coherent light in transmission through a beam splitter, expressed by eqn (5.9.20).

The quantum distribution in eqn (6.10.23) is also evaluated straightforwardly for chaotic light in the limit of an integration time T that is much shorter than the coherence time τ_c. The integrated photon-number operator in eqn (6.10.1) is then approximately

$$\hat{M}(t,T) = \hat{f}(t)T = \hat{a}^\dagger(t)\hat{a}(t)T. \tag{6.10.25}$$

The correlation function in eqn (6.5.2) is approximately

$$\left\langle \hat{a}^\dagger(t)\hat{a}(t') \right\rangle = F\exp[i\omega_0(t-t')], \tag{6.10.26}$$

and the factorization rule in eqn (6.5.4) gives

$$\left\langle \hat{a}^\dagger(t)\hat{a}^\dagger(t')\hat{a}(t')\hat{a}(t) \right\rangle = 2F^2. \tag{6.10.27}$$

Similar factorization rules are used to separate the higher-order expectation values into products of binary correlation functions [2].

Problem 6.16 Prove that the photocount distribution for chaotic light in the limit $T \ll \tau_c = 1/\gamma$ is

$$P_m(T) = \frac{\langle m \rangle^m}{\left(1 + \langle m \rangle\right)^{1+m}}, \tag{6.10.28}$$

where $\langle m \rangle$ is given by eqn (6.10.21).

This is the same geometric photocount distribution as found from the semiclassical theory in eqn (3.9.17). The generation of a geometric distribution in the inefficient photodetection results from preservation of the statistics of quantum chaotic light in transmission through a beam splitter, expressed by eqn (5.9.15).

Despite the similarities in the results of semiclassical and quantum theories for light beams that have adequate classical descriptions, there is an important distinction between the two as regards the origin of the photocount shot noise. Thus, in the discussion of the semiclassical expression (3.9.19) for the variance of the photocount distribution, the origin of the shot noise contribution $\langle m \rangle$ is ascribed to the discrete nature of the photoelectric ionization, there being no analogous term in the intensity variance of the measured light beam. In the quantum theory, however, the electromagnetic field excitation itself has a variance contribution analogous to the shot noise. This is apparent in the photon-number variance of single-mode coherent light, equal to $\langle n \rangle$ in eqn (5.3.19), and in the particle contribution $\langle n \rangle$ to the variance for single-mode chaotic light in eqn (5.4.2). There are corresponding contributions, equal to F, to the photon flux variances for the continuous-mode coherent and chaotic light treated in §§6.4 and 6.5. Thus, in the quantum theory, the shot noise is regarded as already present in the measured coherent or chaotic light beams and the detector transfers this noise to the photocurrent without any addition of noise in the photo-detection process. The role of the detector in transferring noise from light to photocurrent is also evident in the results for a number-state pulse, where the occurrence of a photocount variance in eqn (6.10.15) smaller than the shot noise value of $\langle m \rangle$ reflects the sub-Poissonian photon-number properties of the light itself.

6.11 Homodyne detection

The direct detection treated in the preceding section measures the properties of the incident intensity, or photon flux, in a light beam. The phase properties of the light play no role in the measurement. Homodyne detection measures the electric field, or quadrature-operator expectation values, of the incident light as functions of the measurement phase angle. It is a particularly important technique for the study of squeezed light, where the nonclassical effects are phase dependent.

Figure 6.5 shows the arrangement of components in a homodyne detector. The crucial role is again played by an optical beam splitter, which is here assumed to have 50:50 characteristics, with reflection and transmission coefficients given by

$$|\mathcal{R}| = |\mathcal{T}| = 1/\sqrt{2} \quad \text{and} \quad \phi_{\mathcal{R}} - \phi_{\mathcal{T}} = \pi/2, \tag{6.11.1}$$

as in eqn (3.2.13). The light beam to be measured, the *signal*, is incident in arm 1 of the beam splitter with a destruction operator denoted $\hat{a}(t)$. Arm 2 carries a strong coherent light beam, the *local oscillator*, with a destruction operator denoted $\hat{a}_L(t)$. Both the signal and the local oscillator are assumed to have sufficiently narrow bandwidths that the time-dependent forms of the creation and destruction operators can be safely used. The most sensitive measurements of nonclassical properties are made by *balanced homodyne detection*, where there are photodetectors in both output arms 3 and 4. The quantity to be measured is the *difference* between the numbers of photons arriving at the two detectors during the integration time T, represented by the operator

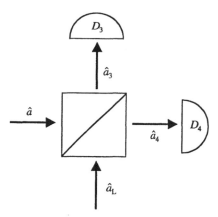

Fig. 6.5. Arrangement of components in a balanced homodyne detector.

$$\hat{M}_-(t,T) = \int_t^{t+T} dt' \left[\hat{f}_3(t') - \hat{f}_4(t') \right] = \int_t^{t+T} dt' \left[\hat{a}_3^\dagger(t')\hat{a}_3(t') - \hat{a}_4^\dagger(t')\hat{a}_4(t') \right]. \quad (6.11.2)$$

It follows from the beam-splitter input–output relations in eqn (6.6.2) and the forms of the beam-splitter coefficients in eqn (6.11.1) that

$$\hat{M}_-(t,T) = i \int_t^{t+T} dt' \left[\hat{a}^\dagger(t')\hat{a}_L(t') - \hat{a}_L^\dagger(t')\hat{a}(t') \right], \quad (6.11.3)$$

where the notation for the input operators is that defined above. The homodyne detector measures photon numbers, or strictly photocounts, but the effect of the local oscillator is to produce measurements proportional to the signal field.

The quantum efficiencies of the two phototubes are generally less than 100% and these are inserted in the manner described in the previous section. It is assumed for simplicity that both detectors have the same quantum efficiency η, which is modelled by the effective beam-splitter arrangement shown in Fig. 6.4. Thus the quantity measured by balanced homodyne detection is represented by the operator

$$\hat{M}_H(t,T) = \int_t^{t+T} dt' \left[\hat{d}_3^\dagger(t')\hat{d}_3(t') - \hat{d}_4^\dagger(t')\hat{d}_4(t') \right], \quad (6.11.4)$$

where the $\hat{d}$ operators for the two detectors are related to the $\hat{a}$ operators in the two output arms and vacuum field operators by relations similar to eqn (6.10.3). The mean homodyne difference photocount is accordingly related to the expectation value of the operator in eqn (6.11.3) by

$$\langle m_- \rangle = \langle \hat{M}_{\mathrm{H}}(t,T) \rangle = \eta \langle \hat{M}_-(t,T) \rangle$$

$$= i\eta \int_t^{t+T} dt' \langle \hat{a}^\dagger(t')\hat{a}_{\mathrm{L}}(t') - \hat{a}_{\mathrm{L}}^\dagger(t')\hat{a}(t') \rangle. \tag{6.11.5}$$

In order to calculate the variance of the difference photocount distribution, we need a relation that can be derived by the use of the standard commutator given in eqn (6.2.14).

Problem 6.17 Prove the relation

$$\left[\hat{M}_{\mathrm{H}}(t,T)\right]^2 = \eta^2 \left[\hat{M}_-(t,T)\right]^2$$

$$+ \eta(1-\eta) \int_t^{t+T} dt' \left[\hat{a}_{\mathrm{L}}^\dagger(t')\hat{a}_{\mathrm{L}}(t') + \hat{a}^\dagger(t')\hat{a}(t')\right]. \tag{6.11.6}$$

The difference photocount variance is thus obtained as

$$(\Delta m_-)^2 = \eta^2 \left\langle \left[\Delta\hat{M}_-(t,T)\right]^2 \right\rangle + \eta(1-\eta) \int_t^{t+T} dt' \left\langle \hat{a}_{\mathrm{L}}^\dagger(t')\hat{a}_{\mathrm{L}}(t') + \hat{a}^\dagger(t')\hat{a}(t') \right\rangle. \tag{6.11.7}$$

These expressions are valid for arbitrary signal and local oscillator beams. It is now assumed that the local oscillator is a 'single-mode' coherent light beam, of the kind described by eqns (6.4.10) to (6.4.13), with a complex amplitude

$$\alpha_{\mathrm{L}}(t) = F_{\mathrm{L}}^{1/2} \exp(-i\omega_{\mathrm{L}}t + i\theta_{\mathrm{L}}). \tag{6.11.8}$$

The local oscillator frequency ω_{L} in homodyne detection is chosen to coincide with the carrier frequency of the signal beam. The expectation value in eqn (6.11.5) is partially evaluated with use of the standard coherent-state eigenvalue property in eqn (6.4.5), and the result is

$$\langle m_- \rangle = i\eta F_{\mathrm{L}}^{1/2} \int_t^{t+T} dt' \left\langle \hat{a}^\dagger(t')\exp(-i\omega_{\mathrm{L}}t' + i\theta_{\mathrm{L}}) - \hat{a}(t')\exp(i\omega_{\mathrm{L}}t' - i\theta_{\mathrm{L}}) \right\rangle. \tag{6.11.9}$$

It is useful to define a *homodyne electric-field operator* as

$$\hat{E}_{\mathrm{H}}(\chi,t,T) = \tfrac{1}{2} T^{-1/2} \int_t^{t+T} dt' \left[\hat{a}^\dagger(t')\exp(-i\omega_{\mathrm{L}}t' + i\chi) + \hat{a}(t')\exp(i\omega_{\mathrm{L}}t' - i\chi)\right], \tag{6.11.10}$$

where

$$\chi = \theta_L + \tfrac{\pi}{2}. \tag{6.11.11}$$

The mean difference photocount from eqn (6.11.9) thus takes the form

$$\langle m_- \rangle = 2\eta (F_L T)^{1/2} \langle \hat{E}_H(\chi,t,T) \rangle. \tag{6.11.12}$$

The variance from eqn (6.11.7) is approximately

$$(\Delta m_-)^2 = \eta F_L T \left\{ 4\eta \langle [\Delta \hat{E}_H(\chi,t,T)]^2 \rangle + 1 - \eta \right\}, \tag{6.11.13}$$

for conditions where the local oscillator flux is much greater than the signal flux, as is the case in most homodyne experiments. The field operator defined by eqn (6.11.10) is analogous to the single-mode field operator defined by eqn (5.1.3). It represents the property of the signal that the balanced homodyne detector seeks to measure.

Consider first the homodyne measurement on a signal that is also in a 'single mode' coherent state. The signal frequency is assumed to equal that of the local oscillator, and its wavepacket amplitude from eqn (6.4.11) is

$$\alpha(t) = F^{1/2} \exp(-i\omega_L t + i\theta). \tag{6.11.14}$$

The coherent signal field, given by the expectation value of the homodyne field operator from eqn (6.11.10), is

$$S = \langle \hat{E}_H(\chi,t,T) \rangle = (FT)^{1/2} \cos(\chi - \theta). \tag{6.11.15}$$

This expression is the same as that derived in eqn (5.3.31) but with the single-mode coherent amplitude replaced by the square root of the mean photon number incident on the detector during the integration time. The variance of the homodyne field operator, or the noise, is readily evaluated as

$$\mathcal{N} = \langle [\Delta \hat{E}_H(\chi,t,T)]^2 \rangle = \tfrac{1}{4}, \tag{6.11.16}$$

which is identical to the single-mode result in eqn (5.3.32). The signal-to-noise ratio as defined in eqn (5.1.12) is

$$\text{SNR} = 4FT \cos^2(\chi - \theta). \tag{6.11.17}$$

The representation of the mean and uncertainty of the coherent-state electric field shown in Fig. 5.4 applies to the above signal and noise of the continuous-mode homodyne field if the amplitude $|\alpha|$ is replaced by $(FT)^{1/2}$.

These signal and noise results are properties of the coherent light beam incident on the detector, but their measured values contain additional factors that result from the quantum efficiencies of the two phototubes. Thus the mean difference photocount obtained from eqn (6.11.12) is

$$\langle m_- \rangle = 2\eta (F_L F)^{1/2} T \cos(\chi - \theta). \tag{6.11.18}$$

The difference photocount variance is obtained by evaluation of eqn (6.11.13), with the result

$$(\Delta m_-)^2 = \eta F_L T. \tag{6.11.19}$$

The homodyne signal-to-noise ratio is defined in terms of the measured coherent signal and noise powers, or photocount rates, in the homodyne detector, equivalent to the definition of eqn (5.1.12) in terms of the field mean and variance, as

$$\text{SNR}_H = \frac{\langle m_- \rangle^2}{(\Delta m_-)^2} = 4\eta FT \cos^2(\chi - \theta) = \eta \text{SNR}. \tag{6.11.20}$$

The ratio is thus degraded by a factor of η compared to that in eqn (6.11.17).

Balanced homodyne detection plays an important role in the measurement of quadrature squeezing, where its sensitivity to the phase of the signal field allows unambiguous identifications of squeezed light to be made. The properties of squeezed-vacuum light derived in §6.9 are mainly expressed in terms of the frequency, rather than the time variable, so that the time-dependent operators used in the homodyne field of eqn (6.11.10) need to be Fourier transformed by use of eqn (6.2.13) before expectation values are determined. It is immediately clear with the use of eqn (6.9.7) that the mean homodyne field vanishes,

$$\langle \hat{E}_H(\chi, t, T) \rangle = 0, \tag{6.11.21}$$

analogous to the single-mode result given in eqn (5.5.22). The mean difference photocount is accordingly zero,

$$\langle m_- \rangle = 0. \tag{6.11.22}$$

The variance of the homodyne field is calculated with the aid of the expectation values given in eqns (6.9.9) and (6.9.11). The local oscillator frequency ω_L is set equal to the central frequency $\omega_p/2$ of the squeezed light and the variance is

$$\langle [\Delta \hat{E}_H(\chi, t, T)]^2 \rangle = \frac{1}{16\pi T} \int_t^{t+T} dt' \int_t^{t+T} dt'' \int_{-\infty}^{\infty} d\omega \exp[-i(\omega - \omega_L)(t' - t'')] \tag{6.11.23}$$

$$\times \{ 2\sinh^2[s(\omega)] + 1 - \exp[2i\chi - i\vartheta(\omega)] \sinh[2s(\omega)] \} + \text{comp. conj.}$$

This expression can in principle be evaluated, at least numerically, for known forms of the functions $s(\omega)$ and $\vartheta(\omega)$ that describe the amplitude and phase of the squeezing, as defined in eqn (6.9.6).

The variance can also be evaluated analytically in an approximation that is valid when the integration time T is sufficiently long that the functions $s(\omega)$ and $\vartheta(\omega)$ vary insignificantly over the frequency range $1/T$. With the use of eqn (2.4.1), the time integrals in eqn (6.11.23) then give an approximate delta-function on ω and ω_L,

$$\frac{1}{T}\int_t^{t+T}\!\!dt'\int_t^{t+T}\!\!dt''\exp\left[-i(\omega-\omega_L)(t'-t'')\right]=\frac{4\sin^2\left[\frac{1}{2}(\omega-\omega_L)T\right]}{(\omega-\omega_L)^2 T}\to 2\pi\delta(\omega-\omega_L).$$

$$(6.11.24)$$

The final result can be arranged as

$$\left\langle\left[\Delta\hat{E}_H(\chi,t,T)\right]^2\right\rangle=\tfrac{1}{4}\left\{\exp\left[2s(\omega_L)\right]\sin^2\left[\chi-\tfrac{1}{2}\vartheta(\omega_L)\right]\right.$$

$$\left.+\exp\left[-2s(\omega_L)\right]\cos^2\left[\chi-\tfrac{1}{2}\vartheta(\omega_L)\right]\right\},$$

$$(6.11.25)$$

which has a form identical to the single-mode result in eqn (5.5.23). The signal field is quadrature squeezed when its variance is smaller than the coherent value of eqn (6.11.16), specified by the range

$$0\le\left\langle\left[\Delta\hat{E}_H(\chi,t,T)\right]^2\right\rangle<\tfrac{1}{4},$$

$$(6.11.26)$$

similar to eqn (5.5.1). The ranges of the signal parameters for which squeezing is observed are determined by essentially the same equations as in Problem 5.13. The experimental points shown in Fig. 5.9 were measured by the method of balanced homodyne detection described in the present section.

In terms of the variance of the difference photocounts, the quadrature squeezing range obtained with the use of eqn (6.11.13) is

$$0\le(\Delta m_-)^2<\eta F_L T,$$

$$(6.11.27)$$

when the local oscillator flux is much larger than the signal flux. The upper limit of the variance range is the detection shot noise associated with the incidence of $F_L T$ photons at a detector of quantum efficiency η. However, the noise described by $(\Delta m_-)^2$ is caused by the beating of the strong coherent field of the local oscillator with the noise in the measured signal, the shot noise of the local oscillator itself being cancelled out by the subtraction inherent in the balanced homodyne detection scheme [13,17]. Thus the detection noise figures smaller than the shot-noise value obtained when eqn (6.11.25) is substituted into eqn (6.11.13) reflect the reduced noise on the squeezed signal. In addition to the squeezed vacuum state considered here, it is also possible to excite continuous-mode squeezed coherent light, analogous to the single-mode states treated in

§5.6. Such squeezed light is potentially useful as a low-noise carrier of optical information.

6.12 The electromagnetic vacuum

The vacuum state $|\{0\}\rangle$ of the electromagnetic field, with no photons excited in any of the modes, is defined by eqns (4.4.24) and (4.4.25). The vacuum-state energy eigenvalue relation is given by eqns (4.4.26) and (4.4.27). These equations refer to a discrete-mode description of the fields but they are readily converted to a continuous-mode description by the replacement of wavevector sums by integrals. The vacuum, or zero-point, energy remains infinite, as mentioned in §6.2.

It is shown in §4.11 that measurements of the intensity of a light beam are insensitive to the vacuum energy and only detect the excitation energy of the electromagnetic field above its (infinite) ground-state value. This happy feature of intensity measurements is ensured by the occurrence of normally-ordered electric-field operators in the representation of the photoionization rate that essentially determines the form of the intensity operator, given by eqn (4.11.6) for a parallel beam of light. There are, however, other forms of measurement that respond more directly to the vacuum energy itself or, usually, to changes in the vacuum energy brought about by changes from the free-space environment assumed in quantization of the electromagnetic field. The aim here is to give a simple example of vacuum-state effects from the rich and sophisticated variety of associated phenomena.

Casimir forces [18] provide a straightforward illustration of vacuum effects. Suppose that a plane parallel cavity made from a pair of perfectly conducting and reflecting mirrors a distance L apart is placed in free space. We consider the change in the vacuum energy brought about by the presence of the cavity. The energy in the free-space regions on both sides of the cavity remains the same but the energy inside the cavity is modified by the change in mode structure from continuous to discrete. The vacuum energy of the standing-wave modes with wavevector perpendicular to the mirrors is obtained from eqn (4.4.27) as

$$\mathcal{E}_{\text{cav}} = \sum_k \hbar c k = (\pi \hbar c / L) \sum_{\nu=1}^{\infty} \nu, \qquad (6.12.1)$$

where the one-dimensional wavevectors are given by eqn (1.1.4) or (1.10.11). The vacuum energy in the same region of space in the absence of the cavity is given by the same expression but with the wavevector replaced by a continuous variable,

$$\mathcal{E}_{\text{free}} = (\pi \hbar c / L) \int_0^{\infty} d\nu \, \nu. \qquad (6.12.2)$$

Both these expressions for the vacuum energy are clearly infinite.

Consider, however, the *change* in energy produced by the presence of the cavity, given by the difference of these expressions,

$$\Delta E = E_{\text{cav}} - E_{\text{free}} = (\pi \hbar c / L) \left\{ \sum_{v=1}^{\infty} v - \int_{0}^{\infty} dv\, v \right\}. \tag{6.12.3}$$

Problem 6.18 Show by the introduction of a convergence factor $e^{-\varepsilon v}$, where $\varepsilon \to 0$, to regularize both the sum and the integral, that

$$\sum_{v=1}^{\infty} v - \int_{0}^{\infty} dv\, v = -1/12. \tag{6.12.4}$$

The energy change from eqn (6.12.3) is therefore finite, with the value

$$\Delta E = -\pi \hbar c / 12 L. \tag{6.12.5}$$

It follows that the energy diminishes with decreasing L, and there is an attractive force between the mirrors given by

$$F = -\partial \Delta E / \partial L = -\pi \hbar c / 12 L^2. \tag{6.12.6}$$

The force is an example of the Casimir forces that act between all bodies placed in the electromagnetic vacuum.

This calculation of the force between two parallel mirrors takes account only of the modes with wavevectors perpendicular to the mirrors. A realistic three-dimensional theory must include the contributions of all wavevector orientations. The principles of the calculation remain the same but the Casimir force per unit area, or pressure, is given by [18-21]

$$P = -\pi^2 \hbar c / 240 L^4. \tag{6.12.7}$$

The force is very small and mirror separations in the micron or submicron range are needed to produce observable values. Thus, for a mirror separation of $0.5\,\mu\text{m}$, the magnitude of the force per unit area, or pressure, is $0.021\,\text{Nm}^{-2}$. It is particularly difficult to measure for the flat-plate geometry considered here, as mirrors of diameter $10\,\text{mm}$ need to be maintained accurately parallel to each other with very small separations. The experimental problems are reduced when one of the flat mirrors is replaced by a curved surface, for example a lens [22] or a sphere [23]. The measurements are in very good agreement with the calculated Casimir force, given by a suitably modified version of eqn (6.12.7) that depends on the radius of curvature of the second surface. These examples of two flat plates or one flat and one curved both produce attractive Casimir forces but some other plate geometries produce replusive forces and it is necessary to calculate the force afresh in each case.

The Casimir force is one manifestation of the electromagnetic vacuum and its zero-point energy. The calculation of its strength for parallel conducting plates followed earlier work on the van der Waals forces between neutral atoms and the Casimir–Polder force between a neutral atom and a flat conducting plate (see [24] for a historical review). Spontaneous emission can also be regarded, to some extent, as stimulated by the electric-field fluctuations in the vacuum state. Modifications in the vacuum fields, and hence in the free-space spontaneous emission rates, are brought about by the proximity of material bodies to the emitting atom. Again, the Lamb shifts in atomic transition frequencies that accompany radiative decay, as discussed in §7.7, can be interpreted as consequences of the vacuum field fluctuations. The electromagetic vacuum forms an important area a study with many subtle aspects [19].

References

[1]   Blow, K.J., Loudon, R., Phoenix, S.J.D. and Shepherd, T.J., Continuum fields in quantum optics, *Phys. Rev. A* **42**, 4102–14 (1990).

[2]   Barnett, S.M. and Radmore, P.M., *Methods in Theoretical Quantum Optics* (Clarendon Press, Oxford, 1997).

[3]   Hong, C.K. and Mandel, L., Experimental realization of a localized one-photon state, *Phys. Rev. Lett.* **56**, 58–60 (1986).

[4]   Teich, M.C. and Saleh, B.E.A., Photon bunching and antibunching, *Prog. Opt.* **26**, 1–104 (1988).

[5]   Zou, X.T. and Mandel, L., Photon antibunching and sub-Poissonian photon statistics, *Phys. Rev. A* **41**, 476–6 (1990).

[6]   Grangier, P., Roger, G. and Aspect, A., Experimental evidence for a photon anticorrelation effect on a beam splitter; a new light on single-photon interferences, *Europhys. Lett.* **1**, 173–9 (1986).

[7]   Campos, R.A., Saleh, B.E.A., and Teich, M.C., Fourth-order interference on joint single-photon wavepackets in lossless optical systems, *Phys. Rev. A* **42**, 4127–37 (1990).

[8]   Brendel, J., Schütrumpf, S., Lange, R., Martienssen, W. and Scully, M.O., A beam splitting experiment with correlated photons, *Europhys. Lett.* **5**, 223–8 (1988).

[9]   Feynman, R.P., Leighton, R.B. and Sands, M., *The Feynman Lectures on Physics* (Addison–Wesley, Reading , Mass, 1965) vol. 3, chaps 3 and 4.

[10]   Hong, C.K., Ou, Z.Y. and Mandel, L., Measurement of subpicosecond time intervals between two photons by interference, *Phys. Rev. Lett.* **59**, 2044–6 (1987).

[11]   Barnett, S.M. and Knight, P.L., Thermofield analysis of squeezing and statistical mixtures in quantum optics, *J. Opt. Soc. Am.* **2**, 467–79 (1985); Squeezing in correlated quantum systems, *J. Mod. Opt.* **34**, 841–53 (1987).

[12]   Burnham, D.C. and Weinberg, D.L., Observation of simultaneity in

parametric production of optical photon pairs, *Phys. Rev. Lett.* **25**, 84–7 (1970).

[13] Yuen, H.P. and Shapiro, J.H., Optical communication with two-photon coherent states – Part III: Quantum measurements realizable with photo-emissive detectors, *IEEE Trans. Inf. Theor.* **26**, 78–92 (1980).

[14] Funk, A.C. and Beck, M., Sub-Poissonian photocurrent statistics: Theory and undergraduate experiment, *Am. J. Phys.* **65**, 492–500 (1997).

[15] Kelley, P.L. and Kleiner, W.H., Theory of electromagnetic field measurement and photoelectron counting, *Phys. Rev. A* **136**, 316–34 (1964).

[16] Chmara, W., A quantum open-systems theory approach to photo-detection, *J. Mod. Opt.* **34**, 455–67 (1987).

[17] Yuen, H.P. and Chan, V.W.S., Noise in homodyne and heterodyne detection, *Opt. Lett.* **8**, 177–9 (1983).

[18] Casimir, H.G.B., On the attraction between two perfectly conducting plates, *Proc. K. Ned. Akad. Wet.* **51**, 793–6 (1948).

[19] Milonni, P.W., *The Quantum Vacuum* (Academic Press, San Diego, 1994).

[20] Milonni, P.W. and Shih, M.-L., Casimir forces, *Contemp. Phys.* **33**, 313–22 (1992).

[21] Mostepanenko, V.M. and Trunov, N.N., *The Casimir Effect and its Applications* (Clarendon Press, Oxford, 1997).

[22] Lamoreaux, S.K., Demonstration of the Casimir force in the 0.6 to 6 μm range, *Phys. Rev. Lett.* **78**, 5–8 (1997); **81**, 5475–6 (1998).

[23] Mohideen, U. and Roy, A., Precision measurement of the Casimir force from 0.1 to 0.9 μm, *Phys. Rev. Lett.* **81**, 4549–52 (1998).

[24] Casimir, H.G.B., Van der Waals forces and zero point energy, in *Physics of Strong Fields*, ed. W. Greiner (Plenum Press, New York, 1987) pp. 957–64.

7 Optical generation, attenuation and amplification

A typical optical experiment consists of three main components. The light is provided by a source, which may be a thermal source that emits chaotic light, but is most often a laser that emits coherent light in quantum-optical experiments. The light is processed in some interaction region of the experiment, where it may, for example, be subjected to attenuation, amplification, beam splitting, or one of the various nonlinear optical processes. The processed light beam is finally detected, for example by the direct or homodyne schemes covered in Chapter 6.

The present chapter considers the light emitted by various kinds of source, together with some examples of the optical processing in the second stage of the typical experiment. The emitting atoms in a gas laser are usually contained in an optical cavity with highly-reflecting mirrors and the emission excites a single standing-wave mode of the cavity for suitable values of the experimental parameters. It is thus appropriate to begin the treatment with a single-mode theory of photon absorption and emission. The theoretical description of the light emitted by a laser requires an extension of the treatment given in §1.10 to include spontaneous emission effects in the interaction of the single-mode field with the atomic transitions.

The light beam that emerges from the laser propagates towards the detector through the interaction region of the experiment. Although this region has a finite extent, it does not consitute a closed optical cavity and the light has the form of a travelling electromagnetic wave. The beam is best described by the continuous-mode theory presented in Chapter 6 and this formalism is also best adapted to the theories of direct and homodyne detection covered in §§6.10 and 6.11. The treatment of optical processing in the present chapter is limited to travelling-wave attenuation and amplification.

The light in these calculations is assumed to take the form of a polarized, plane parallel beam so that a one-dimensional theory is adequate. However, there are also important quantum-optical experiments in which nonclassical light is generated by emission from single driven atoms in free space. The radiation is emitted into all spatial directions with both polarizations and it is necessary to apply a three-dimensional theory to the problem. The light produced by a driven two-level atom, treated in §7.9, shows photon-antibunching characteristics while the light emitted by a driven three-level atom, treated in §8.6, shows nonclassical interbeam correlations.

7.1 Single-mode photon rate equations

Consider an optical cavity of volume V that contains N identical two-level atoms with transition frequency ω_0. We treat the interaction of the atoms with the radiation in a single mode $\mathbf{k}\lambda$ of the cavity, and omit the $\mathbf{k}\lambda$ subscripts from the field variables. The atomic transition is assumed to be collision broadened, with a total linewidth parameter γ whose collisional part γ_{coll} greatly exceeds the radiative part γ_{sp}. The interaction of light with the atoms is then well described by rate equations, as discussed in §2.10, even though the bandwidth of the light may be much smaller than the total atomic linewidth.

The excitation rate, or photon absorption rate, is obtained from eqn (2.10.7). An expression for the orientation-averaged Rabi frequency is provided by eqn (4.10.12) as

$$\overline{\mathcal{V}^2} = \frac{2e^2\omega D_{12}^2}{3\varepsilon_0 \hbar V}n = 4\overline{g^2}n. \tag{7.1.1}$$

Substitution of this averaged squared Rabi frequency in eqn (2.10.7) gives the absorption rate in the form

$$R_{ab} = \frac{e^2\omega D_{12}^2}{3\varepsilon_0 \hbar V}\frac{\gamma}{(\omega_0 - \omega)^2 + \gamma^2}n. \tag{7.1.2}$$

Note that, with the Einstein B coefficient taken from eqn (2.3.20) or (4.10.5), the absorption rate can be written as

$$R_{ab} = B_{12}\frac{\hbar\omega n}{V}\frac{\gamma/\pi}{(\omega_0 - \omega)^2 + \gamma^2} = B_{12}\langle W(\omega)\rangle, \tag{7.1.3}$$

where a form of modified mean radiative energy $\langle W(\omega)\rangle$ per unit volume per unit angular frequency range is here defined as the product of the photon energy density with the normalized lineshape function.

Suppose that the frequency ω of the chosen cavity mode coincides with the centre of the atomic absorption line at ω_0 and define

$$\Gamma_{st} = \frac{e^2\omega_0 D_{12}^2}{3\varepsilon_0 \hbar V\gamma} = \frac{2\pi c^3}{V\omega_0^2}\frac{\gamma_{sp}}{\gamma}, \tag{7.1.4}$$

where the form of the spontaneous emission rate γ_{sp} is taken from eqn (2.3.21) or (4.10.10). The absorption rate becomes simply

$$R_{ab} = \Gamma_{st}n. \tag{7.1.5}$$

The corresponding rate of emission of a photon of frequency $\omega = \omega_0$ is obtained by replacement of n by $n+1$ in eqn (7.1.5) (see the paragraphs that follow eqn (4.10.5)) to give

$$R_{em} = \Gamma_{st}(n+1). \tag{7.1.6}$$

The two terms n and 1 correspond respectively to stimulated and spontaneous emission. The quantity Γ_{st} is the rate of stimulated emission per photon *and* the rate of spontaneous emission into the single cavity mode considered. By contrast, the total rate of spontaneous emission into all modes (including the chosen cavity mode) is determined by the Einstein A coefficient,

$$\Gamma_{sp} = A_{21} = 2\gamma_{sp}, \tag{7.1.7}$$

with a detailed form given by eqn (2.3.21) or (4.10.10). With these developments in notation, the rate equation (1.7.1) for the numbers of atoms in the two levels takes the form

$$dN_1/dt = -dN_2/dt = N_2\Gamma_{sp} + (N_2 - N_1)\Gamma_{st}n, \tag{7.1.8}$$

where the spontaneous emission into the chosen mode is counted once only.

The effect of absorption and emission by the two-level atoms is to cause changes in the photon probability distribution $P(n)$, whose time dependence is described by rate equations. Figure 7.1 shows the energy-level diagram for the chosen cavity mode. The rate of change of the probability for a given n is influenced by four kinds of transition, whose contributions are shown against the four vertical arrows. Thus if n photons are indeed excited, the N_1 atoms in their ground states each absorb photons at the rate shown in eqn (7.1.5) while each of the N_2 excited atoms emits photons at the rate given in eqn (7.1.6). The resulting combined rate of change of $P(n)$ is

$$-N_1\Gamma_{st}nP(n) - N_2\Gamma_{st}(n+1)P(n). \tag{7.1.9}$$

There are also two positive contributions to the rate of change of $P(n)$. Thus if $n-1$ photons are excited, the N_2 excited atoms each emit photons at a rate obtained from eqn (7.1.6) with n replaced by $n-1$, to give

$$N_2\Gamma_{st}nP(n-1). \tag{7.1.10}$$

Again, if $n+1$ photons are excited, the N_1 atoms in their ground states each absorb photons at a rate obtained from eqn (7.1.5) with n replaced by $n+1$, to give

$$N_1\Gamma_{st}(n+1)P(n+1). \tag{7.1.11}$$

The contributions of eqns (7.1.9) to (7.1.11) combine to give the total rate of change of the nth element of the photon probability distribution as

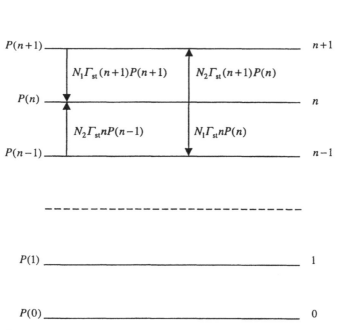

Fig. 7.1. Energy-level diagram for the photons in the chosen cavity mode, showing the contributions to $dP(n)/dt$.

$$dP(n)/dt = \Gamma_{st}\{N_2 nP(n-1) - N_1 nP(n) - N_2(n+1)P(n) + N_1(n+1)P(n+1)\}.$$

$$(7.1.12)$$

Note that only two of the four kinds of transition contribute to the rate for $n = 0$, when eqn (7.1.12) reduces to

$$dP(0)/dt = \Gamma_{st}\{-N_2 P(0) + N_1 P(1)\}. \qquad (7.1.13)$$

The rate of change of the mean number of photons in the chosen mode is given by

$$\frac{d\langle n \rangle}{dt} = \sum_n n \frac{dP(n)}{dt}, \qquad (7.1.14)$$

where the expressions from eqns (7.1.12) and (7.1.13) are to be substituted on the right-hand side. The summation over n is difficult to evaluate, as the atomic-level populations N_1 and N_2 obtained by solution of eqn (7.1.8) are generally functions of n. Thus each level population in eqns (7.1.12) and (7.1.13) should strictly be shown as the value appropriate to the photon number in the element of $P(n)$ that it multiplies. However, it is sometimes a good approximation to replace n by $\langle n \rangle$ in the solutions for the atomic-level populations and this is done for the remainder of the present section.

Problem 7.1   Derive the mean-photon-number rate-equation

$$\mathrm{d}\langle n \rangle / \mathrm{d}t = \Gamma_{\mathrm{st}}\left\{ N_2 + (N_2 - N_1)\langle n \rangle \right\}. \tag{7.1.15}$$

Together with the atomic rate equations (7.1.8), with $\langle n \rangle$ substituted for n, this can in principle be solved for the mean atomic populations and mean photon number as functions of the time for arbitrary initial conditions. However, as the equations contain products of the variables, and are therefore nonlinear, analytic solutions are available only for some simple special cases.

The rate equations display an energy conservation property expressed by

$$\frac{\mathrm{d}N_2}{\mathrm{d}t} + \frac{\mathrm{d}\langle n \rangle}{\mathrm{d}t} + N_2\left(\Gamma_{\mathrm{sp}} - \Gamma_{\mathrm{st}} \right) = 0. \tag{7.1.16}$$

Thus, upon multiplication by $\hbar\omega_0$, the three terms represent the rates of change of the atomic excitation energy, the electromagnetic energy in the chosen cavity mode, and the rate of radiative energy emission into all other modes of the cavity. There are infinitely many of these other modes and they contain all of the energy of the system in the long-time steady state, when $N_2 = 0$ and $\langle n \rangle = 0$.

The equations (7.1.12) and (7.1.13) for the rate of change of the photon probability distribution also provide an equation of motion for the degree of second-order coherence, obtained from eqn (5.1.15) for single-mode excitations.

Problem 7.2   Derive an equation of motion for the second factorial moment of the photon-number distribution in the form

$$\frac{\mathrm{d}}{\mathrm{d}t}\left\{ \langle n(n-1) \rangle - 2\langle n \rangle^2 \right\} =$$
$$- 2\Gamma_{\mathrm{st}}(N_1 - N_2)\left\{ \langle n(n-1) \rangle - 2\langle n \rangle^2 \right\}, \tag{7.1.17}$$

where N_1 and N_2 are again determined by solution of eqn (7.1.8) with n replaced by $\langle n \rangle$.

The solution of this compact equation enables the time dependence of the degree of second-order coherence to be found. The equation is again nonlinear, as the atomic populations are themselves generally time-dependent.

7.2 Solutions for fixed atomic populations

The photon rate equations derived in the previous section are solved quite easily when the atomic populations N_1 and N_2 are assumed to be maintained at fixed values by some external influence. We consider first the steady-state photon-number distribution for $N_2 < N_1$ and then the dynamics of the mean number and degree of second-order coherence for atomic populations in which $N_2 < N_1$ or $N_1 < N_2$. These conditions correspond respectively to attenuation and amplification of the light in the chosen cavity mode. The quantum theory of travelling-wave attenuation and amplification, analogous to the semiclassical theory of §§1.8 and 1.9, is presented in §§7.5 and 7.6.

The steady-state photon distribution is determined by eqns (7.1.12) and (7.1.13) with the rates of change of the $P(n)$ set equal to zero. The resulting chain of equations for the different values of n is solved, beginning with the steady-state solution of eqn (7.1.13) for $n = 0$. This solution is used to simplify eqn (7.1.12) for $n = 1$, which is in turn used to simplify eqn (7.1.12) for $n = 2$, and so on. The general condition for a steady state is

$$N_2 P(n-1) = N_1 P(n) \tag{7.2.1}$$

and the solution of the chain of equations is

$$P(n) = \left(N_2/N_1\right) P(n-1) = \left(N_2/N_1\right)^n P(0). \tag{7.2.2}$$

The remaining unknown, $P(0)$, is determined by the normalization condition

$$\sum_{n=0}^{\infty} P(n) = 1. \tag{7.2.3}$$

The final result,

$$P(n) = \left(1 - \frac{N_2}{N_1}\right)\left(\frac{N_2}{N_1}\right)^n = \frac{\langle n \rangle_{\infty}^n}{\left(1 + \langle n \rangle_{\infty}\right)^{1+n}}, \tag{7.2.4}$$

is the photon distribution for chaotic light with a mean photon number

$$\langle n \rangle_{\infty} = N_2/\left(N_1 - N_2\right), \tag{7.2.5}$$

where the ∞ subscript signifies the photon number achieved after an infinitely-long time interval. The distribution is the same as that found in eqn (1.5.14) for light in interaction with atoms in thermal equilibrium, where the atomic levels are here taken as nondegenerate with $g_1 = g_2 = 1$. The more general derivation here shows that the light generated by atoms maintained at arbitrary levels of excitation has the same chaotic nature as thermal light.

The rate equation (7.1.15) for the mean photon number is readily solved for fixed N_1 and N_2 to give

$$\langle n \rangle = \left(\langle n \rangle_0 - \langle n \rangle_{\infty}\right) \exp\left\{-(N_1 - N_2)\Gamma_{st} t\right\} + \langle n \rangle_{\infty}, \tag{7.2.6}$$

where $\langle n \rangle_0$ is the mean photon number at time $t = 0$. The solution of the equation of motion (7.1.17) for the second factorial moment gives a simple exponential time dependence that can be expressed in terms of the zero time-delay degree of second-order coherence as

$$\langle n \rangle^2 \{ g^{(2)}(0) - 2 \} = \langle n \rangle_0^2 \{ g_0^{(2)}(0) - 2 \} \exp\{ -2\Gamma_{st}(N_1 - N_2)t \}. \tag{7.2.7}$$

The quantities with 0 subscripts are values at time $t = 0$, while the unsubscripted quantities are values at time t.

Equations (7.2.6) and (7.2.7) determine the time developments of the two main indicators of the statistical properties of the light produced by its interaction with the atoms. Figure 7.2 shows the time dependences of the mean photon numbers for various initial values. Note that the mean number of photons at long times always tends to the steady-state value given by eqn (7.2.5). The degree of second-order coherence always tends to the chaotic value of 2 at long times. Light that is initially chaotic remains chaotic at all subsequent times, while light that is initially coherent has a degree of second-order coherence that increases from the initial value of 1 to the chaotic value of 2 in the steady state. As an example of nonclassical initial light, the degree of second-order coherence of a photon number state, given by eqn (5.2.3), also increases to 2 in the steady state.

The statistical properties of the light at times intermediate between 0 and ∞ are those of a superposition of a diminishing amount of light with the arbitrary initial statistical properties and a growing amount of chaotic light. Thus the mean photon number from eqn (7.2.6) can be written

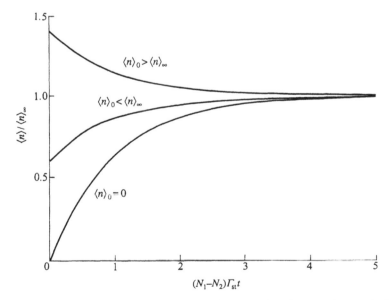

Fig. 7.2. Time dependence of the mean photon number in a cavity mode interacting with atoms, most of which are in their ground states. The three curves correspond to the initial values $\langle n \rangle_0$ of mean photon number indicated.

$$\langle n \rangle = \langle n_{ar} \rangle + \langle n_{ch} \rangle, \tag{7.2.8}$$

where

$$\langle n_{ar} \rangle = \langle n \rangle_0 \exp\{-(N_1 - N_2)\Gamma_{st}t\} \tag{7.2.9}$$

and

$$\langle n_{ch} \rangle = \langle n \rangle_\infty \{1 - \exp[-(N_1 - N_2)\Gamma_{st}t]\}. \tag{7.2.10}$$

The chaotic contribution has the nature of a noise component that is added to the attenuated remnant of an initial signal. The degree of second-order coherence from eqn (7.2.7) can likewise be written

$$g^{(2)}(0) = \frac{\langle n_{ar} \rangle^2 g_0^{(2)}(0) + 4\langle n_{ar} \rangle\langle n_{ch} \rangle + 2\langle n_{ch} \rangle^2}{\left(\langle n_{ar} \rangle + \langle n_{ch} \rangle\right)^2}, \tag{7.2.11}$$

and this expression agrees with the known form for a superposition of chaotic light with light of arbitrary statistics [1]. It reproduces a degree of second-order coherence equal to 2 when the intial light is itself chaotic and it provides a convenient form of the solution for other initial statistics. It is emphasized that the superposition statistics are those of the photons as a whole and there is no implication that individual photons can in some way be identified as having the initial or the chaotic statistics.

The statistical properties of the light are particularly simple in conditions where all N atoms in the cavity are maintained in their ground states, with $N_2 = 0$ and $N_1 = N$. This condition normally applies in thermal equilibrium at ordinary temperatures for atomic transitions in the range of visible frequencies. Then only absorption processes occur, represented by the downwards-pointing arrows in Fig. 7.1, and eqn (7.1.12) reduces to

$$dP(n)/dt = N\Gamma_{st}\{-nP(n) + (n+1)P(n+1)\}. \tag{7.2.12}$$

The form of the photon probability distribution is easily calculated in this case. Consider a photon present in the cavity at time $t = 0$. The probability that it remains unabsorbed at time t is

$$p = \exp(-N\Gamma_{st}t). \tag{7.2.13}$$

Then if l photons are present at $t = 0$, there remain n at time t if n photons are *not* absorbed and $l - n$ photons *are* absorbed, with a probability given by the usual binomial or Bernoulli-sampling distribution

$$P(n) = \sum_{l=n}^{\infty} P_0(l) \frac{l!}{(l-n)!n!}(1-p)^{l-n} p^n. \tag{7.2.14}$$

Problem 7.3   Prove that eqn (7.2.14) is indeed the solution of eqn (7.2.12).

The photon distribution is similar to that derived in eqn (5.9.10) for transmission through a beam splitter and it has the same general properties. There is no chaotic noise contribution, as $\langle n \rangle_\infty$ given by eqn (7.2.5) vanishes for $N_2 = 0$, and the properties of the light are determined at all times by the initial excitation. Thus the mean photon number is

$$\langle n \rangle = p \langle n \rangle_0, \tag{7.2.15}$$

analogous to eqn (5.9.2), and the degree of second-order coherence is

$$g^{(2)}(0) = g_0^{(2)}(0), \tag{7.2.16}$$

analogous to eqn (5.9.13). An initial chaotic distribution remains chaotic, similar to eqn (5.9.15) and an initial coherent distribution remains coherent, similar to eqn (5.9.20). An initial number state retains its degree of second-order coherence but the form of the distribution is changed by the interaction of the light with the ground-state atoms, and Fig. 5.20 provides an example.

Now consider the time-development of the photon statistics for a single resonant mode of the cavity in interaction with atoms whose fixed populations satisfy the inversion condition $N_2 > N_1$. The solution (7.2.8) for the mean photon number remains valid but the two contributions are rewritten as

$$\langle n_{\text{ar}} \rangle = \langle n \rangle_0 \exp\{(N_2 - N_1)\Gamma_{\text{st}}t\} \tag{7.2.17}$$

and

$$\langle n_{\text{ch}} \rangle = [N_2/(N_2 - N_1)]\{\exp[(N_2 - N_1)\Gamma_{\text{st}}t] - 1\} \tag{7.2.18}$$

to emphasize the exponential growth that now occurs. The mean photon numbers increase without limit and there is no steady state. The chaotic, or noise, component survives in conditions of population inversion but its relative contribution is minimized for conditions of perfect atomic inversion, with $N_2 = N$ and $N_1 = 0$. For a general inversion, the ratio of the two contributions at large times is

$$\frac{\langle n_{\text{ar}} \rangle}{\langle n_{\text{ch}} \rangle} = \frac{\langle n \rangle_0 (N_2 - N_1)}{N_2} \quad \text{for} \quad (N_2 - N_1)\Gamma_{\text{st}}t \gg 1. \tag{7.2.19}$$

Thus for a sufficiently strong initial excitation of photons with

$$\langle n \rangle_0 (N_2 - N_1) \gg N_2, \tag{7.2.20}$$

the net effects of stimulated emission and absorption, proportional to the quantity on the left, dominate the effect of spontaneous emission, proportional to the excited-state population on the right. The result at large times is a predominance of light with the arbitrary initial statistics, whose degree of second-order coherence, given by eqn (7.2.11), is little changed from the initial value. On the other hand, for a feeble initial excitation of photons with

$$\langle n \rangle_0 (N_2 - N_1) \ll N_2, \tag{7.2.21}$$

the spontaneous emission dominates, and it leads to predominantly chaotic light at large times, with a degree of second-order coherence equal to 2.

These results show that stimulated emission by a collection of atoms tends to preserve the coherence properties of the stimulating light. Spontaneous emission, on the other hand, generates chaotic light in the chosen cavity mode. It is again stressed that the coherence is a property of the complete photon excitation and there are no distinguishable groups of different kinds of photon. With this proviso, however, the qualitative distinctions between stimulated and spontaneous emission into the given mode provide a simple and useful picture of their different roles in the modification of the initial photon excitation.

7.3 Single-mode laser theory

Atomic population inversion is one ingredient of a laser system but it is also necessary to model the optical cavity that supports the single mode excited by the atoms, and particularly the transmission of light through the end mirrors of the cavity. For the atomic part of the system, the two-level rate equations (7.1.8) must be generalized to a fully-quantized version of the three-level atomic rate equations derived in §1.9. For the radiative part of the system, the rate equation (7.1.12) for the photon probability distribution must be extended to include the loss of radiation through the cavity mirrors, described in classical terms in §1.10. Laser physics is a very extensive field of study [2,3] and the discussion given here is restricted to a simple idealized model of a three-level single-mode gas laser. The theory is approximate but it provides results identical or close to those of the more accurate but much more complex theory.

Figure 7.3 shows the three-level scheme, similar to that in Fig. 1.13 but with the developments in notation for the transition rates between levels 1 and 2 made in §7.1. The pumping rate from the ground state to the upper excited state for each of the N_0 ground-state atoms is again denoted by R. The magnitudes of the various rates are chosen to produce the simplest possible theory consistent with the derivation of a full range of the characteristic laser phenomena. Thus the spontaneous emission rate A_{20} for transitions from level 2 to level 0 is set equal to zero. Furthermore, the spontaneous emission rate from level 1 to level 0 is assumed to be much larger than the pumping rate and the other transition rates. The coefficients therefore satisfy

$$A_{20} = 0 \quad \text{and} \quad A_{10} \gg R, \Gamma_{\text{sp}} \text{ and } \Gamma_{\text{st}} n. \tag{7.3.1}$$

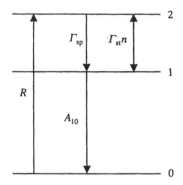

Fig. 7.3. Atomic energy-level scheme for a three-level laser showing the relevant transition rates.

It follows from eqns (1.9.2) and (1.9.3) that

$$N_1 \ll N_2,$$ (7.3.2)

and the population N_1 can be set equal to zero. As in §1.9, N_0 can be replaced by N, the total number of atoms in the cavity, to an excellent approximation, in view of the relatively small excitation of level 2. The quantum version of the atomic rate equation (1.9.2) is then

$$dN_2/dt = NR - N_2\Gamma_{sp} - N_2\Gamma_{st}n.$$ (7.3.3)

The rate of change of the photon probability distribution in the lasing mode has contributions from the interaction of the light with the atoms obtained by setting $N_1 = 0$ in eqn (7.1.12). The terms that remain are represented by the upwards-pointing arrows in Fig. 7.1 and these are reproduced on the left-hand side of Fig. 7.4. The downwards-pointing arrows on the right-hand side of the figure represent the removal of photons from the cavity by transmission through its mirrors. Thus the rate of reflections from the two mirrors together is c/L and a fraction $|\mathcal{T}|^2$ of the radiative energy in the cavity is lost to the outside at each reflection. The cavity loss rate per photon is accordingly

$$\Gamma_{cav} = c|\mathcal{T}|^2/L$$ (7.3.4)

and the resulting contributions are entered on the photon energy-level diagram in Fig. 7.4. The total rate of change of the nth element of the photon probability distribution is thus

$$dP(n)/dt = N_2\Gamma_{st}\{nP(n-1) - (n+1)P(n)\} - \Gamma_{cav}\{nP(n) - (n+1)P(n+1)\},$$ (7.3.5)

and for $n = 0$

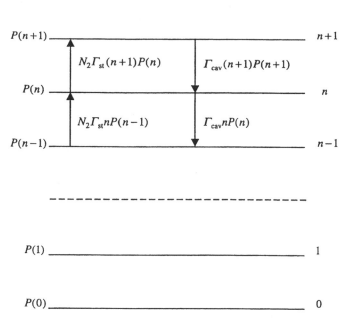

Fig. 7.4. Energy-level diagram for the photons in the lasing mode, showing the contributions to $\mathrm{d}P(n)/\mathrm{d}t$.

$$\mathrm{d}P(0)/\mathrm{d}t = - N_2 \Gamma_{\mathrm{st}} P(0) + \Gamma_{\mathrm{cav}} P(1), \tag{7.3.6}$$

where the level population N_2 should again strictly be shown as the value appropriate to the photon number in the element of $P(n)$ that it multiplies.

The steady-state photon distribution is obtained by setting the time derivatives in eqns (7.3.3), (7.3.5) and (7.3.6) equal to zero. The atomic rate equation gives

$$N_2 = \frac{NR}{\Gamma_{\mathrm{sp}} + \Gamma_{\mathrm{st}} n}, \tag{7.3.7}$$

and the population in the upper level has an inverse dependence on the photon number n. This dependence is crucial to the ability of the laser above threshold to generate coherent light. Thus fluctuations in n above or below the mean value $\langle n \rangle$ have a negative-feedback effect of decreasing or increasing the atomic population, and hence the emission rate, respectively. The breadth of the photon-number distribution is accordingly reduced from its chaotic value to the smaller width characteristic of coherent light. These fluctuation properties are treated in the following section.

The dependence of N_2 on n makes eqn (7.3.5) for the photon probability distribution difficult to solve. However, similar to the two-level theory of §7.1, it is a good approximation to replace the photon number n by its mean value $\langle n \rangle$ when N_2 varies slowly with n. The calculations in the remainder of the present section use this approximation, and eqn (7.3.3) then gives the steady-state condition

$$NR = N_2\Gamma_{sp} + N_2\Gamma_{st}\langle n \rangle. \tag{7.3.8}$$

This equation expresses energy conservation in the atomic part of the laser system, the rate of excitation of level 2 on the left being balanced by the sum on the right of the rates of decay by spontaneous emission into all modes and by stimulated emission into the lasing mode. In the same approximation, a rate equation for the mean photon number is obtained by the method of Problem 7.1 as

$$d\langle n \rangle / dt = \Gamma_{st} N_2 (1 + \langle n \rangle) - \Gamma_{cav}\langle n \rangle, \tag{7.3.9}$$

with the steady-state condition

$$\Gamma_{st} N_2 (1 + \langle n \rangle) = \Gamma_{cav}\langle n \rangle. \tag{7.3.10}$$

This equation expresses energy conservation in the lasing mode, the sum on the left of the rates of generation of photons by spontaneous and stimulated emission being balanced on the right by the rate of loss of photons from the cavity. Elimination of the atomic population from eqns (7.3.8) and (7.3.10) gives

$$\Gamma_{st}\Gamma_{cav}\langle n \rangle^2 - (NR\Gamma_{st} - \Gamma_{sp}\Gamma_{cav})\langle n \rangle - NR\Gamma_{st} = 0. \tag{7.3.11}$$

It is convenient to simplify the notation before solving this equation. Thus the four rates that appear can be reduced to two parameters, which are more transparently related to the performance characteristics of the laser. The cooperation parameter C, defined by eqns (1.9.10) and (1.10.23), reduces with the use of eqn (7.3.4) to

$$C = \frac{NRF(\omega)B_{21}\hbar\omega}{A_{21}V\Gamma_{cav}} \tag{7.3.12}$$

for the conditions in eqn (7.3.1). The lineshape function $F(\omega)$ is assumed to have the Lorentzian form of eqn (2.5.20) with the atomic transition and the cavity mode in resonance, $\omega = \omega_0$. Then, with B_{21} from eqn (2.3.20) and the stimulated and spontaneous transition rates from eqns (7.1.4) and (7.1.7), the definition of the cooperation parameter becomes

$$C = NR\Gamma_{st}/\Gamma_{sp}\Gamma_{cav}. \tag{7.3.13}$$

The laser threshold is defined by the condition $C = 1$, as in the semiclassical theory of §1.10, and it is achieved for the pumping rate given by

$$NR = \Gamma_{sp}\Gamma_{cav}/\Gamma_{st} \quad \text{for} \quad C = 1. \tag{7.3.14}$$

Another useful parameter is the *saturation photon number* n_s of the laser, defined as

$$n_s = \Gamma_{sp}/\Gamma_{st}, \tag{7.3.15}$$

whose role in the characteristics of the laser becomes apparent shortly. With this definition, the form of the atomic population in eqn (7.3.7) resembles the semi-classical forms in §§1.7 and 1.9, with their radiative and saturation energy densities replaced by the corresponding photon-number quantities. The inverse of the saturation photon number, $1/n_s$, known as the *spontaneous emission factor* β, denotes the fraction of the total spontaneous emission that is directed into the lasing mode [4].

The magnitudes of the atomic emission and cavity decay rates determine the approximations that can be made in the solution of the laser equations. We consider only the gas laser, where typical magnitudes of the spontaneous and stimulated emission rates are

$$\Gamma_{sp} \approx 3 \times 10^7 \, \text{s}^{-1} \quad \text{and} \quad \Gamma_{st} \approx 1 \, \text{s}^{-1}. \tag{7.3.16}$$

The saturation photon number is thus

$$n_s \approx 3 \times 10^7. \tag{7.3.17}$$

The cavity decay rate from eqn (7.3.4) for 99% reflectivity mirrors separated by 1 m is

$$\Gamma_{cav} \approx 3 \times 10^6 \, \text{s}^{-1}. \tag{7.3.18}$$

With insertion of the definitions of cooperation parameter and saturation photon number, the quadratic equation (7.3.11) for the mean photon number reduces to [2,4]

$$\langle n \rangle^2 - (C-1)n_s\langle n \rangle - Cn_s = 0, \tag{7.3.19}$$

with positive solution

$$\langle n \rangle = \tfrac{1}{2}\left\{(C-1)n_s + \left[(C-1)^2 n_s^2 + 4Cn_s\right]^{1/2}\right\}. \tag{7.3.20}$$

The mean atomic population from eqn (7.3.8) is

$$N_2 = \frac{\Gamma_{cav}}{\Gamma_{st}} \frac{Cn_s}{n_s + \langle n \rangle}. \tag{7.3.21}$$

For the very large numerical value of n_s given in eqn (7.3.17), the square root in eqn (7.3.20) can be expanded in the form

$$\langle n \rangle = \tfrac{1}{2} \left\{ (C-1)n_s + |C-1|n_s \left[1 + \frac{2C}{(C-1)^2 n_s} + O \left(\frac{1}{n_s^2} \right) \right] \right\} \qquad (7.3.22)$$

to give mean photon numbers, and hence atomic populations, for the various regimes of operation of the laser.

(1) *Laser below threshold*, $C < 1$. Retention only of the lowest-order contribution in $1/n_s$ in eqn (7.3.22) gives the approximate solutions

$$\langle n \rangle \approx \frac{C}{1-C} \quad \text{and} \quad N_2 \approx \frac{\Gamma_{\text{cav}} C}{\Gamma_{\text{st}}}. \qquad (7.3.23)$$

The mean number of photons in the cavity is very small compared to n_s. Note that the steady-state condition (7.3.10) is satisfied exactly but eqn (7.3.8) only in the excellent approximation where $\Gamma_{\text{st}} \langle n \rangle$ is neglected relative to Γ_{sp}. The mean photon number is greater than 1 for $C > \tfrac{1}{2}$ and the rate of stimulated emission thus exceeds the rate of spontaneous emission into the lasing mode for pumping rates above one half the threshold value.

(2) *Laser at threshold*, $C = 1$. It follows directly from eqn (7.3.20) that

$$\langle n \rangle = \sqrt{n_s} \quad \text{and hence} \quad N_2 \approx \Gamma_{\text{cav}}/\Gamma_{\text{st}}. \qquad (7.3.24)$$

The mean photon number is already greatly increased from its below-threshold value.

(3) *Laser above threshold*, $C > 1$. The leading term in eqn (7.3.22) gives the approximate solutions

$$\langle n \rangle \approx (C-1)n_s \quad \text{and} \quad N_2 \approx \Gamma_{\text{cav}}/\Gamma_{\text{st}}. \qquad (7.3.25)$$

The physical significance of the saturation photon number is thus identified as the mean number of photons in the lasing mode of the cavity at twice the threshold pumping rate, $C = 2$. Note that the steady-state condition (7.3.8) is satisfied exactly but eqn (7.3.10) only in the good approximation where 1 is neglected relative to $\langle n \rangle$. The output flux F_{out} from each end of the laser cavity, expressed in photons per second, is defined in the manner outlined in §6.2. It is determined by the transmission through the appropriate mirror at each encounter by the internal photon excitation, so that

$$F_{\text{out}} = \langle n \rangle \left(c|\mathcal{T}|^2/2L \right) = \tfrac{1}{2} \Gamma_{\text{cav}} \langle n \rangle = \tfrac{1}{2} \Gamma_{\text{cav}} (C-1)n_s, \qquad (7.3.26)$$

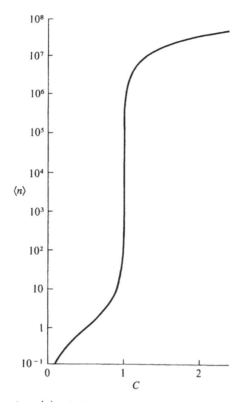

Fig. 7.5. Mean number $\langle n \rangle$ of photons in the lasing mode as a function of the pumping rate C for $n_s = 3 \times 10^7$.

where eqn (7.3.4) is used. Note the similarity of this expression for the mean output photon flux to that for the mean output intensity in the semiclassical theory of the laser, given by eqn (1.10.24).

Figure 7.5 shows the mean photon number in the cavity as a function of the normalized pumping rate C obtained from eqn (7.3.20) for the value of n_s given in eqn (7.3.17). The rapid increase in photon number by about five orders of magnitude in the threshold region is the most striking feature. The logarithmic vertical scale is needed so that the variation of photon number below threshold should be visible at all. The variation of mean atomic population is less dramatic but its linear growth below threshold and constant value above are noteworthy.

Figure 7.6 shows the atomic excitation rate by the pump, the rate of loss of photons into the laser outputs, and the spontaneous emission rate into other modes, as functions of C. The small emission into the lasing mode below threshold is not visible on the linear scale of the figure. Note particularly the locking of the above-threshold rate of spontaneous emission into the other modes to its value at $C = 1$. The additional energy supplied to the lasing transition as C increases above unity is all converted into laser output flux.

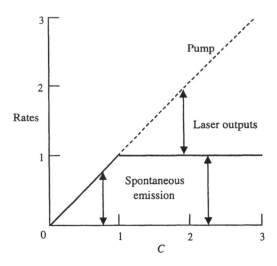

Fig. 7.6. Rates of atomic pumping and of the emission of photons by the laser into the two output beams and into the other modes of the system, in units of $\Gamma_{sp}\Gamma_{cav}/\Gamma_{st}$.

The above calculations are based on a tacit assumption that averages of products of the atomic population and cavity photon number can be factorized into products of the individual averages, denoted N_2 and $\langle n \rangle$ respectively. This factorization is clearly approximate as the atomic population given by eqn (7.3.7) is a function of the photon number n. More precise theories use a joint probability distribution for the atomic population and photon number [4,5]. The resulting expressions for the mean photon numbers at and above threshold are slightly different but those given here are excellent approximations.

7.4 Fluctuations in laser light

The main results of the preceding section are based on the further approximation of replacing the photon-number-dependent atomic population of eqn (7.3.7) by its averaged form obtained from eqn (7.3.8). The approximation produces good results for the mean photon numbers in the different regimes of operation of the laser but a more accurate theory is needed to obtain the fluctuation characteristics that are important in determining the coherence of the laser output.

The chain of rate equations (7.3.5) and (7.3.6) for the photon probability distribution has a steady-state solution for which adjacent elements are related by [5,6]

$$N_2\Gamma_{st}P(n-1) = \Gamma_{cav}P(n),$$ (7.4.1)

similar to eqn (7.2.1). Thus insertion of the atomic population appropriate to the photon number $n-1$ from eqn (7.3.7) gives

$$P(n) = \frac{NR\Gamma_{st}/\Gamma_{cav}}{\Gamma_{sp} + \Gamma_{st}(n-1)} P(n-1) = \frac{Cn_s}{n_s + n - 1} P(n-1), \tag{7.4.2}$$

where the cooperation parameter and saturation photon number are defined by eqns (7.3.13) and (7.3.15). Thus, by iteration,

$$P(n) = \frac{(Cn_s)^n (n_s - 1)!}{(n_s + n - 1)!} P(0). \tag{7.4.3}$$

The remaining unknown element, $P(0)$, is determined by normalization of the distribution. The general result is quite complicated but simple forms are found below and above threshold.

Consider first the laser below threshold, $C < 1$, where the numbers n of excited photons are all very much smaller than n_s. The approximation

$$(n_s + n - 1)! = (n_s + n - 1)(n_s + n - 2)...n_s(n_s - 1)! \approx n_s^n(n_s - 1)! \quad \text{for} \quad n \ll n_s \tag{7.4.4}$$

is very well satisfied and eqn (7.4.3) reduces to

$$P(n) = C^n P(0) = C^n(1 - C), \tag{7.4.5}$$

with the value of $P(0)$ easily determined by normalization. This is a geometric distribution, as in eqn (1.4.2), characteristic of chaotic light with a mean photon number

$$\langle n \rangle = \frac{C}{1 - C}, \tag{7.4.6}$$

identical to the value derived in eqn (7.3.23). The light in the lasing mode below threshold has the usual chaotic-light degree of second-order coherence

$$g^{(2)}(0) = 2. \tag{7.4.7}$$

Now consider the laser above threshold, $C > 1$, where the numbers n of excited photons are of order n_s and the -1 terms in the factorials in eqn (7.4.3) can be ignored, to a very good approximation. The mean photon number is then

$$\langle n \rangle = \sum_{n=0}^{\infty} nP(n) = \sum_{n=1}^{\infty}(n_s + n - n_s)\frac{(Cn_s)^n n_s!}{(n_s + n)!} P(0) \tag{7.4.8}$$
$$= Cn_s - n_s[1 - P(0)] \approx (C - 1)n_s,$$

with $P(0)$ assumed to be negligible for a distribution that encompasses very large photon numbers. This is identical to the value derived in eqn (7.3.25). The approximated distribution from eqn (7.4.3) can be rewritten as

$$P(n) = \frac{(n_s + \langle n \rangle)^n n_s!}{(n_s + n)!} P(0) = \frac{(n_s + \langle n \rangle)^n (n_s + \langle n \rangle)!}{(n_s + \langle n \rangle)^{\langle n \rangle} (n_s + n)!} P(\langle n \rangle),$$ (7.4.9)

and it is expressed as a Gaussian function by methods very similar to those used in Problem 5.6.

Problem 7.4 Prove that the above-threshold normalized photon-number distribution is given by

$$P(n) = \frac{\exp\left\{-(n - \langle n \rangle)^2 / 2(n_s + \langle n \rangle)\right\}}{\left[2\pi(n_s + \langle n \rangle)\right]^{1/2}}$$ (7.4.10)

to a very good approximation.

The distribution resembles that of large-amplitude single-mode coherent light given in eqn (5.3.22), except that its variance has the enhanced value

$$(\Delta n)^2 = n_s + \langle n \rangle = C n_s.$$ (7.4.11)

The contribution n_s represents additional amplitude noise on the coherent excitation of the lasing mode and the distribution tends to that of coherent light in the limit $\langle n \rangle \gg n_s$. The second factorial moment is correspondingly

$$\langle n(n-1) \rangle \approx n_s + \langle n \rangle^2$$ (7.4.12)

and the degree of second-order coherence is

$$g^{(2)}(0) = \frac{n_s}{\langle n \rangle^2} + 1 \approx 1,$$ (7.4.13)

very close to that of coherent light except for values of C very slightly greater than unity.

Figure 7.7 shows a representation of the field in the lasing mode, analogous to the coherent-state representation shown in Fig. 5.4. The single-mode field operators are defined in §5.1, with the field measured in units of $2(\hbar\omega/2\varepsilon_0 V)^{1/2}$. The above calculation provides information only on the photon-number distribution of the light and hence on the *amplitude* of its electric-field. The mean amplitude of the field is $\langle n \rangle^{1/2}$, with $\langle n \rangle$ now a very large number, of order n_s given by eqn (7.3.17). The variance can be written in the manner of eqn (5.3.34) for the pure coherent state as

$$\Delta n = \left(\langle n \rangle^{1/2} + \delta\right)^2 - \left(\langle n \rangle^{1/2} - \delta\right)^2 = 4\delta\langle n \rangle^{1/2}.$$ (7.4.14)

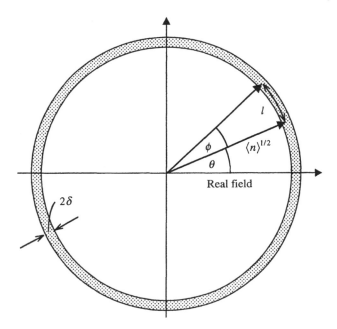

Fig. 7.7. Representation of the field distribution of laser light. The mean amplitude $\langle n \rangle^{1/2}$ of the field and its uncertainty 2δ are shown, with the latter greatly exaggerated. Spontaneous emission causes the tip of the field vector to diffuse around the annular noise region, whose radial width is drawn equal to 2δ.

The expression is consistent with eqn (7.4.11) if

$$\delta = \frac{\Delta n}{4\langle n \rangle^{1/2}} = \tfrac{1}{4}\left(\frac{C}{C-1}\right)^{1/2} \tag{7.4.15}$$

The field-amplitude width 2δ is shown in Fig. 7.7; it reduces to $1/2$ for $C \gg 1$ but it is otherwise somewhat larger than the coherent-state value. The expressions for the amplitude noise obtained in a more accurate theory [4] are broadly similar to the above but they differ in the detailed dependence on C.

It is shown in §7.2 that light of arbitrary initial coherence and photon statistics has its properties preserved by stimulated emission but that spontaneous emission generates chaotic light. The cause of the noise in the laser above threshold is the spontaneous emission of photons into the lasing mode, which occurs at a rate obtained from eqn (7.3.25) as

$$N_2 \Gamma_{st} = \Gamma_{cav}. \tag{7.4.16}$$

As a photon in the lasing mode survives for a time of order $1/\Gamma_{cav}$ before transmission out of the cavity through its end mirrors, there is on average one noise photon present in the mode at all times. This is of course a very small fraction of the total photon number in the lasing mode, given in eqn (7.3.25). The amplitude noise 2δ is correspondingly very much smaller than the mean amplitude $\langle n \rangle^{1/2}$ of the field. However, the noise photons have a more drastic effect on the *phase* of the laser field, which is not constrained to maintain a given mean value. They convert the circular noise disc of coherent light, shown in Fig. 5.4, into the annular noise region of laser light, shown in Fig. 7.7.

The spontaneous emission of a single photon corresponds to the addition of an arrow of unit length and random orientation to the end of the $\langle n \rangle^{1/2}$ field arrow in Fig. 7.7 (see §5.2 for the field properties of a single photon). The repeated emission of spontaneous photons causes the tip of the total field vector to move across the shaded annulus that represents the noise. The contribution of each spontaneous photon to the noise is resolved into two components of equal average magnitude, (a) a radial amplitude component that changes the length of the field vector but not its phase angle θ, and (b) a tangential phase component that changes the angle θ but not the length of the field vector. The amplitude components produced by successive photons are responsible for the amplitude noise derived above and their effect is limited by their small number. The phase component produced by each photon is also small compared to the total field vector but the phase angle θ is unrestricted and successive spontaneous photons have the cumulative effect of causing the tip of the field arrow to diffuse around the noise annulus in Fig. 7.7.

Consider the effect of the Γ_{cav} spontaneous emissions per second on the phase of the laser field. The change ϕ in the phase θ that occurs after a time duration τ is related to the distance l along the noise annulus, through which the tip of the field vector moves, by

$$\langle n \rangle^{1/2} \phi = l. \tag{7.4.17}$$

The mean-square value of l produced by the randomly-oriented unit vector associated with each spontaneous emission event is

$$\overline{l^2} = \tfrac{1}{2}. \tag{7.4.18}$$

The change in θ for each event is equally likely to be positive or negative and the resulting motion of the tip of the arrow is that of a random walk around the annulus with forwards and backwards steps of mean length determined by eqn (7.4.18). According to standard one-dimensional random-walk theory [7], the normalized probability that a distance l is travelled after $\Gamma_{\text{cav}}\tau$ such steps is

$$p(l) = \left(2\pi \overline{l^2} \Gamma_{\text{cav}} \tau\right)^{-1/2} \exp\left(-l^2 / 2\overline{l^2}\Gamma_{\text{cav}}\tau\right). \tag{7.4.19}$$

The probability distribution for l is converted to the equivalent distribution for the change ϕ in phase angle after a time τ by the use of eqns (7.4.17) and (7.4.18),

$$p(\phi) = \left(\langle n \rangle / \pi \Gamma_{\text{cav}} \tau\right)^{1/2} \exp\left(-\langle n \rangle \phi^2 / \Gamma_{\text{cav}} \tau\right). \tag{7.4.20}$$

This one-dimensional random walk differs from the two-dimensional walk used in §3.6 to model the intensity fluctuations of chaotic light.

The probability distribution in eqn (7.4.20) describes a *phase diffusion* of the laser field produced by spontaneous emission into the lasing mode. Its effect is displayed by evaluation of the projection of the electric field at time τ on to the field at time 0,

$$\left\langle \hat{E}^-(0)\hat{E}^+(\tau) \right\rangle = \langle n \rangle \exp(-i\omega_0 \tau)\langle \cos\phi \rangle, \tag{7.4.21}$$

where the field operators are defined in §5.1. The use of a standard integral gives

$$\langle \cos\phi \rangle = \int_{-\infty}^{\infty} \mathrm{d}\phi\, p(\phi)\cos\phi = \exp(-\Gamma_{\mathrm{cav}}\tau/4\langle n \rangle), \tag{7.4.22}$$

and the field correlation function can be written

$$\left\langle \hat{E}^-(0)\hat{E}^+(\tau) \right\rangle = \langle n \rangle \exp(-i\omega_0 \tau - \gamma_{\mathrm{pd}}|\tau|), \tag{7.4.23}$$

where the phase-diffusion damping rate is defined as

$$\gamma_{\mathrm{pd}} = \Gamma_{\mathrm{cav}}/4\langle n \rangle. \tag{7.4.24}$$

It is, however, clear by reference to Fig. 7.7 that the above calculation cannot be entirely correct. The phase angle ϕ is defined modulo 2π, similar to the phase discussed in §5.2; its range extends from 0 to 2π, and not over the infinite range of integration in eqn (7.4.22). The phase distribution must therefore be a periodic function of ϕ and, after the elapse of an infinite time, it must tend to the uniform value of $1/2\pi$, similar to eqn (5.2.22), and not to zero. The more realistic random walk effectively occurs along a one-dimensional axis with periodic boundary conditions or confined between perfectly-reflecting mirrors. In view of these remarks, the probability distribution in eqn (7.4.20) is valid only for times sufficiently short that $p(\phi)$ is negligible for phase changes of order $\pm\pi$. Nevertheless, a more accurate calculation of the phase distribution [8], taking proper account of its periodicity, produces the same result for the field correlation function as in eqn (7.4.23), and the expression in eqn (7.4.24) properly characterizes the phase-diffusion damping.

The correlation function in eqn (7.4.23) is that of light with the Lorentzian spectrum defined by eqn (2.5.20). The full width of the spectrum at half maximum height is

$$2\gamma_{\mathrm{pd}} = \Gamma_{\mathrm{cav}}/2\langle n \rangle = \Gamma_{\mathrm{cav}}^2/4F_{\mathrm{out}}, \tag{7.4.25}$$

where the total laser output flux $2F_{\mathrm{out}}$ is obtained from eqn (7.3.26). This is a form of the *Schawlow–Townes formula* for the laser linewidth. For $C = 2$ and the numerical magnitudes given in eqns (7.3.17) and (7.3.18),

$$2\gamma_{pd} \approx 0.05 \, s^{-1}. \tag{7.4.26}$$

The phase-diffusion linewidth can be measured by a Michelson interferometer with a laser operated slightly above threshold, so that $2\gamma_{pd}$ is significantly larger than the value in eqn (7.4.26). The measurements [9] confirm the Schawlow–Townes formula.

The picture of single-mode laser light provided by the above calculations shows a field excitation that approaches the form of a coherent state as the pumping increases to values well above threshold. The mean amplitude $\langle n \rangle^{1/2}$ of the coherent state is very large, of order 10^4 for the assumed numerical values of the laser parameters, while the amplitude uncertainty 2δ is of order unity. The field thus displays an almost classical behaviour, similar to the stable wave illustrated in Fig. 3.10. However, this picture is valid only over time intervals that are much shorter than the characteristic phase diffusion time, given by the inverse of the spectral width in eqn (7.4.25), and equal to 20 s for the numerical estimate in eqn (7.4.26). The phase of the electric field diffuses away from its initial value for times of this order and the resemblance of the laser excitation to a classical stable wave is lost. The estimated phase diffusion time is very long and the period of stability in a practical laser is often limited by other effects, not included in the idealized model, for example mechanical vibrations of the laser cavity.

The above derivation of the linewidth is based on a somewhat qualitative model of the phase diffusion associated with the spontaneous component of the emission into the lasing mode. A more accurate treatment of phase diffusion [10] uses the equation of motion for the phase distribution described in §5.2. The resulting phase-diffusion rate is similar to eqn (7.4.24) but it is composed of two contributions that take separate accounts of the phase diffusion associated with the emission of photons by the atoms and the loss of photons through the cavity mirrors. This separation has physical consequences in the possibilities for reductions in the laser linewidth. Thus the specification of the pump by a mean rate R of atomic excitation in the model used here produces a Poissonian distribution for the number of excited atoms, which is not explicitly considered in the above calculations. More sophisticated theories of the laser allow for forms of pump statistics that correspond to more regular atomic excitation. The reduction of noise in the atomic pumping leads to reductions in the optical noise component associated with the emission, and it is possible to produce laser light with non-classical or squeezed photon statistics (see [11] and earlier references therein).

7.5 Travelling-wave attenuation

The theories of §§7.1 and 7.2 provide expressions for the *time* dependences of the various statistical properties of light that is attenuated or amplified by interaction with two-level atoms in an optical cavity. The calculations form an important preliminary to the theory of the gas laser, which is the pre-eminent example of a cavity device. The present and following sections are concerned with the attenuation and amplification of a travelling-wave light beam, in contrast to the standing waves of a cavity. The propagation characteristics of a travelling wave are described by the *spatial* dependences of the statistical properties of the beam. The frequency dependences of the refractive index and of the attenuation or gain coefficient of the medium through which the light propagates are allowed for in the calculations that follow; they are important for a realistic description of the spatial development of the beam.

The classical theories of the effects of attenuation and amplification on the mean intensity of a light beam are given in §§1.8 and 1.9. The quantum theory of electromagnetic wave propagation in media that display dispersion and attenuation is most rigorously and completely provided by a formal quantization of the coupled system of the field and the material medium [12]. Such a theory supplies expressions for the dielectric function $\varepsilon(\omega)$, and for the other optical functions of the medium, in terms of its microscopic parameters. The theory also determines the forms of the propagating electromagnetic-field operators. However, some of the main results for the propagation characteristics can be derived by a much simpler calculation based on beam-splitter theory [13] and this approach is followed here. The simpler theory is applicable to a medium whose dielectric function is known and it does not provide any basis for calculating the form of $\varepsilon(\omega)$. The quantization of the electromagnetic field in a material with real dielectric function is treated in §9.2.

It is pointed out in §§1.6 and 1.8 that the scattering associated with spontaneous emission provides a microscopic mechanism for the attenuation of a light beam. Consider the propagation of a signal beam through a medium in which the light is attenuated as a result of scattering by impurity atoms or other inhomogeneities. The scattering centres in an beam path of length L are modelled by the line of beam splitters shown in Fig. 7.8. These are taken to be discrete components in the initial stages of the calculation, but their number

$$N = L/\Delta z \tag{7.5.1}$$

later tends to infinity to model a continuous attenuating medium. The travelling-

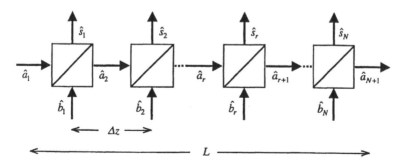

Fig. 7.8. Beam-splitter representation of the scattering in an attenuating medium. The continuous-mode destruction operators all refer to a common frequency ω.

wave signal beam, assumed to have a fixed linear polarization, is described by the formalism of continuous-mode field operators developed in §6.2. The beam splitters are taken to be identical and the input–output relations for the rth element have the forms of eqn (6.6.1),

$$\hat{s}_r(\omega) = \mathcal{R}(\omega)\hat{a}_r(\omega) + \mathcal{T}(\omega)\hat{b}_r(\omega)$$
$$\hat{a}_{r+1}(\omega) = \mathcal{T}(\omega)\hat{a}_r(\omega) + \mathcal{R}(\omega)\hat{b}_r(\omega). \tag{7.5.2}$$

The frequency-dependent reflection and transmission coefficients satisfy the relations (5.7.1) for common frequencies ω.

The operators $\hat{a}_r(\omega)$ and $\hat{a}_{r+1}(\omega)$ refer to the attenuated signal beam before and after the rth scattering centre respectively, while the operator $\hat{s}_r(\omega)$ refers to the light that is scattered *out of* the beam. The pairs of input and output operators satisfy the usual boson commutation relations, for example

$$\left[\hat{a}_r(\omega), \hat{a}_r^\dagger(\omega')\right] = \left[\hat{a}_{r+1}(\omega), \hat{a}_{r+1}^\dagger(\omega')\right] = \left[\hat{s}_r(\omega), \hat{s}_r^\dagger(\omega')\right] = \delta(\omega - \omega'). \tag{7.5.3}$$

The operator $\hat{b}_r(\omega)$ refers to light that is scattered *into* the beam by the scattering centre and its presence is required by the reversibility of the beam-splitter scattering process. These operators represent the addition of noise to the propagating light beam. It is assumed that the noise operators for different scattering centres are independent, with commutation relation

$$\left[\hat{b}_q(\omega), \hat{b}_r^\dagger(\omega')\right] = \delta_{qr}\delta(\omega - \omega'). \tag{7.5.4}$$

A calculation of the effects of attenuation requires an expression for the signal operator after transmission through N beam splitters in terms of the signal operator incident on the first beam splitter. This is readily obtained, by iteration of the second expression in eqn (7.5.2), as

$$\hat{a}_{N+1}(\omega) = [\mathcal{T}(\omega)]^N \hat{a}_1(\omega) + \mathcal{R}(\omega)\sum_{r=1}^{N} [\mathcal{T}(\omega)]^{N-r}\hat{b}_r(\omega). \tag{7.5.5}$$

The beam splitters are now converted to a continuous distribution by taking the limits

$$N \to \infty, \quad \Delta z \to 0 \quad \text{and} \quad |\mathcal{R}(\omega)|^2 \to 0 \tag{7.5.6}$$

in such a way that the *attenuation coefficient* $K(\omega)$, defined by

$$K(\omega) = |\mathcal{R}(\omega)|^2 / \Delta z, \tag{7.5.7}$$

remains finite. With the use of eqn (7.5.1), the usual exponential limit then gives

$$|\mathcal{T}(\omega)|^{2N} = \left[1 - |\mathcal{R}(\omega)|^2\right]^N = \left[1 - \frac{K(\omega)L}{N}\right]^N \rightarrow \exp[-K(\omega)L]. \qquad (7.5.8)$$

The frequency-dependent phase angle of the transmission coefficients in eqn (7.5.5) is chosen to produce the usual propagation phase expressed in terms of the refractive index $\eta(\omega)$ of the attenuating medium, so that

$$[\mathcal{T}(\omega)]^N = \exp\left\{\left[i\eta(\omega)(\omega/c) - \tfrac{1}{2}K(\omega)\right]L\right\}. \qquad (7.5.9)$$

Note that the phase angle of the transmission coefficient $\mathcal{T}(\omega)$ of the individual beam splitter tends to zero in the limit of eqn (7.5.6) and the phase angle of the reflection coefficient $\mathcal{R}(\omega)$ can be chosen as $\pi/2$, in accordance with the general beam-splitter requirements. The attenuation coefficient $K(\omega)$ and the refractive index $\eta(\omega)$ are determined by the dielectric function $\varepsilon(\omega)$ of the medium, which is assumed to be known, by eqns (1.8.3) and (1.8.4).

The noise input operators and their commutators are converted to continuous spatial dependences by replacements analogous to those used in §6.2 to convert discrete-mode to continuous-mode variables,

$$\hat{b}_r(\omega) \rightarrow (\Delta z)^{1/2} \hat{b}(z, \omega) \quad \text{and} \quad \delta_{qr} \rightarrow \Delta z \delta(z - z'). \qquad (7.5.10)$$

The noise operator commutator is thus converted from the discrete form in eqn (7.5.4) to the spatially-continuous form

$$\left[\hat{b}(z, \omega), \hat{b}^\dagger(z', \omega')\right] = \delta(z - z')\delta(\omega - \omega'). \qquad (7.5.11)$$

The noise operators are assigned expectation values

$$\left\langle \hat{b}(z, \omega) \right\rangle = \left\langle \hat{b}^\dagger(z, \omega) \right\rangle = 0 \qquad (7.5.12)$$

and

$$\left\langle \hat{b}^\dagger(z, \omega)\hat{b}(z', \omega') \right\rangle = f_\mathcal{N}(\omega)\delta(z - z')\delta(\omega - \omega'), \qquad (7.5.13)$$

where $f_\mathcal{N}(\omega)$ is a position-independent mean flux of noise photons per unit angular frequency bandwidth.

The summation in eqn (7.5.5) is converted to an integration by the replacement

$$\sum_{r=1}^{N} \rightarrow \frac{1}{\Delta z}\int_0^L dz \qquad (7.5.14)$$

and the input–output relation takes the form

$$\hat{a}_L(\omega) = \exp\left\{\left[i\eta(\omega)(\omega/c) - \tfrac{1}{2}K(\omega)\right]L\right\}\hat{a}_0(\omega)$$

$$+i\sqrt{K(\omega)}\int_0^L dz \exp\left\{\left[i\eta(\omega)(\omega/c) - \tfrac{1}{2}K(\omega)\right](L-z)\right\}\hat{b}(z,\omega). \tag{7.5.15}$$

The discrete beam-splitter subscripts on the signal operators are here replaced by the continuous z coordinate. The input signal operators must satisfy the usual continuous-mode commutation relation of eqn (6.2.5) or (7.5.3),

$$\left[\hat{a}_0(\omega), \hat{a}_0^\dagger(\omega')\right] = \delta(\omega - \omega'). \tag{7.5.16}$$

The output operators are identified by subscripts L that denote the propagation distance, which should not be confused with the local-oscillator L subscripts used in §6.11.

Problem 7.5   Prove that the output signal operators obtained from eqn (7.5.15) also satisfy the standard commutation relation

$$\left[\hat{a}_L(\omega), \hat{a}_L^\dagger(\omega')\right] = \delta(\omega - \omega'). \tag{7.5.17}$$

It is seen from eqn (7.5.15) that the attenuation reduces the input signal component of the output operator and introduces contributions from the noise operators. The input–output relation determines all properties of the output signal when the corresponding properties of the input signal and the noise are known. It has the same overall form as the usual beam-splitter input–output relation and many of the standard beam-splitter results apply to the effects of propagation through an attenuating medium on the properties of the input light.

The Fourier-transformed time-dependent input and output operators are defined by eqn (6.2.13) for sufficiently narrow bandwidths. It is convenient to denote the real part of the signal wavevector in the attenuating medium by

$$k(\omega) = \omega\eta(\omega)/c \tag{7.5.18}$$

and the *signal attenuation* as

$$\mathcal{K}(\omega) = \exp[-K(\omega)L]. \tag{7.5.19}$$

The mean output signal flux, defined as in eqn (6.2.21), is then obtained with the use of eqns (7.5.13) and (7.5.15) as

$$f_S(t) = \left\langle \hat{a}_L^\dagger(t)\hat{a}_L(t) \right\rangle = (2\pi)^{-1}\int d\omega \int d\omega' \left\langle \hat{a}_0^\dagger(\omega)\hat{a}_0(\omega') \right\rangle \sqrt{\mathcal{K}(\omega)\mathcal{K}(\omega')}$$

$$\times \exp\left\{i(\omega - \omega')t - i[k(\omega) - k(\omega')]L\right\} + (2\pi)^{-1}\int d\omega f_{\mathcal{N}}(\omega)\{1 - \mathcal{K}(\omega)\}. \tag{7.5.20}$$

The two terms on the right-hand side represent the attenuated input signal flux and the added noise flux. The integrals can be performed, at least in principle, when the functional forms of the attenuation coefficient, refractive index and noise spectrum are known.

The contributions of the noise operators are negligible in many circumstances. For an attenuating medium in thermal equilibrium, the flux of noise photons has the form given in eqn (1.3.7). For attenuation associated with an atomic transition at the signal frequency ω_0, the resonant noise flux is

$$f_{\mathcal{N}}(\omega_0) = \frac{1}{\exp(\hbar\omega_0/k_B T) - 1} = \frac{N_2}{N_1 - N_2}, \tag{7.5.21}$$

where the expression in terms of two-level atomic populations is obtained with the use of eqn (1.5.10). This expression is also the same as that found in eqn (7.2.5) for the noise photon number in a cavity attenuator. It is very small for atomic transitions in the visible region of the spectrum at ordinary temperatures, where the scattering of extraneous light into the beam is negligible. Thus the fields associated with the noise operators $\hat{b}(z, \omega)$ and their Fourier transforms are taken be in their vacuum states and we set

$$f_{\mathcal{N}}(\omega) = 0 \tag{7.5.22}$$

for the remainder of the present section.

The calculations are greatly simplified when the input signal bandwidth is sufficiently narrow that the attenuation coefficient can be set equal to its value at the central frequency ω_0 of the signal spectrum. The wavevector is expanded to first order in the frequency as

$$k(\omega) = k(\omega_0) + \frac{\omega - \omega_0}{v_G(\omega_0)}, \tag{7.5.23}$$

where $v_G(\omega_0)$ is the usual group velocity at frequency ω_0, defined by

$$\frac{1}{v_G(\omega_0)} = \left. \frac{\partial k(\omega)}{\partial \omega} \right|_{\omega=\omega_0} \tag{7.5.24}$$

The frequency integrals in the expression (7.5.20) for the output flux are easily performed when these approximations are made, with the result

$$f_S(t) = \mathcal{K}(\omega_0) f_0(t_R), \tag{7.5.25}$$

where

$$t_R = t - [L/v_G(\omega_0)]. \tag{7.5.26}$$

This *retarded time* takes account of the propagation of the signal operator through the appropriate distance in the medium. The output flux is determined by the input flux at the retarded time and its magnitude is reduced by the same exponential factor as occurs in the classical expression (1.8.16). The transmission of a narrow bandwidth signal through an attenuating medium thus displays the usual physical attributes of propagation at the group velocity $v_G(\omega_0)$ and spatial decay with coefficient $K(\omega_0)$.

Measurements of the photon-number statistical properties of the light before and after propagation through the medium are made by direct detection, whose theory is outlined in §6.10. The measured property of the light is represented by the operator defined in eqn (6.10.1), whose roles in determining the photocount mean, second factorial moment and variance are described by eqns (6.10.5), (6.10.7) and (6.10.8) respectively. The effects of propagation are very similar to those of the detector efficiency, in that eqn (6.10.3) and the Fourier transform of eqn (7.5.15) both show an output field operator on the left equal to the sum of an appropriate fraction of an input field operator on the right and operators associated with vacuum-state fields. The vacuum field operators do not contribute to normally-ordered expectation values and the fractional quantities cancel in forming the degree of second-order coherence. Thus, analogous to eqn (6.10.9), the degree of second-order coherence is unchanged by propagation through an attenuating medium.

Measurements of the electric field or quadrature-operator properties of the light are made by homodyne detection, whose theory is outlined in §6.11. The measured property of the light is represented by the operator defined in eqn (6.11.4), whose roles in determining the mean and the variance of the difference photocount are described by eqns (6.11.5) to (6.11.7). These measurable quantities are conveniently expressed by eqns (6.11.12) and (6.11.13) in terms of the mean and the variance of the homodyne electric-field operator defined in eqn (6.11.10). The latter quantities are in turn related to the coherent signal S and the noise $\mathcal{N}$ as in the left-hand sides of eqns (6.11.15) and (6.11.16) respectively. The central frequency of the signal spectrum is taken to coincide with the frequency of the local oscillator, $\omega_0 = \omega_L$. The effects of attenuation on the homodyne field operator are readily obtained from the Fourier transform of eqn (7.5.15) when the signal bandwidth is sufficiently narrow.

Problem 7.6   Show that the mean homodyne fields before and after propagation through the medium are related by

$$\left\langle \hat{E}_H(\chi,t,T) \right\rangle_L = \sqrt{K(\omega_0)} \left\langle \hat{E}_H(\chi_R,t_R,T) \right\rangle_0, \qquad (7.5.27)$$

where

$$\chi_R = \chi - k(\omega_0)L \qquad (7.5.28)$$

and the retarded time is defined in eqn (7.5.26).

The noise operators make no contribution to this mean value, on account of the properties in eqn (7.5.12). The noise does, however, contribute to the expectation value of the square of the homodyne field, which contains a term with the operators in the order $\hat{b}\hat{b}^\dagger$. The relation between input and output homodyne field variances is obtained straightforwardly with the use of eqn (7.5.11). It takes a simple form when the integration time T is sufficiently long that the delta-function limit in eqn (2.4.1) is applicable, similar to eqn (6.11.24).

Problem 7.7 Prove that the variances of the homodyne fields before and after propagation though an attenuating medium of length L are related by

$$\left\langle\left[\Delta\hat{E}_{\mathrm{H}}(\chi,t,T)\right]^2\right\rangle_L - \tfrac{1}{4} = \mathcal{K}(\omega_0)\left\{\left\langle\left[\Delta\hat{E}_{\mathrm{H}}(\chi_{\mathrm{R}},t_{\mathrm{R}},T)\right]^2\right\rangle_0 - \tfrac{1}{4}\right\}.$$

(7.5.29)

The homodyne-field variance thus tends to the value 1/4, characteristic of coherent light or the vacuum state, in conditions of severe attenuation. Note that the signal attenuation enters the expressions (7.5.27) and (7.5.29) in similar ways to the occurrence of the detector quantum efficiency in the expressions (6.11.12) and (6.11.13) for the difference-photocount mean and variance in balanced homodyne detection.

These expressions simplify for an input 'single-mode' coherent signal, whose homodyne mean and variance are given by eqns (6.11.15) and (6.11.16). The corresponding output mean, or coherent signal, given by eqn (7.5.27) is

$$S_L = \left\langle\hat{E}_{\mathrm{H}}(\chi,t,T)\right\rangle_L = \left[\mathcal{K}(\omega_0)F_0 T\right]^{1/2}\cos(\chi_{\mathrm{R}} - \theta_0),$$

(7.5.30)

where F_0 is the input photon flux and θ_0 is the input phase. The output variance, or noise, given by eqn (7.5.29),

$$\mathcal{N}_L = \left\langle\left[\Delta\hat{E}_{\mathrm{H}}(\chi,t,T)\right]^2\right\rangle_L = \tfrac{1}{4},$$

(7.5.31)

is unchanged from the input value, and the amplitude uncertainty of the state is thus unaffected by propagation through the attenuating medium. Figure 7.9 shows a representation of the mean and the uncertainty of the homodyne electric field for the input coherent state before and after propagation, similar to Fig. 5.4 for the single-mode coherent state. The phase uncertainty is increased by the propagation according to

$$\Delta\varphi_L = 1/2\left[\mathcal{K}(\omega_0)F_0 T\right]^{1/2} = \Delta\varphi_0/\left[\mathcal{K}(\omega_0)\right]^{1/2},$$

(7.5.32)

where the input phase uncertainty is

$$\Delta\varphi_0 = 1/2\left(F_0 T\right)^{1/2},$$

(7.5.33)

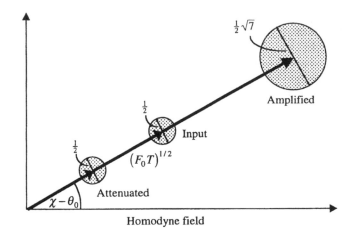

Fig. 7.9. Representations of the means and uncertainties in the homodyne electric fields for an input coherent state and the output states after propagation through an attenuating medium with $\mathcal{K}(\omega_0) = 1/4$ and an amplifying medium with $\mathcal{G}(\omega_0) = 4$. The measurement phase angles χ and χ_R are assumed equal.

analogous to that in eqn (5.3.35). The signal-to-noise ratio, given by

$$\text{SNR}_L = \mathcal{K}(\omega_0)\text{SNR}_0 \tag{7.5.34}$$

for equal measurement phase angles χ and χ_R, is degraded by propagation through the attenuating medium, with the same reduction factor as the beam flux in eqn (7.5.25).

The homodyne-field variance for an input squeezed vacuum state is given by eqn (6.11.25). We consider only the local-oscillator phase angle for which the input variance has its minimum value,

$$\left\langle \left[\Delta\hat{E}_H(\chi_R, t_R, T)\right]^2 \right\rangle_0 = \tfrac{1}{4}\exp[-2s(\omega_L)] \quad \text{for} \quad \chi_R = \tfrac{1}{2}\vartheta(\omega_L). \tag{7.5.35}$$

The output variance given by eqn (7.5.29) is

$$\left\langle \left[\Delta\hat{E}_H(\chi, t, T)\right]^2 \right\rangle_L = \tfrac{1}{4}\left\{1 - \mathcal{K}(\omega_L)\left(1 - \exp[-2s(\omega_L)]\right)\right\}, \tag{7.5.36}$$

where the local oscillator selects the input contribution at frequency ω_L. The output variance is increased above its squeezed input value by the attenuation and the squeezing effect disappears for heavy attenuation when the homodyne variance tends to the coherent-state or vacuum value of 1/4. The advantages of squeezed light for the transmission of information with reduced noise are lost when there is significant attenuation in the system.

7.6 Travelling-wave amplification

The effects of gain on electromagnetic wave propagation in media that display amplification can also be modelled by a line of beam splitters similar to that shown in Fig. 7.8. The beam splitters now represent an inverted population of effectively two-level atoms, similar to those assumed in the amplifier theories of §§1.9 and 7.2. These fictitious beam splitters have to be assigned unusual properties in order that they can model a travelling-wave inverted-population amplifier but the resulting input–output relations are the same as those derived in more realistic theories [14].

It is necessary to modify the input–output relations of the individual beam splitters in Fig. 7.8 in order to produce gain, rather than loss, in the transmitted signal. The amplifying property is achieved by inverting the harmonic oscillators associated with the $\hat{b}_r(\omega)$ input modes. The harmonic oscillators associated with the $\hat{s}_r(\omega)$ output modes must also be inverted for consistency of the model. These inversions have the effect of replacing the corresponding destruction operators by creation operators [15]. The relations (7.5.2) for the output destruction operators in terms of the input operators of the rth beam splitter are thus converted to

$$\hat{s}_r(\omega) = \mathcal{R}(\omega)\hat{a}_r^\dagger(\omega) + \mathcal{T}(\omega)\hat{b}_r(\omega)$$
$$\hat{a}_{r+1}(\omega) = \mathcal{T}(\omega)\hat{a}_r(\omega) + \mathcal{R}(\omega)\hat{b}_r^\dagger(\omega). \tag{7.6.1}$$

The pairs of input and output operators must satisfy the usual continuous-mode boson commutation relations, as in the examples of eqns (7.5.3) and (7.5.4). It follows that

$$|\mathcal{T}(\omega)|^2 - |\mathcal{R}(\omega)|^2 = 1, \tag{7.6.2}$$

and this is the amplifying beam-splitter replacement for the ordinary beam-splitter relations in eqn (5.7.1). The relation makes $|\mathcal{T}(\omega)|$ greater than unity, as required for amplification. The commutator of the two output operators vanishes automatically without any condition on the reflection and transmission coefficients. The noise operators retain their basic forms of commutator given in eqns (7.5.4) and (7.5.11).

The calculation for the amplifier proceeds by the same steps as for the attenuator, but the attenuation coefficient in eqn (7.5.7) is relabelled as a *gain coefficient* $G(\omega)$, defined by

$$G(\omega) = |\mathcal{R}(\omega)|^2/\Delta z. \tag{7.6.3}$$

The change of sign of the second term in eqn (7.6.2) reverses the sign of the exponent in eqn (7.5.8), so that

$$|\mathcal{T}(\omega)|^{2N} \rightarrow \exp[G(\omega)L]. \tag{7.6.4}$$

The conversion of the line of discrete beam splitters to a continuous amplifying medium follows the same steps as in eqns (7.5.10) to (7.5.14) and the input-output relation analogous to eqn (7.5.15) is

$$\hat{a}_L(\omega) = \exp\left\{\left[i\eta(\omega)(\omega/c) + \tfrac{1}{2}G(\omega)\right]L\right\}\hat{a}_0(\omega) + \hat{b}_{\mathcal{N}}(\omega), \tag{7.6.5}$$

where

$$\hat{b}_{\mathcal{N}}(\omega) = i\sqrt{G(\omega)}\int_0^L dz\exp\left\{\left[i\eta(\omega)(\omega/c) + \tfrac{1}{2}G(\omega)\right](L-z)\right\}\hat{b}^\dagger(z,\omega). \tag{7.6.6}$$

It is seen that the amplification has the effects of increasing the input signal component of the output operator and of introducing contributions from the noise operators. All of the properties of the output signal can be determined when the properties of the input signal and the noise are known.

The total output noise is represented by the operator $\hat{b}_{\mathcal{N}}(\omega)$, which is formally written as a destruction operator, despite its definition as an integral over noise creation operators. It has a simple commutation relation.

Problem 7.8   Prove that

$$\left[\hat{b}_{\mathcal{N}}(\omega), \hat{b}^\dagger_{\mathcal{N}}(\omega')\right] = -\left\{\exp[G(\omega)L] - 1\right\}\delta(\omega - \omega'). \tag{7.6.7}$$

This form of expression for the output noise commutator in terms of the gain conforms to a very general requirement that applies to all linear amplifiers [14]. It is readily verified with its use that the total output operator in eqn (7.6.5) continues to satisfy the commutation relation in eqn (7.5.17).

It is convenient to define the *signal gain* as

$$\mathcal{G}(\omega) = \exp[G(\omega)L], \tag{7.6.8}$$

analogous to the signal attenuation defined in eqn (7.5.19). The output signal flux, analogous to eqn (7.5.20), is obtained with the use of eqns (7.6.5) and (7.5.11) as

$$f_S(t) = (2\pi)^{-1}\int d\omega\int d\omega'\left\langle\hat{a}_0^\dagger(\omega)\hat{a}_0(\omega')\right\rangle\sqrt{\mathcal{G}(\omega)\mathcal{G}(\omega')}$$

$$\times\exp\left\{i(\omega - \omega')t - i[k(\omega) - k(\omega')]L\right\} + (2\pi)^{-1}\int d\omega\left[1 + f_{\mathcal{N}}(\omega)\right]\left[\mathcal{G}(\omega) - 1\right]. \tag{7.6.9}$$

The total time-independent output noise flux, given by the final term, is determined by the noise input $f_{\mathcal{N}}(\omega)$ appropriate to the amplification process. The form of this noise function is determined by the inverted harmonic oscillators associated with the $\hat{b}(z,\omega)$ input modes. The inversion corresponds to an effective negative temperature and, for amplification associated with an atomic transition of frequency ω_0, the exponential in eqn (7.5.21) is replaced by N_2/N_1 to give the resonant value

$$1 + f_{\mathcal{N}}(\omega_0) = 1 + \frac{N_1}{N_2 - N_1} = \frac{N_2}{N_2 - N_1}. \tag{7.6.10}$$

This expression is the same as occurs in the noise contribution for a cavity amplifier, given in eqn (7.2.18). When eqn (7.6.10) is substituted in the final term of eqn (7.6.9), the magnitude of the output noise flux satisfies another general requirement of a linear amplification process [14]. There is no way of totally removing the noise added by the amplification but its magnitude is minimized when there is complete population inversion, with $N_2 = N$ and $N_1 = 0$. More generally, the noise flux $f_{\mathcal{N}}(\omega)$ is assumed to vanish in the following calculations, as in eqn (7.5.22), so that the minimum amount of noise is added at all frequencies.

The calculations are again greatly simplified when the input signal has a narrow bandwidth. The frequency integrals in the first term on the right-hand side of eqn (7.6.9) are then easily performed with the use of eqn (7.5.23), and the result is

$$f_S(t) = G(\omega_0) f_0(t_R) + (2\pi)^{-1} \int d\omega [G(\omega) - 1], \tag{7.6.11}$$

where the retarded time is defined in eqn (7.5.26). The output flux is thus determined by the retarded input flux with its magnitude increased by the same factor as occurs in the classical expression (1.9.13). However, in contrast to the classical result, the quantum-mechanical calculation produces an unavoidable noise contribution, whose minimum value is given by the final term in eqn (7.6.11). The noise is associated with the spontaneous emission that must occur for an inverted atomic population.

The output noise flux can be evaluated if the form of the gain coefficient is known. A very simple example is provided by a signal gain that has a constant value over an angular-frequency bandwidth $\mathcal{B}$ and vanishes outside this bandwidth,

$$G(\omega) = \begin{cases} G_0 & \text{for} \quad \omega_0 - \tfrac{1}{2}\mathcal{B} < \omega < \omega_0 + \tfrac{1}{2}\mathcal{B} \\ 0 & \text{for all other frequencies.} \end{cases} \tag{7.6.12}$$

The output flux from eqn (7.6.11) is then

$$f_S(t) = G_0 f_0(t_R) + (\mathcal{B}/2\pi)(G_0 - 1). \tag{7.6.13}$$

Measurements of the signal flux are made by direct detection, whose theory is outlined in §6.10. The noise flux contributes to the measured quantity represented by the operator in eqn (6.10.1) and thus to the mean photocount derived in eqn (6.10.5).

The measured degree of second-order coherence is determined by the usual expectation value of normally-ordered ouput operators, obtained from the Fourier transform of eqn (7.6.5) and its Hermitian conjugate. However, in contrast to the attenuator, the ouput destruction operator now contains not only a contribution proportional to the input destruction operator but also the noise contribution of eqn (7.6.6), proportional to the noise creation operators. The resulting measured degree of second-order coherence thus acquires noise components, whose complicated detailed forms are not considered here [13].

The electric-field properties of the light, measured by homodyne detection, are calculated more easily, as the homodyne electric-field operator defined in eqn (6.11.10) is of first order in the creation and destruction operators, compared to the second-order form of the flux operator. The input signal is again assumed to have a narrow bandwidth, with its central frequency ω_0 equal to that of the local oscillator. The effects of gain on the signal operator are readily obtained from the Fourier transform of eqn (7.6.5). The expectation values of the noise operators continue to vanish, as in eqn (7.5.12), and it follows from eqn (6.11.10) that

$$\left\langle \hat{E}_H(\chi,t,T) \right\rangle_L = \sqrt{G(\omega_0)} \left\langle \hat{E}_H(\chi_R,t_R,T) \right\rangle_0, \tag{7.6.14}$$

where χ_R is defined in eqn (7.5.28). The noise operators make no contribution to the mean homodyne field but they do contribute to the variance. The calculations for the amplifier follow the same steps as those for the attenuator and the result that corresponds to eqn (7.5.29), for a long integration time is,

$$\left\langle \left[\Delta\hat{E}_H(\chi,t,T) \right]^2 \right\rangle_L + \tfrac{1}{4} = G(\omega_0) \left\{ \left\langle \left[\Delta\hat{E}_H(\chi_R,t_R,T) \right]^2 \right\rangle_0 + \tfrac{1}{4} \right\}. \tag{7.6.15}$$

The noise is always increased by amplification, in contrast to the approach of the noise produced by attenuation to the coherent-state value.

The expressions simplify for a 'single-mode' coherent input signal, whose homodyne-field mean and variance are given by eqns (6.11.15) and (6.11.16). The corresponding output mean, or coherent signal, given by eqn (7.6.14) is

$$S_L = \left\langle \hat{E}_H(\chi,t,T) \right\rangle_L = [G(\omega_0)F_0 T]^{1/2} \cos(\chi_R - \theta_0), \tag{7.6.16}$$

where F_0 is the input photon flux and θ_0 is the input phase. The output variance, or noise, given by eqn (7.6.15),

$$\mathcal{N}_L = \left\langle \left[\Delta\hat{E}_H(\chi,t,T) \right]^2 \right\rangle_L = \tfrac{1}{2} G(\omega_0) - \tfrac{1}{4}, \tag{7.6.17}$$

is increased from the input value. The long arrow and the large disc in Fig. 7.9 show the mean and the uncertainty in the homodyne field after propagation through an amplifying medium. The increased amplitude uncertainty in the output is accompanied by an increased phase uncertainty, obtained with the help of the figure as

$$\Delta\varphi_L = \left\{ 2 - \left[G(\omega_0) \right]^{-1} \right\}^{1/2} \Delta\varphi_0, \tag{7.6.18}$$

where the coherent-state input phase-uncertainty is given in eqn (7.5.33). The signal-to-noise ratio, given by

$$\mathrm{SNR}_L = \frac{G(\omega_0)\mathrm{SNR}_0}{2G(\omega_0) - 1} \tag{7.6.19}$$

for equal measurement phase angles χ and χ_R, is degraded by propagation through the amplifying medium, with a reduction factor that tends to 1/2 in the limit of very high gain.

The condition for a squeezed output signal is given by eqn (6.11.26) as

$$0 \leq \left\langle \left[\Delta\hat{E}_H(\chi,t,T) \right]^2 \right\rangle_L < \tfrac{1}{4}. \tag{7.6.20}$$

The corresponding condition on the input variance is obtained from eqn (7.6.15) as

$$1 \leq G(\omega_0) \left\{ 4 \left\langle \left[\Delta\hat{E}_H(\chi_R,t_R,T) \right]^2 \right\rangle_0 + 1 \right\} < 2. \tag{7.6.21}$$

Amplification thus tends to destroy any squeezing present in the input and, for an extremely squeezed input, the condition for some squeezing to survive in the output is

$$G(\omega_0) < 2 \quad \text{when} \quad \left\langle \left[\Delta\hat{E}_H(\chi_R,t_R,T) \right]^2 \right\rangle_0 = 0. \tag{7.6.22}$$

The general condition is illustrated by the input squeezed vacuum state, with a homodyne-field variance given by eqn (7.5.35), where the output variance is

$$\left\langle \left[\Delta\hat{E}_H(\chi,t,T) \right]^2 \right\rangle_L = \tfrac{1}{4}\left\{ G(\omega_L)\left(1 + \exp\left[-2s(\omega_L) \right] \right) - 1 \right\} \tag{7.6.23}$$

and the local oscillator selects the contribution at frequency ω_L. The maximum twofold gain for the preservation of squeezing is a severe restriction on the ability of a travelling-wave amplifier to maintain the reduced noise characteristics of a squeezed input signal.

Problem 7.9   Consider the system of a 'single-mode' coherent input signal of frequency ω_0 that passes through an attenuator and an amplifier in series. The signal attenuation $\mathcal{K}(\omega_0)$ and signal gain $\mathcal{G}(\omega_0)$ of the two components, both of length L, are related by

$$\mathcal{K}(\omega_0)\mathcal{G}(\omega_0) = 1. \tag{7.6.24}$$

Show that the mean homodyne fields of the output and input are equal and that the output field variance is

$$\mathcal{N}_{2L} = \left\langle \left[\Delta\hat{E}_H(\chi,t,T) \right]^2 \right\rangle_{2L} = \tfrac{1}{2}\mathcal{G}(\omega_0) - \tfrac{1}{4} \tag{7.6.25}$$

when the attenuator precedes the amplifier and

$$\mathcal{N}_{2L} = \left\langle \left[\Delta\hat{E}_H(\chi,t,T) \right]^2 \right\rangle_{2L} = \tfrac{3}{4} - \tfrac{1}{2}\mathcal{K}(\omega_0) \tag{7.6.26}$$

when the amplifier precedes the attenuator.

The greater output noise occurs for the system in which the input coherent signal encounters the attenuator before the amplifier.

7.7 Dynamics of the atom–radiation system

The light sources treated in §§7.1 to 7.4 involve atoms in cavities in conditions where only a single mode is excited. The effects of spontaneous emission into all of the spatial modes are included only in so far as they influence emission into the single chosen mode. The single-mode theory provides a realistic model of the laser for appropriate ranges of the parameters. However, other light sources that are important for key experiments in quantum optics involve atoms radiating into free space. The characteristics of the observed emission in these experiments are determined by the nature of the atomic excitation. Their theoretical interpretations need a relation between the time-dependent electromagnetic field and the atomic excitation that generates it.

As a preliminary to the derivation of the required relation in §7.8, we derive the basic equations of motion for the coupled atomic and radiative operators. The time dependence of the coupled system is conveniently obtained by use of the Heisenberg picture. Thus, for the destruction operator of an arbitrary mode, the equation of motion (4.7.12) and the Hamiltonian from eqn (4.9.24) give

$$i\,d\hat{a}_{\mathbf{k}\lambda}(t)/dt = \omega_k \hat{a}_{\mathbf{k}\lambda}(t) - ig_{\mathbf{k}\lambda}\,\hat{\pi}(t)\exp(-i\mathbf{k}\cdot\mathbf{R}), \tag{7.7.1}$$

where the commutation relation from eqn (4.4.5) is used. This equation is formally integrated to give

$$\hat{a}_{\mathbf{k}\lambda}(t) = \exp(-i\omega_k t)\left\{ \hat{a}_{\mathbf{k}\lambda}(0) - g_{\mathbf{k}\lambda}\int_0^t dt'\,\hat{\pi}(t')\exp(-i\mathbf{k}\cdot\mathbf{R} + i\omega_k t') \right\} \tag{7.7.2}$$

and the corresponding creation operator is given by the Hermitian conjugate of this expression. The equation of motion for the atomic transition operator is similarly obtained as

$$i\,d\hat{\pi}(t)/dt = \omega_0 \hat{\pi}(t) - i\sum_{\mathbf{k}}\sum_{\lambda} g_{\mathbf{k}\lambda}\left[2\hat{\pi}^{\dagger}(t)\hat{\pi}(t) - 1\right]\hat{a}_{\mathbf{k}\lambda}(t)\exp(i\mathbf{k}.\mathbf{R}),\tag{7.7.3}$$

where the operator properties from eqns (4.9.16) and (4.9.18) are used. Formal integration gives

$$\hat{\pi}(t) = \exp(-i\omega_0 t)$$
$$\times\left\{\hat{\pi}(0) - \sum_{\mathbf{k}}\sum_{\lambda} g_{\mathbf{k}\lambda}\int_0^t dt'\left[2\hat{\pi}^{\dagger}(t')\hat{\pi}(t') - 1\right]\hat{a}_{\mathbf{k}\lambda}(t')\exp\left(i\mathbf{k}.\mathbf{R} + i\omega_0 t'\right)\right\}\tag{7.7.4}$$

and the corresponding creation operator is again given by the Hermitian conjugate expression.

In the solutions (7.7.2) and (7.7.4) for the field and atomic operators, the first terms on the right-hand sides give the simple time dependences of the Heisenberg operators in the absence of coupling between the atom and the electromagnetic field. The effects of the coupling, contained in the second terms on the right-hand sides, are to produce complicated time dependences, together with mixings of the field and atomic operators. There are no simple explicit solutions for the time-dependent operators of the coupled system. However, the implicit solutions given above can be used to obtain quite simple iterative expressions for the operators in ascending powers of the atom–radiation coupling $g_{\mathbf{k}\lambda}$, and solutions to second order are often sufficient. The commutation properties of the time-dependent operators can also be verified from the above solutions.

Problem 7.10 Assume that the field and atomic operators have their usual commutation properties at time $t = 0$ and hence prove that the destruction and creation operators given by eqn (7.7.2) and its Hermitian conjugate satisfy

$$\left[\hat{a}_{\mathbf{k}\lambda}(t), \hat{a}_{\mathbf{k}'\lambda'}^{\dagger}(t)\right] = \delta_{\mathbf{k},\mathbf{k}'}\delta_{\lambda,\lambda'}\tag{7.7.5}$$

to order $g_{\mathbf{k}\lambda}^2$.

Consider the spontaneous emission by an excited atom as a first example of the iterative solution of the Heisenberg equations of motion. The level of excitation of the atom is represented by the operator $\hat{\pi}^{\dagger}\hat{\pi}$. The equation of motion of this operator is obtained from eqns (4.7.12) and (4.9.24) as

$$d\left[\hat{\pi}^{\dagger}(t)\hat{\pi}(t)\right]/dt = \sum_{\mathbf{k}}\sum_{\lambda} g_{\mathbf{k}\lambda}\left\{\hat{\pi}^{\dagger}(t)\hat{a}_{\mathbf{k}\lambda}(t)\exp(i\mathbf{k}.\mathbf{R}) + \hat{a}_{\mathbf{k}\lambda}^{\dagger}(t)\hat{\pi}(t)\exp(-i\mathbf{k}.\mathbf{R})\right\}.$$
$$\tag{7.7.6}$$

With the atom initially in its excited state and the field in its vacuum state, a solution correct to order $g_{k\lambda}^2$ is obtained by use of expressions for the field operators correct to order $g_{k\lambda}$. It is thus sufficient to use the zero-order solution

$$\hat{\pi}(t') = \hat{\pi}(t)\exp\left[i\omega_0(t - t')\right] \tag{7.7.7}$$

in the integrand of eqn (7.7.2), which becomes

$$\hat{a}_{k\lambda}(t) = \exp(-i\omega_k t)$$

$$\times \left\{ \hat{a}_{k\lambda}(0) - g_{k\lambda}\,\hat{\pi}(t)\exp(-i\mathbf{k}.\mathbf{R} + i\omega_0 t)\int_0^t dt'\exp\left[i(\omega_k - \omega_0)t'\right]\right\}. \tag{7.7.8}$$

The corresponding approximate solution for the creation operator is given by the Hermitian conjugate of this expression, and substitution for the field operators in eqn (7.7.6) gives

$$d\left[\hat{\pi}^\dagger(t)\hat{\pi}(t)\right]\big/dt$$

$$= \hat{\pi}^\dagger(t)\sum_k \sum_\lambda g_{k\lambda}\left\{\hat{a}_{k\lambda}(0)\exp(i\mathbf{k}.\mathbf{R} - i\omega_k t) - g_{k\lambda}\,\hat{\pi}(t)\int_0^t dt'\exp\left[i(\omega_k - \omega_0)(t' - t)\right]\right\}$$

$$+ \text{Hermitian conjugate.} \tag{7.7.9}$$

The expectation values of the field operators vanish for the initial vacuum state of the electromagnetic field, denoted $|\{0\}\rangle$ as before. The vacuum state expectation value of eqn (7.7.9) is therefore

$$\frac{d\langle\{0\}|\hat{\pi}^\dagger(t)\hat{\pi}(t)|\{0\}\rangle}{dt} = -\langle\{0\}|\hat{\pi}^\dagger(t)\hat{\pi}(t)|\{0\}\rangle\sum_k \sum_\lambda g_{k\lambda}^2 \int_0^t dt'\exp\left[i(\omega_k - \omega_0)(t' - t)\right]$$

$$+ \text{complex conjugate}$$

$$= -\langle\{0\}|\hat{\pi}^\dagger(t)\hat{\pi}(t)|\{0\}\rangle\sum_k \sum_\lambda 2g_{k\lambda}^2 \sin\left[(\omega_k - \omega_0)t\right]\big/(\omega_k - \omega_0). \tag{7.7.10}$$

The final factors in this expression form a representation of the delta-function for sufficiently large t, as in eqn (2.4.8). The summation in eqn (7.7.10) is then identical to that in eqn (4.10.6), and

$$d\langle\{0\}|\hat{\pi}^\dagger(t)\hat{\pi}(t)|\{0\}\rangle\big/dt = -2\gamma_{sp}\langle\{0\}|\hat{\pi}^\dagger(t)\hat{\pi}(t)|\{0\}\rangle, \tag{7.7.11}$$

with the solution

$$\langle\{0\}|\hat{\pi}^\dagger(t)\hat{\pi}(t)|\{0\}\rangle = \langle\{0\}|\hat{\pi}^\dagger(0)\hat{\pi}(0)|\{0\}\rangle\exp(-2\gamma_{sp}t). \tag{7.7.12}$$

The atomic excitation decays at a rate $2\gamma_{sp}$, and eqn (7.7.12) is the Heisenberg-picture description of the spontaneous emission process, equivalent to the phenomenological result in eqn (1.7.11). The above calculation thus confirms a known result with the use of the Heisenberg picture.

A similar procedure is used to put the equation of motion (7.7.3) for the atomic transition operator into a more instructive form. Thus, its vacuum expectation value correct to second order in the atom–radiation coupling is given with the use of eqn (7.7.8) as

$$\frac{d\langle\{0\}|\hat{\pi}(t)|\{0\}\rangle}{dt} = -\langle\{0\}|\hat{\pi}(t)|\{0\}\rangle\left\{i\omega_0 + \sum_k\sum_\lambda g_{k\lambda}^2\int_0^t dt'\exp[i(\omega_k - \omega_0)(t' - t)]\right\}.$$

(7.7.13)

The integral is more difficult to evaluate than that in eqn (7.7.10) because of the absence of the complex conjugate. By straightforward integration

$$\int_0^t dt'\exp[i(\omega_k - \omega_0)(t' - t)] = \frac{1 - \cos[(\omega_k - \omega_0)t]}{i(\omega_k - \omega_0)} + \frac{\sin[(\omega_k - \omega_0)t]}{\omega_k - \omega_0}.$$ (7.7.14)

The real term on the right-hand side is equivalent to a delta function for large t, and only this part contributes to eqn (7.7.10). However, there is now also an imaginary term, which is related to the standard definition of the *principal part* of an integral [16] by

$$\frac{\mathcal{P}}{\omega_k - \omega_0} = \underset{t\to\infty}{Lt}\frac{1 - \cos[(\omega_k - \omega_0)t]}{\omega_k - \omega_0}.$$

(7.7.15)

When this function occurs in an integral over ω_k, the rapidly-oscillating cosine makes no contribution, except at the value $\omega_k = \omega_0$, where it cancels the 1 in the numerator to give zero. The principal part thus behaves everywhere like the inverse of $\omega_k - \omega_0$ except for the removal of a small range of the integration symmetrically placed around $\omega_k = \omega_0$. Equation (7.7.14) now takes the form

$$\int_0^t dt'\exp[i(\omega_k - \omega_0)(t' - t)] = -i\frac{\mathcal{P}}{\omega_k - \omega_0} + \pi\delta(\omega_k - \omega_0).$$

(7.7.16)

The equation of motion (7.7.13) is converted by this substitution to

$$\frac{d\langle\{0\}|\hat{\pi}(t)|\{0\}\rangle}{dt} = -\langle\{0\}|\hat{\pi}(t)|\{0\}\rangle\left\{i\omega_0 - i\frac{V}{(2\pi)^3}\sum_\lambda\mathcal{P}\int dk\frac{g_{k\lambda}^2}{\omega_k - \omega_0} + \gamma_{sp}\right\},$$

(7.7.17)

where the summation over $\mathbf{k}$ is converted to an integration by eqn (4.5.6) and the radiative decay rate is obtained from the delta-function part as in eqn (7.7.11). According to eqn (4.9.20), the time dependence of the atomic transition dipole moment is determined by eqn (7.7.17) and its Hermitian conjugate. The first two terms in the large bracket give the oscillation frequency of the dipole, while the third term gives its decay rate. The second term is a correction to the oscillation frequency ω_0 brought about by the coupling of the atom to the vacuum radiation field.

The calculation of radiative corrections to the energy-level separations of atoms forms an extensive field of study [17]. The integral given in eqn (7.7.17) is not adequate on its own for such calculations as there is a further contribution to the level shifts from the final term in eqn (4.8.28), which is neglected here. Calculations for the hydrogen atom show that the shift vanishes unless one of the states in the transition is an S state. The shift is always very small and, for example, its value for the hydrogen $2^2S_{1/2}$ state is about 10^9 Hz, roughly six orders of magnitude smaller than the $n = 2$ excitation energy. The existence of level shifts was first demonstrated by Lamb and Retherford in experiments on transitions between the $2^2S_{1/2}$ state and the unshifted $2^2P_{1/2}$ state. The splitting between these states is known as the *Lamb shift*. The radiative level shifts are ignored in the calculations that follow, where they are assumed to be included in the basic transition frequencies.

7.8 The source-field expression

A useful expression for the electric field radiated by an excited atom is obtained straightforwardly by a calculation in the Heisenberg picture [17]. Consider the time dependence of the electric field operator at a point $\mathbf{r}$ some distance from the radiating atom at position $\mathbf{R}$. The electric field is divided into two parts, as in eqn (4.4.13), where the Heisenberg operator analogous to eqn (4.4.14) is

$$\hat{\mathbf{E}}_T^+(\mathbf{r},t) = i\sum_k \sum_\lambda \mathbf{e}_{k\lambda} (\hbar\omega_k/2\varepsilon_0 V)^{1/2} \hat{a}_{k\lambda}(t)\exp(i\mathbf{k}.\mathbf{r}). \qquad (7.8.1)$$

The Heisenberg destruction operator given by eqn (7.7.2) has two parts, an initial-value or freely-propagating part that is independent of the atom and a part that represents the field radiated by the atomic light source. The initial-value part makes no contributions to the expectation values of normally-ordered operators for an electromagnetic field in its vacuum state at time $t = 0$. Thus, with the assumption that the field operators are to be used only in such expectation-value calculations, the initial-value part can be dropped. Substitution of the source-field part of eqn (7.7.2) into eqn (7.8.1) gives

$$\hat{\mathbf{E}}_{sf}^+(\mathbf{r},t) = -\frac{ie}{16\pi^3\varepsilon_0}\int d\mathbf{k}\sum_\lambda \omega_k \mathbf{e}_{k\lambda}(\mathbf{e}_{k\lambda}.\mathbf{D}_{12})\exp\{i\mathbf{k}.(\mathbf{r}-\mathbf{R})-i\omega_k t\}$$

$$\times \int_0^t dt'\hat{\pi}(t')\exp(i\omega_k t'), \qquad (7.8.2)$$

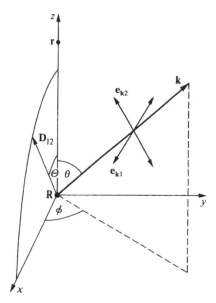

Fig. 7.10. Orientations of the vectors used in the evaluation of the electric field at **r** produced by radiation from an atom at **R**.

with $g_{k\lambda}$ substituted from eqn (4.9.22) and the wavevector sum converted to an integral by eqn (4.5.6).

Figure 7.10 shows the orientations of the various vectors that enter eqn (7.8.2). The line joining the atom to the observation point is taken as the z axis of Cartesian and polar coordinates, with the atom as origin. The polar coordinates of the wavevector **k** and the two independent polarizations $\mathbf{e}_{k1}$ and $\mathbf{e}_{k2}$ are given in eqns (4.5.2) to (4.5.4), while the coordinates of the dipole moment $\mathbf{D}_{12}$ are given in eqn (4.10.7). It is clear from symmetry that the radiated field has a zero y component. The x and z components are found by evaluation of the integrals in eqn (7.8.2). Just as in the corresponding classical theory of radiation by a dipole [18], the field at **r** has nonzero x and z components close to the dipole but, at distances such that

$$k|\mathbf{r} - \mathbf{R}| \gg 1, \tag{7.8.3}$$

the field consists of transverse waves polarized parallel to the x axis. We consider only the far field of the dipole and assume that the inequality (7.8.3) is satisfied for the wavevectors that contribute most strongly to the emission.

The angular part of the wavevector integration in eqn (7.8.2) is

$$\int_0^\pi d\theta \int_0^{2\pi} d\phi \sin\theta \sum_\lambda \mathbf{e}_{k\lambda} (\mathbf{e}_{k\lambda} \cdot \mathbf{D}_{12}) \exp\{ik|\mathbf{r} - \mathbf{R}|\cos\theta\}$$

$$\approx -\frac{2\pi i D_{12} \sin\Theta}{k|\mathbf{r} - \mathbf{R}|} \{\exp(ik|\mathbf{r} - \mathbf{R}|) - \exp(-ik|\mathbf{r} - \mathbf{R}|)\}\mathbf{i}, \tag{7.8.4}$$

where $\mathbf{i}$ is a unit vector parallel to the x axis. The integral is evaluated by substitution of the explicit forms of the polarization and dipole vectors, and some simple angular integrations produce terms in the inverse first, second and third powers of $|\mathbf{r} - \mathbf{R}|$. Only the term in the inverse first power is retained in the second line of eqn (7.8.4) in view of the inequality (7.8.3). The radial part of the wavevector integration in eqn (7.8.2) can now be performed after substitution of the angular part from eqn (7.8.4). With $\omega_k = ck$, the k-dependent terms produce a radial integral

$$
\int_{-\infty}^{\infty} dk\, k^2 \{\exp(ik|\mathbf{r} - \mathbf{R}|) - \exp(-ik|\mathbf{r} - \mathbf{R}|)\} \exp\{ick(t' - t)\}
$$
$$
= -\frac{2\pi}{c^3}\left(\frac{\partial}{\partial t'}\right)^2 \left\{\delta\left(t' - t + \frac{|\mathbf{r} - \mathbf{R}|}{c}\right) - \delta\left(t' - t - \frac{|\mathbf{r} - \mathbf{R}|}{c}\right)\right\},
$$

(7.8.5)

where the range of integration is extended to $-\infty$, as explained after eqn (7.8.8) below. The double time derivative is inserted before integration to generate the k^2 factor in the integrand, and the delta-functions then follow from eqn (2.4.9).

The derivation is now completed by substitution of eqn (7.8.5) and the remaining factors from eqn (7.8.4) into eqn (7.8.2). The time derivatives that act on the delta-functions are transferred to the function $\hat{\pi}(t')$ by integration by parts [16]. The second delta-function makes no contribution for the given range from 0 to t of the t' integration and the first delta-function contributes only for

$$t > |\mathbf{r} - \mathbf{R}|/c.$$

(7.8.6)

The final result is

$$
\hat{E}_{sf}^+(\mathbf{r},t) = \frac{eD_{12}\sin\Theta}{4\pi\varepsilon_0 c^2|\mathbf{r} - \mathbf{R}|}\, i\left(\frac{d}{dt}\right)^2 \hat{\pi}\left(t - \frac{|\mathbf{r} - \mathbf{R}|}{c}\right).
$$

(7.8.7)

The field is thus proportional to the acceleration of the effective charges that constitute the transition dipole-moment operator, similar to the classical theory of radiation by moving charges. Its form is indeed identical to that which relates the classical field radiated by a dipole to the classical dipole moment [18]. Note that the field operator at time t is determined by the dipole operator at a retarded time, thus making proper allowance for the travel time of the electromagnetic waves from $\mathbf{R}$ to $\mathbf{r}$. The component $\hat{E}_{sf}^-(\mathbf{r},t)$ of the field operator is given by the conjugate expression and the total operator represents the field at point $\mathbf{r}$ in terms of the dipole operator $e\hat{\mathbf{D}}$ of the radiating source atom at point $\mathbf{R}$.

The time dependence of $\hat{\pi}(t)$ is obtained in leading order from eqn (7.7.7), and eqn (7.8.7) becomes

$$
\hat{E}_{sf}^+(\mathbf{r},t) = -\frac{e\omega_0^2 D_{12}\sin\Theta}{4\pi\varepsilon_0 c^2|\mathbf{r} - \mathbf{R}|}\, i\hat{\pi}\left(t - \frac{|\mathbf{r} - \mathbf{R}|}{c}\right),
$$

(7.8.8)

where the inequality (7.8.6) is again assumed to be satisfied. This form of the source-field expression is used in §§7.9, and 8.2 to 8.4 to determine the coherence properties of the radiation fields emitted by atoms whose dipole moments have known time dependences. It is emphasized that the relation can be used only for the evaluation of expectation values of normally-ordered operators when the distance and time satisfy the inequalities (7.8.3) and (7.8.6). It should also be cautioned that the derivation is based on the atom–field interaction Hamiltonian from eqn (4.9.24), in which the rotating-wave approximation is made. The extension of the range of integration to $-\infty$ in eqn (7.8.5) artificially restores the effects of the omitted terms. Derivations that use the full interaction Hamiltonian [17,19,20] put the relations (7.8.7) and (7.8.8) on a firmer foundation.

Problem 7.11 Show that the integral over all directions of the intensity at radius r radiated by an atom placed at the origin of coordinates is

$$r^2 \int d\Omega \left\langle \hat{I}(\mathbf{r},t) \right\rangle = 2\gamma_{\mathrm{sp}} \hbar \omega_0 \left\langle \hat{\pi}^\dagger \left(t - \frac{r}{c} \right) \hat{\pi} \left(t - \frac{r}{c} \right) \right\rangle, \tag{7.8.9}$$

where the intensity operator is defined in §4.11 and $d\Omega$ is an element of solid angle.

The right-hand side of eqn (7.8.9) is the same as the rate of spontaneous emission of energy per excited atom given by $(N_2/N)\hbar\omega_0 A$ in the Einstein theory of §1.7.

7.9 Emission by a driven atom

The coherence properties of the light emitted by a single atom in free space are conveniently calculated with use of the source-field expression. Suppose that the emitted light is observed at position $\mathbf{r}$ and that measurements are made on the light polarized parallel to the x axis. The field operators can then be treated as scalars and the degrees of first and second-order coherence are obtained from eqns (4.12.3), (4.12.7) and (7.8.8) as

$$g^{(1)}(\tau) = \frac{\left\langle \hat{E}_{\mathrm{sf}}^-(\mathbf{r},t) \hat{E}_{\mathrm{sf}}^+(\mathbf{r},t+\tau) \right\rangle}{\left\langle \hat{E}_{\mathrm{sf}}^-(\mathbf{r},t) \hat{E}_{\mathrm{sf}}^+(\mathbf{r},t) \right\rangle} = \frac{\left\langle \hat{\pi}^\dagger(t) \hat{\pi}(t+\tau) \right\rangle}{\left\langle \hat{\pi}^\dagger(t) \hat{\pi}(t) \right\rangle} \tag{7.9.1}$$

and

$$g^{(2)}(\tau) = \frac{\left\langle \hat{E}_{\mathrm{sf}}^-(\mathbf{r},t) \hat{E}_{\mathrm{sf}}^-(\mathbf{r},t+\tau) \hat{E}_{\mathrm{sf}}^+(\mathbf{r},t+\tau) \hat{E}_{\mathrm{sf}}^+(\mathbf{r},t) \right\rangle}{\left\langle \hat{E}_{\mathrm{sf}}^-(\mathbf{r},t) \hat{E}_{\mathrm{sf}}^+(\mathbf{r},t) \right\rangle^2}$$

$$= \frac{\left\langle \hat{\pi}^\dagger(t) \hat{\pi}^\dagger(t+\tau) \hat{\pi}(t+\tau) \hat{\pi}(t) \right\rangle}{\left\langle \hat{\pi}^\dagger(t) \hat{\pi}(t) \right\rangle^2}, \tag{7.9.2}$$

where the time delay τ is assumed positive. Note that the normal-ordering requirement for the validity of eqn (7.8.8) is satisfied by the expectation values in the degrees of coherence. The time t in these relations is assumed to be sufficiently long for the atoms and the field to have arrived at steady states. The degrees of coherence are then independent of t and, in addition, the fixed retardation time difference $|\mathbf{r} - \mathbf{R}|/c$ between the field and the atomic operators in eqn (7.8.8) can be ignored. The prefactors of the atomic operators conveniently cancel in the degrees of coherence.

The calculation of the coherence properties of the emitted light is thus converted to a problem of determining steady-state expectation values of products of atomic transition operators. The zero time-delay degrees of coherence are obtained very simply. It is obvious from eqn (7.9.1) that

$$g^{(1)}(0) = 1. \tag{7.9.3}$$

For the degree of second-order coherence, the properties (4.9.17) and (4.9.18) apply at any common time and it follows that

$$g^{(2)}(0) = 0. \tag{7.9.4}$$

The light emitted by a single driven atom is thus nonclassical in accordance with the criterion in eqn (5.10.4).

The atomic expectation values needed for the complete time dependences of the degrees of coherence can be calculated in any of the pictures outlined in §§4.7 and 4.9 and it is convenient here to use the interaction picture. It follows from the definitions in eqns (4.9.11) and (4.9.14) with the use of eqns (4.9.32) and (4.9.38) that

$$\langle \hat{\pi}(t) \rangle = \mathrm{Tr}\{\hat{\rho}_I(t)\hat{\pi}\} \exp(-i\omega_0 t)$$
$$= \rho_{21}(t)\exp(-i\omega_0 t) = \tilde{\rho}_{21}(t)\exp(-i\omega t), \tag{7.9.5}$$

$$\langle \hat{\pi}^\dagger(t) \rangle = \rho_{12}(t)\exp(i\omega_0 t) = \tilde{\rho}_{12}(t)\exp(i\omega t) \tag{7.9.6}$$

and

$$\langle \hat{\pi}^\dagger(t)\hat{\pi}(t) \rangle = \rho_{22}(t) = \tilde{\rho}_{22}(t), \tag{7.9.7}$$

where the $\tilde{\rho}_{ij}(t)$ are the modified density matrix elements defined in eqn (2.7.9). The time dependences of the density matrix elements can in principle be calculated using, where appropriate, rate equations or the interaction-picture Bloch equations. However, the above expectation values involve only a single time t, whereas the degrees of coherence involve expectation values of products of operators evaluated at different times, t and $t + \tau$. It is necessary to generalize the above expressions.

Consider first the relation between density matrix elements at different times.

As all three independent elements are coupled by the equations of motion, the elements at time $t + \tau$ are expressed in the general forms

$$\tilde{\rho}_{21}(t + \tau) = \alpha_1(\tau) + \alpha_2(\tau)\tilde{\rho}_{21}(t) + \alpha_3(\tau)\tilde{\rho}_{12}(t) + \alpha_4(\tau)\tilde{\rho}_{22}(t), \qquad (7.9.8)$$

with the complex conjugate expression for $\tilde{\rho}_{12}(t + \tau)$, and

$$\tilde{\rho}_{22}(t + \tau) = \beta_1(\tau) + \beta_2(\tau)\tilde{\rho}_{21}(t) + \beta_3(\tau)\tilde{\rho}_{12}(t) + \beta_4(\tau)\tilde{\rho}_{22}(t). \qquad (7.9.9)$$

The coefficients $\alpha_i(\tau)$ and $\beta_i(\tau)$ are determined by solution of the equations of motion. They must clearly satisfy the initial conditions

$$\alpha_2(0) = \beta_4(0) = 1$$
$$\alpha_1(0) = \alpha_3(0) = \alpha_4(0) = \beta_1(0) = \beta_2(0) = \beta_3(0) = 0. \qquad (7.9.10)$$

In addition, it is assumed that the density matrix elements are independent of their values at time t after a very long additional time τ has elapsed, so that

$$\alpha_2(\infty) = \alpha_3(\infty) = \alpha_4(\infty) = \beta_2(\infty) = \beta_3(\infty) = \beta_4(\infty) = 0. \qquad (7.9.11)$$

The forms of solution (7.9.8) and (7.9.9) for the density matrix elements are re-expressed in terms of atomic-operator expectation values with the use of eqns (7.9.5) to (7.9.7),

$$\langle \hat{\pi}(t + \tau) \rangle \exp\{i\omega(t + \tau)\} = \alpha_1(\tau) + \alpha_2(\tau)\langle \hat{\pi}(t) \rangle \exp(i\omega t)$$
$$+ \alpha_3(\tau)\langle \hat{\pi}^\dagger(t) \rangle \exp(-i\omega t) + \alpha_4(\tau)\langle \hat{\pi}^\dagger(t)\hat{\pi}(t) \rangle \qquad (7.9.12)$$

and

$$\langle \hat{\pi}^\dagger(t + \tau)\hat{\pi}(t + \tau) \rangle = \beta_1(\tau) + \beta_2(\tau)\langle \hat{\pi}(t) \rangle \exp(i\omega t)$$
$$+ \beta_3(\tau)\langle \hat{\pi}^\dagger(t) \rangle \exp(-i\omega t) + \beta_4(\tau)\langle \hat{\pi}^\dagger(t)\hat{\pi}(t) \rangle. \qquad (7.9.13)$$

In view of eqn (7.9.11), the steady-state expectation values are given by

$$\underset{t \to \infty}{\mathrm{Lt}} \langle \hat{\pi}(t) \rangle \exp(i\omega t) = \alpha_1(\infty) \qquad (7.9.14)$$

and

$$\langle \hat{\pi}^\dagger(\infty)\hat{\pi}(\infty) \rangle = \beta_1(\infty). \qquad (7.9.15)$$

The relations (7.9.12) and (7.9.13) are used to obtain the expectation values for the degrees of coherence with the help of the *quantum regression theorem* [16,21]. Suppose that the expectation value of any operator $\hat{A}$ at time $t + \tau$ is related to the expectation values of a set of operators $\hat{A}_i$ at an earlier time t according to

$$\left\langle \hat{A}(t+\tau) \right\rangle = \sum_i \alpha_i(\tau) \left\langle \hat{A}_i(t) \right\rangle. \tag{7.9.16}$$

The regression theorem proves that

$$\left\langle \hat{B}(t)\hat{A}(t+\tau)\hat{C}(t) \right\rangle = \sum_i \alpha_i(\tau) \left\langle \hat{B}(t)\hat{A}_i(t)\hat{C}(t) \right\rangle, \tag{7.9.17}$$

where the capital letters denote any operators. The theorem thus expresses double-time expectation values in terms of single-time expectation values in a manner that can be applied immediately to the present problem. Thus it follows from eqns (7.9.12) and (7.9.13) with the use of eqns (4.9.17) and (4.9.18) that

$$\left\langle \hat{\pi}^\dagger(t)\hat{\pi}(t+\tau) \right\rangle = \exp\{-i\omega(t+\tau)\}$$
$$\times \left\{ \alpha_1(\tau)\left\langle \hat{\pi}^\dagger(t) \right\rangle + \alpha_2(\tau)\left\langle \hat{\pi}^\dagger(t)\hat{\pi}(t) \right\rangle \exp(i\omega t) \right\} \tag{7.9.18}$$

and

$$\left\langle \hat{\pi}^\dagger(t)\hat{\pi}^\dagger(t+\tau)\hat{\pi}(t+\tau)\hat{\pi}(t) \right\rangle = \beta_1(\tau)\left\langle \hat{\pi}^\dagger(t)\hat{\pi}(t) \right\rangle, \tag{7.9.19}$$

where the first terms on the right-hand sides of eqns (7.9.12) and (7.9.13) correspond to contributions in eqn (7.9.16) in which the operator $\hat{A}_i$ is an algebraic function independent of t.

The degrees of coherence are now obtained by substitution into eqns (7.9.1) and (7.9.2). As t is assumed sufficiently large for steady-state conditions to be established, the expressions (7.9.14) (and its conjugate) and (7.9.15) should be used for the single-time expectation values. Thus

$$g^{(1)}(\tau) = \exp(-i\omega\tau)\left\{ \frac{\alpha_1(\tau)\alpha_1^*(\infty)}{\beta_1(\infty)} + \alpha_2(\tau) \right\} \tag{7.9.20}$$

and

$$g^{(2)}(\tau) = \beta_1(\tau)/\beta_1(\infty). \tag{7.9.21}$$

The infinite time-delay degrees of coherence are obtained with the help of eqn (7.9.11) as

$$\left| g^{(1)}(\infty) \right| = \left| \alpha_1(\infty) \right|^2 / \beta_1(\infty) \tag{7.9.22}$$

and

$$g^{(2)}(\infty) = 1. \tag{7.9.23}$$

The complete time dependences of the degrees of coherence are determined by $\alpha_1(\tau)$, $\alpha_2(\tau)$, and $\beta_1(\tau)$, and the remaining coefficients are not needed. Thus the off-diagonal density matrix element in eqn (7.9.8) must be calculated for arbitrary values of $\tilde{\rho}_{21}(t)$ and $\tilde{\rho}_{12}(t)$, but $\tilde{\rho}_{22}(t)$ can be set equal to zero. The diagonal matrix element in eqn (7.9.9) need only be calculated for the condition

$$\tilde{\rho}_{21}(t) = \tilde{\rho}_{12}(t) = \tilde{\rho}_{22}(t) = 0. \tag{7.9.24}$$

A simple example of these methods of calculation is provided by a system in which the atomic motion is adequately described by a rate equation of the kind shown in eqn (2.10.6),

$$d\tilde{\rho}_{22}(t)/dt = R - 2\gamma_{\text{sp}}\tilde{\rho}_{22}(t). \tag{7.9.25}$$

As explained in §2.10, this simple equation of motion is valid when $\mathcal{V}^2 \ll \gamma\gamma_{\text{sp}}$, with weak broad-band incident light, or large collision or Doppler broadening. It may also be valid for methods of atomic excitation other than optical pumping, such as electron bombardment. Only the diagonal density matrix element appears in the rate equation and we consider only the degree of second-order coherence of the emitted light. The general solution of eqn (7.9.25) in the form of eqn (7.9.9) is

$$\tilde{\rho}_{22}(t + \tau) = \left(R/2\gamma_{\text{sp}}\right)\left[1 - \exp\left(-2\gamma_{\text{sp}}\tau\right)\right] + \tilde{\rho}_{22}(t)\exp\left(-2\gamma_{\text{sp}}\tau\right). \tag{7.9.26}$$

Thus

$$\beta_1(\tau) = \left(R/2\gamma_{\text{sp}}\right)\left[1 - \exp\left(-2\gamma_{\text{sp}}\tau\right)\right] \tag{7.9.27}$$

and the degree of second-order coherence from eqn (7.9.21) is

$$g^{(2)}(\tau) = 1 - \exp\left(-2\gamma_{\text{sp}}\tau\right). \tag{7.9.28}$$

This function is plotted in Fig. 7.11 and it clearly violates the classical inequalities (3.7.6) and (3.7.11). The light emitted by a single driven atom has a degree of second-order coherence that satisfies both the quantum inequality of eqn (6.5.9), for sub-Poissonian statistics, and the quantum inequality of eqn (6.5.8), for photon antibunching.

The zero value of the degree of second-order coherence at zero time delay, established in general in eqn (7.9.4), reflects the inability of the single two-level atom to emit a pair of photons simultaneously. Successive photon emissions are separated by the time needed for the atom arriving in its ground state to be re-excited and emit a second photon. The emitted photons are thus 'spaced out', with a minimum separation of the order of the radiative lifetime τ_{R}, or $1/2\gamma_{\text{sp}}$.

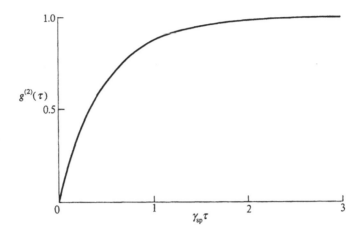

Fig. 7.11. Dependence on the time delay τ of the degree of second-order coherence of the light emitted by a single driven atom.

The degree of second-order coherence in eqn (7.9.28) and Fig. 7.11 should be compared with that of Gaussian–Lorentzian chaotic light given by eqn (3.7.19) and illustrated in Fig. 3.12. The two expressions are essentially reflections of each other in a horizontal line at $g^{(2)}(\tau) = 1$, and the photon bunching property of chaotic light mirrors the photon antibunching property of the light radiated by the driven atom.

These nonclassical effects survive direct detection of the light radiated by the atom. It follows from eqns (6.2.18) and (7.8.8) that the continuous-mode destruction operator $\hat{a}(t)$ is proportional to the atomic transition operator $\hat{\pi}(t)$, with appropriate retardation. The total steady-state rate of radiation of photons, integrated over all directions, is obtained from the final term in eqn (7.9.25) as

$$2\gamma_{sp}\tilde{\rho}_{22}(\infty) = 2\gamma_{sp}\beta_1(\infty) = R, \tag{7.9.29}$$

with the obvious interpretation that the rate of atomic decay must equal the rate of atomic excitation. The mean photocount $\langle m \rangle$ is obtained from eqns (6.10.1) and (6.10.5) on multiplication of R by a proportionality factor that includes the quantum efficiency, the integration time, and allowances for the orientation and cross section of the photodetector. The proportionality factors cancel in the measured degree of second-order coherence.

Problem 7.12   Use the quantum theory of direct detection to show that the measured degree of second-order coherence of the light radiated by a single driven atom for zero time delay is

$$g_D^{(2)}(0) = \frac{1 - 2\gamma_{sp}T + 2\gamma_{sp}^2T^2 - \exp(-2\gamma_{sp}T)}{2\gamma_{sp}^2T^2}. \tag{7.9.30}$$

This expression resembles the degree of second-order coherence of chaotic light, given essentially in eqn (3.8.7), but with the reflection property mentioned above. Thus $g_D^{(2)}(0)$ vanishes for $T \to 0$ and it tends to unity for $T \to \infty$.

Figure 6.1(c) shows a representation of the time series of photocounts, with the rate of occurrence of closely-spaced counts depressed below the rate in the random series for coherent light shown in Fig. 6.1(b). The detected Mandel Q parameter defined in eqn (6.10.11) is

$$Q_D = \langle m \rangle \frac{1 - 2\gamma_{sp}T - \exp(-2\gamma_{sp}T)}{2\gamma_{sp}^2 T^2},\tag{7.9.31}$$

where the mean count $\langle m \rangle$ includes the appropriate proportionality factors. This again resembles the corresponding expression for chaotic light, given by eqn (6.10.22), but with changes of sign that ensure negative values of Q_D. The sub-Poissonian statistics of the radiation from a single driven atom are thus reflected in the photocount characteristics measured in direct detection.

More favourable conditions for the experimental observation of photon anti-bunching and sub-Poissonian statistics occur for an atom driven by narrow-band light, when the above treatment by rate equations does not apply. This topic is covered in §§8.2 to 8.4.

References

[1]   Loudon, R. and Shepherd, T.J., Properties of the optical quantum amplifier, *Optica Acta* **31**, 1243–69 (1984).

[2]   Siegman, A.E., *Lasers* (University Science Books, Mill Valley, CA, 1986).

[3]   Milonni, P.W. and Eberly, J.H., *Lasers* (Wiley, New York, 1988).

[4]   Rice, P.R. and Carmichael, H.J., Photon statistics of a cavity-QED laser: A comment on the laser-phase-transition analogy, *Phys. Rev. A* **50**, 4318–29 (1994).

[5]   Scully, M.O. and Lamb, W.E., Quantum theory of an optical maser. I. General theory, *Phys. Rev.* **159**, 208–26 (1967).

[6]   Sargent, M., Scully, M.O. and Lamb, W.E., *Laser Physics* (Addison–Wesley, Reading, MA, 1974).

[7]   Goodman, J. W., *Statistical Optics* (Wiley, New York, 1985).

[8]   Louisell, W.H., *Quantum statistical properties of radiation* (Wiley, New York, 1973).

[9]   Gerhardt, H., Welling, H. and Güttner, A., Observation of quantum-phase and quantum-amplitude noise for a laser below and above threshold, *Phys. Lett. A* **40**, 191–3 (1972).

[10]   Barnett, S.M., Stenholm, S. and Pegg, D.T., A new approach to optical phase diffusion, *Opt. Comm.* **73**, 314–8 (1989).

[11]   Kolobov, M.I., Davidovich, L., Giacobino, E. and Fabre, C., Role of

pumping statistics and dynamics of atomic polarization in quantum fluctuations of laser sources, *Phys. Rev. A* **47**, 1431–46 (1993).

[12]   Huttner, B. and Barnett, S.M., Quantization of the electromagnetic field in dielectrics, *Phys. Rev. A* **46**, 4306–22 (1992).

[13]   Jeffers, J., Imoto, N. and Loudon, R., Quantum optics of travelling-wave attenuators and amplifiers, *Phys. Rev. A* **47**, 3346–59 (1993).

[14]   Caves, C.M., Quantum limits on noise in linear amplifiers, *Phys. Rev. D* **26**, 1817–39 (1982).

[15]   Glauber, R.J., Amplifiers, attenuators and the quantum theory of measurement, in *Frontiers in Quantum Optics*, eds E.R. Pike and S. Sarkar (Adam Hilger, Bristol, 1986) pp. 534–82.

[16]   Barnett, S.M. and Radmore, P.M., *Methods in Theoretical Quantum Optics* (Oxford University Press, Oxford, 1997).

[17]   Milonni, P.W., *The Quantum Vacuum* (Academic Press, San Diego, 1994).

[18]   Jackson, J.D., *Classical Electrodynamics*, 3rd edn (Wiley, New York, 1999).

[19]   P.W. Milonni and P.L. Knight, Retardation in the resonant interaction of two identical atoms, *Phys. Rev. A* **10**, 1096–108 (1974).

[20]   G. Compagno, R. Passante and F. Persico, Virtual photons, causality and atomic dynamics in spontaneous emission, *J. Mod. Opt.* **37**, 1377–82 (1990).

[21]   Mandel, L. and Wolf, E., *Optical Coherence and Quantum Optics* (Cambridge University Press, Cambridge, 1995) chap. 17.

8 Resonance fluorescence and light scattering

The basic components of a light-scattering experiment include a scattering region, with the atoms of interest illuminated by a parallel beam of light, and a detector that measures the scattered intensity at some finite angle to the direction of the incident beam. The scattering at small angles is usually difficult to resolve from the transmitted incident beam itself, and it is convenient in many experiments to measure the light scattered at right angles. In high-resolution studies, the atoms may be injected into the scattering region in the form of a beam in the third mutually perpendicular direction, to minimize the Doppler broadening of the scattered light.

In terms of the quantum theory of light, the scattering involves the destruction of a photon of energy $\hbar\omega$ from the incident beam and the creation of a photon of energy $\hbar\omega_{sc}$ in the scattered beam. The light scattering is thus a second-order process, as two interactions of the radiation field and the atomic electrons take place. Scattering occurs for all values of the incident frequency relative to the transition frequencies of the scattering atoms. The intensity of the scattering is, however, particularly strong in conditions where the incident frequency lies close to that of an atomic transition.

Figure 8.1 shows some atomic energy levels, with the magnitude of the incident frequency represented by the arrow on the left. The remaining arrows represent components that occur in the scattered spectrum. The *elastic* or *Rayleigh* contribution to the scattering includes light whose frequency ω_{sc} equals the incident frequency ω. The *inelastic* contributions include all of the remaining components in the scattered light, with ω_{sc} different from ω. The inelastic components are subdivided according to whether or not the atom makes a transition to a final state that differs from its initial state before the scattering event. We consider only atoms that are initially in their ground states.

Suppose first that the atom is again in its ground state after the scattering event. Energy conservation suggests that the scattered photon must have $\omega_{sc} = \omega$ and therefore contribute only to the elastic or Rayleigh component. However, when ω is close to resonance with a homogeneously-broadened atomic excitation frequency ω_l, the scattered light has additional inelastic components with the same central frequency and linewidth as the atomic transition concerned. These are *fluorescent* contributions in which the atom reradiates by ordinary emission after excitation by the incident light; the energy difference between $\hbar\omega$ and $\hbar\omega_{sc}$ is taken up within the energy uncertainty associated with the line-broadening processes. These Rayleigh and fluorescent components are represented by the two arrows on the left of the energy-level diagram in Fig. 8.1.

Now suppose that the atom is left in an excited state of energy $\hbar\omega_f$ after the scattering. The scattering is entirely inelastic, and the *Raman* contribution includes

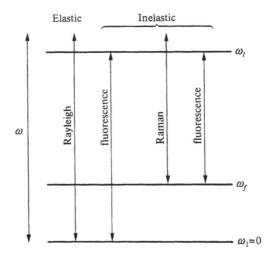

Fig. 8.1. Schematic energy-level diagram showing the incident frequency ω close to resonance with an atomic transition ω_l and the four kinds of contribution to the scattered frequencies ω_{sc}.

light whose frequency ω_{sc} satisfies the energy-conservation equality with $\omega - \omega_f$. There is, in addition, a further resonance fluorescence component associated with radiative decay to the final-state energy level $\hbar\omega_f$ when the incident frequency lies close to an atomic excitation frequency ω_l. These Raman and excited-state fluorescent components are represented by the two arrows on the right of Fig. 8.1. There are similar pairs of contributions for each atomic energy level whose excitation frequency ω_f lies below the incident frequency ω.

The calculations of the present chapter fall into two main parts. We first consider the scattering by a two-level atom, where the atomic state after the scattering can only be the same as the initial ground state; it is possible to give a detailed but simple theory of the spectral and coherence properties of the Rayleigh-scattered and resonance-fluorescent light. The additional optical effects of quantum jumps and cascade emission occur when the atomic transition scheme is extended to include a third level. The second main calculation refers to the light scattered by a multilevel atom, where a straightforward derivation provides the intensities of the Rayleigh and Raman components in the scattered spectrum.

8.1 The scattering cross-section

It is convenient to preface the more specific calculations by some general remarks on light scattering, which apply irrespective of the distribution of scattered frequencies ω_{sc}. Classical electromagnetic theory is used in the present section, but the definitions and the main results apply to the quantum calculations in the remainder of the chapter.

Consider first the geometrical arrangement of a scattering experiment. The scattering atom is placed at the origin of the coordinate system shown in Fig. 8.2. The parallel beam of incident light has its electric vector $\mathbf{E}$ parallel to the x axis and its propagation direction parallel to the z axis. The cycle-averaged Poynting vector of a classical coherent incident beam is

$$\bar{I} = \tfrac{1}{2}\varepsilon_0 c |E|^2,$$ (8.1.1)

where E is the complex amplitude of the field vector.

The scattered light is usually distributed over all directions in space. Consider the scattered light with electric vector $\mathbf{E}_{sc}$ at position $\mathbf{r}$. Its averaged Poynting vector is

$$\bar{I}_{sc} = \tfrac{1}{2}\varepsilon_0 c |E_{sc}|^2.$$ (8.1.2)

The rate at which the atom scatters electromagnetic energy into an elementary solid angle $d\Omega$ in the direction of the vector $\mathbf{r}$ is obtained on multiplication of the Poynting vector by $r^2 d\Omega$. The total scattered energy includes the contributions of both polarizations of the light at position $\mathbf{r}$.

The strength of the scattering is conveniently expressed in terms of the *differential scattering cross-section* $d\sigma(\omega)/d\Omega$. This is defined as the rate at which energy is removed from the incident beam by the scattering into unit solid angle in the given direction, divided by the rate at which energy in the incident beam crosses a unit area perpendicular to its propagation direction. For elastic scattering, all of the energy removed from the incident beam is converted into

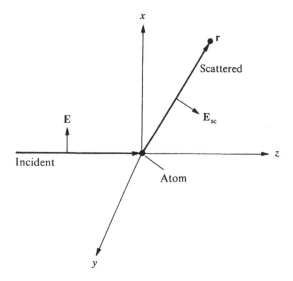

Fig. 8.2. Geometrical arrangement of a light-scattering experiment. The scattered light has two independent polarization directions.

scattered energy. However, this is not the case for inelastic scattering, where some of the energy removed from the incident beam is given to the atom, and only a fraction ω_{sc}/ω appears in the scattered light. The differential cross-section for scattering of light of frequency ω into light of frequency ω_{sc} is thus,

$$d\sigma(\omega)/d\Omega = r^2\omega\bar{I}_{sc}/\omega_{sc}\bar{I}. \tag{8.1.3}$$

The differential cross-section clearly has the dimensions of area and it is independent of r, as $\bar{I}_{sc}$ falls off with the inverse square of the distance from the atom.

The *scattering cross-section* is defined in the same way as the differential cross-section except that it refers to the total rate at which energy is removed from the incident beam by scattering in all directions. It is obtained by integration of eqn (8.1.3) over the complete 4π steradian solid angle,

$$\sigma(\omega) = \int d\Omega \left(r^2\omega\bar{I}_{sc}/\omega_{sc}\bar{I} \right). \tag{8.1.4}$$

The cross-section also has the dimensions of area and is independent of r.

The cross-sections so defined refer to the light scattering by a single atom. Some experiments do observe single-atom light scattering but the scattering is more generally produced by a large number N of atoms, and one must consider how to add their contributions. For the scattering of visible light by a gas, the mean separation between atoms is very small compared to the wavelength of the radiation, provided that the gas pressure at room temperature is greater than about 100 Pa. The total electric vector of the scattered light is given by a linear superposition of the contributions of the N atoms.

Consider first the scattering by a homogeneous distribution of atoms. The incident light is assumed to be a classical electromagnetic wave of well-defined amplitude and phase. The components of the scattered field that arise from atoms contained within a half-wavelength of the incident radiation are in phase with each other, but are out of phase with the fields arising from the atoms that lie within adjacent half-wavelengths. For a uniform distribution of atoms, the contributions to the total scattered field from the groups of atoms in the different half-wavelengths exactly cancel by destructive interference, except in the forward direction, $\theta = 0$. Thus no scattering occurs and the intensity of the incident light beam remains unchanged. There is, for example, no Rayleigh scattering by a perfect crystal in which the atoms are fixed on a regular array of lattice sites.

The same simple model also shows how light scattering is restored by the presence of spatial inhomogeneities in the atomic distribution. Thus for the N-atom gas, the total volume V that it occupies is divided into smaller regions that contribute fields of the same phase to the scattered light in a given direction. Suppose that a total of N_+ atoms lie in regions that contribute a scattered field of positive phase factor and N_- in regions that contribute a field of negative phase factor, with

$$N_+ + N_- = N. \tag{8.1.5}$$

It is assumed that the atoms in the gas are randomly distributed in space such that each atom is equally likely to lie in one region or the other.

Problem 8.1   Prove that the mean scattered intensity in the given direction, proportional to the square of the field, is

$$\left\langle (N_+ - N_-)^2 \right\rangle \bar{I}_{sc} = N\bar{I}_{sc}, \tag{8.1.6}$$

where the angle brackets here denote an average over the random distribution of the atoms between regions of opposite phase and $\bar{I}_{sc}$ is the mean scattered intensity for a single atom.

The scattering cross-section for the N atoms is thus equal to that for a single atom multiplied by N. It is assumed that the refractive indices of the gas at the incident and scattered frequencies are close to unity and need not appear in the Poynting vectors. A similar result applies to glassy solids, where the static density fluctuations give rise to Rayleigh scattering, which is an important source of attenuation in optical fibres.

The inhibition of light scattering for a uniform spatial distribution of atoms and its restoration for a random spatial distribution also has implications for the radiative broadening of atomic transitions. Thus in the theory of the linear susceptibility given in §2.5, the polarization of a gas of N atoms is assumed in eqns (2.5.15) and (2.5.16) to equal the dipole moment of a single atom multiplied by the gas density, with the same damping γ_{sp}. The radiative damping relies on the presence of scattering of light out of the incident beam by spontaneous emission. The calculation of §2.5 is thus correct for a random distribution of the N atoms, but it does not apply to a uniform distribution, where the radiative damping vanishes.

It is emphasized in §§1.6 and 1.8 that the attenuation and scattering of light are different aspects of the same process; the light removed from a beam by absorption reappears as scattered radiation. From the viewpoint of attenuation, the rate of removal of energy from the incident beam in a volume V is given by eqn (1.8.2) as

$$-V\,\partial\bar{I}/\partial z = VK(\omega)\bar{I}, \tag{8.1.7}$$

where $K(\omega)$ is the attenuation coefficient for a light beam assumed sufficiently weak to avoid saturation. From the viewpoint of scattering, the cross-section $\sigma(\omega)$ is defined by eqn (8.1.4) such that the rate of removal of energy by N atoms is

$$N\sigma(\omega)\bar{I}. \tag{8.1.8}$$

These expressions describe the same physical quantity, so that

$$\sigma(\omega) = (V/N)K(\omega) = (2V\omega/Nc)\kappa(\omega), \tag{8.1.9}$$

where eqn (1.8.3) is used. This relation is known as the *optical theorem*. It is derived on the basis of general considerations of energy conservation, and it is valid for both classical and quantum-mechanical treatments of light scattering. Similar optical theorems hold for a wide variety of scattering problems [1,2]. The application of the theorem to Rayleigh and Raman scattering is discussed in §§8.8 and 8.9 respectively. The derivation given above assumes that all of the light removed from the incident beam reappears as scattered radiation. A more general form of the optical theorem, not considered here, applies when some of the light is removed from the beam by nonradiative processes.

8.2 Resonance fluorescence

The simplest scattering system is a two-level atom illuminated by single-mode coherent incident light. The two-level theory developed below [3,4] also applies to the Rayleigh and ground-state fluorescent scattering by real multilevel atoms when the exciting frequency ω lies close to only a single atomic transition frequency. Multilevel atoms show the additional excited-state scattering represented on the right of Fig. 8.1. It is possible to treat the two-level scattering with a greater degree of detail than can easily be achieved for scattering by multilevel atoms. We use the methods outlined in §§7.7 to 7.9 to relate the properties of the emitted light to the time development of the atomic state.

The behaviour of a two-level atom subjected to coherent incident light is described by the interaction-picture optical Bloch equations. The semiclassical theory derived in Chapter 2 remains valid when the classical driving field is replaced by a quantum coherent-state light beam. The required equations of motion are thus identical to the semiclassical forms in eqns (2.8.1) and (2.9.4),

$$d\tilde{\rho}_{22}/dt = -\tfrac{1}{2}i\mathcal{V}(\tilde{\rho}_{12} - \tilde{\rho}_{21}) - 2\gamma_{sp}\tilde{\rho}_{22} \tag{8.2.1}$$

and

$$d\tilde{\rho}_{12}/dt = \tfrac{1}{2}i\mathcal{V}(\tilde{\rho}_{11} - \tilde{\rho}_{22}) + [i(\omega_0 - \omega) - \gamma]\tilde{\rho}_{12}, \tag{8.2.2}$$

where $\mathcal{V}$, taken to be real throughout, is the Rabi frequency. Here γ_{sp} is the radiative decay rate defined in eqn (2.5.11) and γ, defined in eqn (2.9.5), is the sum of γ_{sp} and the elastic collisional decay rate γ_{coll}. The remaining density matrix elements are determined as usual from

$$\tilde{\rho}_{11} + \tilde{\rho}_{22} = 1 \quad \text{and} \quad \tilde{\rho}_{21} = \tilde{\rho}_{12}^*. \tag{8.2.3}$$

The motions of the expectation values of the atomic transition operators are obtained from eqns (7.9.5) to (7.9.7) when the time dependences of the density matrix elements are known. The expectation values of the radiated-field operators are then determined by the source-field expression in the form derived in eqn (7.8.8) for a transition dipole of frequency ω_0. The required connections between field expectation values and atomic density-matrix elements are thus

$$\left\langle \hat{E}_{sc}^+(\mathbf{r},t)\right\rangle = -\frac{e\omega_0^2 \mathbf{e}_{sc}\cdot\mathbf{D}_{12}}{4\pi\varepsilon_0 c^2 r}\tilde{\rho}_{21}\left(t-\frac{r}{c}\right)\exp\left\{-i\omega\left(t-\frac{r}{c}\right)\right\}, \tag{8.2.4}$$

with the conjugate relation for $\hat{E}_{sc}^-$, and

$$\left\langle \hat{E}_{sc}^-(\mathbf{r},t)\hat{E}_{sc}^+(\mathbf{r},t)\right\rangle = \left(\frac{e\omega_0^2 \mathbf{e}_{sc}\cdot\mathbf{D}_{12}}{4\pi\varepsilon_0 c^2 r}\right)^2 \tilde{\rho}_{22}\left(t-\frac{r}{c}\right). \tag{8.2.5}$$

The atom is placed at the coordinate origin, sc subscripts distinguish scattered-field quantities, and $\mathbf{e}_{sc}$ is the polarization vector of the scattered light (parallel to the x axis in Fig. 7.10). These relations are used to determine the intensity, spectral properties and coherence of the scattered field. Only steady-state properties are considered, with t taken infinite.

The steady-state scattered intensity obtained from eqn (8.2.5) is

$$\bar{I}_{sc} = \left\{e^2\omega_0^4(\mathbf{e}_{sc}\cdot\mathbf{D}_{12})^2/8\pi^2\varepsilon_0 c^3 r^2\right\}\tilde{\rho}_{22}(\infty), \tag{8.2.6}$$

which is essentially the same as the quantity integrated over solid angle in eqn (7.8.9). The steady-state degree of atomic excitation given by eqn (2.9.6) is

$$\tilde{\rho}_{22}(\infty) = \frac{(\gamma/4\gamma_{sp})\mathcal{V}^2}{(\omega_0-\omega)^2 + \gamma^2 + (\gamma/2\gamma_{sp})\mathcal{V}^2}. \tag{8.2.7}$$

The rate of production of scattered photons by a single two-level atom is equal to $2\gamma_{sp}\tilde{\rho}_{22}(\infty)$, as in eqn (7.9.29), and the mean time interval between scattering events is given by the inverse of this quantity. The scattering rate is thus proportional to $\mathcal{V}^2$ for weak incident light and it has a limiting value of γ_{sp} for very strong incident light with a saturated atomic transition. The mean incident intensity in a single-mode light beam is given by eqn (4.11.7) as

$$\bar{I} = c\hbar\omega\langle n\rangle/V. \tag{8.2.8}$$

The squared Rabi frequency that appears in eqn (8.2.7) can thus be written with the use of eqn (4.10.11) as

$$\mathcal{V}^2 = \frac{2e^2}{\varepsilon_0 c\hbar^2}(\mathbf{e}\cdot\mathbf{D}_{12})^2\bar{I}, \tag{8.2.9}$$

where $\mathbf{e}$ is the polarization vector of the incident light. The differential cross-section (8.1.3) for the scattering by a single atom is therefore

$$\frac{d\sigma(\omega)}{d\Omega} = \frac{e^4 \omega_0^4 \omega}{16\pi^2 \varepsilon_0^2 \hbar^2 c^4 \omega_{sc}} \frac{(\mathbf{e}_{sc} \cdot \mathbf{D}_{12})^2 (\mathbf{e} \cdot \mathbf{D}_{12})^2 \, \gamma/\gamma_{sp}}{(\omega_0 - \omega)^2 + \gamma^2 + (\gamma/2\gamma_{sp})\mathcal{V}^2},$$ (8.2.10)

and this is multiplied by N for the scattering by N identical atoms. Apart from the ω factor in the numerator, the cross-section has a Lorentzian dependence on the incident frequency, centred on the atomic transition frequency ω_0 and with a width that combines the effects of radiative, power, and collision broadening in the same manner as eqn (2.9.9). The cross-section is essentially independent of the incident intensity for low intensities, where the Rabi frequency is small. However, the cross-section falls off inversely with incident intensity for high intensities, where the $\mathcal{V}^2$ term in the denominator is largest. This is the scattering analogue of the saturation effect on the attenuation coefficient discussed in §1.8.

Information on the first-order coherence properties of the scattered light is obtained from the expectation values of the field operators in eqn (8.2.4) and its conjugate. The sum of these expectation values has a sinusoidal variation with position and time, similar to the mean field of a single-mode coherent state in eqn (5.3.31). The ratio of the part of the scattered intensity that has this first-order coherent behaviour to the total scattered intensity is expressed as [3]

$$\frac{\bar{I}_{sc}^{coherent}}{\bar{I}_{sc}} = \frac{\langle \hat{E}_{sc}^-(\mathbf{r},t) \rangle \langle \hat{E}_{sc}^+(\mathbf{r},t) \rangle}{\langle \hat{E}_{sc}^-(\mathbf{r},t) \hat{E}_{sc}^+(\mathbf{r},t) \rangle}$$ (8.2.11)

This ratio is unity for a purely coherent excitation of the scattered field but it vanishes for chaotic light. The steady-state ratio is

$$\bar{I}_{sc}^{coherent}/\bar{I}_{sc} = |\tilde{\rho}_{21}(\infty)|^2 / \tilde{\rho}_{22}(\infty)$$ (8.2.12)

for the expectation values given in eqns (8.2.4) and (8.2.5). It follows from the general property of the density matrix elements quoted in eqn (2.8.18) that

$$|\tilde{\rho}_{21}|^2 / \tilde{\rho}_{22} \leq \tilde{\rho}_{11}$$ (8.2.13)

and the ratio in eqn (8.2.12) is accordingly always less than or equal to unity. It is shown below that the degree of second-order coherence of the 'coherent' part the scattered light does not at all resemble that of a coherent state, and use of the word is here limited to the description of first-order properties.

The coherent fraction of eqn (8.2.12) can be expressed in various forms. It follows with the use of eqns (7.9.8), (7.9.9) and (7.9.11) that

$$\bar{I}_{sc}^{coherent}/\bar{I}_{sc} = |\alpha_1(\infty)|^2 / \beta_1(\infty) = |g^{(1)}(\infty)|,$$ (8.2.14)

with the infinite time-delay degree of first-order coherence from eqn (7.9.22). Alternatively, with the steady-state solutions of the optical Bloch equations from eqns (2.9.6) and (2.9.7), eqn (8.2.12) becomes

$$\frac{\bar{I}_{sc}^{coherent}}{\bar{I}_{sc}} = \frac{\left[(\omega_0 - \omega)^2 + \gamma^2\right](\gamma_{sp}/\gamma)}{(\omega_0 - \omega)^2 + \gamma^2 + (\gamma/2\gamma_{sp})\mathcal{V}^2}. \tag{8.2.15}$$

The scattered light is therefore almost entirely first-order coherent when the collision broadening is small ($\gamma_{coll} \ll \gamma_{sp}$, $\gamma \approx \gamma_{sp}$) and the incident beam is weak ($\mathcal{V} \ll \gamma_{sp}$). The coherent part of the scattered light falls to a very small fraction in the opposite limits of large collision broadening ($\gamma_{coll} \gg \gamma_{sp}$, $\gamma \gg \gamma_{sp}$) or an intense incident beam ($\mathcal{V} \gg \gamma_{sp}$). The expressions for the coherent fraction remain valid for the scattering by N atoms.

More complete information on the nature of the light produced in resonant scattering conditions is obtained from its degrees of first and second-order coherence. These are determined from the solutions of the optical Bloch equations by the procedure described in §7.9. The complete time dependences are particularly instructive when the Bloch equations have reasonably simple solutions, as in the cases of a weak incident beam or of zero detuning treated in §§8.3 and 8.4 respectively. For the scattering by a single atom, only the limiting values

$$g^{(1)}(0) = 1, \tag{8.2.16}$$

$$g^{(2)}(0) = 0 \quad \text{and} \quad g^{(2)}(\infty) = 1 \tag{8.2.17}$$

derived in §7.9 apply for general magnitudes of the parameters. The vanishing of the degree of second-order coherence for zero time delay shows that the scattered light has sub-Poissonian statistics, in accordance with eqn (5.10.5). The light is thus nonclassical even when its coherent fraction is large.

The absolute strength of the scattering is given by the differential cross-section in eqn (8.2.10) but the relative sizes of the contributions are conveniently shown by the normalized spectrum of the scattered light. The spectrum is obtained from the degree of first-order coherence by eqn (3.5.10) or (3.5.11) as

$$F(\omega_{sc}) = \frac{1}{2\pi} \int_{-\infty}^{\infty} d\tau g^{(1)}(\tau) \exp(i\omega_{sc}\tau) = \frac{1}{\pi} \mathrm{Re} \int_{0}^{\infty} d\tau g^{(1)}(\tau) \exp(i\omega_{sc}\tau). \tag{8.2.18}$$

It contains two distinct contributions, corresponding to the parts of the degree of first-order coherence that decay to zero as τ tends to infinity and to the constant-amplitude part,

$$\exp(-i\omega\tau)|g^{(1)}(\infty)|, \tag{8.2.19}$$

that remains at infinite time delay. The constant contribution to the degree of first-order coherence is given by eqn (7.9.22) and the decaying contributions are given by the remainder of the large bracket in eqn (7.9.20) when this constant term is subtracted.

The Fourier transform of the decaying parts of the degree of first-order coherence generates a spectrum at frequencies ω_{sc} that differ from the frequency ω of the incident light; this is the inelastic or fluorescent part of the scattered spectrum. On the other hand, substitution of the constant-amplitude part (8.2.19) of the degree of first-order coherence into eqn (8.2.18) gives the elastic or Rayleigh part of the scattered spectrum as

$$F(\omega_{sc})_{\text{elastic}} = \left|g^{(1)}(\infty)\right|\delta(\omega_{sc} - \omega). \tag{8.2.20}$$

It is seen by comparison with eqn (8.2.14) that the strength of the elastic part of the spectrum is equal to the coherent fraction of the scattered light, and the elastic and inelastic contributions are sometimes called the 'coherent' and 'incoherent' parts, respectively. However, as shown in the present and next sections, the elastically-scattered light has a degree of second-order coherence that differs from the usual unit value for coherent light. The expression for the coherent fraction given in eqn (8.2.15) shows that the elastic contribution is significant only when the incident beam is weak and the collision broadening is small. These properties of the spectrum remain valid for the scattering by an arbitrary number N of identical atoms.

8.3 Weak incident beam

The incident beam is assumed in the present section to be so weak that the Rabi frequency is much smaller than the radiative decay rate,

$$\mathcal{V} \ll \gamma_{sp}. \tag{8.3.1}$$

Then eqns (8.2.14) and (8.2.15) give

$$\bar{I}_{sc}^{\text{coherent}}/\bar{I}_{sc} = \left|g^{(1)}(\infty)\right| = \gamma_{sp}/\gamma = \gamma_{sp}/(\gamma_{sp} + \gamma_{coll}); \tag{8.3.2}$$

the scattered light is entirely coherent and elastic in the absence of any collision broadening. The complete cross-section is given by eqn (8.2.10) with the final term in the denominator neglected.

The solutions of the optical Bloch equations (8.2.1) and (8.2.2) in the limit of a weak incident beam are given by eqns (2.10.2) and (2.10.3). The degrees of coherence of the scattered light are obtained by the method of §7.9. The coefficients needed for the off-diagonal density-matrix element of eqn (7.9.8) are identified as

$$\alpha_1(\tau) = \frac{i\mathcal{V}/2}{i(\omega_0 - \omega) + \gamma}\left(\exp\left\{-\left[i(\omega_0 - \omega) + \gamma\right]\tau\right\} - 1\right) \tag{8.3.3}$$

and

$$\alpha_2(\tau) = \exp\left\{-\left[i(\omega_0 - \omega) + \gamma\right]\tau\right\}, \tag{8.3.4}$$

where the time delay τ is again assumed to be positive. The coefficient $\beta_1(\tau)$ in the diagonal density-matrix element in eqn (7.9.9) is given by the entire expression from eqn (2.10.2).

The degree of first-order coherence is obtained from eqn (7.9.20) as

$$g^{(1)}(\tau) = \frac{\gamma_{coll}}{\gamma}\exp(-i\omega_0\tau - \gamma\tau) + \frac{\gamma_{sp}}{\gamma}\exp(-i\omega\tau). \tag{8.3.5}$$

The exponential in the first term is identical to the usual degree of first-order coherence for collision-broadened chaotic light for an atomic transition frequency ω_0. The second exponential is the same as the degree of first-order coherence for coherent light of frequency ω.

The normalized spectrum derived from the degree of first-order coherence by means of eqn (8.2.18) is [5]

$$F(\omega_{sc}) = \frac{\gamma_{coll}}{\gamma}\frac{\gamma/\pi}{(\omega_0 - \omega_{sc})^2 + \gamma^2} + \frac{\gamma_{sp}}{\gamma}\delta(\omega_{sc} - \omega). \tag{8.3.6}$$

This simple form of scattered spectrum is illustrated in Fig. 8.3. It includes an inelastic part whose strength is directly proportional to the collision-broadening decay rate. The inelastic part of the spectrum has the same lineshape as the

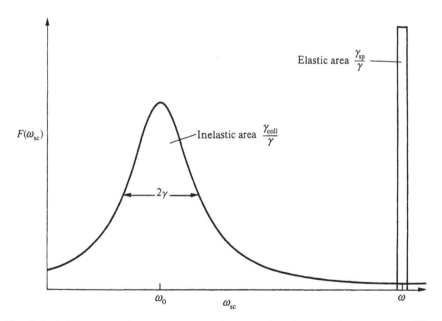

Fig. 8.3. Light-scattering spectrum of a two-level atom for $\omega > \omega_0$. The Rayleigh delta-function component at frequency ω and the fluorescent Lorentzian centred on the atomic transition frequency ω_0 correspond to the two arrows in the left of the energy-level diagram in Fig. 8.1.

absorption spectrum of the atom, and the scattering is equivalent to absorption of the incident light followed by fluorescent emission. The difference between the energies of the incident and scattered photons is assimilated by the energy uncertainty of the excited state that results from the collisions, and the effect is sometimes called *collisional redistribution* of the energy. The remainder of the scattered intensity lies in the elastic Rayleigh part represented by the delta-function term in eqn (8.3.6). Both contributions to the spectrum are observed experimentally, for example, in the light scattering by Sr vapour [6].

The properties of the first-order coherence suggest that the scattered light contains a sum of ordinary coherent and chaotic components. However, this simple picture is modified by an evaluation of the second-order coherence. The degree of second-order coherence is obtained by substitution of the expression for $\beta_1(\tau)$ given by the entire eqn (2.10.2) into eqn (7.9.21), with the result

$$g^{(2)}(\tau) = 1 + \frac{\left[(2\gamma_{sp} - \gamma)/\gamma\right]\left[(\omega_0 - \omega)^2 + \gamma^2\right]}{(\omega_0 - \omega)^2 + (2\gamma_{sp} - \gamma)^2} \exp(-2\gamma_{sp}\tau) - 2\gamma_{sp}\exp(-\gamma\tau) \times$$

$$\frac{\left[(\omega_0 - \omega)^2 + \gamma(2\gamma_{sp} - \gamma)\right]\cos[(\omega_0 - \omega)\tau] + 2(\omega_0 - \omega)(\gamma - \gamma_{sp})\sin[(\omega_0 - \omega)\tau]}{\gamma\left[(\omega_0 - \omega)^2 + (2\gamma_{sp} - \gamma)^2\right]}.$$

$$(8.3.7)$$

This complicated expression is best appreciated by consideration of a couple of limiting cases.

Suppose first that there is no collision broadening, when the spectrum contains only the elastic Rayleigh component with the single scattered frequency $\omega_{sc} = \omega$, and eqn (8.3.7) reduces to

$$g^{(2)}(\tau) = 1 + \exp(-2\gamma_{sp}\tau) - 2\cos[(\omega_0 - \omega)\tau]\exp(-\gamma_{sp}\tau) \quad (\gamma_{coll} = 0). \quad (8.3.8)$$

The simplest possible case is that of zero detuning, where [7]

$$g^{(2)}(\tau) = \left[1 - \exp(-\gamma_{sp}\tau)\right]^2 \quad (\gamma_{coll} = 0, \ \omega = \omega_0 = \omega_{sc}); \quad (8.3.9)$$

the degree of second-order coherence increases significantly from its zero initial value only after a time delay τ of the order of the radiative lifetime of the atomic transition. The variations of the degree of second-order coherence for several values of the detuning are shown by the full curves in Fig. 8.4. For comparison, the same limit of eqn (8.3.5) gives

$$g^{(1)}(\tau) = \exp(-i\omega\tau) \quad (\gamma_{coll} = 0), \quad (8.3.10)$$

the ordinary result for coherent light. However, the degree of second-order coher-

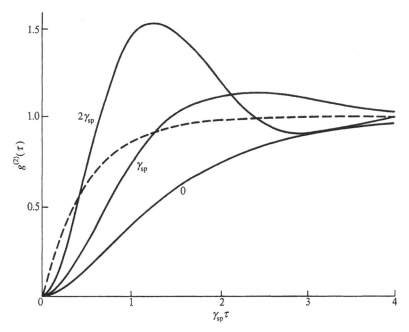

Fig. 8.4. Dependence of the degree of second-order coherence of the scattered light on the time delay τ. Full curves: weak incident beam with no collision broadening and the detunings $|\omega_0 - \omega|$ shown. Broken curve: large collision broadening.

ence for ordinary coherent light equals unity for all τ, unlike the expression (8.3.8) obtained for the light scattered in resonance fluorescence. It is clear from Fig. 8.4 that the fluorescent light has a degree of second-order coherence that violates the classical inequalities (3.7.6) and (3.7.11). The elastically-scattered light is antibunched for the qualitative reasons discussed after eqn (7.9.28). It also has sub-Poissonian statistics, or photon-number squeezing. These results for zero collision broadening apply to experiments on the scattering by a single atom, where there are no other atoms with which to collide. Single-atom scattering is treated further in §8.4.

The second simple limit is that of very large collision broadening, when the spectrum contains only the inelastic fluorescent component. The second exponential in eqn (8.3.7) now decays to zero at much shorter time delays τ than does the first exponential. Thus for τ sufficiently large that the final term is negligible, eqn (8.3.7) reduces to

$$g^{(2)}(\tau) = 1 - \exp(-2\gamma_{sp}\tau) \quad \left(\gamma_{coll} \gg \gamma_{sp}, \ \gamma_{coll}\tau \gg 1\right). \tag{8.3.11}$$

The conditions assumed here are those in which the time dependences of the diagonal elements of the atomic density matrix are described by rate equations, as discussed in §2.10. The result (8.3.11), illustrated by the broken curve in Fig.

8.4, is the same as that derived in eqn (7.9.28) and illustrated in Fig. 7.11. The corresponding degree of first-order coherence is obtained from eqn (8.3.5) as

$$g^{(1)}(\tau) = \exp(-i\omega_0\tau - \gamma_{coll}\tau) \quad (\gamma_{coll} \gg \gamma_{sp}), \tag{8.3.12}$$

the ordinary result for chaotic light. This degree of first-order coherence reduces to zero for the time scale assumed in eqn (8.3.11). However, the existence of a large collision broadening implies a significant atomic density and, as discussed in the following section, the lack of correlation between the photons emitted by different atoms increases the degree of second-order coherence above the value given in eqn (8.3.11). The conditions assumed for this expression are not of great practical importance for studies of nonclassical light.

As is explained above, the inelastic conversion of some of the incident photons of frequency ω into scattered photons of different frequencies ω_{sc} depends upon the collisional uncertainty in the atomic excited-state energy. Doppler broadening results from the thermal distribution of atomic velocities, and it causes no additional uncertainty in the energy of a particular atom. The Doppler effect changes the spectral distribution of the inelastically-scattered light but it does not affect its total strength.

8.4 Single-atom resonance fluorescence

The scattering theory is here specialized to a single atom, with the collision and Doppler broadening mechanisms ignored. The calculations are further simplified when the incident coherent light is exactly resonant with the atomic transition, and ω_0 is set equal to ω throughout the present section. Equations (8.2.14) and (8.2.15) then give

$$\frac{\bar{I}_{sc}^{coherent}}{\bar{I}_{sc}} = \left|g^{(1)}(\infty)\right| = \frac{2\gamma_{sp}^2}{2\gamma_{sp}^2 + V^2}. \tag{8.4.1}$$

The coherent fraction of the scattered light is thus simply determined by the strength of the incident beam. The calculations that follow are concerned with the ways in which the scattered spectrum and degrees of coherence vary with the incident beam strength.

The optical Bloch equations (8.2.1) and (8.2.2) are partially solved for zero collision broadening and zero detuning in eqn (2.8.10). This diagonal density-matrix element provides the coefficient $\beta_1(\tau)$ in eqn (7.9.9). The corresponding solution for the off-diagonal density matrix element is obtained straightforwardly but with lengthy and tedious algebra [7]. The first two coefficients in eqn (7.9.8) are found to be

$$\alpha_1(\tau) = -\frac{i\mathcal{V}}{2\gamma_{sp}^2 + \mathcal{V}^2}\left\{\gamma_{sp} - \left[\exp(-3\gamma_{sp}\tau/2)/16i\lambda\right]\right.$$

$$\left. \times\left(\left[4(\gamma_{sp} + i\lambda)^2 - \gamma_{sp}^2\right]\exp(i\lambda\tau) - \text{comp.conj.}\right)\right\}$$

(8.4.2)

and

$$\alpha_2(\tau) = \tfrac{1}{2}\exp(-\gamma_{sp}\tau) + \frac{\exp(-3\gamma_{sp}\tau/2)}{8i\lambda}$$

$$\times\left[(\gamma_{sp} + 2i\lambda)\exp(i\lambda\tau) - (\gamma_{sp} - 2i\lambda)\exp(-i\lambda\tau)\right],$$

(8.4.3)

where

$$\lambda = \left(\mathcal{V}^2 - \tfrac{1}{4}\gamma_{sp}^2\right)^{1/2}.$$

(8.4.4)

These expressions are valid for both $\mathcal{V} > \gamma_{sp}/2$ and $\mathcal{V} < \gamma_{sp}/2$. It is convenient to put

$$\lambda = i\left(\tfrac{1}{4}\gamma_{sp}^2 - \mathcal{V}^2\right)^{1/2} = i|\lambda|$$

(8.4.5)

in the latter regime.

The degree of first-order coherence obtained by insertion of these expressions into eqn (7.9.20) is [7]

$$g^{(1)}(\tau) = \exp(-i\omega\tau)\left\{\frac{2\gamma_{sp}^2}{2\gamma_{sp}^2 + \mathcal{V}^2} + \tfrac{1}{2}\exp(-\gamma_{sp}\tau)\right.$$

$$\left. + \frac{(\gamma_{sp} - 2i\lambda)^2\exp\left[-i\lambda\tau - (3\gamma_{sp}\tau/2)\right]}{8i\lambda(3\gamma_{sp} + 2i\lambda)} + (\lambda \to -\lambda)\right\},$$

(8.4.6)

where the fourth term in the large bracket is the same as the third term with the sign of λ reversed. The degree of coherence consists of the sum of a contribution with the usual coherent time dependence and three contributions with chaotic-light time dependences. The coherent fraction in the first term agrees with the expression in eqn (8.4.1).

Consider first the regime of weaker incident beams, where the replacement in eqn (8.4.5) is made and the spectrum is given by eqn (8.2.18) as [3]

$$F(\omega_{sc}) = \frac{2\gamma_{sp}^2}{2\gamma_{sp}^2 + \mathcal{V}^2} \delta(\omega_{sc} - \omega) + \frac{\gamma_{sp}/2\pi}{(\omega - \omega_{sc})^2 + \gamma_{sp}^2}$$

$$- \frac{1}{16\pi|\lambda|} \frac{(\gamma_{sp} + 2|\lambda|)^2}{(\omega - \omega_{sc})^2 + [(3\gamma_{sp}/2) - |\lambda|]^2} - (|\lambda| \to -|\lambda|)$$

(8.4.7)

for $\mathcal{V} < \gamma_{sp}/2$. For a very weak incident beam with $\mathcal{V} = 0$, the elastic part of the spectrum reduces to the same unit-strength delta-function as given by eqn (8.3.6) when the collision broadening decay rate γ_{coll} is set equal to zero.

Problem 8.2   Show that for small but nonzero $\mathcal{V}$, the leading term in the inelastic part of the spectrum is [8]

$$F(\omega_{sc}) = \frac{\gamma_{sp}\mathcal{V}^2/\pi}{\left[(\omega - \omega_{sc})^2 + \gamma_{sp}^2\right]^2}.$$

(8.4.8)

Show that the width of the spectrum is

$$\text{FWHM} = 2\gamma_{sp}(\sqrt{2} - 1)^{1/2}.$$

(8.4.9)

The width of this squared Lorentzian lineshape is thus reduced to 64% of that for an ordinary Lorentzian with the same damping parameter. The reduction in the width is associated with a moderate quadrature squeezing that occurs in the scattered light [8], in addition to the photon-number squeezing produced in resonance fluorescence. The degree of second-order coherence for weaker incident beams is obtained from eqns (2.8.10) and (7.9.21) as

$$g^{(2)}(\tau) = 1 - \left[\cosh(|\lambda|\tau) + (3\gamma_{sp}/2|\lambda|)\sinh(|\lambda|\tau)\right]\exp(-3\gamma_{sp}\tau/2).$$

(8.4.10)

The expression reduces to eqn (8.3.9) in the weak-beam limit and, in general, the degree of second-order coherence shows a monotonic increase from 0 to 1 as τ increases from 0 to ∞.

Now consider the regime of stronger incident beams with $\mathcal{V} > \gamma_{sp}/2$. The spectrum obtained by insertion of eqn (8.4.6) into eqn (8.2.18) is [3]

$$F(\omega_{sc}) = \frac{2\gamma_{sp}^2}{2\gamma_{sp}^2 + \mathcal{V}^2} \delta(\omega_{sc} - \omega) + \frac{\gamma_{sp}/2\pi}{(\omega - \omega_{sc})^2 + \gamma_{sp}^2}$$

$$+ \frac{3\gamma_{sp}\lambda(\mathcal{V}^2 - 2\gamma_{sp}^2) + \gamma_{sp}(5\mathcal{V}^2 - 2\gamma_{sp}^2)(\omega + \lambda - \omega_{sc})}{8\pi\lambda(2\gamma_{sp}^2 + \mathcal{V}^2)\left[(\omega + \lambda - \omega_{sc})^2 + (3\gamma_{sp}/2)^2\right]} + (\lambda \to -\lambda).$$

(8.4.11)

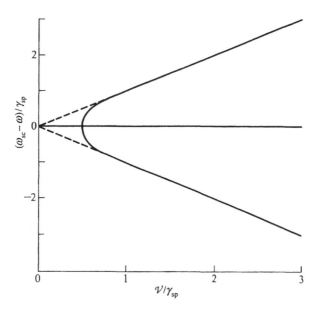

Fig. 8.5. Variation with Rabi frequency $\mathcal{V}$ of the central frequencies of the inelastic spectral lines.

The four contributions consist of an elastic delta-function at $\omega_{sc} = \omega$ and three inelastic Lorentzians centred on the frequencies $\omega_{sc} = \omega$, $\omega + \lambda$ and $\omega - \lambda$ respectively. Figure 8.5 shows how these central frequencies vary with the Rabi frequency $\mathcal{V}$, which is proportional to the electric field of the incident beam. The separations between the frequencies are very close to $\mathcal{V}$ for $\mathcal{V} > \gamma_{sp}$.

Problem 8.3   Obtain the integrated strengths of the four contributions to the spectrum and verify that they sum to unity, in accordance with the normalization of $F(\omega_{sc})$.

The spectral contributions in eqn (8.4.11) overlap to give a complicated dependence on the frequency when $\mathcal{V}$ is only a little larger than $\gamma_{sp}/2$. The spectrum takes a much simpler form for larger values of $\mathcal{V}$, when the elastic Rayleigh contribution is negligible and the inelastic contributions reduce to

$$F(\omega_{sc}) = \frac{3\gamma_{sp}/8\pi}{\left(\omega - \mathcal{V} - \omega_{sc}\right)^2 + \left(3\gamma_{sp}/2\right)^2} + \frac{\gamma_{sp}/2\pi}{\left(\omega - \omega_{sc}\right)^2 + \gamma_{sp}^2}$$
$$+ \frac{3\gamma_{sp}/8\pi}{\left(\omega + \mathcal{V} - \omega_{sc}\right)^2 + \left(3\gamma_{sp}/2\right)^2},$$

(8.4.12)

now written in order of ascending central frequency. The three peaks have integrated strengths of 1/4, 1/2 and 1/4, and radiative linewidths of $3\gamma_{sp}$, $2\gamma_{sp}$, and

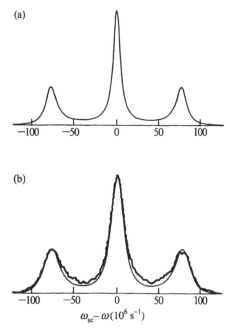

Fig. 8.6. Resonance fluorescence spectrum of Na for the transition between levels $3^2S_{1/2}$ (F = 2, m_F = 2) and $3^2P_{3/2}$ (F = 3, m_F = 3) with $\mathcal{V} = 7.8 \times 10^7$ s^{-1} and $\gamma_{sp} = 5 \times 10^6$ s^{-1} showing (a) theory and (b) experiment compared to a convolution of the instrumental response and the theoretical lineshape. (After [10])

$3\gamma_{sp}$ respectively. The triple-peaked spectrum is a striking characteristic of the resonance fluorescence, which is observed experimentally [9,10]. Figure 8.6(a) shows the spectrum of scattered light predicted by eqn (8.4.12) for $\mathcal{V} = 15.6\gamma_{sp}$ and Fig. 8.6(b) shows a measured spectrum for the corresponding strength of incident beam. The two-level theory is valid for the interpretation of the results in the conditions of this experiment.

The occurrence of triple-peaked spectra in the light scattered by a two-level atom at high incident intensities is understood qualitatively in terms of the dynamic Stark effect, mentioned in §2.7 [11,12]. Consider a state of the coupled atom and incident radiation in which $n + 1$ quanta are excited. There are two states of this kind; either the atom is in its ground state, accompanied by $n + 1$ photons or the atom is in its excited state, accompanied by n photons. The coupled system oscillates between the two states in the manner shown in Fig. 2.5. In terms of stationary states, the effect of the coupling is to split the two $(n + 1)$-quantum states by an energy $\hbar\mathcal{V}$, as shown in Fig. 8.7. The n-quantum states are split in a similar fashion. The four downward transitions indicated in the figure occur when the $(n + 1)$-quantum states decay into n-quantum states by fluorescent emission of a photon into the scattered beam. There are clearly three distinct transition frequencies separated by the Rabi frequency $\mathcal{V}$, and the central peak acquires contri-

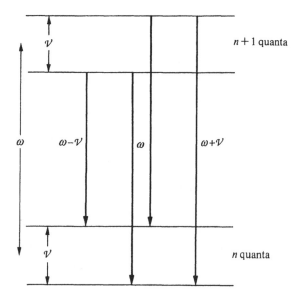

Fig. 8.7. Formation of a three-peaked emission spectrum by radiative decay between levels split by the dynamic Stark effect.

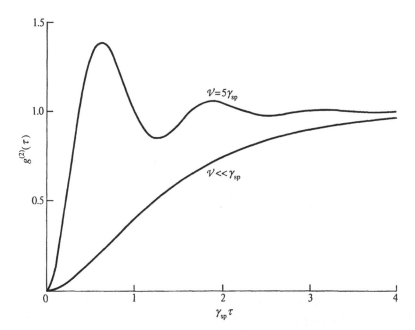

Fig. 8.8. Dependence of the degree of second-order coherence of the scattered light on the time delay τ for zero detuning and the values of Rabi frequency indicated.

butions from two kinds of transition, thus gaining twice the strength of each outer peak.

The degree of second-order coherence of the scattered light for $V > \gamma_{sp}/2$ is obtained from eqns (2.8.10) and (7.9.21) as [7]

$$g^{(2)}(\tau) = 1 - \left[\cos(\lambda\tau) + \left(3\gamma_{sp}/2\lambda\right)\sin(\lambda\tau)\right]\exp\left(-3\gamma_{sp}\tau/2\right). \qquad (8.4.13)$$

The degree of second-order coherence now has an oscillatory dependence on τ. An example is shown in Fig. 8.8, with the monotonic dependence of eqn (8.3.9) for comparison. The degree of second-order coherence again violates the classical inequalities for small $\gamma_{sp}\tau$ and the scattered light is antibunched in accordance with eqn (6.5.8).

The scattered light also has sub-Poissonian photon statistics with $Q < 0$, as defined in eqn (5.10.6), and with $g^{(2)}(0) < 1$, as defined in eqn (6.5.9). The variance in the number of scattered photons is accordingly smaller than their mean value. This reduction in the variance below the Poisson value is the cause of the decrease in the breadth of the distribution of atomic deflections produced by interaction with a light beam, shown in Fig. 1.18. As outlined in §1.11, the mean deflection is proportional to the mean number of absorption–spontaneous emission events, and the different numbers of events experienced by different atoms cause broadening of the distribution. A reduction in the spread of spontaneously-emitted photon number thus produces a corresponding reduction in the spread of deflection angle. The dashed line in the figure shows the distribution calculated on the assumption of Poisson statistics for the scattered photons, with $Q = 0$, while the continuous line shows the distribution calculated for the value $Q = -0.58$ appropriate to the experiment.

Measurements of the degree of second-order coherence can be made by observations of the resonance fluorescence from atoms in the form of a beam [13]. The atoms are excited by a perpendicular light beam and the scattered light is detected in the third mutually-perpendicular direction. The single-atom resonance fluorescence is not observed directly with this arrangement because two or more atoms are sometimes simultaneously in the field of view. An average of the number of atoms over a Poisson distribution (see Problem 8.7) produces a measured degree of second-order coherence greater than unity. The sub-Poissonian character of the scattered light is therefore lost, although the antibunching property survives. The single-atom coherence is extracted from the measured data by theoretical analysis.

The more recent availability of ion traps allows the direct measurement of single-atom resonance fluorescence [14]. Some examples of the observed degree of second-order coherence of the light scattered by $^{24}\text{Mg}^+$ are shown in Fig. 8.9. The frequency ω of the incident light in the measurements is smaller than the atomic transition frequency ω_0, because of a need to avoid heating of the atom, and the zero-detuning theory of the present section does not apply. However, the single-atom experiments have the advantage of directly observing degrees of second-order coherence smaller than unity. The continuous lines in Fig. 8.9 show the calculated degrees of second-order coherence obtained from a more general

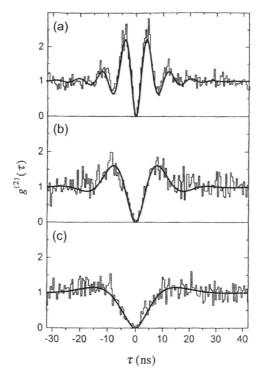

Fig. 8.9. Measured degrees of second-order coherence for a single ^{24}Mg$^+$ ion for detunings $(\omega_0 - \omega)/\gamma_{sp}$ and Rabi frequencies $\mathcal{V}/\gamma_{sp}$ given respectively by (a) 4.6 and 5.6 (b) 2.2 and 2.0 (c) 1.0 and 1.2. (After [14])

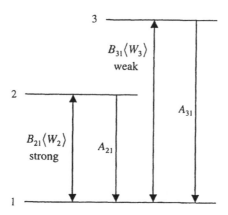

Fig. 8.10. Three-level V configuration used in the observation of quantum jumps.

theory that includes the effects of nonzero detuning [13]. The agreement between experiment and theory is seen to be very good.

8.5 Quantum jumps

The light emitted in resonance fluorescence shows further nonclassical properties when the atomic transition scheme is extended to include a third participating level. Figure 8.10 shows a three-level scheme with the important transitions marked. The transition between levels 1 and 2 is strongly allowed, and it is strongly driven by an incident light beam, while the much weaker transition between levels 1 and 3 is weakly driven. Transitions between levels 2 and 3 are forbidden. Light is scattered from the strong transition in accordance with the theory presented in preceding sections. However, the weak transition occasionally occurs, and the flow of strongly scattered light is interrupted for the duration of the excitation to level 3. The time dependence of the resonance fluorescence from the strong transition thus shows a pattern of continuous periods of bright scattered light separated by periods of darkness, whose random occurrences and durations are governed by the relevant transition rates. The interruptions are experimentally observable [15,16] and Fig. 8.11 shows an example of the random behaviour. Level 3 is known as a *shelving state* of the atom. Its effect in switching off the strong-transition resonance fluorescence provides the means for direct observation of the *quantum jumps* from level 1 to level 3 and thus for the detection of weak transitions in single-atom spectroscopy [17-19].

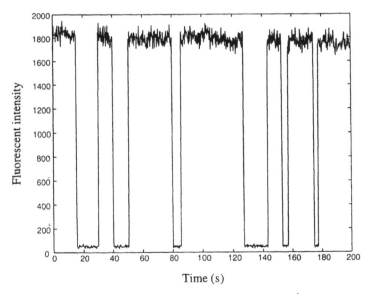

Fig. 8.11. Resonance fluorescence intensity from a single Ba$^+$ ion, expressed in photocounts per 1/5 s, as a function of the time. (R. Blatt, unpublished)

The arrangement of the three levels and transitions shown in Fig. 8.10 is known as the *V configuration*. The transition scheme differs from the *cascade configuration* for three-level laser theory shown in Figs. 1.13 and 7.3 but similar theoretical techniques can be used. The experiments are usually performed with coherent incident light beams but a rate-equation analysis, appropriate to broadband incident light, is simpler and it provides the required insight into the nature of the quantum-jump process. The three levels are assumed to be nondegenerate and, analogous to eqns (1.9.2) to (1.9.4), the rate equations for the level populations are

$$d\rho_{33}/dt = -\rho_{33}A_{31} + (\rho_{11} - \rho_{33})B_{31}\langle W_3 \rangle, \tag{8.5.1}$$

$$d\rho_{22}/dt = -\rho_{22}A_{21} + (\rho_{11} - \rho_{22})B_{21}\langle W_2 \rangle \tag{8.5.2}$$

and

$$d\rho_{11}/dt = \rho_{22}A_{21} + \rho_{33}A_{31} - (\rho_{11} - \rho_{22})B_{21}\langle W_2 \rangle - (\rho_{11} - \rho_{33})B_{31}\langle W_3 \rangle, \tag{8.5.3}$$

where the single-atom populations are expressed in terms of the diagonal density-matrix elements. The Einstein coefficients for the two transitions are as defined in §1.5. The sum of the three density-matrix elements is clearly conserved by these equations of motion.

The steady-state solutions are readily derived and they have simple forms in the limit

$$B_{21}\langle W_2 \rangle \gg A_{21} \gg A_{31} \text{ and } B_{31}\langle W_3 \rangle \tag{8.5.4}$$

that represents the conditions in quantum-jump experiments.

Problem 8.4   Obtain the solutions of the rate equations (8.5.1) to (8.5.3) in the steady state and show that they reduce to

$$\rho_{33}(\infty) = \frac{B_{31}\langle W_3 \rangle}{2A_{31} + 3B_{31}\langle W_3 \rangle} \tag{8.5.5}$$

and

$$\rho_{22}(\infty) = \rho_{11}(\infty) = \frac{A_{31} + B_{31}\langle W_3 \rangle}{2A_{31} + 3B_{31}\langle W_3 \rangle}, \tag{8.5.6}$$

when the inequalities (8.5.4) are satisfied.

The transition between levels 1 and 2 is fully saturated and the equal mean populations in these levels are always greater than that in level 3. The atom is in level 3 for a small fraction of the time when $B_{31}\langle W_3\rangle$ is much smaller than A_{31}, and the quantum jumps occur infrequently in this case. The steady-state population of level 3 increases with growing $B_{31}\langle W_3\rangle$ and, for $B_{31}\langle W_3\rangle$ much greater than A_{31}, the atom is in its shelving state for almost one third of the time. The quantity $\rho_{33}(\infty)$ determines the fraction of the time for which the resonance fluorescence from the strong transition between levels 1 and 2 is quenched.

A more complete account of the properties of the scattered light requires the relations between the diagonal density-matrix elements, or atomic populations, at different times. These relations have the forms of eqn (7.9.9), generalized to the three-level system. The equations of motion (8.5.1) to (8.5.3) can be solved exactly but the general solutions are complicated. The calculations of degrees of second-order coherence that follow need only the terms analogous to $\beta_1(\tau)$ in eqn (7.9.9); the remaining terms are here ingored by temporarily setting the populations of levels 2 and 3 equal to zero at time t. Further simplifications result when the inequality (8.5.4) is assumed at the outset. A somewhat tedious calculation then leads to [20]

$$\rho_{33}(\tau) = \frac{B_{31}\langle W_3\rangle}{2A_{31} + 3B_{31}\langle W_3\rangle}\left\{1 - \exp\left[-\left(A_{31} + \tfrac{3}{2}B_{31}\langle W_3\rangle\right)\tau\right]\right\} \tag{8.5.7}$$

and

$$\rho_{22}(\tau) = \frac{A_{31} + B_{31}\langle W_3\rangle}{2A_{31} + 3B_{31}\langle W_3\rangle} + \frac{B_{31}\langle W_3\rangle/2}{2A_{31} + 3B_{31}\langle W_3\rangle}\exp\left[-\left(A_{31} + \tfrac{3}{2}B_{31}\langle W_3\rangle\right)\tau\right]$$
$$-\tfrac{1}{2}\exp\left[-\left(A_{21} + 2B_{21}\langle W_2\rangle + \tfrac{1}{2}B_{31}\langle W_3\rangle\right)\tau\right]. \tag{8.5.8}$$

The time dependences of the upper-level populations are illustrated by the continuous curves in Fig. 8.12 and their steady-state values are shown by the dashed lines. The parameters are chosen for ease of illustration and they only weakly satisfy the inequalities in eqn (8.5.4). The population of level 2 increases rapidly from zero to a value close to the steady-state population of 5/11, obtained from the two-level theory of eqn (1.7.4) with the relative values of the parameters A_{21} and $B_{21}\langle W_2\rangle$ given in the caption. The presence of level 3 is felt at longer times, when transitions into the shelving state begin to take a significant share of the unit total population, and $\rho_{22}(\tau)$ falls towards its steady state value of 3/8. The fraction of the time for which the strong-transition resonance fluorescence is quenched grows monotonically towards its steady-state value of $\rho_{33}(\infty) = 1/4$.

The pairs of transition operators for the three-level V configuration shown in Fig. 8.10 are defined by

$$\begin{array}{ll}
\hat{\pi}_2^\dagger = |2\rangle\langle 1| & \hat{\pi}_3^\dagger = |3\rangle\langle 1| \\
\text{and} & \\
\hat{\pi}_2 = |1\rangle\langle 2| & \hat{\pi}_3 = |1\rangle\langle 3|,
\end{array} \tag{8.5.9}$$

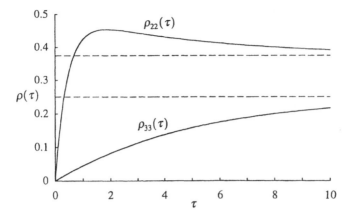

Fig. 8.12. Time dependences of the populations of levels 2 and 3 for the para-
meter values $A_{31} = 0.05$, $B_{31}\langle W_3 \rangle = 0.1$ and $A_{21} = 0.2$, all expressed in units of
$B_{21}\langle W_2 \rangle$. The dashed lines show the steady-state values.

similar to eqn (4.9.11), and the products of operators for the same transition
satisfy relations similar to eqns (4.9.14) to (4.9.18). The products of the different
creation operators or of the different destruction operators vanish by orthogonality
of the atomic states. The density-matrix elements are related to expectation values
of transition operators by

$$\rho_{33}(t) = \left\langle \hat{\pi}_3^\dagger(t)\hat{\pi}_3(t) \right\rangle \quad \text{and} \quad \rho_{22}(t) = \left\langle \hat{\pi}_2^\dagger(t)\hat{\pi}_2(t) \right\rangle, \tag{8.5.10}$$

similar to eqn (7.9.7). The radiated fields are related to the appropriate transition
operators by the source-field expression (7.8.8).

The nature of the light emitted by the three-level atom is clarified by consideration of its degrees of second-order coherence. The rate equation theory outlined
above provides expressions for the intrabeam degrees of second-order coherence
of the strong fluorescent emission from the transition between levels 1 and 2, and
also for the weaker fluorescent emission from the transition between levels 1 and
3. The correlations between photons with the different emission frequencies are
determined by the interbeam degrees of coherence, defined classically in eqn
(3.7.26). There thus are four distinct degrees of second-order coherence in total.
Their quantum forms are expressed in terms of the transition-operator expectation
values by a generalization of eqn (7.9.2) to

$$g_{l,m}^{(2)}(\tau) = \frac{\left\langle \hat{\pi}_l^\dagger(t)\hat{\pi}_m^\dagger(t+\tau)\hat{\pi}_m(t+\tau)\hat{\pi}_l(t) \right\rangle}{\left\langle \hat{\pi}_l^\dagger(t)\hat{\pi}_l(t) \right\rangle \left\langle \hat{\pi}_m^\dagger(t)\hat{\pi}_m(t) \right\rangle} \quad l,m = 2,3. \tag{8.5.11}$$

This expression is valid when the fields are detected at equal linear distances from
the atom so that the retardation time differences between field and atomic operators

in eqn (7.8.8) can be ignored. With the usual assumptions of stationary and ergodic statistics, the averages in the degrees of second-order coherence refer to a long series of measurements on the light radiated by the single atom considered.

The expectation values in the numerators of the degrees of second-order coherence are calculated with use of the quantum regression theorem of eqn (7.9.17) and the method follows closely that given in §7.9 for the emission by a driven two-level atom.

Problem 8.5   With the shorthand notation

$$\langle l,m \rangle \equiv \langle \hat{\pi}_l^\dagger(t)\hat{\pi}_m^\dagger(t+\tau)\hat{\pi}_m(t+\tau)\hat{\pi}_l(t) \rangle, \tag{8.5.12}$$

verify that the results given in eqns (8.5.7) and (8.5.8) are sufficient to determine all four of these functions. Show that

$$\langle l,m \rangle \equiv \langle \hat{\pi}_l^\dagger(t)\hat{\pi}_l(t) \rangle \rho_{mm}(\tau) \quad l,m = 2,3. \tag{8.5.13}$$

The four degrees of second-order coherence are therefore

$$g_{3,3}^{(2)}(\tau) = g_{2,3}^{(2)}(\tau) = 1 - \exp\left[-\left(A_{31} + \tfrac{3}{2}B_{31}\langle W_3 \rangle\right)\tau\right] \tag{8.5.14}$$

and

$$
\begin{aligned}
g_{2,2}^{(2)}(\tau) = g_{3,2}^{(2)}(\tau) &= 1 + \frac{B_{31}\langle W_3 \rangle/2}{A_{31} + B_{31}\langle W_3 \rangle}\exp\left[-\left(A_{31} + \tfrac{3}{2}B_{31}\langle W_3 \rangle\right)\tau\right] \\
&\quad - \frac{2A_{31} + 3B_{31}\langle W_3 \rangle}{2\left(A_{31} + B_{31}\langle W_3 \rangle\right)}\exp\left[-\left(A_{21} + 2B_{21}\langle W_2 \rangle + \tfrac{1}{2}B_{31}\langle W_3 \rangle\right)\tau\right],
\end{aligned}
\tag{8.5.15}
$$

where the subscripts on $g^{(2)}(\tau)$ denote light from the strong and weak transitions by labels 2 and 3 respectively. The degrees of coherence are thus identical in pairs and their time dependences are illustrated in Fig. 8.13 for the same parameter values as in Fig. 8.12. Their functional forms are in fact normalized versions of the level populations given in eqns (8.5.7) and (8.5.8). The intrabeam expressions show that the light emitted from both transitions is antibunched and has sub-Poissonian statistics. The properties of the fluorescent light are similar to those found in §7.9 from the rate-equation analysis of emission by a driven two-level atom.

The interbeam degrees of second-order coherence $g_{l,m}^{(2)}(\tau)$ describe measurements of correlations in which a detection of photon m occurs a time τ after a detection of photon l. It is seen from Fig. 8.13 that the correlations are quite different for the two time-orders of detection of the distinct frequencies of photon. Thus, for short time delays τ, the second detected photon is much more likely to originate from the strong transition than from the weak transition, and the normalized correlation, or degree of second-order coherence, is indeed independent of the

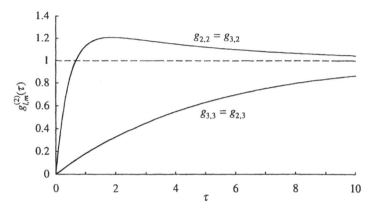

Fig. 8.13. Time dependences of the four degrees of second-order coherence of the two emitted frequencies for the same parameter values as in Fig. 8.12.

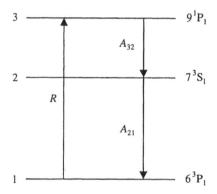

Fig. 8.14. Three-level configuration used for the observation of two-photon cascade emission. The mercury-level designations refer to experiment [21].

nature of the first photon. In addition to the violation of the classical inequalities in eqns (3.7.6) and (3.7.11) by the intrabeam degrees of coherence, the interbeam degree of coherence $g_{3,2}^{(2)}$ violates the inequality in eqn (3.7.27). The light emitted in three-level resonance fluorescence displays the characteristic phenomenon of the quantum jump, with a directness that is difficult to achieve in other forms of experiment, and it also provides light with nonclassical coherence and statistics.

8.6 Two-photon cascade emission

The present section is concerned with a process that is not strictly resonance fluorescence or light scattering but whose theory closely parallels the quantum jump analysis of the preceding section. Figure 8.14 shows the three-level atomic energy-level scheme for two-photon cascade emission. The configuration is the same as that of the three-level laser shown in Fig. 7.3 but the conditions of the experiment are different. Atoms are excited into level 3 at a rate R by electron bombardment, whence they decay to level 2 by emission of a photon of frequency ω_{32} at a spontaneous rate A_{32}. They subsequently decay to level 1 with emission of a second photon of frequency ω_{21} at a spontaneous rate A_{21}. In contrast to the laser, there is no build-up of a significant number of the ω_{32} photons. The individual atoms are weakly excited and the observations are made on a gas of similar atoms to achieve detectable amounts of light at the two frequencies [21]. It is possible to investigate the degrees of intrabeam and interbeam coherence.

The conditions of the light source, with large collision and Doppler broadening, ensure that the time developments of the atomic density-matrix elements are adequately described by rate equations, analogous to eqn (2.10.6) or eqns (8.5.1) to (8.5.3). The excitation of the upper levels is assumed to be sufficiently weak that the population ρ_{11} of the lowest level is very close to unity. The appropriate rate equations for the scheme in Fig. 8.14 are then

$$d\rho_{33}/dt = R - A_{32}\rho_{33}, \tag{8.6.1}$$

$$d\rho_{22}/dt = A_{32}\rho_{33} - A_{21}\rho_{22} \tag{8.6.2}$$

and

$$d\rho_{11}/dt = A_{21}\rho_{22} - R. \tag{8.6.3}$$

The sum of the three diagonal density-matrix elements is conserved by these equations of motion.

The rate equations are relatively simple but, in contrast to the quantum-jump process, calculations of the degrees of second-order coherence need expressions for the diagonal density-matrix elements analogous to eqn (7.9.9) for arbitrary values of $\rho_{22}(t)$. The value of $\rho_{33}(t)$ is also taken as arbitrary for completeness.

Problem 8.6   Obtain the general solutions

$$\rho_{33}(t+\tau) = (R/A_{32})\left[1 - \exp(-A_{32}\tau)\right] + \rho_{33}(t)\exp(-A_{32}\tau) \tag{8.6.4}$$

and

$$\rho_{22}(t+\tau) = \frac{R\left\{A_{21}\left[1 - \exp(-A_{32}\tau)\right] - A_{32}\left[1 - \exp(-A_{21}\tau)\right]\right\}}{A_{21}(A_{21} - A_{32})}$$
$$+ \rho_{22}(t)\exp(-A_{21}\tau) - \rho_{33}(t)\frac{A_{32}\left[\exp(-A_{21}\tau) - \exp(-A_{32}\tau)\right]}{A_{21} - A_{32}}. \tag{8.6.5}$$

These solutions enable the calculation of two-time correlation functions by the method of §7.9.

The pairs of transition operators for the cascade scheme shown in Fig. 8.14 are defined by

$$\hat{\pi}_2^\dagger = |2\rangle\langle 1| \qquad \hat{\pi}_3^\dagger = |3\rangle\langle 2|$$
$$\text{and}$$
$$\hat{\pi}_2 = |1\rangle\langle 2| \qquad \hat{\pi}_3 = |2\rangle\langle 3|, \tag{8.6.6}$$

similar to eqn (8.5.9), and the products of operators for the same transition satisfy relations similar to eqns (4.9.14) to (4.9.18). It should be noted however, in contrast to the corresponding operators for the V configuration, that one ordered product of different creation or different destruction operators does not vanish,

$$\hat{\pi}_3^\dagger \hat{\pi}_2^\dagger = |3\rangle\langle 1| \quad \text{and} \quad \hat{\pi}_2 \hat{\pi}_3 = |1\rangle\langle 3|. \tag{8.6.7}$$

The density-matrix elements are related to expectation values of transition operators by eqn (8.5.10). The four degrees of second-order coherence are obtained from these and the fourth-order expectation values as in eqn (8.5.11). With the shorthand notation of eqn (8.5.12), the required expectation values are straightforwardly calculated as

$$\langle 3,3 \rangle = \left\langle \hat{\pi}_3^\dagger(t)\hat{\pi}_3(t) \right\rangle (R/A_{32})\left[1 - \exp(-A_{32}\tau)\right], \tag{8.6.8}$$

$$\langle 2,2 \rangle = \left\langle \hat{\pi}_2^\dagger(t)\hat{\pi}_2(t) \right\rangle \frac{R\left\{A_{21}\left[1 - \exp(-A_{32}\tau)\right] - A_{32}\left[1 - \exp(-A_{21}\tau)\right]\right\}}{A_{21}(A_{21} - A_{32})}, \tag{8.6.9}$$

$$\langle 3,2 \rangle = \left\langle \hat{\pi}_3^\dagger(t)\hat{\pi}_3(t) \right\rangle$$
$$\times \left\{ \frac{R\left\{A_{21}\left[1 - \exp(-A_{32}\tau)\right] - A_{32}\left[1 - \exp(-A_{21}\tau)\right]\right\}}{A_{21}(A_{21} - A_{32})} + \exp(-A_{21}\tau) \right\} \tag{8.6.10}$$

and

$$\langle 2,3 \rangle = \left\langle \hat{\pi}_2^\dagger(t)\hat{\pi}_2(t) \right\rangle (R/A_{32})\left[1 - \exp(-A_{32}\tau)\right], \tag{8.6.11}$$

where the final term in the large bracket of eqn (8.6.10) is contributed by the expectation value of operators in the order given in eqn (8.6.7).

The degrees of second-order coherence are now easily obtained as

$$g_{3,3}^{(2)}(\tau) = g_{2,3}^{(2)}(\tau) = 1 - \exp(-A_{32}\tau), \tag{8.6.12}$$

$$g_{2,2}^{(2)}(\tau) = \frac{A_{21}\left[1 - \exp(-A_{32}\tau)\right] - A_{32}\left[1 - \exp(-A_{21}\tau)\right]}{A_{21} - A_{32}} \tag{8.6.13}$$

and

$$g_{3,2}^{(2)}(\tau) = g_{2,2}^{(2)}(\tau) + (A_{21}/R)\exp(-A_{21}\tau), \tag{8.6.14}$$

where the steady-state populations of levels 3 and 2 are set equal to R/A_{32} and R/A_{21} respectively. Note that these upper-level populations must be much smaller than unity in order so satisfy the requirement that ρ_{11} remains very close to unity. The subscripts 3 and 2 on $g^{(2)}(\tau)$ denote the emitted light with frequencies ω_{32} and ω_{21} respectively. The intrabeam degrees of second-order coherence in eqns (8.6.12) and (8.6.13) resemble the expression (7.9.28) for single-photon emission by a driven two-level atom. The light at the two emitted frequencies is again antibunched and it has sub-Poissonian statistics.

The main feature of interest in the two-photon cascade emission is the form of the 3,2 interbeam degree of second-order coherence in eqn (8.6.14). The first term on the right-hand side vanishes at $\tau = 0$ but the second term does not. With τ positive, this degree of coherence describes observations in which photon ω_{21} follows photon ω_{32} in time. The second term is the contribution from pairs of photons produced in a single cascade emission. It can be much larger than the first term, which arises from photons produced in different cascade emissions, if the excitation rate R is sufficiently small. There is, of course, no similar contribution to the 2,3 interbeam degree of coherence in eqn (8.6.12) because, by the nature of a single cascade emission, photon ω_{32} must precede photon ω_{21}.

For comparison with experiment, the single-atom degrees of coherence must be generalized to obtain the properties of the light emitted by a large and fluctuating number of similar atoms. The classical relations between the correlation functions for the light from a single atom and from ν similar atoms are given in eqns (3.4.2) and (3.7.14). The corresponding quantum relations are obtained by the replacement of classical field variables by electric-field operators and the results are identical.

Problem 8.7   Assume that the fluctuations in the number of observed atoms are governed by a Poisson distribution with mean $\bar{\nu}$. Show that the degrees of coherence G for the collection of atoms are related to the single-atom degrees of coherence g by

$$G^{(1)}(\tau) = g^{(1)}(\tau) \tag{8.6.15}$$

and

$$G^{(2)}(\tau) = \left(g^{(2)}(\tau)/\bar{\nu}\right) + 1 + \left|g^{(1)}(\tau)\right|^2. \tag{8.6.16}$$

The many-atom degree of second-order coherence reproduces the relation (3.7.16) when $\bar{\nu}$ is taken to be very large. In the conditions of two-photon cascade emission, the large collision and Doppler broadenings produce a negligibly small degree of first-order coherence for times τ of the order of the radiative lifetimes. It thus follows from eqns (8.6.12) and (8.6.13) that

$$G_{3,3}^{(2)}(\tau) = G_{2,3}^{(2)}(\tau) = G_{2,2}^{(2)}(\tau) = 1 \qquad (8.6.17)$$

for $\bar{v} \gg 1$, and the antibunching property of the single-atom emission is lost when the contributions of many atoms are combined. However, it follows from eqn (8.6.14) that

$$G_{3,2}^{(2)}(\tau) = 1 + \left(A_{21}/R\bar{v}\right)\exp\left(-A_{21}\tau\right), \qquad (8.6.18)$$

and the additional interbeam correlation has a significant magnitude when $R\bar{v}$ is smaller than or comparable to A_{21}.

Figure 8.15 shows experimental results [21] for the square of the interbeam degree of second-order coherence in cascade emission from mercury, using the three energy levels shown in Fig. 8.14. Positive τ corresponds to the detection of photon ω_{21} after photon ω_{32} and negative τ to the reverse order of detection. The solid curves are the theoretical predictions from eqns (8.6.17) and (8.6.18) for the numerical values

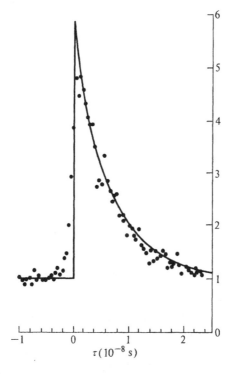

Fig. 8.15. Comparison of experimental points [21] with theoretical results for the square of the interbeam degree of second-order coherence.

$$A_{21} = 1.19 \times 10^8 \, \text{s}^{-1} \quad \text{and} \quad R\bar{v} = 8.38 \times 10^7 \, \text{s}^{-1}. \tag{8.6.19}$$

The radiative lifetime is measured independently but the excitation rate $R\bar{v}$, not available from the experiment, is chosen to give the best fit to the measurements. The agreement between theory and experiment is seen to be very good. The measurements also confirm the unit intrabeam degrees of second-order coherence calculated in eqn (8.6.17).

The physical origin of the interbeam correlation is obvious in the quantum picture of cascade emission, and it is also clear that it must be enhanced relative to the intrabeam correlations at low excitation rates R; the mean interval $1/R\bar{v}$ between successive photons in the same beam is then long but emission of the photon pair, one into each beam, always occurs within a much shorter time interval of order $1/A_{21}$. There is no possibility of a classical description of this correlation effect and the classical inequality (3.7.27) is seriously violated by the experimental results [21].

The two-photon cascade-emission source is an important element in the single-photon Brown–Twiss and Mach–Zehnder interference experiments [22] described in §5.8. The isolation of two-photon pairs for low excitation rates R allows the use of the first photon of the pair to activate an electronic gate, which then has a low probability of admitting any photon other than the second of the same pair into the interferometers. The photon pair states are described by the formalism of §6.7.

Problem 8.8   Consider a single two-photon cascade-emission event that follows the excitation of an atom to level 3 at time $t = 0$. Assume that the emitted light is rendered unidirectional by a suitable arrangement of mirrors and lenses. Find the appropriate two-photon wavepacket amplitude $\beta(t,t')$ in the pair creation operator of eqn (6.7.19). Show that the photon fluxes at the two emitted frequencies are

$$f_3(t) = A_{32} \exp(-A_{32}t) \tag{8.6.20}$$

and

$$f_2(t) = \frac{A_{32}A_{21}}{A_{21} - A_{32}} \left[\exp(-A_{32}t) - \exp(-A_{21}t) \right]. \tag{8.6.21}$$

The single-photon state generated by detection of the other photon of the pair is obtained by the procedure of eqns (6.7.30) to (6.7.32). The experimental points and theoretical curve in Fig. 5.18 confirm the success of the technique for isolating single-photon states when R, and hence $\langle n \rangle$, are sufficiently small.

The degree of second-order coherence for cascade-emission light in eqn (8.6.18) is qualitatively similar to that given in eqn (6.9.25) for the light generated by parametric down-conversion. The two processes generate photon pairs that are relatively well separated from successive pairs when $A_{21} \gg R\bar{v}$ and $\Delta \gg F$ respectively. The nonlinear parametric process has the advantage over cascade emission that the light is generated in well-defined beams, instead of in random directions. Parametric down-conversion is further treated in §9.4.

8.7 The Kramers–Heisenberg formula

The treatment of light scattering so far in the present chapter is concerned with atomic transition schemes that use two, or at most three, energy levels. We now consider a more realistic model of the scattering atoms that allows for an unlimited number of levels. In order to keep the treatment reasonably compact, the collision broadening is assumed negligible and the incident beam is assumed to be weak, with the relevant Rabi frequencies much smaller than the radiative linewidths. The scattering by a two-level atom, derived in §8.3, is extremely simple in this case, with all of the scattered intensity in the elastic Rayleigh component. The treatment that follows extends this result to the multilevel atom. Total strengths of the various components in the scattered light are derived but not their detailed spectral distributions or coherence properties. It is possible to extend the two-level Bloch-equation method to multilevel atoms but the required results are obtained more straightforwardly by transition-rate theory.

All light-scattering processes include the destruction of an incident photon of frequency ω and the creation of a scattered photon of frequency ω_{sc}. They fall into the category of *second-order* radiative transitions, in contrast to the *first-order* radiative transitions represented by the absorption and emission rates calculated in §2.3. Thus, for example, the Einstein B and A coefficients in eqns (2.3.20) and (2.3.21) include the square of the transition dipole moment D_{12}, but the scattering cross-section in eqn (8.2.10) includes the fourth power of the dipole moment. Fermi's golden rule from eqn (2.4.15) is generalized to apply to second-order processes as [2]

$$\frac{1}{\tau} = \frac{2\pi}{\hbar^2} \sum_f \left| \langle f|\hat{\mathcal{H}}_{ED}|i\rangle + \frac{1}{\hbar}\sum_l \frac{\langle f|\hat{\mathcal{H}}_{ED}|l\rangle\langle l|\hat{\mathcal{H}}_{ED}|i\rangle}{\omega_i - \omega_l} \right|^2 \delta(\omega_f - \omega_i). \tag{8.7.1}$$

The first term in the modulus signs reproduces the Fermi golden rule expression and it provides the rate of first-order direct transitions from initial state $|i\rangle$ to final state $|f\rangle$. The second term provides the rate of second-order indirect transitions from state $|i\rangle$ to state $|f\rangle$ via a range of *virtual intermediate states* $|l\rangle$. The delta-function imposes conservation of energy between the initial and final states of the transition but there are no restrictions on the energies $\hbar\omega_l$ of the intermediate states. However, the *energy denominator* $\hbar(\omega_i - \omega_l)$ in the second term reduces the contributions to the transition rate for states whose energies $\hbar\omega_l$ differ greatly from $\hbar\omega_i$. The system is considered to pass through the intermediate states in a virtual sense that does not require energy to be conserved until the final state is reached.

The terms in the modulus signs of eqn (8.7.1) are the first two contributions in a series that extends to higher orders. The subsequent terms contain two or more intermediate states with their corresponding energy denominators. The magnitudes of the terms normally diminish sufficiently rapidly that only the lowest-order contributing term for a given radiative process need be retained. This is the second term for light scattering, where one of the electric-dipole matrix elements is associated with destruction of the incident photon and the other with creation of the scattered photon. Following the energy-level scheme shown in Fig. 8.1, the atom is assumed to be in its ground state $|1\rangle$ at the beginning of the scattering event and it is left in some state $|f\rangle$ at the end of the scattering. The scattering cross-section is independent of the statistical and coherence properties of the incident light. The incident beam is thus simply assumed to contain n photons of frequency ω and wavevector $\mathbf{k}$ initially, with no scattered photons excited. At the end of the scattering there are $n-1$ incident photons and a single scattered photon with frequency ω_{sc} and wavevector $\mathbf{k}_{sc}$. The scattering transition rate is obtained from eqn (8.7.1) as

$$\frac{1}{\tau} = \frac{2\pi}{\hbar^4} \sum_f \sum_{\mathbf{k}_{sc}} \left| \sum_l \frac{\langle n-1,1,f|\hat{\mathcal{H}}_{ED}|l\rangle\langle l|\hat{\mathcal{H}}_{ED}|n,0,1\rangle}{n\omega - \omega_l} \right|^2 \delta(\omega_f + \omega_{sc} - \omega), \quad (8.7.2)$$

where the entries in the bras and kets refer respectively to the states of the incident photons, the scattered photons and the atom. The l summation runs over all states of the combined atom and radiation systems, and the intermediate-state energy $\hbar\omega_l$ here includes the contributions of both systems. The sum over final states in front of the square modulus takes separate account of all possible atomic states $|f\rangle$ and scattered photon modes $\mathbf{k}_{sc}$.

Figure 8.16 shows diagrammatic representations of the two possible contributions to the scattering matrix element, following the conventions of Fig. 4.2. The l summation now refers only to the atomic part of the coupled system and the state of the radiation is accounted for separately. Diagram (a) shows an event in

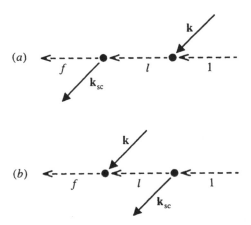

Fig. 8.16. Diagrammatic representations of the two contributions to the light-scattering cross-section.

which an incident photon is destroyed in the excitation of the atom from state $|1\rangle$ to state $|l\rangle$ and a scattered photon is created in the atomic transition from state $|l\rangle$ to state $|f\rangle$; the denominator in eqn (8.7.2) equals $\omega - \omega_l$ for this contribution. Diagram (b) shows an event in which a scattered photon is created in the excitation of the atom from state $|1\rangle$ to state $|l\rangle$ and an incident photon is destroyed in the atomic transition from state $|l\rangle$ to state $|f\rangle$; the denominator in eqn (8.7.2) equals $-\omega_{sc} - \omega_l$ for this contribution. The first step in the latter process uses the non-energy-conserving diagram (d) in Fig. 4.2 but it is possible for the scattering event as a whole to conserve energy. The matrix elements are evaluated with the use of the full form of electric-dipole interaction Hamiltonian given in eqn (4.9.9). The transition rate from eqn (8.7.2) takes the form

$$
\frac{1}{\tau} = \sum_f \sum_{\mathbf{k}_{sc}} \frac{\pi e^4 \omega \omega_{sc} n}{2\varepsilon_0^2 \hbar^2 V^2} \left| \sum_l \left(\frac{(\mathbf{e}_{sc} \cdot \mathbf{D}_{fl})(\mathbf{e} \cdot \mathbf{D}_{l1})}{\omega_l - \omega} + \frac{(\mathbf{e} \cdot \mathbf{D}_{fl})(\mathbf{e}_{sc} \cdot \mathbf{D}_{l1})}{\omega_l + \omega_{sc}} \right) \right|^2
$$
$$
\times \delta(\omega_f + \omega_{sc} - \omega), \tag{8.7.3}
$$

where $\mathbf{e}$ and $\mathbf{e}_{sc}$ are again the unit polarization vectors of the incident and scattered photons. In contrast to eqn (8.7.2), the label l in eqn (8.7.3) refers to the intermediate states of the atom alone.

The scattering rate is best expressed as a cross-section, following the definitions outlined in §8.1. Thus the rate of removal of energy from the incident beam divided by the rate at which energy in the beam passes through a unit area perpendicular to its propagation direction is

$$
\sigma(\omega) = \frac{\hbar\omega/\tau}{cn\hbar\omega/V} = \frac{V}{cn\tau}. \tag{8.7.4}
$$

The summation over scattered wavevectors in eqn (8.7.3) is converted into an integration over frequency and solid angle $d\Omega$ by a development of eqn (4.5.6) as

$$
\sum_{\mathbf{k}_{sc}} \rightarrow \frac{V}{(2\pi)^3} \iint d\omega_{sc} d\Omega \frac{\omega_{sc}^2}{c^3}. \tag{8.7.5}
$$

With this substitution in eqn (8.7.3), the frequency integration is readily performed with use of the delta-function. The further restriction to scattered wave-vector directions that lie within the element of solid angle $d\Omega$ produces the differential light-scattering cross-section as

$$
\frac{d\sigma(\omega)}{d\Omega} = \sum_f^{\omega_f \le \omega} \frac{e^4 \omega (\omega - \omega_f)^3}{16\pi^2 \varepsilon_0^2 \hbar^2 c^4} \left| \sum_l \left(\frac{(\mathbf{e}_{sc} \cdot \mathbf{D}_{fl})(\mathbf{e} \cdot \mathbf{D}_{l1})}{\omega_l - \omega} + \frac{(\mathbf{e} \cdot \mathbf{D}_{fl})(\mathbf{e}_{sc} \cdot \mathbf{D}_{l1})}{\omega_l + \omega_{sc}} \right) \right|^2. \tag{8.7.6}
$$

The summation over f is restricted because the delta-function in eqn (8.7.3) can be satisfied only for atomic final states whose excitation energy is smaller than $\hbar\omega$. The complete differential cross-section is obtained by summation of the contributions of the two independent polarizations for the scattered photon.

Equation (8.7.6) is the Kramers–Heisenberg formula for the differential cross-section. It is the basic expression of quantum-mechanical light-scattering theory. The cross section includes both elastic Rayleigh scattering, corresponding to $f = 1$ and $\omega_f = 0$, and inelastic Raman scattering, corresponding to all the remaining contributions in the summation over f. The radiative broadening of the transitions is not obtained from the second-order transition rate and its effects are inserted 'by hand' in the following sections.

8.8 Elastic Rayleigh scattering

Consider first the elastic-scattering component, where the atom returns to its ground state $|1\rangle$ at the conclusion of the scattering event. The differential cross-section from eqn (8.7.6) reduces to

$$\frac{d\sigma(\omega)}{d\Omega} = \frac{e^4\omega^4}{16\pi^2\varepsilon_0^2\hbar^2 c^4}\left|\sum_l\left(\frac{(\mathbf{e}_{sc}\cdot\mathbf{D}_{1l})(\mathbf{e}\cdot\mathbf{D}_{l1})}{\omega_l-\omega}+\frac{(\mathbf{e}\cdot\mathbf{D}_{1l})(\mathbf{e}_{sc}\cdot\mathbf{D}_{l1})}{\omega_l+\omega}\right)\right|^2 \tag{8.8.1}$$

It is instructive to examine some limiting cases of this general expression before considering its overall form.

Suppose first that ω is very much larger than all of the intermediate-state excitation frequencies ω_l. The frequency denominators in eqn (8.8.1) are expanded according to

$$\frac{1}{\omega_l\pm\omega}=\frac{1}{\omega}\left(\pm1-\frac{\omega_l}{\omega}\pm\ldots\right) \tag{8.8.2}$$

The first terms produce a zero contribution when they are inserted into eqn (8.8.1) and a closure relation similar to that in eqn (4.9.2) is applied to the sum over intermediate states. The contributions of the second terms are written in a compact form by the use of a standard sum rule that can be proved for the energy levels and wavefunctions of any atom [2], given by

$$\sum_l\omega_l(\mathbf{e}_{sc}\cdot\mathbf{D}_{1l})(\mathbf{e}\cdot\mathbf{D}_{l1})=(Z\hbar/2m)\mathbf{e}\cdot\mathbf{e}_{sc}, \tag{8.8.3}$$

where Z is the number of bound electrons in the atom. The cross-section in eqn (8.8.1) is converted by this procedure to

$$d\sigma/d\Omega = Z^2 r_e^2(\mathbf{e}\cdot\mathbf{e}_{sc})^2 \quad \omega\gg\omega_l, \tag{8.8.4}$$

where

$$r_e = e^2/4\pi\varepsilon_0 mc^2 = 2.8 \times 10^{-15} \text{ m} \tag{8.8.5}$$

is called the classical electron radius.

The scattering of high-frequency incident light, with differential cross-section given by eqn (8.8.4), is known as *Thomson scattering*. The cross-section does not depend on the atomic wavefunctions or energy levels and, indeed, the same result is obtained for scattering by free electrons. Note the proportionality to the *square* of the number of electrons, despite the linear dependence of scattered intensity on the number of scatterers proved in eqn (8.1.6). The Z electrons in the same atom have very small separations compared to the wavelength of the light and they therefore radiate with the same phase, in the kind of classical calculation made in Problem 8.1, to give a cross-section that is Z^2 times that for a single-electron atom. The above theory breaks down for incident frequencies so high that the wavelength is comparable to the atomic radius and the electric-dipole approximation becomes invalid. A more comprehensive breakdown occurs when $\hbar\omega$ is comparable to the relativistic rest-mass energy mc^2 of the electron. A significant amount of energy is then transferred to the electron and the scattering process becomes inelastic with scattered frequencies smaller than the incident frequency. This is the regime of *Compton scattering*, not treated here.

Consider now the opposite limit, where ω is very much smaller than all of the intermediate-state excitation frequencies ω_l. The elastic differential cross-section from eqn (8.8.1) becomes

$$\frac{d\sigma(\omega)}{d\Omega} = \frac{e^4\omega^4}{16\pi^2\varepsilon_0^2\hbar^2 c^4}\left|\sum_l \frac{(\mathbf{e}_{sc}\cdot\mathbf{D}_{1l})(\mathbf{e}\cdot\mathbf{D}_{l1}) + (\mathbf{e}\cdot\mathbf{D}_{1l})(\mathbf{e}_{sc}\cdot\mathbf{D}_{l1})}{\omega_l}\right|^2 \quad \omega \ll \omega_l. \tag{8.8.6}$$

The cross-section is thus proportional to the fourth power of the frequency. The Rayleigh scattering of light by atoms whose main absorption frequencies lie in the ultraviolet is thus greatest at the violet end of the visible spectrum and it accounts for the blue of the sky and the red of the sunset.

The summation in eqn (8.8.6) can be evaluated for the hydrogen atom, where only the P intermediate states have nonzero matrix elements. The members of each trio of P states make equal contributions and eqn (8.8.6) reduces to

$$\frac{d\sigma(\omega)}{d\Omega} = \frac{e^4\omega^4}{4\pi^2\varepsilon_0^2\hbar^2 c^4}\left|\sum_l \frac{X_{1l}^2}{\omega_l}\right|^2 (\mathbf{e}\cdot\mathbf{e}_{sc})^2 \quad \omega \ll \omega_l. \tag{8.8.7}$$

The value of the summation is [23]

$$\sum_l \frac{X_{1l}^2}{\omega_l} = \frac{9\hbar}{16m\omega_R^2}, \tag{8.8.8}$$

and with use of the Rydberg frequency from eqn (2.2.15) and the electron radius from eqn (8.8.5), the cross-section simplifies to

$$\frac{d\sigma(\omega)}{d\Omega} = \frac{81 r_e^2}{64} \left(\frac{\omega}{\omega_R}\right)^4 (\mathbf{e}.\mathbf{e}_{sc})^2 \quad \omega \ll \omega_l. \tag{8.8.9}$$

The cross-section also simplifies for resonance scattering with an incident frequency ω that lies very close to one of the atomic excitation frequencies ω_l. The first term in the large bracket of eqn (8.8.1) dominates the second in the vicinity of a resonance and it is a good approximation to neglect all contributions in the first term except the one that corresponds to the resonant intermediate state. An infinite cross-section occurs for exact resonance with $\omega = \omega_l$. The occurrence of the infinity is a consequence of the neglect of radiative damping in the derivation of the general cross-section in eqn (8.7.6). A more complete calculation shows that the damping adds an imaginary part to ω in the denominators of eqn (8.8.1), similar to the imaginary parts in the linear susceptibility of eqn (2.5.16). The spontaneous emission from an excited state of a multilevel atom occurs by transitions to all levels of lower energy, with rates denoted by the Einstein coefficients A_{lm}. The coefficients for nondegenerate levels are determined by a generalization of eqn (2.3.21) or (4.10.10) to

$$A_{lm} \equiv 2\gamma_{lm} = e^2 (\omega_l - \omega_m)^3 D_{lm}^2 / 3\pi\varepsilon_0 \hbar c^3 . \tag{8.8.10}$$

The total linewidth or damping parameter for level l is thus made up from a sum of the partial contributions γ_{lm},

$$\gamma_l = \sum_m^{\omega_m < \omega_l} \gamma_{lm} . \tag{8.8.11}$$

The resonant cross-section obtained from eqn (8.8.1) is thus

$$\frac{d\sigma(\omega)}{d\Omega} = \frac{e^4 \omega_l^4}{16\pi^2 \varepsilon_0^2 \hbar^2 c^4} \frac{(\mathbf{e}_{sc}.\mathbf{D}_{1l})^2 (\mathbf{e}.\mathbf{D}_{l1})^2}{(\omega_l - \omega)^2 + \gamma_l^2} \quad \omega \approx \omega_l. \tag{8.8.12}$$

This agrees with the two-level cross-section in eqn (8.2.10) in the conditions of a weak incident beam and no collision broadening assumed in the present analysis.

The total resonant cross-section is obtained from the differential cross-section by integration over the wavevector directions of the scattered photon and addition of the contributions of its two independent polarizations. The calculation is essentially the same as that in eqn (4.10.8). Most experiments observe the simultaneous scattering by many randomly-oriented atoms and it is convenient to take a further average of the cross-section over atomic orientations. The orientation average is the same as that in eqn (2.3.19). The total averaged cross-section is thus

$$\sigma(\omega) = \frac{e^4 \omega_l^4}{18\pi\varepsilon_0^2\hbar^2 c^4} \frac{D_{1l}^4}{\left(\omega_l - \omega\right)^2 + \gamma_l^2} \quad \omega \approx \omega_l, \tag{8.8.13}$$

and this is multiplied by N to obtain the cross-section for scattering by N atoms.

Consider the resonant cross-section for the first excited state, denoted $|2\rangle$. The summation in eqn (8.8.11) reduces to a single term and the radiative damping constant γ_2 is the same quantity as γ_{sp} defined in eqn (2.5.11) or (4.10.10). Thus eqn (8.8.13) is written

$$\sigma(\omega) = \frac{e^2 \omega_0 D_{12}^2}{3\varepsilon_0\hbar c} \frac{\gamma_{sp}}{\left(\omega_0 - \omega\right)^2 + \gamma_{sp}^2} \quad \omega \approx \omega_0, \tag{8.8.14}$$

where the transition frequency of the first excited state is here denoted ω_0 for comparison of the cross-section with the attenuation coefficient from eqn (2.5.19). These expressions correctly satisfy the optical theorem of eqn (8.1.9).

For exact resonance, the cross-section reduces, with further use of eqn (4.10.10) or (8.8.10), to

$$\sigma(\omega) = 2\pi c^2/\omega^2 = \lambda^2/2\pi \quad \omega = \omega_0, \tag{8.8.15}$$

where λ is the wavelength of the incident light. This strikingly simple result shows that the cross-section for exact resonance with the first excited state of an atom depends only on the transition frequency and it is independent of all other atomic properties. For a resonance in the visible region of the spectrum, $\sigma(\omega)$ given by eqn (8.8.15) is of order 6×10^{-14} m^2. Away from resonance, on the other hand, the earlier discussion shows that the cross-section is typically of order r_e^2 or 8×10^{-30} m^2. The resonant enhancement of the cross-section is thus of enormous proportions. If $|l\rangle$ is a higher excited state of the scattering atom, the cross-section for exact resonance is given by a generalization of eqn (8.8.15) to

$$\sigma(\omega) = \frac{\lambda^2}{2\pi}\left(\frac{\gamma_{l1}}{\gamma_l}\right)^2 \quad \omega = \omega_l, \tag{8.8.16}$$

· where γ_{l1} is a partial contribution to the linewidth, as defined in eqn (8.8.10). The resonant cross-section thus becomes smaller for the more highly excited states.

The complete expression (8.8.1) for the differential cross-section must be used when ω is neither very high, nor very low, nor very close to one of the atomic excitation frequencies. Numerical results for the elastic scattering by hydrogen [24] are shown in Fig. 8.17, where the vertical broken lines indicate the positions of the resonant excitation frequencies. The excitation frequencies crowd together when ω is close to ω_R, and the resonances are shown only up to the $n = 5$ excited state. For $\omega \gg \omega_R$, the scaled differential cross-section tends to the unit value for Thomson scattering, indicated by the horizontal broken line. Inclusion of the radiative damping removes the resonant infinities from the cross-section, as in

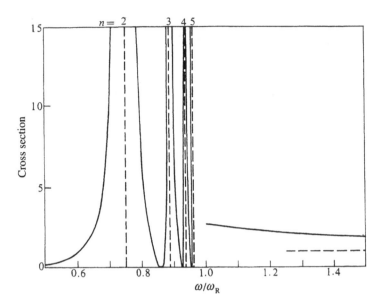

Fig. 8.17. Frequency dependence of the differential cross-section for elastic scattering by atomic hydrogen calculated numerically [24]. The magnitudes shown should be multiplied by $r_e^2 (\mathbf{e} \cdot \mathbf{e}_{sc})^2$.

the expressions given above, and it also removes the zeros between the infinities; these changes are imperceptible on the scale of Fig. 8.17.

Problem 8.9   Prove that the differential cross-section at the $n = 2$ resonance has a peak value of order $10^{14} r_e^2$ when radiative damping is included.

8.9 Inelastic Raman scattering

Incident light beams whose frequencies ω are higher than the lowest atomic excitation frequency are subjected to inelastic Raman-scattering processes in which the final atomic state is not the ground state. The differential cross-section is given by eqn (8.7.6) with the elastic $f = 1$ contribution excluded from the summation over f. The scattered radiation generally has frequency components determined by the delta-function in eqn (8.7.3) as

$$\omega_{sc} = \omega - \omega_f, \tag{8.9.1}$$

with a different frequency ω_{sc} for each distinct energy level f whose frequency ω_f is smaller than ω. The inelastic scattering is illustrated by the 'Raman' arrow in Fig. 8.1.

The cross-section again has simple properties in the limits of high or low incident frequencies.

Problem 8.10 Consider the analogue of the Thompson-scattering regime for inelastic scattering. Prove that the inelastic contributions to the cross-section all tend to zero in the limit of high incident frequencies, $\omega \gg \omega_l$, in contrast to the high-frequency limit of the elastic contribution given by eqn (8.8.4). The proof needs a generalization of the sum rule in eqn (8.8.3) following the method given in [2].

The cross-section is, of course, also zero in the opposite extreme of low incident frequencies, $\omega \ll \omega_l$, because ω must be greater than the smallest atomic excitation frequency for any inelastic scattering to occur.

The most striking feature of the inelastic cross-section is again the occurrence of a resonance when ω is close to one of the atomic transition frequencies ω_l. The resonances arise from the first term in the large bracket of eqn (8.7.6) and it is a good approximation to retain only the resonant term in the summation over l. The total resonant cross-section is obtained in exactly the same way as the purely elastic contribution, by the introduction of the linewidth γ_l defined in eqn (8.8.11). The integration and the orientation average that lead to eqn (8.8.13) then produce a generalization of the cross-section that includes the inelastic scattering,

$$\sigma(\omega) = \sum_f^{\omega_f < \omega_l} \frac{e^4 \omega_l (\omega_l - \omega_f)^3}{18 \pi \varepsilon_0^2 \hbar^2 c^4} \frac{D_{fl}^2 D_{l1}^2}{(\omega_l - \omega)^2 + \gamma_l^2} \qquad \omega \approx \omega_l. \qquad (8.9.2)$$

The elastic cross-section in eqn (8.8.13) is the $f = 1$ term from this expression. It is seen that enhancements of the cross-section occur for all scattered frequencies $\omega_{sc} = \omega_l - \omega_f$ when ω resonates with an atomic excitation frequency ω_l.

The total resonant cross-section can be written as a sum of contributions,

$$\sigma(\omega) = \sum_f \sigma_f(\omega), \qquad (8.9.3)$$

where $\sigma_f(\omega)$ is the cross-section for scattering with emission of light of frequency $\omega - \omega_f$, corresponding to a given atomic final state $|f\rangle$. According to eqn (8.9.2), and with the use of eqn (8.8.10),

$$\sigma_f(\omega) = \frac{2\pi c^2}{\omega_l^2} \frac{\gamma_{l1} \gamma_{lf}}{(\omega_l - \omega)^2 + \gamma_l^2} \qquad \omega \approx \omega_l. \qquad (8.9.4)$$

Thus, when the incident light resonates with the excitation frequency of state $|l\rangle$, the intensity of the scattered light of frequency $\omega - \omega_f$ is proportional to the radiative decay rate from state $|l\rangle$ to state $|f\rangle$. For exact resonance,

$$\sigma_f(\omega) = \frac{2\pi c^2}{\omega_l^2} \frac{\gamma_{l1} \gamma_{lf}}{\gamma_l^2} \qquad \omega = \omega_l; \qquad (8.9.5)$$

the ratio of damping parameters is less than unity, in view of eqn (8.8.11), except when $|l\rangle$ is the first excited state. The inelastic resonant cross-section is therefore smaller than the elastic value in eqn (8.8.15), but it is still usually greatly enhanced by comparison with the nonresonant cross-section.

Summation of the cross-section in eqn (8.9.4) over f to form the total cross-section of eqn (8.9.3) gives

$$\sigma(\omega) = \frac{2\pi c^2}{\omega_l^2} \frac{\gamma_{l1}\gamma_l}{\left(\omega_l - \omega\right)^2 + \gamma_l^2} \quad \omega \approx \omega_l, \tag{8.9.6}$$

with the use of eqn (8.8.11). For exact resonance

$$\sigma(\omega) = \frac{\lambda^2}{2\pi} \frac{\gamma_{l1}}{\gamma_l} \quad \omega = \omega_l, \tag{8.9.7}$$

which again reproduces the elastic value in eqn (8.8.15) when $|l\rangle$ is the first excited state.

Apart from special cases where the expression for the cross-section simplifies, it is generally the case that several states $|l\rangle$ in the intermediate-state summation of eqn (8.7.6) make comparable contributions. The frequency dependence of the cross-section is then determined by a detailed calculation in terms of the atomic wavefunctions and energy eigenvalues. The general appearance of the frequency variation for inelastic scattering is qualitatively similar to that of elastic scattering, shown in Fig. 8.17, but with a low-frequency cut-off at the first atomic excitation frequency and a fall-off to zero at high frequencies. The infinities at the exact resonances that occur in the cross-section of eqn (8.7.6) are again removed by the insertion of radiative damping, as in the resonant cross-section expressions given above.

The optical theorem of eqn (8.1.9) applies generally to the attenuation and scattering of light by a multilevel atom, where both elastic and inelastic processes contribute. It is not possible to demonstrate the theorem completely with results derived here, because we do not have a sufficiently general expression for the attenuation coefficient that includes resonant, nonresonant, and cross terms analogous to those in the square modulus in the cross-section of eqn (8.7.6). Attenuation coefficients are usually measured only close to resonance of the incident light with an atomic transition, the attenuation far from resonance being often too small for detection. On the other hand, it is commonly possible to detect nonresonant scattering (as in a visual inspection of the blue sky!), as the effect produces light with frequencies and directions that would not otherwise be present. Thus, even an elementary theory of scattering needs to include the resonant and nonresonant contributions from the various excited states of a multilevel atom, as in eqn (8.7.6), whereas a useful expression for the attenuation lineshape often need include only the two states involved in the resonant transition, as in eqn (2.5.19).

The above calculations provide only limited information on the spectral distribution of the scattered light. We conclude the section with some qualitative remarks on the spectra produced in resonance scattering of a weak incident beam of coherent light by atoms subjected to radiative and collision broadening. Figure 8.1 shows the kinds of component that occur when ω is close to ω_l. An atom that returns to its ground state $|1\rangle$ at the end of the scattering process produces the two components represented on the left of the figure. The form of the spectrum is expressed in eqn (8.3.6) and illustrated in Fig. 8.3. There is an elastic Rayleigh line of zero width and an inelastic fluorescent line with the same Lorentzian distribution as the transition from state $|1\rangle$ to state $|l\rangle$. The strength of the inelastic component is proportional to the collision-broadening parameter of state $|l\rangle$.

Qualitatively similar components, represented on the right of Fig. 8.1, occur for an atom that is left in an excited state $|f\rangle$ at the end of the scattering event. The Raman component has a frequency given by eqn (8.9.1); however, because the final atomic state is now an excited one, the Raman-scattered light has a Lorentzian distribution around $\omega - \omega_f$ with the linewidth of state $|f\rangle$. There is also a fluorescent component whose strength is proportional to the collision-broadening parameter of state $|l\rangle$ and whose linewidth is a sum of contributions from the two states $|l\rangle$ and $|f\rangle$ involved in the transition. The combined spectrum of these two contributions is similar to that shown in Fig. 8.3 except that the frequencies are reduced by ω_f and both components gain an additional width from the final state $|f\rangle$. The cross-sections derived in the present section refer to the sums of the two contributions integrated over the scattered frequency.

References

[1]   Newton, R.G., Optical theorem and beyond, *Am. J. Phys.* **44**, 639–42 (1976).

[2]   Merzbacher, E., *Quantum Mechanics*, 3rd edn (Wiley, New York, 1998).

[3]   Mollow, B.R., Power spectrum of light scattered by two-level systems, *Phys. Rev.* **188**, 1969–75 (1969).

[4]   Knight, P.L. and Milonni, P.W., The Rabi frequency in optical spectra, *Phys. Reps.* **66**, 21–107 (1980).

[5]   Huber, D.L., Resonant scattering of monochromatic light in gases, *Phys. Rev.* **178**, 93–102 (1969).

[6]   Carlston, J.L., Szöke, A. and Raymer, M.G., Collisional redistribution and saturation of near-resonance scattered light, *Phys. Rev. A* **15**, 1029–45 (1977).

[7]   Carmichael, H.J. and Walls, D.F., A quantum-mechanical master equation treatment of the dynamical Stark effect, *J. Phys. B* **9**, 1199–219 (1976).

[8]   Rice, P.R. and Carmichael, H.J., Nonclassical effects in optical spectra, *J. Opt. Soc. Am. B* **5**, 1661–8 (1988).

[9]   Schuda, F., Stroud, C.R. and Hercher, M., Observation of the resonant Stark effect at optical frequencies, *J. Phys. B* **7**, L198–202 (1974).

[10]   Grove, R.E., Wu, F.Y. and Ezekiel, S., Measurement of the spectrum of resonance fluorescence from a two-level atom in an intense monochromatic field, *Phys. Rev. A* **15**, 227–33 (1977).

[11]   Cohen-Tannoudji, C. and Reynaud, S., Atoms in strong light fields:

photon antibunching in single atom fluorescence, *Phil. Trans. R. Soc. A* **293**, 223–37 (1979).

[12]   Compagno, G., Passante, R. and Persico, F., *Atom–Field Interactions and Dressed Atoms* (Cambridge University Press, Cambridge, 1995).

[13]   Dagenais, M. and Mandel, L., Investigation of two-time correlations in photon emissions from a single atom, *Phys. Rev. A* **18**, 2217–28 (1978).

[14]   Höffges, J.T., Baldauf, H.W., Eichler, T., Helmfrid, S.R. and Walther, H., Heterodyne measurement of the fluorescent radiation of a single trapped ion, *Opt. Comm.* **133**, 170–4 (1997).

[15]   Nagourney, W., Sandberg, J. and Dehmelt, H., Shelved optical electron amplifier: observation of quantum jumps, *Phys. Rev. Lett.* **56**, 2797–9 (1986).

[16]   Sauter, Th., Neuhauser, W., Blatt, R. and Toschek, P.E., Observation of quantum jumps, *Phys. Rev. Lett.* **57**, 1696–8 (1986).

[17]   Dehmelt, H.G., Proposed 10^{14} $\Delta v > v$ laser fluorescence spectroscopy on Tl^+ mono-ion oscillator II, *Bull. Am. Phys. Soc.* **20**, 60 (1975).

[18]   Cook, R.J. and Kimble, H.J., Possibility of direct detection of quantum jumps, *Phys. Rev. Lett.* **54**, 1023–6 (1985).

[19]   Cook, R.J., Quantum jumps, *Prog. Opt.* **28**, 361–416 (1990).

[20]   Pegg, D.T., Loudon, R. and Knight, P.L., Correlations in light emitted by three-level atoms, *Phys. Rev. A* **33**, 4085–91 (1986).

[21]   Clauser, J.F., Experimental distinction between the quantum and classical field-theoretic predictions for the photoelectric effect, *Phys. Rev. D* **9**, 853–60 (1974).

[22]   Grangier, P., Roger, G. and Aspect, A., Experimental evidence for a photon anticorrelation effect on a beam splitter: a new light on single-photon interferences, *Europhys. Lett.* **1**, 173–9 (1986).

[23]   Bethe, H.A. and Salpeter, E.E., *Quantum Mechanics of One- and Two-Electron Atoms* (Springer-Verlag, Berlin, 1957) §61.

[24]   M. Gavrila, Elastic scattering of photons by a hydrogen atom, *Phys. Rev.* **163**, 147–55 (1967).

9 Nonlinear quantum optics

The book so far is almost entirely concerned with linear optical processes, where the response of a material to incident light is described by its linear susceptibility and the rates of optical processes are proportional to the incident intensity. However, examples of nonlinear behaviour occur in the saturation of atomic transitions, discussed in §1.7, and in the power-broadened susceptibility, derived in eqn (2.8.6). Atomic saturation is also an important component in the laser theory of §§1.10 and 7.3. The expressions derived in these examples contain the electromagnetic energy density or photon number in denominators whose expansions include terms proportional to all positive powers of the optical energy. The higher-order, or nonlinear, terms are normally much smaller than the linear term for conventional light sources but their influence is often significant in experiments that use laser sources. The present chapter treats the effects of the nonlinear terms in modifying the properties of incident light beams.

The effects of linear absorption and emission in changing the initial statistical properties of light beams are treated in §§7.2, 7.5 and 7.6. It is there shown that any nonclassical features in the incident light tend to disappear with increasing time or propagation distance. By contrast, some nonlinear processes have the ability to change the initial degrees of coherence so as to introduce quantum-optical features into light that previously had entirely classical properties. Further, they have the additional attribute that the rates of the nonlinear processes themselves generally depend on the degrees of coherence of the participating beams. The rates thus change with time as the nonlinear process itself changes the initial coherence and photon statistics of the light.

Nonlinear optics embraces a wide variety of phenomena and it forms an extensive field of study. Its theoretical treatment was initiated by Bloembergen in the early 1960s (his classic monograph is reprinted [1]). Nonlinear optics is also covered in more recent textbooks and monographs [2–5]. The aim of the present chapter is to describe the principles and outline the theories of a selection of some of the lower-order nonlinear optical processes that are particularly important in the practical realization of nonclassical light. The emphasis is thus on quantum effects in nonlinear optics, which may be treated by discrete or continuous-mode theory. It is not possible to include a comprehensive account of the theoretical background of semiclassical nonlinear optics in the available space and the reader is assumed to be familiar with its main aspects, as presented in the above references.

9.1 The nonlinear susceptibility

The linear frequency-dependent susceptibility of a medium, for example the expression derived for an atomic gas in §2.5, provides a complete description of the linear propagation of electromagnetic waves. In a similar way, the propagation of electromagnetic waves through a medium in which there are significant non-linear processes is described by its *nonlinear susceptibility*. The nonlinear suscep-tibility is calculated by retaining the terms of higher order in the electric field of the radiation that are neglected in the derivation of the linear susceptibility. Thus the expressions (2.5.12) and (2.5.13) for the C coefficients are just the first terms in power series in the atom–radiation interaction. The series for $C_1(t)$ contains all even powers of the electric field (or equivalently $\mathcal{V}$), while that for $C_2(t)$ contains all odd powers. The polarization formed from eqns (2.5.5) and (2.5.15) thus contains terms proportional to all odd powers of the electric field, and the semiclassical relation between polarization and electric field is generalized to the schematic form

$$P = \varepsilon_0\left(\chi^{(1)}E + \chi^{(3)}E^3 + \chi^{(5)}E^5 + \ldots\right). \tag{9.1.1}$$

Here $\chi^{(1)}$ is the same as the linear susceptibility $\chi(\omega)$ derived in §2.5 while $\chi^{(3)}$, $\chi^{(5)}$, and so on, are components of the nonlinear susceptibility. These compo-nents are functions of the frequencies and polarization vectors of the electric fields that multiply them.

The treatment of the susceptibility in §2.5, on which the above remarks are based, applies to a two-level atom in free space. The system is invariant under the operation of spatial inversion in the atomic nucleus and the atomic wavefunctions have the well-defined parities that are used, for example, in the discussion of the interaction Hamiltonian in §2.2. Such a system is said to be *centrosymmetric*. The inversion operation changes the signs of both P and E, and it is clear that, with no even powers of E on the right-hand side, eqn (9.1.1) is correctly invariant under spatial inversion. The form of this expression remains valid when the treat-ment is extended to multilevel atoms, as must be done in the derivation of realistic expressions for the nonlinear contributions to the susceptibility. The multilevel generalization of the two-level form of the linear susceptibility from eqn (2.5.16), without the average over atomic orientations, is

$$\chi^{(1)}(\omega;\omega) = \frac{Ne^2}{\varepsilon_0\hbar V}\sum_l (e.\mathbf{D}_{1l})^2\left\{\frac{1}{\omega_l - \omega - i\gamma_l} + \frac{1}{\omega_l + \omega + i\gamma_l}\right\}, \tag{9.1.2}$$

where the damping parameter γ_l is defined in eqns (8.8.10) and (8.8.11). The modified notation for $\chi^{(1)}$ is consistent with that of the nonlinear susceptibilities treated below. It indicates that an incident light beam with a field E of frequency ω induces an oscillating linear polarization at the same frequency ω. Figure 9.1 shows diagrammatic representations of the two contributions to the linear suscepti-bility, following the conventions of Fig. 4.2.

Many materials important in nonlinear quantum optics are not invariant under spatial inversion, for example, anisotropic molecules and many crystal structures. Such systems are said to be *noncentrosymmetric*. The system wavefunctions in these cases do not have well-defined parities and transition schemes are possible in

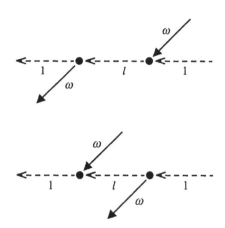

Fig. 9.1. Diagrammatic representations of the pairs of electric-dipole interactions in the two contributions to the linear susceptibility.

which odd numbers of electric-dipole interactions restore the system to its initial ground state $|1\rangle$. The relation (9.1.1) between polarization and field is then generalized by the addition of even powers of the field,

$$P = P_L + P_{NL} = \varepsilon_0 \chi^{(1)} E + \varepsilon_0 \left(\chi^{(2)} E^2 + \chi^{(3)} E^3 + ... \right). \tag{9.1.3}$$

The linear polarization P_L is the same as before and the linear susceptibility retains the explicit form of frequency dependence shown in eqn (9.1.2). The nonlinear polarization P_{NL} now begins with the term in E^2, which is the first even contribution in a series that contains all positive integer powers of E. Laser sources are available in practice for which successive terms in the series have comparable magnitudes, but such high-power light beams tend to damage the nonlinear medium and they are not needed for the production of useful and interesting nonclassical light. The strengths of the beams used in the experiments to be considered here are sufficiently low that only the terms shown explicitly in eqn (9.1.3) need be retained. Our treatment is accordingly restricted to the nonlinear susceptibility components $\chi^{(2)}$ and $\chi^{(3)}$.

The effect of a nonzero linear susceptibility is a modification from the free-space propagation characteristics of light caused by the presence of a medium, as described in earlier chapters. The behaviour of the nonlinear terms is more complicated. Consider the second-order term; if the electric field of the incident light contains two distinct frequency components,

$$E(t) = E(\omega_1) \exp(-i\omega_1 t) + E(\omega_2) \exp(-i\omega_2 t) + \text{comp. conj.}, \tag{9.1.4}$$

then E^2 contains components having the 5 distinct frequencies

$$\omega_1 + \omega_2 \qquad \omega_1 - \omega_2$$

$$2\omega_1 \qquad 2\omega_2$$

$$0. \tag{9.1.5}$$

According to eqn (9.1.3), the field $E(t)$ induces components of the polarization at all 5 frequencies. The oscillating polarization in turn acts as a source of electromagnetic waves at these new frequencies. The frequencies in the first line are the sum and difference of the incident frequencies, those in the second line are their second harmonics, and the third line represents a static component. The corresponding nonlinear optical processes associated with $\chi^{(2)}$ are thus sum and difference frequency generation, second-harmonic generation, and optical rectification. The magnitude of the nonlinear susceptibility depends on the frequencies of the electric-field components that it multiplies and it has different forms for the different $\chi^{(2)}$ processes. In addition, the nonlinear susceptibility depends, via its electric-dipole matrix elements, on the polarization vectors of the field components and also on the direction of the induced polarization. Thus $\chi^{(2)}$ is a function of two frequencies and three spatial directions.

The general quantum-mechanical forms of the various nonlinear components of the susceptibility are given by quite cumbersome expressions that can be found in books on nonlinear optics [1-4]. Thus, for example, the form of $\chi^{(2)}$ appropriate to generation of the sum frequency $\omega_1 + \omega_2$ in eqn (9.1.5) is

$$\chi^{(2)}(\omega_1 + \omega_2; \omega_1, \omega_2) = \frac{Ne^3}{2\varepsilon_0 \hbar^2 V} \sum_{l,m} \left\{ \frac{(\mathbf{e}.\mathbf{D}_{1l})(\mathbf{e}_2.\mathbf{D}_{lm})(\mathbf{e}_1.\mathbf{D}_{m1})}{(\omega_l - \omega_2 - \omega_1)(\omega_m - \omega_1)} \right.$$

$$\left. + \frac{(\mathbf{e}_2.\mathbf{D}_{1l})(\mathbf{e}.\mathbf{D}_{lm})(\mathbf{e}_1.\mathbf{D}_{m1})}{(\omega_l + \omega_2)(\omega_m - \omega_1)} + 4 \text{ more terms} \right\}, \tag{9.1.6}$$

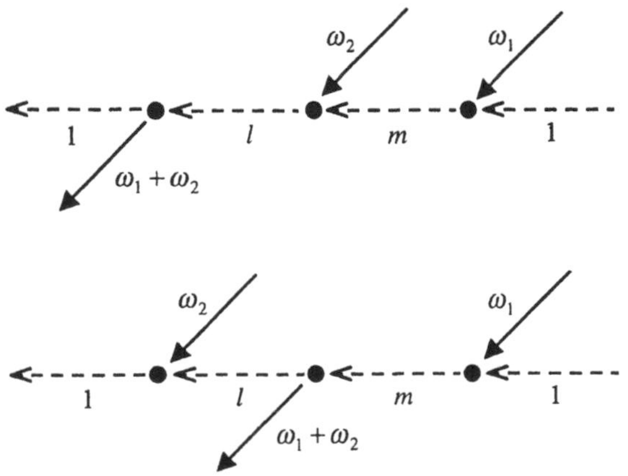

Fig. 9.2. Diagrammatic representations of the electric-dipole interactions in two of the six contributions to the nonlinear susceptibility $\chi^{(2)}$ for sum-frequency generation.

where $\mathbf{e}$, $\mathbf{e}_1$, and $\mathbf{e}_2$ are unit vectors in the directions of the induced polarization and the electric fields of the incident beams of frequencies ω_1 and ω_2 respectively. The notation for $\chi^{(2)}$ shows the frequencies of the induced polarization and of the incident beams that combine to produce it. The nonlinear susceptibility is generally complex, with damping parameters γ_l and γ_m in the denominators similar to those in the linear susceptibility of eqn (9.1.2). Damping is important in removing the resonant infinities in $\chi^{(2)}$ but it is here omitted for simplicity. The diagrams in Fig. 9.2 represent the two terms shown explicitly in eqn (9.1.6).

The third-order term applies in general to field contributions with three distinct frequencies in the incident light, similar to eqn (9.1.4) but with an additional term in ω_3, so that E^3 contains the 22 distinct frequency components

$$
\begin{array}{llll}
\omega_1 + \omega_2 + \omega_3 & \omega_1 + \omega_2 - \omega_3 & \omega_1 - \omega_2 + \omega_3 & -\omega_1 + \omega_2 + \omega_3 \\
2\omega_1 \pm \omega_2 & 2\omega_1 \pm \omega_3 & & \\
2\omega_2 \pm \omega_1 & 2\omega_2 \pm \omega_3 & & \\
2\omega_3 \pm \omega_1 & 2\omega_3 \pm \omega_2 & & \\
3\omega_1 & 3\omega_2 & 3\omega_3 & \\
\omega_1 & \omega_2 & \omega_3. & \quad (9.1.7)
\end{array}
$$

Here the first four lines contain the sums and differences of three or two of the frequencies in the incident light, the fifth line contains their third-harmonic frequencies, and the final line contains the incident frequencies. The $\chi^{(3)}$ nonlinearity thus embraces the generation of a wealth of new frequencies from those in the incident light. It is not intended to consider all of these in detail, but the terms in ω_1 are of particular interest here as they represent the processes of self-phase modulation and two-photon absorption, both of which are important in the production of nonclassical light. The explicit form of $\chi^{(3)}$ is not shown here but it follows the pattern established by the expressions for $\chi^{(1)}$ and $\chi^{(2)}$; in general, $\chi^{(i)}$ is given by an expression similar to eqns (9.1.2) and (9.1.6) but with products of $i+1$ electric-dipole matrix elements in the numerators and products of i frequency factors in the denominators. Comparison of eqns (9.1.5) and (9.1.7) illustrates the rapid escalation in the number of distinct processes with the order of the nonlinearity. The $\chi^{(2)}$ and $\chi^{(3)}$ nonlinearities are fortunately sufficient for discussions of the main nonclassical effects.

Problem 9.1   Compare the magnitudes of the susceptibilities $\chi^{(i)}$ and $\chi^{(i+1)}$ for incident beams that are detuned from the atomic resonances by frequencies in the visible region of the spectrum. Show that a beam intensity of order 10^{18} Wm^{-2} is needed to equalize the ith and $(i+1)$th terms on the right-hand side of eqn (9.1.3). Such intensities are produced only by the highest-power laser sources.

The form of the linear susceptibility of a dielectric medium in eqn (9.1.2) is not changed by the removal of the 1/2 factors between the relations in eqns (2.5.1) and (2.5.2) and those in eqns (9.1.3) and (9.1.4). It is a complex function of ω, which is conveniently expressed by eqns (1.8.4) and (1.8.5) in terms of the refractive index $\eta(\omega)$ and the extinction coefficient $\kappa(\omega)$ as

$$\chi^{(1)}(\omega;\omega) = [\eta(\omega) + i\kappa(\omega)]^2 - 1. \qquad (9.1.8)$$

The optical dispersion governed by $\eta(\omega)$ and the loss governed by $\kappa(\omega)$ are intimately related by the Kramers–Kronig relations [6], such that a dielectric whose loss vanishes at all frequencies must have a constant unit refractive index. However, the losses often occur in frequency ranges that are well removed from those of interest for nonlinear-optical processes and indeed the relevant experiments are usually designed to minimize the losses by appropriate choices of operating frequencies and material. Thus, with the loss neglected, eqn (9.1.8) reduces to

$$\chi^{(1)}(\omega;\omega) = [\eta(\omega)]^2 - 1 \quad \text{for} \quad \kappa(\omega) = 0. \qquad (9.1.9)$$

The calculations that follow are restricted to frequencies for which the loss is negligible, so that the damping parameters can be removed from eqn (9.1.2) and the linear susceptibility is a real function.

The susceptibilities are functions of the frequency and the effects of the nonlinear polarization are accordingly best treated in terms of Fourier-transformed fields. For plane waves propagating parallel to the z axis, the positive-frequency parts of the fields are expressed as

$$E(z,t) = (2\pi)^{-1/2} \int_0^\infty d\omega\, E(z,\omega) \exp[-i\omega t + ik(\omega)z] \qquad (9.1.10)$$

and

$$P(z,t) = (2\pi)^{-1/2} \int_0^\infty d\omega\, P(z,\omega) \exp(-i\omega t), \qquad (9.1.11)$$

where the real wavevector $k(\omega)$ is given by eqn (7.5.18). The *linear* polarization is related to the electric field by

$$P_L(z,t) = \varepsilon_0 (2\pi)^{-1/2} \int_0^\infty d\omega\, \chi^{(1)}(\omega;\omega) E(\omega) \exp[-i\omega t + ik(\omega)z], \qquad (9.1.12)$$

where the field amplitude is independent of z in the absence of nonlinearity. The Fourier component of the linear polarization is obtained by comparison with eqn (9.1.11) as

$$P_L(z,\omega) = \varepsilon_0 \chi^{(1)}(\omega;\omega) E(\omega) \exp[ik(\omega)z]. \qquad (9.1.13)$$

It is shown in textbooks on nonlinear optics [1-5] that the Fourier components of the electric field and the *nonlinear* polarization that generates it are related by

$$\frac{\partial E(z,\omega)}{\partial z} = \frac{i\omega}{2\varepsilon_0 c\eta(\omega)} P_{\text{NL}}(z,\omega)\exp[-ik(\omega)z]. \tag{9.1.14}$$

The relation is valid in the *slowly-varying amplitude approximation*, which holds for almost all nonlinear optical processes. It is used in subsequent sections to determine the electromagnetic waves generated by nonlinear processes

The above discussion is based on a semiclassical theory, with quantum-mechanical susceptibility components but classical fields. The calculations for specific nonlinear processes that follow use a fully quantized theory, as is essential for the derivation of nonclassical properties. However, the formalism of the present section continues to apply, with straightforward conversion of classical fields to quantum field operators. The treatments of individual nonlinear processes are preceded by an account of the electromagnetic field quantization in dielectric media.

9.2 Electromagnetic field quantization in media

The procedure of quantization of the electromagnetic field in free space is treated in §4.4. In the subsequent calculations of the effects of interaction of the field with atomic media, it has been assumed that the materials are sufficiently dilute for the quantized field to retain its free-space form. This is no longer a valid assumption in the dense media used for many experiments in nonlinear optics. We therefore need to consider the modifications of the quantized field operators brought about by the presence of the medium. It is assumed that the optical nonlinearities of the medium do not significantly affect the quantization and only the effects of the linear terms are included in the determination of the field operators. Many of the important nonlinear-optical materials are crystalline, but dielectric anisotropy does not play an essential role in the quantum effects and it is ignored here.

The quantization of the electromagnetic field in a dielectric is most fundamentally carried out by an extension of the methods of Chapter 4 to diagonalize the complete Hamiltonian of the system of electromagnetic field, dense atomic medium, and electric-dipole interaction between the two. The diagonalization can be performed explicitly [7] and it provides expressions for the quantized field operators and for the linear susceptibility of the medium. On a more pheno-menological level, equivalent expressions for the field operators are obtained from a theory in which the linear susceptibility is assumed to have a known form, for example that given in eqn (9.1.2), and the field operators are obtained by solution of the operator Maxwell equations for the medium [8,9]. The electromagnetic field quantization is greatly simplified if the linear susceptibility is again taken to be real, with $\kappa(\omega) = 0$, and the treatment here is restricted to frequencies for which this is a reasonable approximation. The resulting field operators are therefore valid only for frequencies in the transparent regions of the dielectric spectrum.

The light beams that participate in nonlinear optical processses generally propagate in different spatial directions and a full three-dimensional quantization is necessary in principle. However, the beams are often approximately collinear and the quantum effects to be treated here are not significantly affected by the exact configurations of the beams. We therefore consider the modifications of the one-dimensional continuous-mode field operators in eqns (6.2.6) and (6.2.7) brought about by the presence of the dielectric [10]. Propagation parallel to the z axis is assumed, with the electric and magnetic fields oriented parallel to the x and y axes, respectively. The quantization is performed in essentially the same manner as in the free-space theory of §4.4, by requiring the energy operator of the field, or the field Hamiltonian, to take the form of an integral over harmonic-oscillator contributions.

We assume that the electric-field operator has a form similar to the classical expression of eqn (9.1.10), or to the field operator in eqn (6.2.6) but with a wave-vector given by eqn (7.5.18). With the effects of the nonlinearities ignored, the amplitude of the Fourier component of the electric field operator is independent of position and we can write

$$
\hat{E}_T^+(z,t) = (2\pi)^{-1/2} \int_0^\infty d\omega \hat{E}_T^+(\omega) \exp[-i\omega t + ik(\omega)z]
$$

$$
= i \int_0^\infty d\omega f(\omega) \hat{a}(\omega) \exp[-i\omega t + ik(\omega)z],
$$

(9.2.1)

where the normalization function $f(\omega)$ is to be determined. The displacement operator in a dispersive material is obtained from the electric-field operator as

$$
\hat{D}^+(z,t) = i\varepsilon_0 \int_0^\infty d\omega f(\omega) [\eta(\omega)]^2 \hat{a}(\omega) \exp[-i\omega t + ik(\omega)z],
$$

(9.2.2)

where $\hat{D}$ must not be confused with the transition dipole-moment operator $\hat{\mathbf{D}}$ defined in §4.9, whose matrix elements occur in the susceptibility expressions given in §9.1. The form of the magnetic-field operator is obtained from Maxwell's equations as

$$
\hat{B}^+(z,t) = i \int_0^\infty d\omega f(\omega) [\eta(\omega)/c] \hat{a}(\omega) \exp[-i\omega t + ik(\omega)z].
$$

(9.2.3)

Only the parts of the fields that correspond to propagation in the positive z direction are shown in these expressions.

It is convenient to consider the normal-order parts of the field energy-density and Poynting-vector operators, denoted $\hat{W}(z,t)$ and $\hat{I}(z,t)$ respectively. These operators represent the excitation energy of the electromagnetic field in the dielectric above its ground-state, or zero-point, value. It follows from the operator

Maxwell equations that the field energy satisfies the equation of continuity

$$\partial \hat{W}(z,t)/\partial t = -\partial \hat{I}(z,t)/\partial z, \tag{9.2.4}$$

similar to the classical energy-continuity equation [11]. Note that this relation has a different sign from that in eqn (1.8.11), which relates two ways of expressing the loss rate in an attenuating medium. The time derivative of the normally-ordered energy operator in eqn (9.2.4) has the explicit form

$$
\frac{\partial \hat{W}(z,t)}{\partial t} = \hat{E}_T^-(z,t)\frac{\partial \hat{D}^+(z,t)}{\partial t} + \frac{\partial \hat{D}^-(z,t)}{\partial t}\hat{E}_T^+(z,t)
$$
$$
+ \frac{1}{\mu_0}\hat{B}^-(z,t)\frac{\partial \hat{B}^+(z,t)}{\partial t} + \frac{1}{\mu_0}\frac{\partial \hat{B}^-(z,t)}{\partial t}\hat{B}^+(z,t), \tag{9.2.5}
$$

obtained by conversion of the corresponding classical expression [11]. An expression for the energy operator $\hat{W}(z,t)$ itself cannot be obtained by integration of this relation as its right-hand side is not in general a perfect time-derivative in the presence of the dielectric. The Poynting vector operator is

$$\hat{I}(z,t) = \varepsilon_0 c^2 \left\{ \hat{E}_T^-(z,t)\hat{B}^+(z,t) + \hat{B}^-(z,t)\hat{E}_T^+(z,t) \right\}, \tag{9.2.6}$$

with the same form as in eqn (4.11.5) for the assumed orientations of the electric and magnetic vectors.

Despite the difficulty of integration of eqn (9.2.5) in general, the right-hand side *can* be integrated if the expressions (9.2.1) to (9.2.3) for the field operators are substituted first. A straightforward calculation then gives

$$
\hat{W}(z,t) = \varepsilon_0 c \int_0^\infty d\omega \int_0^\infty d\omega' f(\omega) f(\omega') \frac{k(\omega)-k(\omega')}{\omega-\omega'}[\eta(\omega)+\eta(\omega')]
$$
$$
\times \hat{a}^\dagger(\omega)\hat{a}(\omega')\exp\{i(\omega-\omega')t - i[k(\omega)-k(\omega')]z\}, \tag{9.2.7}
$$

where $f(\omega)$ is assumed to be real. The form of this function is found by insisting that the total normally-ordered energy operator in a region of space of area A in the xy plane should equal that of a continuous-mode set of harmonic oscillators,

$$A \int_{-\infty}^\infty dz\, \hat{W}(z,t) = \int_0^\infty d\omega\, \hbar\omega\, \hat{a}^\dagger(\omega)\hat{a}(\omega). \tag{9.2.8}$$

This expression is the same as the integral on the right-hand side of eqn (6.2.8).

Problem 9.2   Prove that the normalization function is

$$f(\omega) = \left(\hbar\omega/4\pi\varepsilon_0 cA\eta(\omega)\right)^{1/2}, \tag{9.2.9}$$

where the delta-function of eqn (2.4.9) is used and the frequency integration is accomplished by a change of variable from ω to $k(\omega)$.

The Fourier-transform electric-field operator is accordingly

$$\hat{E}_T^+(\omega) = i\left(\frac{\hbar\omega}{2\varepsilon_0 cA\eta(\omega)}\right)^{1/2} \hat{a}(\omega), \qquad (9.2.10)$$

and the time-dependent electric and magnetic field operators are obtained from eqns (9.2.1) and (9.2.3) as

$$\hat{E}_T^+(z,t) = i\int_0^\infty d\omega \left(\frac{\hbar\omega}{4\pi\varepsilon_0 cA\eta(\omega)}\right)^{1/2} \hat{a}(\omega)\exp[-i\omega t + ik(\omega)z] \qquad (9.2.11)$$

and

$$\hat{B}^+(z,t) = i\int_0^\infty d\omega \left(\frac{\hbar\omega\eta(\omega)}{4\pi\varepsilon_0 c^3 A}\right)^{1/2} \hat{a}(\omega)\exp[-i\omega t + ik(\omega)z]. \qquad (9.2.12)$$

It is emphasized that these forms of the field operators are valid only in frequency ranges where negligible attenuation occurs. They reduce to the free-space field operators in eqns (6.2.6) and (6.2.7) when $\eta(\omega)$ is set equal to unity.

The energy of a travelling-wave field is most conveniently expressed in terms of its intensity or Poynting vector, as is discussed in §6.2. The Poynting vector operator obtained by substitution of the quantized fields in eqn (9.2.6) is

$$\hat{I}(z,t) = \frac{\hbar}{4\pi A}\int_0^\infty d\omega \int_0^\infty d\omega' \left(\frac{\omega\omega'}{\eta(\omega)\eta(\omega')}\right)^{1/2} [\eta(\omega) + \eta(\omega')]$$
$$\times \hat{a}^\dagger(\omega)\hat{a}(\omega')\exp\{i(\omega - \omega')t - i[k(\omega) - k(\omega')]z\}, \qquad (9.2.13)$$

which again reduces to its free-space form in eqn (6.2.10) when $\eta(\omega) = 1$. This complicated operator has simple integrated properties. Thus the integral over all time gives the total energy that flows through a plane of constant z as

$$A\int_{-\infty}^\infty dt\hat{I}(z,t) = \int d\omega \hbar\omega \hat{a}^\dagger(\omega)\hat{a}(\omega), \qquad (9.2.14)$$

which equals the total field energy of eqn (9.2.8). This result is unchanged from that in free space, given by eqn (6.2.11), and it retains the physical significance that all of the field energy must, in the fullness of time, pass each point on the z axis. The total energy flow over the entire length of the z axis at a given time t is

$$A \int_{-\infty}^{\infty} dz \hat{I}(z.t) = \int d\omega \hbar \omega v_G(\omega) \hat{a}^\dagger(\omega) \hat{a}(\omega). \qquad (9.2.15)$$

The contribution of each frequency component to the flow is now weighted by the appropriate group velocity $v_G(\omega)$, defined in eqn (7.5.24), instead of the uniform velocity c that appears in the free-space result of eqn (6.2.12).

For a light beam of narrow bandwidth centred on frequency ω_0, is it often a good approximation to take the frequency and refractive-index factors outside the integrals in the above expressions for the field operators. The Poynting vector operator from eqns (9.2.6) and (9.2.13) can then be written in the equivalent forms

$$\hat{I}(z,t) = 2\varepsilon_0 c \eta(\omega_0) \hat{E}_T^-(z,t) \hat{E}_T^+(z,t) = (\hbar\omega_0/A) \hat{a}^\dagger(t_R) \hat{a}(t_R), \qquad (9.2.16)$$

where the time-dependent destruction operator is defined in eqn (6.2.13) and the retarded time is

$$t_R = t - [z/v_G(\omega_0)], \qquad (9.2.17)$$

similar to that in eqn (7.5.26). The final expression in eqn (9.2.16) is a generalization of that in eqn (6.2.19).

The restriction of the field operators in eqns (9.2.11) and (9.2.12) to propagation parallel to the z axis enables them to be written as intregrals over the frequency. More general expressions that include all directions of propagation must be expressed as sums or integrals over the three-dimensional wavevector $\mathbf{k}$, similar to the free-space field operators in eqns (4.4.14) and (4.4.17). The corresponding field operators in dielectric media [10,12] are not considered here.

9.3 Second-harmonic generation

Second-harmonic generation was the first nonlinear optical process to be observed experimentally [13] and it forms a convenient starting point to illustrate some of the principal features of nonlinear quantum optics. A single narrow-band incident beam at a fundamental frequency ω_f is assumed and the basic quantum event consists of the simultaneous destruction of two photons of frequency ω_f and the creation of one second-harmonic photon of frequency $2\omega_f$. The process is formally described by the nonlinear susceptibility $\chi^{(2)}$, whose explicit form is given by eqn (9.1.6) with the number of distinct terms reduced to three when both incident frequencies are set equal to ω_f. The resulting nonlinear polarization in eqn (9.1.3) acts as a source for the electromagnetic waves of frequency $2\omega_f$ that form the second-harmonic beam.

In the calculation that follows, the incident beam is assumed to be sufficiently intense that its properties are unchanged by the modest amount of harmonic generation. The symbolic second-order term in the polarization of eqn (9.1.3) is then converted to quantum-mechanical form by the replacement of the classical fields by the field operators from eqn (9.2.10) or (9.2.11). The incident light beam is assumed to have a small spread of frequencies around ω_f in the initial stages of the calculation. The positive-frequency part of the nonlinear polarization is given by

$$\hat{P}_{NL}^+(z,t) = \frac{\varepsilon_0}{2\pi} \int_0^\infty d\omega' \int_0^\infty d\omega'' \chi^{(2)}(\omega' + \omega''; \omega', \omega'') \hat{E}_T^+(\omega') \hat{E}_T^+(\omega'')$$

$$\times \exp\{-i(\omega' + \omega'')t + i[k(\omega') + k(\omega'')]z\}, \tag{9.3.1}$$

where the frequencies ω' and ω'' refer to the incident light, whose narrow-band spectrum is specified by an appropriate choice of input state. Insertion of the form of the field operator from eqn (9.2.10) gives

$$\hat{P}_{NL}^+(z,t) = -\frac{\hbar}{4\pi cA} \int_0^\infty d\omega' \int_0^\infty d\omega'' \left(\frac{\omega'\omega''}{\eta(\omega')\eta(\omega'')} \right)^{1/2} \chi^{(2)}(\omega' + \omega''; \omega', \omega'')$$

$$\times \hat{a}(\omega')\hat{a}(\omega'') \exp\{-i(\omega' + \omega'')t + i[k(\omega') + k(\omega'')]z\}. \tag{9.3.2}$$

The component of the nonlinear polarization at frequency ω is obtained from the Fourier transform of this expression, in accordance with eqn (9.1.11), as

$$\hat{P}_{NL}^+(z,\omega) = -\frac{\hbar}{(2\pi)^{1/2}2cA} \int_0^\infty d\omega' \int_0^\infty d\omega'' \left(\frac{\omega'\omega''}{\eta(\omega')\eta(\omega'')} \right)^{1/2} \chi^{(2)}(\omega' + \omega''; \omega', \omega'')$$

$$\times \hat{a}(\omega')\hat{a}(\omega'') \exp\{i[k(\omega') + k(\omega'')]z\} \delta(\omega - \omega' - \omega''). \tag{9.3.3}$$

The growth of the second-harmonic electric field over a propagation distance L is obtained by integration of the quantized version of eqn (9.1.14). The presence of the nonlinearity restores the position dependence of the Fourier component of the electric-field operator, which is given by

$$\hat{E}_T^+(L,\omega) = \frac{i\omega}{2\varepsilon_0 c\eta(\omega)} \int_0^L dz \, \hat{P}_{NL}^+(z,\omega) \exp[-ik(\omega)z]. \tag{9.3.4}$$

The time-dependent electric-field operator is therefore

$$\hat{E}_T^+(L,t) = \frac{i}{(2\pi)^{1/2}2\varepsilon_0 c} \int_0^\infty d\omega \int_0^L dz \, \frac{\omega}{\eta(\omega)} \hat{P}_{NL}^+(z,\omega) \exp\{-i\omega t + ik(\omega)(L - z)\}, \tag{9.3.5}$$

and the mean value of the narrow-bandwidth second-harmonic intensity at position

L is obtained in accordance with eqn (9.2.16) as

$$\langle \hat{I}_h(L,t) \rangle = 2\varepsilon_0 c \eta(2\omega_f) \langle \hat{E}_T^-(L,t) \hat{E}_T^+(L,t) \rangle, \tag{9.3.6}$$

where the expectation value is taken with respect to the unchanged state of the incident light.

The characteristics of the second-harmonic light are obtained by successive substitutions of eqn (9.3.3) into eqn (9.3.5) and then into eqn (9.3.6). Consider first the spatial dependence of the second-harmonic intensity. With narrow-band incident light, the frequencies ω' and ω'' in eqn (9.3.3) are essentially equal to ω_f and ω is equal to $2\omega_f$. The spatial integration in eqn (9.3.5) thus produces the factor

$$\left| \int_0^L dz \exp(-i\Delta k z) \right|^2 = \frac{\sin^2(\Delta k L/2)}{(\Delta k/2)^2} \tag{9.3.7}$$

in eqn (9.3.6), where

$$\Delta k = k(2\omega_f) - 2k(\omega_f) = (2\omega_f/c)[\eta(2\omega_f) - \eta(\omega_f)] \tag{9.3.8}$$

is known as the *phase mismatch* of the second-harmonic process. In the absence of dispersion, where the refractive indices at frequencies $2\omega_f$ and ω_f are equal, the fundamental and harmonic beams are *perfectly phase-matched* with $\Delta k = 0$, and the spatial factor in eqn (9.3.6) reduces to L^2. In the presence of dispersion, however, the spatial factor oscillates with propagation distance and its maximum value is $(2/\Delta k)^2$. The effect of the phase mismatch is thus to reduce the mean second-harmonic intensity substantially when $\Delta k L \gg 2$. The phase mismatch is assumed to be zero in the remainder of the present derivation.

The general form of the second-harmonic intensity obtained from eqn (9.3.6) is complicated but the main quantum effects survive a quite drastic approximation, in which the integrated incident frequencies are all set equal to ω_f, except in the time-dependent exponent.

Problem 9.3 Prove that the second-harmonic intensity is given approximately by

$$\langle \hat{I}_h(L,t) \rangle = \frac{\hbar^2}{2\varepsilon_0 c^3 A^2} \frac{\omega_f^4 L^2}{[\eta(\omega_f)]^2 \eta(2\omega_f)}$$

$$\times \left| \chi^{(2)}(2\omega_f; \omega_f, \omega_f) \right|^2 \langle \hat{a}^\dagger(t) \hat{a}^\dagger(t) \hat{a}(t) \hat{a}(t) \rangle, \tag{9.3.9}$$

where the time-dependent operators are defined by eqn (6.2.13) for narrow-band incident light.

The main feature of interest for quantum-optical effects is the proportionality of the mean intensity to the second-order correlation function of the incident beam. For stationary incident light at the fundamental frequency ω_f, the correlation is related to the beam flux defined in eqn (6.2.24) and its degree of second-order coherence from eqn (6.2.26), identified by f,f subscripts,

$$\left\langle \hat{a}^\dagger(t)\hat{a}^\dagger(t)\hat{a}(t)\hat{a}(t)\right\rangle = g_{f,f}^{(2)}(0)\left\langle \hat{a}^\dagger(t)\hat{a}(t)\right\rangle^2 = g_{f,f}^{(2)}(0)F^2. \tag{9.3.10}$$

The second-harmonic intensity is therefore proportional to the degree of second-order coherence and to the square of the mean photon flux of the incident fundamental light. It follows, for example, that incident chaotic light is twice as effective in the generation of second-harmonic light as is incident coherent light of the same mean photon flux.

The proportionality of the harmonic intensity to the degree of second-order coherence is readily understood in qualitative terms. The process depends on the simultaneous absorption of two fundamental photons for the creation of each second-harmonic photon. The degree of second-order coherence itself provides a measure of the two-photon absorption rate, as explained in the justification of the quantum-mechanical form of $g^{(2)}(\tau)$ in §4.12. The twofold increase in the generation rate for chaotic light over that for coherent light reflects the greater availability of photon pairs in the photon bunches of the former, as is discussed in §6.5 and illustrated in parts (a) and (b) of Fig. 6.1. By contrast, the second-harmonic generation rate for incident light that is antibunched, as illustrated in part (c) of Fig. 6.1, is diminished below the rate for coherent light. The generation rate falls to zero for incident light whose degree of second-order coherence vanishes; a trivial example is provided by a single-photon wavepacket of the form treated in §6.3, where the absence of a second fundamental photon removes any possibility of second-harmonic generation. Strictly, the rate is determined by the zero time-delay degree of second-order coherence in eqn (9.3.10) and thus by the super-Poissonian, Poissonian or sub-Poissonian statistics of the fundamental. However, the bunching or antibunching is accompanied by super- or sub-Poissonian statistics respectively for the examples of light beam considered here, despite the lack of a general correspondence between the two nonclassical characteristics [14].

The intrabeam degree of second-order coherence of the second harmonic is calculated from the definition in eqn (4.12.9) with the use of the field operator from eqn (9.3.5). An expectation value of the product of four field operators occurs in the numerator of the degree of coherence, while the expectation value from eqn (9.3.6) appears squared in the denominator. The degree of coherence can be calculated in the same approximation as is used in the derivation of the mean intensity in eqn (9.3.9) and, without detailed derivation, it is seen to have the form

$$g_{h,h}^{(2)}(0) = \frac{\left\langle \hat{a}^\dagger(t)\hat{a}^\dagger(t)\hat{a}^\dagger(t)\hat{a}^\dagger(t)\hat{a}(t)\hat{a}(t)\hat{a}(t)\hat{a}(t)\right\rangle}{\left\langle \hat{a}^\dagger(t)\hat{a}^\dagger(t)\hat{a}(t)\hat{a}(t)\right\rangle^2} = \frac{g_{f,f}^{(4)}(0)}{\left[g_{f,f}^{(2)}(0)\right]^2}, \tag{9.3.11}$$

where the expression on the right contains the degrees of second- and fourth-order coherence of the fundamental. It follows from the coherent-state property in eqn (6.4.5) that the second-harmonic degree of second-order coherence equals unity for incident coherent light and the same result follows from the classical degrees of coherence in §3.7. The corresponding value for incident chaotic light is 6.

The above calculations are based on the assumption that the properties of the incident beam are unchanged over the propagation distance L and this is valid for a small fractional conversion of fundamental to second harmonic. The assumption becomes invalid for propagation distances L over which a significant fraction of the fundamental intensity is converted to the second harmonic. The field amplitudes of both fundamental and second harmonic must then be treated as variables, whose coupled motion is described by a pair of equations similar to eqn (9.1.14). The semiclassical solutions of the equations of motion in conditions of perfect phase matching [1–5] show a complete conversion of incident fundamental intensity into second-harmonic intensity. These intensities are proportional respectively to the squares of the hyperbolic secant and tangent of a normalized propagation distance. The L^2 dependence of the second-harmonic intensity in eqn (9.3.9) thus continues to hold at sufficiently short distances but the harmonic intensity approaches a value equal to the incident fundamental intensity at large distances.

There are, however, additional variations in the intensities of the two beams caused by changes in the intrabeam degrees of second-order coherence and in the interbeam correlations that occur with increasing propagation distance. Consider first the effects of the second-harmonic generation on the degree of second-order coherence of the fundamental beam. For the examples shown in Fig. 6.1, the preferential removal of pairs of coincident or nearly-coincident photons from the incident beam tends to convert the patterns of part (a) to part (b) and of part (b) to part (c). The fundamental degree of second-order coherence thus decreases with increasing conversion to second harmonic and there is a corresponding reduction in the conversion rate. The harmonic degree of second-order coherence also falls from its initial value as the expectation values in eqn (9.3.11) are evaluated for fundamental states of increasingly sub-Poissonian and antibunched natures. Finally, the interbeam degree of second-order coherence falls from its initial value of unity, characteristic of uncorrelated beams, as the second-harmonic photons tend to coincide with dark sections of the fundamental beam left by the removal of the pairs of photons that produced them. The conversion of classical to nonclassical light is most readily achieved for incident coherent light, whose statistical properties lie adjacent to the nonclassical ranges of values.

The inclusion of changes in the coherence and correlation properties of the fundamental and harmonic with propagation distance leads to quite complicated equations of motion that are best solved numerically. Figure 9.3 shows some calculated variations of the intrabeam and interbeam degrees of second-order coherence for coherent incident light as functions of a normalized propagation distance. Substantial reductions in all three degrees of coherence are apparent. The theory outlined above applies to travelling-wave second-harmonic generation but similar effects occur with the nonlinear material placed in an optical cavity [15], and this configuration is used for many experimental studies. The production of sub-Poissonian photon statistics is related to the development of amplitude quadrature

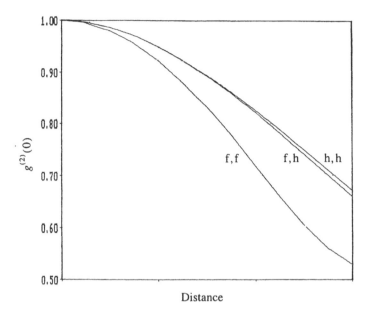

Fig. 9.3. Developments of the intrabeam (f,f and h,h) and interbeam (f,h) degrees of second-order coherence of an intially-coherent fundamental and initially-unexcited second harmonic as functions of a normalized propagation distance. (H.D. Simaan, unpublished)

squeezing, as in the example of amplitude-squeezed coherent light considered in eqn (5.6.19). Much experimental effort is devoted to the study of quadrature squeezing in second-harmonic generation [16] but this is not considered here.

9.4 Parametric down-conversion

The process of parametric down-conversion, or parametric splitting, provides an effective experimental technique for the generation of the single-mode and two-mode squeezed vacuum states described in §6.9. The theory of the process resembles that of second-harmonic generation but now the basic quantum event consists of the destruction of a photon of frequency ω_p from a single narrow-band incident pump beam and the simultaneous creation of two photons with frequencies ω and $\omega_p - \omega$. The two emitted photons are described as forming the *signal* and *idler* beams, respectively, although they occur symmetrically in the theory of the down-conversion process. By convention, the higher-frequency beam is identified as the signal. Both the signal and idler fields are initially in their vacuum states and the down-converted photons are generated by spontaneous emission.

The overall form of the nonlinear susceptibility $\chi^{(2)}$ is given by eqn (9.1.6) with the specializations of the frequencies to the parametric process shown in the following expressions. Figure 9.4 shows diagrammatic representations of two of the six terms. The single-mode quadrature-squeezed vacuum state is generated when the pairs of photons are emitted into the same continuous-mode field; the signal and idler beams are then indistinguishable and the process is known as *degenerate* parametric down-conversion. This process is the inverse of second-harmonic generation. The two-mode quadrature-squeezed vacuum state is generated when the signal and idler beams are distinguishable, for example by their propagation directions or polarizations, and the process is then known as *non-degenerate* parametric down-conversion.

The appropriate component of the nonlinear susceptibility couples the pump and signal fields to produce a nonlinear polarization at the idler frequency, which in turn acts as the source for a field at the idler frequency. As in the above theory of second-harmonic generation, the incident pump beam is assumed to be sufficiently intense that its field strength is almost unchanged by the parametric process, and it can be written in the form of eqn (9.2.10) or (9.2.11). The signal and idler fields, however, are changed significantly and it is essential to retain the z dependences of their amplitudes. The form of the nonlinear polarization, analogous to eqn (9.3.1), is accordingly

$$\hat{P}_{NL}^{+}(z,t) = \frac{\varepsilon_0}{2\pi}\int_0^\infty d\omega' \int_0^\infty d\omega'' \chi^{(2)}(-\omega'+\omega'';-\omega',\omega'')\hat{E}_T^-(z,\omega')\hat{E}_T^+(\omega'')$$
$$\times \exp\{i(\omega'-\omega'')t - i[k(\omega')-k(\omega'')]z\}, \tag{9.4.1}$$

where the frequency ω' refers to the signal and ω'' to the pump. The changes of sign with respect to eqn (9.3.1) reflect the replacement of the destruction operation by the creation operation for the photon of frequency ω'. The component of the nonlinear polarization at the idler frequency $\omega_p - \omega$ is obtained from the Fourier transform of this expression, in accordance with eqn (9.1.11), as

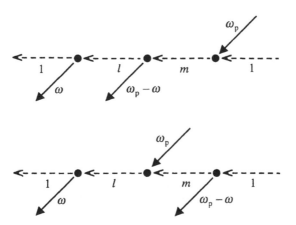

Fig. 9.4. Diagrammatic representations of the electric-dipole interactions in two of the six contributions to the nonlinear susceptibility $\chi^{(2)}$ for parametric down-conversion.

$$\hat{P}_{NL}^{+}(z,\omega_p-\omega) = \frac{\varepsilon_0}{(2\pi)^{1/2}} \int_0^\infty d\omega' \int_0^\infty d\omega'' \chi^{(2)}(-\omega'+\omega'';-\omega',\omega'') \hat{E}_T^-(z,\omega') \hat{E}_T^+(\omega'')$$

$$\times \exp\{-i[k(\omega')-k(\omega'')]z\} \delta(\omega_p-\omega+\omega'-\omega'').$$

$$(9.4.2)$$

The strong pump beam is now assumed to be in a continuous-mode coherent state with an eigenvalue relation of the form of eqn (6.4.4),

$$\hat{a}(\omega'')|\{\alpha_p\}\rangle = \alpha_p(\omega'')|\{\alpha_p\}\rangle. \qquad (9.4.3)$$

The narrow-band pump is taken to have a spectral amplitude with the 'single-mode' form of eqn (6.4.10),

$$\alpha_p(\omega'') = (2\pi F_p)^{1/2} \exp(i\theta_p)\delta(\omega''-\omega_p), \qquad (9.4.4)$$

where F_p is the mean photon flux of the pump beam and θ_p is the phase of the pump field. The nonlinear polarization is replaced by its expectation value for the single-mode pump with the use of eqns (9.2.10) and (9.4.4). The integrand in eqn (9.4.2) now contains two delta-functions and the integrals are easily evaluated to give

$$\hat{P}_{NL}^{+}(z,\omega_p-\omega) = i\left(\frac{\varepsilon_0 \hbar \omega_p F_p}{2cA\eta(\omega_p)}\right)^{1/2} \chi^{(2)}(\omega_p-\omega;-\omega,\omega_p)\hat{E}_T^-(z,\omega)$$

$$\times \exp\{i\theta_p + i[k(\omega_p)-k(\omega)]z\}. \qquad (9.4.5)$$

In physical terms, this relation describes the beating of the pump and signal, with frequencies ω_p and ω, to produce a polarization at the idler frequency $\omega_p-\omega$ by a difference-frequency generation process.

The growth of the electric field at the idler frequency is determined by solution of eqn (9.1.14), with the nonlinear polarization from eqn (9.4.5) substituted on its right-hand side, to give

$$\frac{\partial \hat{E}_T^+(z,\omega_p-\omega)}{\partial z} = -\frac{\omega_p-\omega}{\eta(\omega_p-\omega)} \left(\frac{\hbar \omega_p F_p}{8\varepsilon_0 c^3 A\eta(\omega_p)}\right)^{1/2} \chi^{(2)}(\omega_p-\omega;-\omega,\omega_p)$$

$$\times \hat{E}_T^-(z,\omega)\exp\{i\theta_p + i[k(\omega_p)-k(\omega)-k(\omega_p-\omega)]z\}. \qquad (9.4.6)$$

The total field consists of the three contributions of pump, signal and idler, and the pump also beats with the idler to produce a polarization at the signal frequency by another difference-frequency generation process. In view of the symmetrical occurrences of the signal and idler variables in the theory, the spatial development of the signal field is also described by eqn (9.4.6) but with the replacement $\omega \to \omega_p - \omega$. The Hermitian conjugate of the resulting equation is

$$\frac{\partial \hat{E}_T^-(z,\omega)}{\partial z} = -\frac{\omega}{\eta(\omega)} \left(\frac{\hbar \omega_p F_p}{8\varepsilon_0 c^3 A \eta(\omega_p)} \right)^{1/2} \chi^{(2)} \left(\omega; -\omega_p + \omega, \omega_p \right)^*$$

$$\times \hat{E}_T^+ \left(z, \omega_p - \omega \right) \exp \left\{ -i\theta_p - i \left[k(\omega_p) - k(\omega) - k(\omega_p - \omega) \right] z \right\},$$

(9.4.7)

which, together with eqn (9.4.6), provides a pair of coupled equations for the two fields $\hat{E}_T^-(z,\omega)$ and $\hat{E}_T^+\left(z, \omega_p - \omega\right)$.

The position-dependent phase factor in the two equations expresses the mismatch in the phases of pump, signal and idler beams in the down-conversion process. The effects of the phase mismatch are similar to those in second-harmonic generation, discussed in §9.3. It plays no essential role in the quantum effects associated with the nonlinear process and it is assumed henceforth that the beams are perfectly phased-matched with

$$k\left(\omega_p\right) = k(\omega) + k\left(\omega_p - \omega\right).$$

(9.4.8)

The pump, signal and idler beams often propagate in different directions, a feature that greatly facilitates the separation of the two outputs in experiments on non-degenerate parametric down-conversion. A three-dimensional theory is then necessary for the detailed description of experimental results. In terms of the phase matching, such a theory produces a condition similar to that derived here, except that eqn (9.4.8) is replaced by the corresponding vector relation between the wavevectors of the three beams. The one-dimensional theory is retained for the calculations that follow.

The electric-field operators are replaced by continuous-mode photon creation and destruction operators as in eqn (9.2.10) and its Hermitian conjugate, except that the position dependences of the photon operators must now be included. We first assume distinguishable fields for the signal and idler, with different photon creation and destruction operators. Thus, similar to the notation used in §7.5, we put

$$\hat{E}_T^+(z,\omega) = i \left(\frac{\hbar \omega}{2\varepsilon_0 c A \eta(\omega)} \right)^{1/2} \hat{a}_z(\omega)$$

(9.4.9)

for the signal field and

$$\hat{E}_T^+\left(z, \omega_p - \omega\right) = i \left(\frac{\hbar\left(\omega_p - \omega\right)}{2\varepsilon_0 c A \eta\left(\omega_p - \omega\right)} \right)^{1/2} \hat{b}_z\left(\omega_p - \omega\right).$$

(9.4.10)

for the idler field. The second-order nonlinear susceptibilities that occur in eqns (9.4.6) and (9.4.7) are in fact equal [4] and it is convenient to define dimensionless amplitude and phase functions $s(\omega)$ and $\vartheta(\omega)$ by

$$
s(\omega)\exp[i\vartheta(\omega)] = -\left(\frac{\hbar\omega_p\omega(\omega_p-\omega)F_p}{8\varepsilon_0c^3A\eta(\omega_p)\eta(\omega)\eta(\omega_p-\omega)}\right)^{1/2}
$$
$$
\times\chi^{(2)}(\omega_p-\omega;-\omega,\omega_p)\exp(i\theta_p)L,
\tag{9.4.11}
$$

where L is the total propagation distance in the nonlinear material. This expression is unchanged when ω is replaced by $\omega_p-\omega$. The field propagation equations (9.4.6) and (9.4.7) now become

$$
\frac{\partial\hat{b}_z(\omega_p-\omega)}{\partial z} = -\frac{s(\omega)}{L}\exp[i\vartheta(\omega)]\hat{a}_z^\dagger(\omega)
\tag{9.4.12}
$$

and

$$
\frac{\partial\hat{a}_z^\dagger(\omega)}{\partial z} = -\frac{s(\omega)}{L}\exp[-i\vartheta(\omega)]\hat{b}_z(\omega_p-\omega).
\tag{9.4.13}
$$

Problem 9.4   Show that the solutions of eqns (9.4.12) and (9.4.13) are

$$
\hat{a}_L^\dagger(\omega) = \hat{a}_0^\dagger(\omega)\cosh[s(\omega)]
$$
$$
-\hat{b}_0(\omega_p-\omega)\exp[-i\vartheta(\omega)]\sinh[s(\omega)]
\tag{9.4.14}
$$

and

$$
\hat{b}_L(\omega_p-\omega) = \hat{b}_0(\omega_p-\omega)\cosh[s(\omega)]
$$
$$
-\hat{a}_0^\dagger(\omega)\exp[i\vartheta(\omega)]\sinh[s(\omega)].
\tag{9.4.15}
$$

Verify that independent boson commutation relations for the pairs of operators at $z=0$ are preserved for the operators at $z=L$.

These relations, which describe the process of nondegenerate parametric down-conversion, are identical in form to the two-mode squeezing transformations in eqns (6.9.15) and (6.9.16).

The process of degenerate parametric down-conversion is described by the same basic theory, but with the b operators replaced by a operators. Equations (9.4.14) and (9.4.15) then essentially reduce to the single relation

$$
\hat{a}_L(\omega) = \hat{a}_0(\omega)\cosh[s(\omega)] - \hat{a}_0^\dagger(\omega_p-\omega)\exp[i\vartheta(\omega)]\sinh[s(\omega)].
\tag{9.4.16}
$$

This is equivalent to the single-mode squeezing transformation given in eqn (6.9.7) and the validity of the transformation equivalent to that in eqn (6.9.8) is ensured by the invariance of $s(\omega)$ and $\vartheta(\omega)$, defined in eqn (9.4.11), under the replacement $\omega \rightarrow \omega_p - \omega$.

The parametric down-converter is the main experimental tool for the generation of quadrature-squeezed vacuum states, either the single-mode or the two-mode varieties. The states display all of the nonclassical properties derived in §6.9. The main interest in the two-mode squeezed vacuum states produced by nondegenerate down-conversion lies in the relations between the intrabeam and interbeam properties, as described in the later part of §6.9. With the assumption of a vacuum-state field at $z = 0$, the expectation values of single signal and idler operators at $z = L$ vanish. The second-order correlation functions within the signal and idler beams are obtained from eqns (9.4.14) and (9.4.15) as

$$\left\langle \hat{a}_L^\dagger(\omega)\hat{a}_L(\omega') \right\rangle = \left\langle \hat{b}_L^\dagger\!\left(\omega_p - \omega\right)\hat{b}_L\!\left(\omega_p - \omega'\right) \right\rangle = \sinh^2[s(\omega)]\delta(\omega - \omega'). \quad (9.4.17)$$

The beams are stationary, in accordance with eqn (6.2.23), and the mean photon fluxes in the signal and idler beams at $z = L$ follow from eqn (6.2.21) as

$$f_S(t) = f_I(t) = (2\pi)^{-1}\int d\omega \sinh^2[s(\omega)], \quad (9.4.18)$$

where $s(\omega)$ is defined in eqn (9.4.11). The equal beam fluxes thus begin to grow from their initial zero values with a proportionality to the square of the propagation distance, although the validity of the theory given here is limited to values of L such that there is negligible reduction in the intensity of the pump beam. The interbeam photon correlations have the properties evaluated and discussed on the basis of eqns (6.9.21) to (6.9.25). They provide the optical characteristics needed for a wide range of experiments on the fundamental properties of light including, for example, production of single-photon states (§6.7), single-photon interference (§5.8) and two-photon interference (§6.8). The two-photon correlations are also useful in studies of Bell's inequalities [17], related to the Einstein–Podolsky–Rosen paradox [18], and quantum teleportation [19]. More practical applications of the states include their uses in quantum cryptography [20] and absolute optical metrology [21].

The quadrature squeezing itself is the main focus of interest for the single-mode states produced by degenerate parametric down-conversion. The expectation values of the field operators at $z = L$ continue to vanish when the field at $z = 0$ is in its vacuum state. The second-order correlation functions are obtained from eqn (9.4.16) as

$$\left\langle \hat{a}_L^\dagger(\omega)\hat{a}_L(\omega') \right\rangle = \sinh^2[s(\omega)]\delta(\omega - \omega') \quad (9.4.19)$$

and

$$\left\langle \hat{a}_L(\omega)\hat{a}_L(\omega') \right\rangle = -\tfrac{1}{2}\exp[i\vartheta(\omega)]\sinh[2s(\omega)]\delta(\omega + \omega' - \omega_p). \quad (9.4.20)$$

The mean photon flux is given by the same expression as in eqn (9.4.18) and the correlation functions are identical to eqns (6.9.9) and (6.9.11). Measurements on the squeezed vacuum state are made by balanced homodyne detection, whose theory is covered in §6.11. The calculations of the present section refer to the travelling-wave parametric down-converter and they complement the theory of the single-mode squeezed vacuum given in §5.5, which is more appropriate for experiments where the nonlinear material is placed in an optical cavity. Comparison of the results shows that essentially the same nonclassical phenomena occur in the travelling-wave and cavity configurations.

9.5 Parametric amplification

The parametric down-conversion treated in the previous section is essentially a spontaneous generation process, as the signal and idler beams are initially in their vacuum states. The same device acts as an amplifier, or attenuator, when the signal beam incident at $z = 0$ is excited above the vacuum level. We consider first the *nondegenerate* parametric amplifier, where the signal and idler beams are in distinguishable modes, and then the *degenerate* parametric amplifier, where both beams excite the same mode and are therefore indistinguishable.

The solutions for the parametric down-conversion given in eqns (9.4.14) and (9.4.15) and their Hermitian conjugates have the forms of input–output relations for the nondegenerate parametric amplifier. They are analogous to eqns (7.6.5) and (7.6.6) for the inverted-population amplifier. Thus, with the signal beam in an arbitrary state at $z = 0$ but the idler beam in its vacuum state, the mean output amplitude of the signal mode is

$$\langle \hat{a}_L(\omega) \rangle = \cosh[s(\omega)]\langle \hat{a}_0(\omega) \rangle \tag{9.5.1}$$

and the mean output amplitude of the idler mode is

$$\langle \hat{b}_L(\omega_p - \omega) \rangle = -\sinh[s(\omega)]\exp[i\vartheta(\omega)]\langle \hat{a}_0^\dagger(\omega) \rangle. \tag{9.5.2}$$

The corresponding second-order correlation functions within the signal and idler beams are

$$\langle \hat{a}_L^\dagger(\omega)\hat{a}_L(\omega') \rangle = \langle \hat{a}_0^\dagger(\omega)\hat{a}_0(\omega') \rangle \cosh[s(\omega)]\cosh[s(\omega')]$$
$$+ \sinh^2[s(\omega)]\delta(\omega - \omega') \tag{9.5.3}$$

and

$$\langle \hat{b}_L^\dagger(\omega_p - \omega)\hat{b}_L(\omega_p - \omega') \rangle = \langle \hat{a}_0^\dagger(\omega)\hat{a}_0(\omega') \rangle \exp[-i\vartheta(\omega) + i\vartheta(\omega')]$$
$$\times \sinh[s(\omega)]\sinh[s(\omega')] + \sinh^2[s(\omega)]\delta(\omega - \omega'). \tag{9.5.4}$$

The correlation functions reduce to those given in eqn (9.4.17) in the absence of an input signal.

The beam fluxes are obtained from the definition in eqn (6.2.21), except that the wavevector propagation factors from the integrand of eqn (9.2.11) should be inserted. The mean output signal flux is thus given by

$$f_S(t) = (2\pi)^{-1} \int d\omega \int d\omega' \langle \hat{a}_L^\dagger(\omega) \hat{a}_L(\omega') \rangle \exp\{i(\omega - \omega')t - i[k(\omega) - k(\omega')]L\}.$$

$$(9.5.5)$$

It is convenient to define the signal gain as

$$G(\omega) = \cosh^2[s(\omega)], \tag{9.5.6}$$

and this quantity is always greater than unity. The output flux obtained by substitution of eqn (9.5.3) into eqn (9.5.5) is then

$$f_S(t) = (2\pi)^{-1} \int d\omega \int d\omega' \langle \hat{a}_0^\dagger(\omega) \hat{a}_0(\omega') \rangle \sqrt{G(\omega) G(\omega')}$$
$$\times \exp\{i(\omega - \omega')t - i[k(\omega) - k(\omega')]L\} + (2\pi)^{-1} \int d\omega[G(\omega) - 1].$$

$$(9.5.7)$$

This expression is identical to that derived in eqn (7.6.9) for the inverted population amplifier when the latter has the minimum zero value for its noise input flux $f_\mathcal{N}(\omega)$, which corresponds to conditions of complete population inversion. The first term on the right-hand side of eqn (9.5.7) represents the output flux produced by the input signal, while the second term is a time-independent noise flux. The output noise flux has the minimum value allowed by the general theory of linear amplification [22].

The equivalence of the nondegenerate parametric amplifier to the zero-noise-input inverted-population amplifer enables its main properties to be deduced from those derived in §7.6. Thus, for example, the output flux generated by a narrow-band input flux $f_0(t)$ has a form identical to that given in eqn (7.6.11) as

$$f_S(t) = G(\omega_0) f_0(t_R) + (2\pi)^{-1} \int d\omega[G(\omega) - 1], \tag{9.5.8}$$

where the retarded time is defined in eqn (7.5.26). The main features of homodyne detection of the amplified signal are provided by eqns (7.6.14) to (7.6.23) and there is no need to repeat these. For a 'single-mode' coherent-state input signal with a gain of $G(\omega_0) = 4$, the graphical representation of the homodyne electric fields shown in Fig. 7.9 remains valid for the nondegenerate parametric amplifier.

It is apparent from the form of eqn (9.5.4) that the idler flux also acquires an additional contribution in the presence of a nonzero input signal. This contribution is an attenuated or amplified version of the input, depending on the value of the hyperbolic sine function. For experiments that measure either the signal or the idler output alone, the characteristics of the nondegenerate parametric amplifier are

independent of the phase of the input signal and no nonclassical effects are introduced by the amplification process. However, various nonclassical correlation and
squeezing effects occur in measurements that detect both signal and idler fields
[23,24], and some of these are treated in the discussion of two-mode squeezed
states in §6.9.

We now consider the degenerate parametric amplifier, where both photons are
emitted into the same continuous-mode field. The mean output signal amplitude is
obtained from the expectation value of eqn (9.4.16) as

$$\langle \hat{a}_L(\omega) \rangle = \langle \hat{a}_0(\omega) \rangle \cosh[s(\omega)] - \langle \hat{a}_0^\dagger(\omega_p - \omega) \rangle \exp[i\vartheta(\omega)]\sinh[s(\omega)]. \quad (9.5.9)$$

We assume that the frequency spread of the input signal is much narrower than the
amplifier bandwidth. It is instructive to evaluate the output properties for a 'single-
mode' coherent-state input of the form defined in §6.4 with its frequency equal to
one half that of the pump, $\omega_0 = \omega_p/2$. Then

$$\hat{a}_0(\omega)|\{\alpha_0\}\rangle = \alpha_0(\omega)|\{\alpha_0\}\rangle, \quad (9.5.10)$$

where

$$\alpha_0(\omega) = (2\pi F_0)^{1/2} \exp(i\theta_0)\delta(\omega - \tfrac{1}{2}\omega_p), \quad (9.5.11)$$

F_0 is the mean photon flux of the input signal and θ_0 is its phase.

The nature of the output signal from the degenerate parametric amplifier
remains to be determined but we may reasonably assume that its complex amplitude contains a coherent signal component, with a form similar to that in eqn
(9.5.11),

$$\alpha_S(\omega) = (2\pi F_S)^{1/2} \exp(i\theta_S)\delta(\omega - \tfrac{1}{2}\omega_p), \quad (9.5.12)$$

where F_S is the mean photon flux of the coherent output signal and θ_S is its
phase. The relation between input and output parameters obtained with the use of
eqn (9.5.9) is

$$F_S^{1/2} \exp(i\theta_S) = F_0^{1/2}\{\exp(i\theta_0)\cosh s_0 - \exp[-i\theta_0 + i\vartheta_0]\sinh s_0\}, \quad (9.5.13)$$

where s_0 and ϑ_0 are the amplifier parameters defined in eqn (9.4.11), evaluated
at the signal frequency $\omega_0 = \omega_p/2$. The relation is rewritten as

$$F_S^{1/2} \exp(i\theta_S) = F_0^{1/2} \exp(\tfrac{1}{2}i\vartheta_0)$$
$$\times \{i\sin(\theta_0 - \tfrac{1}{2}\vartheta_0)\exp(s_0) + \cos(\theta_0 - \tfrac{1}{2}\vartheta_0)\exp(-s_0)\}. \quad (9.5.14)$$

The amplitude and phase of the output clearly depend on the phase angle θ_0 of the

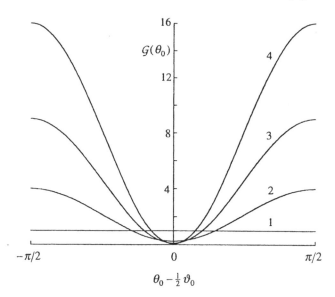

Fig. 9.5. Variation of the gain of the degenerate parametric amplifier with the phase θ_0 of the input signal for the values of $\exp(s_0)$ indicated. (After [25])

input signal. The degenerate parametric amplifier is a *phase-sensitive* device, in contrast to the *phase-insensitive* properties of the inverted population amplifier of §7.6 and the nondegenerate parametric amplifier treated earlier in the present section.

The gain of the amplifier is defined as the ratio of the coherent output flux to the input flux, given by

$$G(\theta_0) = F_S/F_0$$
$$= \sin^2\left(\theta_0 - \tfrac{1}{2}\vartheta_0\right)\exp(2s_0) + \cos^2\left(\theta_0 - \tfrac{1}{2}\vartheta_0\right)\exp(-2s_0). \tag{9.5.15}$$

Figure 9.5 shows the variation of the amplifier gain with the input phase θ_0 for several values of the parametric pumping rate, expressed in terms of $\exp(s_0)$. The gain has a maximum value of

$$G_{max} = \exp(2s_0) \quad \text{for} \quad \theta_0 - \tfrac{1}{2}\vartheta_0 = \pm\pi/2, \tag{9.5.16}$$

for which the coherent output signal flux and phase are

$$F_S = G_{max}F_0 = \exp(2s_0)F_0 \quad \text{and} \quad \theta_S = \theta_0. \tag{9.5.17}$$

The gain is reduced for other input phase angles and it falls to values smaller than unity, corresponding to attenuation rather than amplification, for θ_0 sufficiently close to $\tfrac{1}{2}\vartheta_0$. The gain has a minimum value of

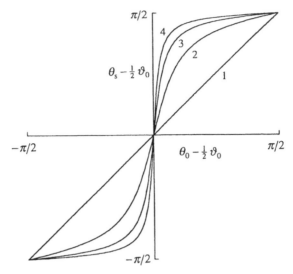

Fig. 9.6. Same as Fig. 9.5 but showing the variation of the phase of the coherent output signal. (After [25])

$$\mathcal{G}_{\min} = \exp(-2s_0) \quad \text{for} \quad \theta_0 - \tfrac{1}{2}\vartheta_0 = 0, \tag{9.5.18}$$

for which the attenuated coherent output flux and the output phase are

$$F_S = \mathcal{G}_{\min} F_0 = \exp(-2s_0)F_0 \quad \text{and} \quad \theta_S = \theta_0. \tag{9.5.19}$$

Figure 9.6 shows the variation of the phase of the coherent output signal with the input phase for the same values of the parametric pumping rate as Fig. 9.5. In conditions of maximum gain, where eqns (9.5.16) and (9.5.17) apply, a large change in input phase produces a relatively small change in the output phase. Conversely, in conditions of maximum attenuation, where eqns (9.5.18) and (9.5.19) apply, a small change in input phase produces a relatively large change in output phase.

The higher-order output correlation functions are similarly obtained by use of eqn (9.4.16). Thus with the state of the input signal defined in eqns (9.5.10) and (9.5.11), the second-order correlation functions are

$$\langle \hat{a}_L^{\dagger}(\omega)\hat{a}_L(\omega')\rangle = \alpha_S^*(\omega)\alpha_S(\omega') + \sinh^2[s(\omega)]\delta(\omega - \omega') \tag{9.5.20}$$

and

$$\langle \hat{a}_L(\omega)\hat{a}_L(\omega')\rangle = \alpha_S(\omega)\alpha_S(\omega') - \tfrac{1}{2}\exp[i\vartheta(\omega)]\sinh[2s(\omega)]\delta(\omega + \omega' - \omega_p). \tag{9.5.21}$$

The first terms on the right-hand sides of these expressions are coherent contributions from the amplified signal, while the second terms, identical to the squeezed vacuum-state results in eqns (9.4.19) and (9.4.20), represent noise components. The total output flux, obtained from eqn (6.2.21) with the use of eqns (9.5.12) and (9.5.20), is

$$f_S(t) = F_S + (2\pi)^{-1}\int d\omega \sinh^2[s(\omega)], \tag{9.5.22}$$

a simple sum of the coherent signal and noise components.

The homodyne detections of the coherent input signal and of the amplified output are treated by the theory outlined in §6.11. The mean and variance of the homodyne field for the 'single-mode' coherent state are given by eqns (6.11.15) and (6.11.16). The coherent input signal and noise are accordingly

$$S_0 = (F_0 T)^{1/2} \cos(\chi - \theta_0) \tag{9.5.23}$$

and

$$\mathcal{N}_0 = \tfrac{1}{4}. \tag{9.5.24}$$

Figure 5.4 shows a pictorial representation of these expressions for the coherent-state electric field and the same representation of the input state is included in Fig. 9.7. The maximum value of the input signal-to-noise ratio,

$$\text{SNR}_0 = 4F_0 T, \tag{9.5.25}$$

is achieved for $\chi = \theta_0$.

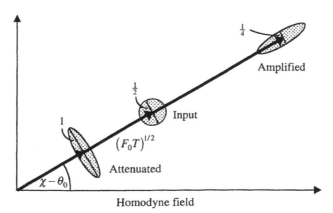

Fig. 9.7. Representations of the means and uncertainties in the homodyne electric fields for an input coherent state and for the output states after propagation through a degenerate parametric amplifier with $\exp(s_0) = 2$ and input phases $\theta_0 = \tfrac{1}{2}\vartheta_0$ (attenuation) and $\tfrac{1}{2}(\vartheta_0 + \pi)$ (amplification). The measurement phase angles χ and χ_R are assumed equal.

The mean homodyne field of the amplifier output state similarly gives a coherent signal

$$S_L = \left(F_S T\right)^{1/2} \cos(\chi_R - \theta_S), \tag{9.5.26}$$

where the retarded detection phase angle χ_R, defined in eqn (7.5.28), takes account of the linear change in signal phase angle on propagation through the amplifier. The optimum value of the output signal-to-noise ratio is thus achieved for $\chi_R = \theta_S$. The homodyne-field variance of the output state is the same as that of the squeezed vacuum state, derived in eqn (6.11.25), and it gives an output noise

$$\mathcal{N}_L = \tfrac{1}{4}\left\{\sin^2\left(\chi_R - \tfrac{1}{2}\vartheta_0\right)\exp(2s_0) + \cos^2\left(\chi_R - \tfrac{1}{2}\vartheta_0\right)\exp(-2s_0)\right\}. \tag{9.5.27}$$

Problem 9.5   Consider the conditions of maximum gain and maximum attenuation for the parametric amplifier and show that the optimum signal-to-noise ratio of the output in both cases satisfies

$$\mathrm{SNR}_L = \mathrm{SNR}_0. \tag{9.5.28}$$

Show more generally that it is not possible for the degenerate parametric amplifier to increase the optimum signal-to-noise ratio of the output above that for the input in eqn (9.5.25) for any choices of the angles θ_0 and χ_R.

The preservation of signal-to-noise ratio that is possible for the phase-sensitive degenerate parametric amplifier contrasts with the inevitable reductions found for the phase-insensitive travelling-wave attenuator and amplifier in eqns (7.5.34) and (7.6.19). Similar reductions in SNR occur for the nondegenerate parametric amplifier treated earlier in the present section. It can be shown that the different noise characteristics of the two classes of device are direct consequences of the different natures of their phase sensitivities [22].

Figure 9.7 includes graphical representations of the output fields of the degenerate parametric amplifier in conditions of maximum attenuation and maximum amplification. It should be compared with the analogous Fig. 7.9, which illustrates the effects of phase-insensitive attenuation and amplification on the same coherent input state. The attenuated output state in Fig. 9.7 has the nature of the amplitude-squeezed coherent state illustrated in Fig. 5.11, while the amplified output state has the nature of the phase-squeezed coherent state of Fig. 5.13. These squeezing effects of the degenerate parametric amplifier are closely related to the corresponding gain and phase variations shown in Figs. 9.5 and 9.6.

The cavity parametric amplifier with a sufficiently weak pump beam has input–output properties very similar to those of the travelling-wave amplifier treated here [25]. However, the system of a gain medium placed in a Fabry–Perot cavity has the same construction as the laser treated in §§1.10 and 7.3. Thus, as the pump intensity is increased, the cavity parametric amplifier displays a lasing threshold. In the regime above threshold, the device provides an output beam in the absence of any input signal, and it transforms into the self-sustaining *optical parametric oscillator*. The properties of the oscillator resemble those of the travelling-wave amplifier only in the below-threshold range of pump intensities.

9.6 Self-phase modulation

The processes of second-harmonic generation, covered in §9.3, and parametric down-conversion, treated in §9.4, are governed by $\chi^{(2)}$ contributions to the nonlinear susceptibility and they occur only in noncentrosymmetric materials. All materials have nonzero $\chi^{(3)}$ components of the susceptibility and the associated nonlinear processes occur universally. As is shown by eqn (9.1.1), they are the lowest order of nonlinear process in centrosymmetric materials. They are important, for example, in optical fibres made of glassy materials, where the generally small values of $\chi^{(3)}$ are compensated by the availability of very long propagation distances, of the order of kilometers. Equation (9.1.7) shows the extensive collection of frequencies that are generated in $\chi^{(3)}$ processes, but the discussions here are restricted to a small selection chosen for their importance in the production of nonclassical light. The theory given below does not include the wide range of interesting phenomena associated with soliton formation in optical pulse propagation through $\chi^{(3)}$ materials [26].

Self-phase modulation occurs in the presence of a single light beam and it corresponds to the frequency ω_1 in eqn (9.1.7), as generated by incident light of the same frequency. More generally, the required nonlinear polarization analogous to eqn (9.3.1) is

$$\hat{P}_{NL}^+(z,t) = \frac{3\varepsilon_0}{(2\pi)^{3/2}} \int_0^\infty d\omega' \int_0^\infty d\omega'' \int_0^\infty d\omega'''$$
$$\times \chi^{(3)}(-\omega' + \omega'' + \omega''';-\omega',\omega'',\omega''')\hat{E}_T^-(z,\omega')\hat{E}_T^+(z,\omega'')\hat{E}_T^+(z,\omega''')$$
$$\times \exp\{i(\omega' - \omega'' - \omega''')t - i[k(\omega') - k(\omega'') - k(\omega''')]z\},$$

$$(9.6.1)$$

where the factor of 3 results from the cubing of the sum of positive- and negative-frequency field operators. The signs of the frequencies in $\chi^{(3)}$ indicate that a photon of frequency ω' is created, while photons of frequencies ω'' and ω''' are destroyed. The incident light is assumed to have a narrow bandwidth centred on frequency ω_0 and it is a good approximation to take the third-order susceptibility out of the integrand. We define the shorthand notation

$$\chi^{(3)}(\omega_0;\omega_0) \equiv \chi^{(3)}(\omega_0;-\omega_0,\omega_0,\omega_0) \qquad (9.6.2)$$

and use of eqn (9.1.10) then converts eqn (9.6.1) to

$$\hat{P}_{NL}^+(z,t) = 3\varepsilon_0\chi^{(3)}(\omega_0;\omega_0)\hat{E}_T^-(z,t)\hat{E}_T^+(z,t)\hat{E}_T^+(z,t). \tag{9.6.3}$$

The sum of linear and nonlinear contributions to the polarization generated by light of frequency ω_0 is

$$\hat{P}^+(z,t) = \varepsilon_0\left\{\chi^{(1)}(\omega_0;\omega_0) + 3\chi^{(3)}(\omega_0;\omega_0)\hat{E}_T^-(z,t)\hat{E}_T^+(z,t)\right\}\hat{E}_T^+(z,t). \tag{9.6.4}$$

The susceptibilities $\chi^{(1)}$ and $\chi^{(3)}$ are real functions of ω_0 for frequencies such that neither linear nor nonlinear attenuation processes occur. The nonlinear term can then be regarded as modifying the linear refractive index, now denoted $\eta_0(\omega_0)$. With use of eqn (9.1.9), the *nonlinear refraction coefficient* $\eta_2(\omega_0)$ is defined by

$$\begin{aligned}
\left[\eta(\omega_0)\right]^2 &= 1 + \chi^{(1)}(\omega_0;\omega_0) + 3\chi^{(3)}(\omega_0;\omega_0)\langle\hat{E}^-\hat{E}^+\rangle \\
&\approx \left[\eta_0(\omega_0) + \eta_2(\omega_0)\langle\hat{E}^-\hat{E}^+\rangle\right]^2,
\end{aligned} \tag{9.6.5}$$

where the expectation value refers to the state of the incident light. Thus for a relatively small nonlinear term, the nonlinear refraction coefficient is given by

$$\eta_2(\omega_0) = 3\chi^{(3)}(\omega_0;\omega_0)/2\eta_0(\omega_0). \tag{9.6.6}$$

The refractive index of an intense light beam is thus modified by a term proportional to its own intensity. The phenomenon is known as the *optical Kerr effect* and it has several important consequences. For example, the transverse distribution of intensity across a laser beam typically has a maximum at the centre and falls off towards the periphery. If $\chi^{(3)}$ is positive, the beam experiences a higher refractive index at its centre than at its edges and the phase velocity at the centre is correspondingly smaller. The result is a self-focusing effect in which the beam converges inwards.

Transverse variations of the field are ignored in the following treatment of self-phase modulation. It is convenient to convert from electric field operators to photon operators with the use of eqn (9.4.9), where the narrow-band assumption allows all the frequencies in the square-root factors to be set equal to ω_0. The Fourier transform of the nonlinear polarization obtained in accordance with the inverse of eqn (9.1.11) is then

$$\begin{aligned}
\hat{P}_{NL}^+(z,\omega) = {}&\frac{3i\varepsilon_0}{2\pi}\left(\frac{\hbar\omega_0}{2\varepsilon_0cA\eta_0(\omega_0)}\right)^{3/2}\chi^{(3)}(\omega_0;\omega_0) \\
&\times \int_0^\infty d\omega'\int_0^\infty d\omega''\int_0^\infty d\omega'''\,\hat{a}_z^\dagger(\omega')\hat{a}_z(\omega'')\hat{a}_z(\omega''') \\
&\times \exp\{-i[k(\omega') - k(\omega'') - k(\omega''')]z\}\delta(\omega + \omega' - \omega'' - \omega''').
\end{aligned} \tag{9.6.7}$$

The propagation equation is obtained from the quantum-mechanical form of the Fourier transform of eqn (9.1.14), with further use of eqn (9.4.9).

Problem 9.6 Prove that the propagation of the beam destruction operator is described by

$$\partial \hat{a}_z(t)/\partial z = i\mu(\omega_0)\hat{a}_z^\dagger(t)\hat{a}_z(t)\hat{a}_z(t),$$ (9.6.8)

where

$$\mu(\omega_0) = \frac{3\hbar\omega_0^2\chi^{(3)}(\omega_0;\omega_0)}{4\varepsilon_0 c^2 A\left[\eta_0(\omega_0)\right]^2}.$$ (9.6.9)

Note that phase matching does not play any role in self-phase modulation, unlike the processes of second-harmonic generation and parametric down-conversion.

It is readily verified that the solution of eqn (9.6.8) for propagation over a distance L is

$$\hat{a}_L(t) = \exp\left\{i\mu(\omega_0)\hat{a}_0^\dagger(t)\hat{a}_0(t)L\right\}\hat{a}_0(t),$$ (9.6.10)

where $\hat{a}_0(t)$ denotes the destruction operator at $z = 0$, the beginning of the non-linear medium. The solution displays the basic nature of the self-phase modulation effect, with the phase of the beam changed by an amount proportional to its own flux. The beam flux operator, as defined in eqn (6.2.20), is unchanged by the process, as

$$\hat{a}_L^\dagger(t)\hat{a}_L(t) = \hat{a}_0^\dagger(t)\hat{a}_0(t).$$ (9.6.11)

The quantity $\mu(\omega_0)L$ represents the nonlinear phase shift per unit photon flux and its magnitude for a typical optical fibre is of order

$$\mu(\omega_0)L \approx 10^{-18} \text{ s} \quad \text{for} \quad L = 1 \text{ km}.$$ (9.6.12)

Consider the self-phase modulation of a 'single-mode' coherent input beam whose wavepacket amplitude from eqn (6.4.11) has the form

$$\alpha_0(t) = F^{1/2}\exp(-i\omega_0 t),$$ (9.6.13)

where F is the input photon flux and the input phase is taken to be zero for simplicity. The homodyne field properties of the input state are represented in Fig. 5.4, as is explained in §6.11, and the representation is reproduced in Fig. 9.8 for a measurement phase angle χ that is also set equal to zero. The destruction operator is depicted according to the prescription

$$\hat{a}_0(t)T^{1/2} \to (FT)^{1/2} + \tfrac{1}{4}e^{i\xi},$$ (9.6.14)

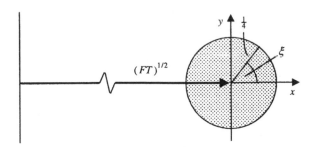

Fig. 9.8. Graphical representation of the homodyne field mean and variance of the input coherent state showing the coordinates used for the noise contour.

where T is the integration time of the detector. The time dependence in $\alpha_0(t)$ is removed by the homodyne detection process and the mean field $(FT)^{1/2}$ is represented by the heavy arrow. The noise circle from the second term in eqn (9.6.14) is specified by

$$\left. \begin{array}{l} x = \tfrac{1}{4}\cos\xi \\ y = \tfrac{1}{4}\sin\xi \end{array} \right\} \quad x^2 + y^2 = \tfrac{1}{16} \tag{9.6.15}$$

in terms of the Cartesian coordinates defined in the figure. For a light beam of intensity $0.1\,\mathrm{W}$ and a detection time T of order $10^{-9}\,\mathrm{s}$,

$$F \approx 3 \times 10^{17}\,\mathrm{s}^{-1} \quad \text{and} \quad (FT)^{1/2} \approx 10^4. \tag{9.6.16}$$

The mean amplitude in eqn (9.6.14) is therefore very much larger than the noise term.

The effects of the self-phase modulation on the homodyne field mean and variance are obtained by a simple calculation based on the graphical representation [27]. Thus replacement of the input operators in eqn (9.6.10) by the prescription in eqn (9.6.14) gives an expression for the output operator as

$$\begin{aligned} \hat{a}_L(t)T^{1/2} &\to \exp\left\{ i\frac{\mu(\omega_0)}{T}\left|(FT)^{1/2} + \tfrac{1}{4}e^{i\xi}\right|^2 L\right\}\left[(FT)^{1/2} + \tfrac{1}{4}e^{i\xi}\right] \\ &\approx \exp\left(i\theta_{\mathrm{SPM}}\right)\left\{(FT)^{1/2} + \tfrac{1}{4}e^{i\xi} + \tfrac{1}{2}i\theta_{\mathrm{SPM}}\cos\xi\right\}, \end{aligned} \tag{9.6.17}$$

where

$$\theta_{\mathrm{SPM}} = \mu(\omega_0)FL \tag{9.6.18}$$

is of order unity or less for the numerical values in eqns (9.6.12) and (9.6.16). Terms of smaller order than this are neglected. It is seen that the mean field vector is rotated by the angle θ_{SPM} but that its magnitude is not changed by the self-phase modulation.

The noise is represented by the second and third terms in the large bracket of eqn (9.6.17). With the coordinates defined in Fig. 9.8, the new noise contour has the form of an ellipse, given by

$$\left. \begin{aligned} x &= \tfrac{1}{4}\cos\xi \\ y &= \tfrac{1}{4}\sin\xi + \tfrac{1}{2}\theta_{SPM}\cos\xi \end{aligned} \right\} \quad x^2 + \left(y - 2\theta_{SPM}x\right)^2 = \tfrac{1}{16}. \tag{9.6.19}$$

The noise circle of the input is distorted into an ellipse because a field fluctuation to a value larger than the mean produces a phase change larger than θ_{SPM}, while a fluctuation to a value smaller than the mean produces a phase change smaller than θ_{SPM}. The form of the noise ellipse is illustrated in Fig. 9.9 for $\theta_{SPM} = 0.5$, together with a selection of the displacements $2\theta_{SPM}x$ that shear the input circle into the output ellipse. The noise ellipse should be shown as attached to the tip of the mean field vector, in the usual way, and it is seen from eqn (9.6.17) that the ellipse as a whole is displaced and rotated through the same angle θ_{SPM} as the mean field. The additional displacement and rotation are omitted from Fig. 9.9 in order to display the shearing effect more clearly. Also, the straight shear lines should strictly be shown as arcs of circles centred on the origin of the mean field vector, in accordance with the conservation of the flux operator in eqn (9.6.11). A curvature of the shear can lead to the replacement of the ellipse by crescent-shaped contours for suitable values of the parameters [28,29]. However, the representation in Fig. 9.9 is a good approximation for the assumed modest value of θ_{SPM} and the very large mean field estimated in eqn (9.6.16).

The occurrence of a noise ellipse suggests that the self-phase modulation converts the input coherent state into a quadrature-squeezed state, and this is indeed the case.

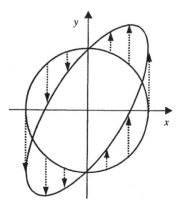

Fig. 9.9. Graphical representation of the self-phase-modulation shearing of the input noise circle into the output elliptical noise contour for $\theta_{SPM} = 0.5$.

Problem 9.7   Prove that the ellipse derived in eqn (9.6.19) has ϑ and s parameters, with their usual significances shown in Fig. 5.7, given by

$$\tan \vartheta = -1/\theta_{SPM}, \tag{9.6.20}$$

$$e^{2s} = 1 + 2\theta_{SPM}^2 + 2\theta_{SPM}\left(1 + \theta_{SPM}^2\right)^{1/2}, \tag{9.6.21}$$

and

$$e^{-2s} = 1 + 2\theta_{SPM}^2 - 2\theta_{SPM}\left(1 + \theta_{SPM}^2\right)^{1/2}, \tag{9.6.22}$$

where ϑ is taken in the fourth quadrant in accordance with Fig. 9.9. It is easily verified that the last two expressions are indeed inverses of each other.

The homodyne field variance is obtained by substitution of these expressions into eqn (6.11.25) and the output state has all of the usual squeezed-state properties. Note, however, that the orientation $\vartheta/2$ of the ellipse has the initial value of $-\pi/4$ for small propagation distances L such that $\theta_{SPM} \ll 1$, and the output is intermediate between the amplitude-squeezed and phase-squeezed coherent states shown in Figs. 5.11 and 5.13 respectively. The output tends to an amplitude-squeezed coherent state for long propagation distances L such that $\theta_{SPM} \gg 1$. The magnitude of the output noise, as measured by the minimum variance $\exp(-2s)/4$, diminishes with increasing propagation distance and it can in principle be reduced to an arbitrarily small value. However, these conclusions for large propagation distances are modified by the accompanying conversion of the elliptical noise contours into crescents, which has the effect of removing the squeezing. The field variance thus has a minimum value at an optimum propagation distance, beyond which it increases again [27]. It should also be mentioned that real optical fibres exhibit substantial attenuation over distances of the order of tens of kilometers. The loss tends to remove nonclassical effects from the light, including the more exotic breakups into superpositions of various numbers of coherent states that are predicted to occur at much larger propagation distances than those considered here [29,30].

The $\pi/4$ quadrature-squeezed state generated in the early stages of self-phase modulation is not as useful for applications as the amplitude-squeezed state. Also, the conservation of the flux operator ensures that the photocount statistics of the light are unchanged from their input values by the self-phase modulation. The state is thus not suitable for observations of reduced noise in either homodyne or direct detection. This situation can be remedied by combination of the output from the self-phase modulation with an additional coherent state, for example by insertion of the nonlinear medium in one arm of a Mach–Zehnder interferometer fed with a coherent-state input [28,29]. The overall effect, for appropriate choices of the parameters, is a rotation of the mean field vector to a $\pi/2$ angle with the major axis of the noise ellipse. The resulting state has both amplitude squeezing and sub-Poissonian photocount statistics.

9.7 Single-beam two-photon absorption

The nonlinear processes covered so far conserve energy within the radiation field and there is no net transfer of energy from field to medium. Some of the other nonlinear effects are associated with a transfer of energy by excitation or deexcitation of the medium. A simple example is provided by single-beam two-photon absorption and emission, as only one light beam is involved. Two-photon absorption resembles single-photon absorption except that the process now excites the medium from its ground state to a higher state by removal of two photons from the beam. Whereas the self-phase modulation of §9.6 relies on the real part of $\chi^{(3)}$, two-photon absorption is controlled by its imaginary part.

Consider first the single-photon attenuation coefficient, which is related to the imaginary part of the linear susceptibility by eqn (2.5.18). Conversion to the notation of the present chapter puts this relation in the form

$$K_0(\omega) = [\omega/c\eta_0(\omega)]\text{Im}\,\chi^{(1)}(\omega;\omega). \tag{9.7.1}$$

The incident light beam is again assumed to have a narrow bandwidth centred on frequency ω_0 and its intensity operator is approximately

$$\hat{I}(z,t) = 2\varepsilon_0 c\eta_0(\omega_0)\hat{E}_T^-(z,t)\hat{E}_T^+(z,t), \tag{9.7.2}$$

similar to eqn (9.2.16). The mean rate of removal of energy from the incident light by the linear attenuation is

$$\begin{aligned} K_0(\omega_0)\langle\hat{I}(z,t)\rangle &= 2\varepsilon_0\omega_0\,\text{Im}\,\chi^{(1)}(\omega_0;\omega_0)\langle\hat{E}_T^-(z,t)\hat{E}_T^+(z,t)\rangle \\ &= 2\omega_0\,\text{Im}\langle\hat{E}_T^-(z,t)\hat{P}_L^+(z,t)\rangle, \end{aligned} \tag{9.7.3}$$

where the form of the linear polarization is taken from eqn (9.6.4). The field expectation values in this and subsequent equations are taken with respect to the state of the incident light.

A similar procedure is used to determine the rate of nonlinear attenuation. Suppose now that the imaginary part of the linear susceptibility $\chi^{(1)}$ at frequency ω_0 is negligible, so that no linear attenuation occurs, but that $\text{Im}\,\chi^{(3)}$ is significant. The expression for the rate of removal of energy from the incident light in the second line of eqn (9.7.3) remains valid when the linear polarization is replaced by the nonlinear polarization from eqn (9.6.4), to give

$$\begin{aligned} 4\omega_0\,&\text{Im}\langle\hat{E}_T^-(z,t)\hat{P}_{\text{NL}}^+(z,t)\rangle \\ &= 12\varepsilon_0\omega_0\,\text{Im}\,\chi^{(3)}(\omega_0;\omega_0)\langle\hat{E}_T^-(z,t)\hat{E}_T^-(z,t)\hat{E}_T^+(z,t)\hat{E}_T^+(z,t)\rangle, \end{aligned} \tag{9.7.4}$$

where an additional factor of 2 occurs because two photons of frequency ω_0 are removed from the beam in each absorption event. The refractive index is assumed to be essentially unchanged from its linear value. The second-order field correlation function is calculated by use of the density operator that describes the state of the incident light.

It is instructive to recalculate the two-photon absorption by second-order transition-rate theory, similar to the calculation of the light-scattering cross-section in §8.7. We use a discrete-mode formalism, as in the theory of single-photon absorption and emission in §§7.1 and 7.2. The photon probability distribution for the initial excitation of the light beam is denoted $P(n)$. The transition rate for an n-photon initial excitation is obtained from eqn (8.7.1) as

$$\frac{1}{\tau} = \frac{2\pi}{\hbar^4} \sum_f \left| \sum_l \frac{\langle n-2, f | \hat{\mathcal{H}}_{\text{ED}} | n-1, l \rangle \langle n-1, l | \hat{\mathcal{H}}_{\text{ED}} | n, 1 \rangle}{\omega_0 - \omega_l} \right|^2 \delta(\omega_f - 2\omega_0). \quad (9.7.5)$$

Here f, l and 1 refer respectively to the final, intermediate and initial atomic states, as shown in the diagrammatic representation of the transition in Fig. 9.10. Only the photon-destructive part of the electric-dipole interaction contributes to the matrix elements shown explicitly in eqn (9.7.5). Thus, with the matrix elements factorized into atomic and radiative parts, similar to the Heisenberg-picture treatment of the photon intensity operator in §4.11, and with the photon probability distribution included, the radiative part contributes a factor

$$\sum_n P(n) \left| \langle n-2 | \hat{E}_T^+(z,t) | n-1 \rangle \langle n-1 | \hat{E}_T^+(z,t) | n \rangle \right|^2$$

$$= \left\langle \hat{E}_T^-(z,t) \hat{E}_T^-(z,t) \hat{E}_T^+(z,t) \hat{E}_T^+(z,t) \right\rangle. \quad (9.7.6)$$

The transition rate from eqn (9.7.5) thus becomes

$$\frac{1}{\tau} = \frac{2\pi e^4}{\hbar^4} \sum_f \left| \sum_l \frac{(\mathbf{e}.\mathbf{D}_{fl})(\mathbf{e}.\mathbf{D}_{l1})}{\omega_0 - \omega_l} \right|^2 \delta(\omega_f - 2\omega_0) \left\langle \hat{E}_T^-(z,t) \hat{E}_T^-(z,t) \hat{E}_T^+(z,t) \hat{E}_T^+(z,t) \right\rangle,$$

$$(9.7.7)$$

where $\mathbf{e}$ is the unit polarization vector of the incident light.

The transition rate shows the same proportionality to the second-order field

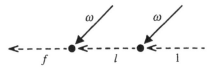

Fig. 9.10. Diagrammatic representation of the electric-dipole interactions in two-photon absorption.

correlation function as the rate of energy removal in eqn (9.7.4) and it can be used to obtain an expression for the appropriate component of the nonlinear susceptibility. Thus, for a medium that consists of a gas of N randomly-oriented atoms in a volume V, the rate of energy removal is given by

$$(2N\hbar\omega_0/V)\overline{(1/\tau)}, \tag{9.7.8}$$

where the overbar denotes an orientation average. Comparison with the expression in eqn (9.7.4) gives

$$\text{Im}\chi^{(3)}(\omega_0;-\omega_0,\omega_0,\omega_0) = \frac{\pi Ne^4}{3\varepsilon_0\hbar^3 V}\sum_f\left|\sum_l\frac{\overline{(\mathbf{e}.\mathbf{D}_{fl})(\mathbf{e}.\mathbf{D}_{l1})}}{\omega_0-\omega_l}\right|^2\delta(\omega_f-2\omega_0), \tag{9.7.9}$$

where the full notation for the susceptibility is restored. This expression for the imaginary part of $\chi^{(3)}$ agrees with the form obtained by direct calculation for the choice of frequencies appropriate to two-photon absorption [1–4].

The transition rates derived above ignore the line-broadening processes for the atomic levels, described in Chapter 2. Homogeneous processes have the effect of adding imaginary parts to the optical frequency and they are easily included in the transition-rate and susceptibility expressions. The delta-function is replaced by a Lorentzian lineshape,

$$\delta(\omega_f-2\omega_0) \to \frac{\gamma_f/\pi}{(\omega_f-2\omega_0)^2+\gamma_f^2}, \tag{9.7.10}$$

where γ_f includes the usual radiative and collisional contributions to the final-state linewidth. In accordance with the discussion in §2.10, the rate-equation treatment of two-photon absorption is valid for narrow-band incident light only when the collisional width is much larger than the radiative width.

The second-order correlation function in the transition rate of eqn (9.7.7) is proportional to the zero time-delay degree of second-order coherence of the incident light,

$$\left\langle \hat{E}_T^-(z,t)\hat{E}_T^-(z,t)\hat{E}_T^+(z,t)\hat{E}_T^+(z,t)\right\rangle = g^{(2)}(0)\left\langle \hat{E}_T^-(z,t)\hat{E}_T^+(z,t)\right\rangle^2, \tag{9.7.11}$$

in accordance with the definition in eqn (4.12.8). The reasons for the proportionality are the same as those in second-harmonic generation, discussed after eqn (9.3.10). Both processes depend on the simultaneous removal of two photons from the same light beam and both are facilitated by the abundance of photon pairs in light with super-Poissonian statistics or inhibited by their rarity in light with sub-Poissonian statistics. The zero time-delay degree of second-order coherence of the incident light provides the appropriate measure of its effectiveness for the two nonlinear processes.

Two-photon absorption also shares with second-harmonic generation the property of smoothing the photon-number fluctuations in the incident light, as described in the last two paragraphs of §9.3. The basic reason is again the same, as the two-photon process preferentially removes pairs of coincident photons and the statistics accordingly tend towards sub-Poissonian form. The development of the two-photon absorption in time must be determined by the use of a transition rate that diminishes with time because of reductions in both the degree of second-order coherence and the squared intensity on the right-hand side of eqn (9.7.11). The main effects are demonstrated in the remainder of the present section by a simple rate-equation treatment of the process for single-mode incident light, similar to the theory for single-photon absorption outlined in §7.1.

For a single-mode excitation with n photons, the second-order correlation function in eqn (9.7.6) is

$$\left\langle \hat{E}_T^-(z,t)\hat{E}_T^-(z,t)\hat{E}_T^+(z,t)\hat{E}_T^+(z,t) \right\rangle = \left[\hbar\omega_0/2\varepsilon_0 V\eta_0(\omega_0)\right]^2 n(n-1), \qquad (9.7.12)$$

where the relation of field to photon operators is taken from eqn (5.1.1), with insertion of an additional factor for the linear refractive index of the medium. Thus, for an n-photon initial state, the two-photon absorption rate is

$$1/\tau_{ab} = \mathcal{K}n(n-1), \qquad (9.7.13)$$

where $\mathcal{K}$ includes all the n-independent factors from eqns (9.7.7) and (9.7.12). The corresponding rate of emission for an initially-excited atom has the same form, except that the photon destruction events are replaced by creation events to give

$$1/\tau_{em} = \mathcal{K}(n+1)(n+2), \qquad (9.7.14)$$

where both spontaneous and stimulated emission are included.

It is useful to consider two-photon transitions in a medium that has fixed numbers of atoms N_1 in the ground state and N_f in the excited state. The effect of two-photon absorption and emission is to cause changes in the photon probability distribution $P(n)$. Figure 9.11 shows the energy-level diagram for the radiative mode, analogous to Fig. 7.1 for single-photon processes, with the rates of change for the four two-photon processes that affect $P(n)$. The resulting rate of change of this element of the photon probability distribution is

$$dP(n)/dt = \mathcal{K}\{N_f(n-1)nP(n-2) - N_1n(n-1)P(n)$$
$$-N_f(n+1)(n+2)P(n) + N_1(n+2)(n+1)P(n+2)\}. \qquad (9.7.15)$$

The equations separate into two sets that couple all the even-numbered or all the odd-numbered elements of the distribution. The first two terms on the right-hand side of eqn (9.7.15) are absent for the lowest levels in the two sets, which have $n=0$ and $n=1$ respectively.

The steady-state photon distribution, obtained by setting the rates in eqn

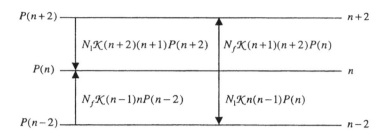

Fig. 9.11. Photon energy-level diagram for a single mode subjected to two-photon absorption and emission, showing the contributions to $dP(n)/dt$.

(9.7.15) equal to zero for all n, has

$$P(n) = \left(N_f/N_1\right)P(n-1) = \left(N_f/N_1\right)^{n/2}P(0) \qquad n \text{ even}$$
$$= \left(N_f/N_1\right)^{(n-1)/2}P(1) \quad n \text{ odd.} \tag{9.7.16}$$

The sums of the probability elements over all even or all odd n are constants of the motion, so that

$$\overset{\text{even}}{\underset{n}{\sum}} P(n) = \left[1 - \left(N_f/N_1\right)\right]^{-1}P(0) = \overset{\text{even}}{\underset{n}{\sum}} P_0(n) \tag{9.7.17}$$

and

$$\overset{\text{odd}}{\underset{n}{\sum}} P(n) = \left[1 - \left(N_f/N_1\right)\right]^{-1}P(1) = \overset{\text{odd}}{\underset{n}{\sum}} P_0(n), \tag{9.7.18}$$

where $P_0(n)$ is the initial probability distribution and N_f is assumed to be smaller than N_1. The complete steady-state distribution is thus easily found for any given initial conditions.

Consider first an initial state in which no photons are excited. The steady-state distribution produced by the two-photon emission is then

$$P(n) = \left(1 - \frac{N_f}{N_1}\right)\left(\frac{N_f}{N_1}\right)^{n/2} \qquad n \text{ even}$$
$$= 0 \qquad\qquad\qquad n \text{ odd,} \tag{9.7.19}$$

similar to the chaotic-light probability distribution of eqn (1.5.14) or (7.2.4) but involving only the even number states.

Problem 9.8   Prove that the steady-state mean photon number and the degree of second-order coherence of the light generated by two-photon emission are respectively [31]

$$\langle n \rangle_\infty = 2N_f \big/ \big(N_1 - N_f \big) \qquad (9.7.20)$$

and

$$g_\infty^{(2)}(0) = \frac{N_1 + 3N_f}{2N_f} = 2 + \frac{1}{\langle n \rangle_\infty}, \qquad (9.7.21)$$

where the ∞ subscripts indicate steady-state quantities achieved after an infinitely-long time interval.

The emitted light thus has a degree of second-order coherence larger than that of ordinary chaotic light, the enhanced fluctuations being caused by the emission of photons in correlated pairs.

   Another simple example is that of initial single-mode coherent light of mean photon number $\langle n \rangle_0$, where the steady-state distribution is

$$
\begin{aligned}
P(n) &= \tfrac{1}{2}\left(1 - \frac{N_f}{N_1}\right)\left(\frac{N_f}{N_1}\right)^{n/2}\left(1 + e^{-2\langle n \rangle_0}\right) & n \text{ even} \\[2mm]
&= \tfrac{1}{2}\left(1 - \frac{N_f}{N_1}\right)\left(\frac{N_f}{N_1}\right)^{(n-1)/2}\left(1 - e^{-2\langle n \rangle_0}\right) & n \text{ odd.}
\end{aligned}
\qquad (9.7.22)
$$

Problem 9.9   Prove that the steady-state mean photon number and the degree of second-order coherence generated by two-photon absorption and emission for initially coherent light with $\langle n \rangle_0 \gg 1$ are approximately [31]

$$\langle n \rangle_\infty = \big(N_1 + 3N_f \big) \big/ 2 \big(N_1 - N_f \big) \qquad (9.7.23)$$

and

$$g_\infty^{(2)}(0) = \frac{16N_f \big(N_1 + N_f \big)}{\big(N_1 + 3N_f \big)^2} = 2 - \frac{1}{2\langle n \rangle_\infty^2}. \qquad (9.7.24)$$

The light now has a degree of second-order coherence smaller than that of chaotic light and it has sub-Poissonian statistics, or photon-number squeezing, for $N_f < 0.094 N_1$, when the enhancement in fluctuations caused by the emission of correlated photon pairs is overcome by the reduction in fluctuations caused by the removal of coincident photon pairs by the absorption.

These steady-state distributions are useful for understanding the effects of the two-photon absorption and emission on the photon statistics but they cannot easily be accessed in experiments, where the amount of nonlinear absorption is usually small. A general explicit solution for the time dependence of the photon probability distribution can be found when the fractional number of excited atoms is negligible, and eqn (9.7.15) reduces to

$$dP(n)/dt = N\mathcal{K}\{-n(n-1)P(n)+(n+2)(n+1)P(n+2)\}. \tag{9.7.25}$$

The rate of change of the mean photon number is then

$$d\langle n\rangle/dt = -2N\mathcal{K}\langle n(n-1)\rangle = -2N\mathcal{K}g^{(2)}(0)\langle n\rangle^2 \tag{9.7.26}$$

and that of the second factorial moment of the photon distribution is

$$d\langle n(n-1)\rangle/dt = -2N\mathcal{K}\{2\langle n(n-1)(n-2)\rangle + \langle n(n-1)\rangle\}, \tag{9.7.27}$$

both obtained from eqn (9.7.25). The rate of change of each moment depends upon the next higher moment of the distribution, in contrast to eqns (7.1.15) and (7.1.17) for single-photon absorption and emission.

Consider first the solutions correct to first order in t, valid for short times. The degree of second-order coherence in eqn (9.7.26) is taken equal to its initial value, and the solution is

$$\langle n\rangle = \langle n\rangle_0 - 2N\mathcal{K}t\langle n(n-1)\rangle_0 = \langle n\rangle_0 - 2N\mathcal{K}tg_0^{(2)}(0)\langle n\rangle_0^2, \tag{9.7.28}$$

where the 0 subscripts indicate the initial values.

Problem 9.10   Derive the short-time approximations

$$g^{(2)}(0) = 1 - 2N\mathcal{K}t \tag{9.7.29}$$

for initially coherent light and

$$g^{(2)}(0) = 2 - 4N\mathcal{K}t\big(2\langle n\rangle_0 + 1\big) \tag{9.7.30}$$

for initially chaotic light.

The short-time absorption of chaotic light proceeds at twice the rate for coherent light but the short-time reduction in the degree of second-order coherence for intense chaotic light with $\langle n\rangle_0 \gg 1$ is very much faster than the reduction for coherent light.

The general solutions for the distribution and the moments [32] are not repro-

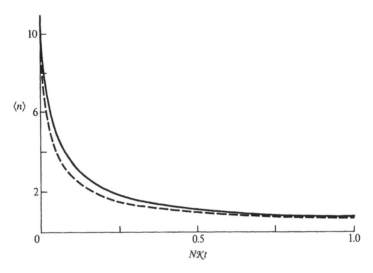

Fig. 9.12. Time dependences of the mean photon numbers for light beams subjected to two-photon absorption. The beams are initially coherent (full curve) and chaotic (broken curve) with mean photon numbers equal to 10. (After [32])

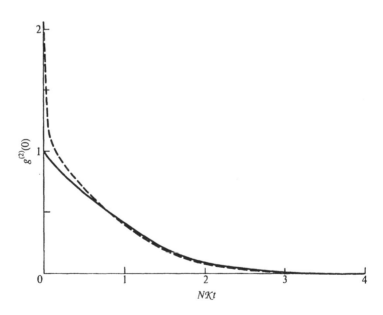

Fig. 9.13. Time dependences of the degrees of second-order coherence for the same light beams as in Fig. 9.12, initially coherent (full curve) and chaotic (broken curve). (After [32])

duced here, but Figs. 9.12 and 9.13 show the calculated time dependences of the mean photon number and degree of second-order coherence for initial beams of chaotic and coherent light with mean photon number equal to 10. The factor of 2 difference between the initial absorption rates for the two cases is clearly visible in Fig 9.12. However, the absorption rates tend to equality at longer times as the degrees of second-order coherence of the remnants of the initial beams themselves tend to equality, as shown in Fig. 9.13. This equality at longer times is achieved by the very much more rapid fall-off in the degree of second-order coherence of initially-chaotic light at short times, even for the modest number of photons assumed in Fig. 9.13. Initially-coherent light immediately begins to acquire sub-Poissonian statistics, or photon-number squeezing, and this has been observed experimentally [33]. Initially-chaotic light also eventually acquires sub-Poissonian statistics, but only after a substantial amount of two-photon absorption, which makes any experimental observation difficult. The development of photon-number squeezing is accompanied by the growth of quadrature amplitude squeezing over a limited range of times [34]. This is another example of the occurrence of the two kinds of squeezing in tandem, mentioned in §§5.6 and 5.10.

9.8 Conclusion

The calculations outlined in the present chapter show how the nonlinear inter-actions of light with matter can change the initial statistical properties of the light. The quadrature variances and degrees of second-order coherence of the individual light beams and the correlations between different beams may be increased or decreased, depending on the nature of the nonlinear process. The newly-created light that is generated in some processes may have statistical properties quite different from those of the initiating light beams.

These effects occur in essentially all nonlinear optical processes but the emphasis here is on processes that convert classical incident light, usually coherent light from a laser source, into light that displays nonclassical properties. Thus second-harmonic generation and two-photon absorption produce light with sub-Poissonian statistics and amplitude squeezing, parametric down-conversion gene-rates quadrature-squeezed single-mode or two-mode vacuum states, with striking interbeam correlations in the latter, parametric amplification produces quadrature-squeezed coherent states, and self-phase modulation converts coherent light into quadrature-squeezed light intermediate between the amplitude and phase varieties. The nonlinear processes provide important experimental techniques for the pro-duction of nonclassical light. They greatly augment the possibilities of nonclassical light generation by the linear processes of resonance fluorescence and two-photon cascade emission, discussed in Chapter 8.

Although it remains the case that the vast majority of light sources, and the optical experiments performed with them, are describable by classical theory, this theory suffers from basic limitations in its range of application. The predictions of the quantum theory of light parallel those of the classical theory in the ranges of validity of the latter, but the quantum theory appears to provide accurate descriptions of the properties of all kinds of light, without any restrictions. Nonclassical light is important in a growing variety of applications, where its specific quantum properties provide advantages, for example, in secure optical communications or noise reduction. It is also important in tests and demonstrations of some of the fundamental concepts of quantum mechanics itself. These include the aspects of quantum measurement theory discussed in §5.10, the distinctions between classical and Bose–Einstein statistics discussed in §6.8, and the vacuum-field fluctuations considered repeatedly, particularly in connection with their modifications in the various kinds of squeezing experiment. Other basic quantum-mechanical phenomena studied by quantum-optical experiments include Bell's inequalities and the Einstein–Podolsky–Rosen paradox [17,18].

In summary, the quantum theory provides the most reliable and universal descriptions and interpretations of optical experiments, despite the extensive regimes of applicability of classical theory. Growing ranges of observed optical phenomena can only be understood in terms of the quantum theory. Finally, calculations in quantum optics are often elegantly simple and direct. It is hoped that the present account has conveyed some of this simplicity and directness.

References

[1]   Bloembergen, N., *Nonlinear Optics* (World Scientific Publishing, Singapore, 1996) (reprint of book first published in 1965).

[2]   Shen, Y.R., *The Principles of Nonlinear Optics* (Wiley, New York, 1984).

[3]   Schubert, M. and Wilhelmi, B., *Nonlinear Optics and Quantum Electronics* (Wiley, New York, 1986).

[4]   Butcher, P.N. and Cotter, D., *The Elements of Nonlinear Optics* (Cambridge University Press, Cambridge, 1990).

[5]   Mills, D.L., *Nonlinear Optics* (Springer-Verlag, Berlin, 1991).

[6]   Nussenzveig, H.M., *Causality and Dispersion Relations* (Academic Press, New York, 1972) §1.9.

[7]   Huttner, B. and Barnett, S.M., Quantization of the electromagnetic field in dielectrics, *Phys. Rev. A* **46**, 4306–22 (1992).

[8]   Matloob, R., Loudon, R., Barnett, S.M. and Jeffers, J., Electromagnetic field quantization in absorbing dielectrics, *Phys. Rev. A* **52**, 4823–38 (1995).

[9]   Gruner, T. and Welsch, D.–G., Green-function approach to the radiation-field quantization for homogeneous and inhomogeneous Kramers–Kronig dielectrics, *Phys. Rev. A* **53**, 1818–29 (1996).

[10]   Blow, K.J., Loudon, R., Phoenix, S.J.D. and Shepherd, T.J., Continuum fields in quantum optics, *Phys. Rev. A* **42**, 4102–14 (1990).

[11]   Jackson, J.D., *Classical Electrodynamics*, 3rd edn. (Wiley, New York, 1999).

[12]   Milonni, P.W., Field quantization and radiative processes in dispersive dielectric media, *J. Mod. Opt.* **42**, 1191–2004 (1995).

[13] Franken, P., Hill, A.E., Peters, C.W. and Weinreich, G., Generation of optical harmonics, *Phys. Rev. Lett.* **7**, 118–9 (1961).

[14] Zou, X.T. and Mandel, L., Photon antibunching and sub-Poissonian photon statistics, *Phys. Rev. A* **41**, 475–6 (1990).

[15] Horowicz, R.J., Quantum correlation between fundamental and second harmonic in SHG, *Europhys. Lett.* **10**, 537–42 (1989).

[16] Bachor, H.-A., *A Guide to Experiments in Quantum Optics* (Wiley–VCH, Weinheim, 1998).

[17] Rarity, J.G. and Tapster, P.R., Experimental violation of Bell's inequality based on phase and momentum, *Phys. Rev. Lett.* **64**, 2495–8 (1990).

[18] Bell, J.S., *Speakable and Unspeakable in Quantum Mechanics* (Cambridge University Press, Cambridge, 1987).

[19] Zeilinger, A., Fundamentals of quantum information, *Physics World* **11**, 35–40 (March 1998).

[20] Phoenix, S.J.D. and Townsend, P.D., Quantum cryptography: how to beat the code breakers using quantum mechanics, *Contemp. Phys.* **36**, 165–95 (1995).

[21] Migdall, A., Correlated-photon metrology without absolute standards, *Physics Today* **52**, 41–6 (January 1999).

[22] Caves, C.M., Quantum limits on noise in linear amplifiers, *Phys. Rev. D* **26**, 1817–39 (1982).

[23] Ekert, A.K. and Knight, P.L., Correlations and squeezing of two-mode oscillations, *Am. J. Phys.* **57**, 692–7 (1989).

[24] Townsend, P.D. and Loudon, R., Quantum noise reduction at frequencies up to 0.5 GHz using pulsed parametric amplification, *Phys. Rev. A* **45**, 458–67 (1992).

[25] Collett, M.J. and Loudon, R., Output properties of parametric amplifiers in cavities, *J. Opt. Soc. Am. B* **4**, 1525–34 (1987).

[26] Newell, A.C. and Moloney, J.V., *Nonlinear Optics* (Addison–Wesley, Redwood City, 1992).

[27] Blow, K.J., Loudon, R. and Phoenix, S.J.D., Graphical representation of self-phase modulation noise, *J. Mod. Opt.* **40**, 2515–24 (1993).

[28] Kitagawa, M. and Yamamoto, Y., Number–phase minimum-uncertainty state with reduced number uncertainty in a Kerr nonlinear interferometer, *Phys. Rev. A* **34**, 3974–88 (1986).

[29] Sizmann, A. and Leuchs, G., The optical Kerr effect and quantum optics in fibres, *Prog. Opt.* **39**, 369–465 (1999).

[30] Buzek, V. and Knight, P.L., Quantum interference, superposition states of light, and nonclassical effects, *Prog. Opt.* **34**, 1–158 (1995).

[31] McNeil. K.J. and Walls, D.F., Possibility of observing enhanced photon bunching from two-photon emission, *Phys. Lett.* **51**A, 233–4 (1975).

[32] Simaan, H.D. and Loudon, R., Quantum statistics of single-beam two-photon absorption, *J. Phys. A* **8**, 539–54 (1975).

[33] Ispasoiu, R.G. and Goodson III, T., Photon-number squeezing by two-photon absorption in an organic polymer, *Opt. Comm.* **178**, 371–6 (2000).

[34]   Gilles, L., and Knight, P.L., Two-photon absorption and nonclassical states of light, *Phys. Rev. A* **48**, 1582–93 (1993).

Index

A coefficient, *see* Einstein *A* coefficient
absorption of photons, *see* Einstein *B* coefficient; photon
amplification
  inverted population 31–5, 319–24
  laser 39–40
  parametric 404–11
  travelling-wave
    quantum theory 319–24
    semiclassical theory 31–5
amplitude-squeezed state, *see* squeezed coherent state
antibunching, *see* photon antibunching
atom–radiation dynamics 324–8
atomic
  angular momentum 158
  density matrix
    definition 68
    properties 68, 71–2, 76, 346
    related to transition operators 332–4, 363
    solutions with collision broadening 79
    solutions with radiative damping 73–5
    steady-state solutions 72–3, 77–8
    three-level atom 361–3, 366
  electric dipole moment 50, 157
    second quantization 163–5
  electric quadrupole moment 158
  energy eigenstates 47
  excitation 23–7
  fluorescence, *see* fluorescent emission; resonance fluorescence
  Hamiltonian 160, 162
    second quantization 162–5
  hydrogen, *see* hydrogen atom
  level degeneracy 16–19
  magnetic dipole moment 158
  multipole moments 157–8
  orientation average 56, 63
  potential energy 157–8
  rate equation, *see* rate equation
  saturation 24
  transition
    operator 163–4
    rate, *see* transition rate
  velocity distribution 66
  wavefunction parity 50

attenuation
  coefficient
    beam-splitter expression 312
    effect of saturation 31
    related to *B* coefficient 30
    related to scattering cross-section 343, 380
    related to susceptibility 63, 417
    semiclassical theory 27–31, 63
  travelling-wave
    effect on coherent signal 317–18
    effect on homodyne field 316–18
    effect on squeezed vacuum 318
    quantum theory 310–18
    semiclassical theory 27–31

B coefficient, *see* Einstein *B* coefficient
balanced homodyne detection, *see* homodyne detection
beam splitter
  amplifier model 319–20
  arbitrary single-arm input 221–7
  attenuator model 311–13
  input–output relations
    classical theory 88–91
    field 215
    photon number 216
    quantum theory 212–16
  matrix 89–90
  output
    field correlation 218
    noise 217
    photon correlation 218
    single-photon input 216–21
Bell inequalities 403, 426
Bernoulli sampling 224, 295
black-body radiation 5
bleaching 31
Bloch equations, *see* optical Bloch equations
Bohr radius 49
Boltzmann distribution 18, 31
Brown–Twiss interferometer 3
  classical theory 114–17
  quantum theory 117, 218, 220–3, 252–3
Brownian motion 87

Cascade emission, *see* two-photon cascade
    emission
Casimir force 284–6
Cauchy's inequality 108
cavity, *see* Fabry–Perot cavity; optical
    cavity
centrosymmetric medium 384
chaotic light
    amplitude distribution 105–6
    coherence length 88
    coherence time 86, 97
    collision-broadened source 83–7
    density operator 200
    Doppler-broadened source 87–8
    electric field 83–6, 106, 200
    first-order coherence 94–9, 235–6, 248
    intensity distribution 107
    intensity fluctuations 86–8, 103–7
    interaction with two-level atoms 293–4,
        296
    laser below threshold 305
    multimode 235
    phase
        distribution 105–6
        fluctuations 84–5
    photocount distribution 122–3, 276–7
    photon bunching 249
    photon-number distribution 14, 199
    quadrature uncertainty 200
    rth-order coherence 114
    second-order coherence 109–11, 200,
        229, 249
    time dependence 83–8
    two-photon absorption 423–5
    unpolarized 114
charge density 126, 156
classical electron radius 375
classical stable wave
    definition 99
    first-order coherence 99
    intensity fluctuations 107
    photocount distribution 120–2
    rth-order coherence 114
    second-order coherence 111
closure theorem 149, 162
coherence
    length 88
    spatial 101, 117
    time 86–8, 95, 97
    see also degree of first-order coherence;
        degree of rth order coherence;
        degree of second-order coherence
coherent light
    conditions for 98, 112
    laser above threshold 306–7, 310
    two-photon absorption 422–5
    see also coherent state
coherent signal
    coherent state 195, 409

definition 182
degenerate parametric amplifier 406–10
effect of amplification 322
effect of attenuation 316–18
number state 186
squeezed coherent state 209
squeezed vacuum 204
coherent state
    continuous-mode 245–8
        degrees of coherence 247
        eigenvalue relation 246
        intensity 246
        photocount distribution 275–7
        photon flux 247
        'single-mode' 247–8, 280, 317–18,
            322–4, 400, 413
    multimode 234–5
    single-mode
        comparison with classical stable wave
            194
        definition 190
        displacement operator 192
        eigenvalue relation 191
        electric-field mean 195
        graphical representation 196–7
        interaction with two-level atoms 294,
            296
        noise 195, 213
        number uncertainty 193
        overlap 191
        phase uncertainty 196–9
        quadrature mean and uncertainty 195
        second-order coherence 194, 229
        signal-to-noise ratio 195
        two-photon absorption 422–5
collision broadening 76–9, 83–4
complex error function 67
composite lineshape 67–8, 103
Compton scattering 40, 375
continuous-mode field
    degrees of coherence 241
    field operators 238, 240
    Hamiltonian 238
    intensity operator 239–40
    number operator 238
    photon flux operator 240–1
correlation function
    first-order
        chaotic light 94–7
        classical stable wave 99
        definition 92–3
        laser light 309
        related to first-order coherence 93,
            176–7
    second-order
        chaotic light 110
        definition 108
        quantum jump 364
        related to first-order correlation 110

related to second-order coherence 108, 177–8
two-photon cascade emission 367
Coulomb gauge 127
creation operator, *see* harmonic oscillator; number state; photon
cross-section
 light scattering 340–4, 373
  related to attenuation coefficient 343, 380
 resonance fluorescence 346
current density 126, 156

de Broglie relation 40
degeneracy of atomic levels 16–19
degree of first-order coherence
 classical definition 93, 98, 101
 classical stable wave 99–100
 collision-broadened chaotic light 94–6
 composite chaotic light 100, 103
 Doppler-broadened chaotic light 96–7
 emission by driven atom 331–2, 334
 general properties 94, 98
 multimode chaotic light 235–6
 multimode coherent state 235
 number state 244–5
 quantum definition 176–7, 241
 related to spectral distribution 102–3
 resonance fluorescence 331–2, 334, 347–9, 352–3
 single-mode state 183
degree of fourth-order coherence 396
degree of *r*th-order coherence
 chaotic beam 114, 236
 classical definition 114
 classical stable wave 114
 multimode coherent state 236
degree of second-order coherence
 chaotic–arbitrary superposition 295
 chaotic light 109–11, 229
 classical definition 108, 112-13
 classical inequalities 108–9, 113, 249
 classical stable wave 111
 coherent state 194, 229, 249
 effect of amplification 322
 effect of attenuation 316
 emission by driven atom 331–2, 334–7
 Gaussian chaotic light 111
 interbeam 113–14, 178, 258, 270, 363–5, 370, 397–8
 laser 305–6
 Lorentzian chaotic light 111, 235–6
 multimode coherent state 235
 nonclassical range 228–9, 250
 number state 185, 229, 245
 pair state 256–8
 preservation by beam splitter 225
 quantum definition 177–8, 241
 quantum inequality 178

quantum jump 363–5
related to first-order coherence 110, 236
resonance fluorescence 331–2, 334–7, 347, 350–1, 354, 357–60
second-harmonic generation 396–8
single-mode state 183–4
squeezed vacuum 203, 229
time dependence 294–7
two-photon absorption 419–20, 422–5
two-photon cascade emission 367–9
unpolarized light 114
density matrix, *see* atomic density matrix
density of modes in a cavity 4–7
density operator
 definition 149
 general properties 150
 Heisenberg picture 155
 interaction picture 168
 multimode chaotic light 152
 multimode thermal excitation 152
 pure states 150–1
 Schrödinger picture 154
 single-mode thermal excitation 151–2
destruction operator
 see harmonic oscillator; photon
dielectric function 28
differential cross-section
 light scattering 341–2
  elastic 374–6
  inelastic 378
  Kramers–Heisenberg formula 373
  Thomson 884–5
 resonance fluorescence 346
Dirac delta-function 57–9
 definition 58
 homogeneous broadening 419
 properties 59
 related to Kronecker delta 237
 representations 58–9
direct detection
 quantum theory 271–8
 semiclassical theory 117–23
Doppler broadening 65–8, 78–9
dynamic Stark effect 70, 356–7

Einstein
 A coefficient
  definition 17
  fluorescent emission 26
  multilevel atom 376
  numerical value 57
  quantized field calculation 170–1
  quantum-mechanical expression 57, 171
  related to *B* coefficient 19
  related to radiative lifetime 26, 57, 62, 171
 B coefficient
  definition 17–18

Einstein, *B* coefficient *cont'd*
    orientation average 56
    quantum-mechanical expression 52–7,
        170
    related to *A* coefficient 19
    related to attenuation coefficient 30
Einstein–Podolsky–Rosen paradox 403,
    426
elastic light scattering
    hydrogen atom 375–8
    quantum theory 374–8
    resonant cross-section 376–7
    Thomson scattering 375
electric
    dipole moment 50, 157
    permittivity 8
    quadrupole moment 158
    susceptibility, *see* linear susceptibility;
        nonlinear susceptibility
electric-dipole approximation 161
electric-field operator
    continuous-mode field 238, 240
    discrete-mode field 141–2, 153
    single-mode field 181
    source-field expression 330
electromagnetic field, classical theory
    Coulomb gauge 127
    energy 132–3
    field equations 127–8
    gauge transformation 127
    harmonic oscillator coordinates 131
    interaction with atom 156–9
    longitudinal part 128
    Maxwell equations 5, 8, 126, 129
    mode 4–7, 130–1
    potential theory 126–7
    transition to quantum theory 141
    transverse part 128
electromagnetic field, quantum theory, *see*
        quantized radiation field
electromagnetic vacuum 143–4, 284–6
electromagnetic wave
    attenuation coefficient 27–31, 63
    classical stable wave 99, 107, 114, 120–2
    equation 5
    extinction coefficient 28
    intensity 27–8
        related to energy density 29
    Poynting vector 27
    refractive index 28
    see also quantized radiation field
emission of photons, *see* Einstein; photon
ensemble
    average 13–14, 149
        related to time average 13, 93
    chaotic light beams 107
    definition 13
entangled state 217, 224, 253, 263, 265–6

ergodic theorem 13, 93
extinction coefficient 28, 388

Fabry–Perot cavity 4, 35–6
Fermi's golden rule 60
fine-structure constant 159
first-order
    coherence, *see* degree of first-order
        coherence,
    correlation function, *see* correlation
        function, first-order
fluorescent emission 339–40
    lifetime 26
    see also resonance fluorescence
Fock state, *see* number state
fractal 87

gain coefficient
    inverted-population amplifier 33
    travelling-wave amplifier 319
gauge transformation 127
Gaussian
    approximation to photon-number
        distribution
        coherent state 193
        laser 306
    lineshape 64, 66–7, 87–8
    pulse 242–3
Gaussian–Gaussian light 106, 111
Gaussian–Lorentzian light 106, 111, 116,
    122–3
geometric distribution 15, 199
group velocity 315–16, 393

harmonic oscillator 133–9
    creation operator 134, 137
    destruction operator 133, 137
    energy 1, 8–10, 136–7
    ground-state condition 136
    Hamiltonian 133–4, 138
    normalization 137
    number operator 137
    quadrature operators 138
Heisenberg picture 154–5
Helmholtz theorem 128
Hermitian operator 138
homodyne detection 278–84
    balanced 278
    degenerate parametric amplifier 409–10
    difference photocount 278–9
        mean and variance 280–1
    electric-field operator 280
        effect of amplification 322–4
        effect of attenuation 316–18
    local oscillator 278
    'single-mode' 280
    self-phase modulation 413–16

signal-to-noise ratio 281–2
squeezed light 282–4
    effect of amplification 323
    effect of attenuation 318
homogeneous line broadening 65
hydrogen atom
    Bohr radius 49
    elastic light scattering 375–6
    radiative level shift 328
    radiative lifetime of 2P state 57

ideal squeezed state, *see* squeezed coherent
    state
idler 398
inelastic light scattering
    Compton scattering 40, 375
    quantum theory 378–81
    resonant cross-section 379–80
inhomogeneous line-broadening 67, 78
integration related to summation
    one dimension 235, 237, 313
    three dimensions 7, 145, 373
integration time 116, 271, 278, 414
intensity interference, *see* Brown–Twiss
    interferometer
interaction Hamiltonian
    diagrammatic representation 166
    electric-dipole 49–51, 161–9
    Heisenberg picture 166
    interaction picture 168
    minimal-coupling form 159
    multipolar form 161
    parity 50
    rotating-wave approximation 53, 69, 165
    Schrödinger picture 163, 165
    second quantization 163–6
    unitary transformation 160–1
interaction picture 166–8
interbeam second-order coherence
    classical definition 113–14
    quantum definition 178
    violation of classical inequality 258, 270,
        365, 370
interference, *see* Brown–Twiss interferome-
    ter; Mach–Zehnder interferometer;
    photon interference; two-photon
    interference
intermediate state 371
inverted-population amplifier
    related to parametric amplifier 405
    semiclassical theory 31–5
    signal-to-noise ratio 322–3
    travelling-wave 319–24

Jaynes–Cummings model 172

Kramers–Heisenberg formula 371–4
Kramers–Kronig relations 388

Lamb shift 286, 328
laser
    amplitude noise 306–7
    atomic rate equation 32, 298
    cavity loss rate 298, 301, 307
    comparison with coherent light 306–7,
        310
    cooperation parameter 39, 300
    differential gain 39–40
    electric-field fluctuation 306–9
    invention 3
    mean photon number 301–4
    output flux 302
    phase diffusion 308–10
    photon-number distribution 305–6
    saturation photon number 301
    Schawlow–Townes formula 309–10
    second-order coherence 305–6
    single-pass gain 37
    spontaneous emission factor 301
    three-level theory 31–40, 297–304
    threshold condition 38, 300–1
    threshold pumping rate 301
light scattering
    cross-section 340–4
    diagrammatic representation 372
    distribution of atoms 342–3
    Kramers–Heisenberg formula 373
    optical theorem 343–4, 380
    quantum theory 371–4
    related to attenuation 22–3
    see also elastic light scattering; inelastic
        light scattering
linear susceptibility
    collision broadening 78
    definition 28, 61
    power broadening 73
    radiative broadening 60–5, 384
    semiclassical theory 60–5
lineshape function
    composite 67–8
    Gaussian 64, 66–7
    homogeneous 65
    inhomogeneous 67
    Lorentzian 63–5
    related to first-order coherence 102–3
    squared Lorentzian 354
    Voigt 68
linewidth
    collision 78
    Doppler 66–7
    Gaussian lineshape 66
    Lorentzian lineshape 64
    natural 65
    power 73
    radiative 64
local oscillator 278
longitudinal

longitudinal *cont'd*
  delta-function 147
  vector field 128, 145–7
Lorentzian lineshape 63–5, 86–7
  squared 354

Mach–Zehnder interferometer 1
  classical theory 91–4, 100–1
  fringe visibility 100-1
  independent sources 227
  quantum theory 176–7, 218–21, 223–4,
    251–2
  single-photon input 218–21, 252
magnetic-field operator
  continuous-mode field 238
  discrete-mode field 142
magnetic permeability 8
magnetization 156–7
Mandel Q parameter 230, 274, 337
Maxwell
  equations 5, 8, 126, 129
  velocity distribution 66
measurement-conditioned state
  beam-splitter output 217
  single-photon wavepacket 259–60
mode of radiation field
  density of states 4–7
  polarization 6, 131, 144
  wavevector 6, 130–1, 144

negative-temperature condition 31
noise
  amplifier 320–1
  attenuator 313–15, 317–18
  coherent state 195, 409
  definition 183
  number state 186
  self-phase modulation 414–15
  squeezed coherent state 209
  squeezed vacuum 204
noncentrosymmetric medium 384
nonclassical light 180, 227–31, 250
  see also photon antibunching; photon pair
    states; squeezed coherent state;
    squeezed vacuum; sub-Poissonian
    statistics
nonlinear refraction coefficient 412
nonlinear susceptibility 73, 78, 383–9
  definition 384–5
  frequency components 385–7
  intensity-dependent refractive index 412
  numerical estimate 387
  parametric down-conversion 386, 398–
    401
  second-harmonic generation 386, 393
  second-order 385–7
  self-focusing 412
  self-phase modulation 411

third-order 387
two-photon absorption 419
normal ordering 175
number state
  continuous-mode 242–5, 253
    eigenvalue relation 244
    first-order coherence 244
    generation from cascade emission 220,
      370
    generation from pair state 259–60
    intensity and photon flux 244
    second-order coherence 245
    wavepacket creation operator 243
  multimode 139–40
  single-mode
    creation operator 139
    effect of beam splitter 224, 227–8
    electric-field expectation values 186–7
    phase properties 187–90
    quadrature expectation values 185
    second-order coherence 185, 229
    signal-to-noise ratio 186

optical beam splitter, *see* beam splitter
optical Bloch equations
  collision broadening 77, 344
  comparison with rate equations 79–81
  derivation 68–9
  Doppler broadening 78–9, 81
  rotating-wave approximation 69
  spontaneous emission 72
  see also atomic density matrix
optical cavity
  density of modes 4–7, 131
  see also Fabry–Perot cavity; mode of
    radiation field
optical detection, *see* direct detection;
    homodyne detection
optical excitation of atoms 23–7
optical Kerr effect 412
  see also self-phase modulation
optical theorem 343–4, 380
optical tweezers 44
orientation average 56, 63

pair state, *see* photon pair state
parametric amplifier 404–11
  degenerate 406–11
    phase sensitivity 407–8
    signal-to-noise ratio 409–10
  nondegenerate 404–6
    related to inverted-population amplifier
      405
parametric down-conversion 398–404
  degenerate 402–4
    quadrature squeezing 403–4
  nondegenerate 399–402
    two-mode squeezed vacuum 402–3

nonlinear susceptibility 386, 398–401
  phase matching 401
parity
  interaction Hamiltonian 50
  wavefunction 50, 384
particle fluctuations 199
partition noise 216, 274
phase diffusion 308–10
phase matching 395, 401
phase, quantum-mechanical 187–9
  coherent state 196–9
  distribution 187, 198
  number–phase uncertainty product 196,
    210
  number state 189–90
  operator 187
  squeezed state 210–12
  state 188
photocount distribution
  chaotic light 122–3, 276–7
  classical stable wave 120–2
  classical theory 118–23
  coherent light 275-7
  Mandel formula 120
  Mandel Q parameter 274, 337
  mean and variance 273
  number state 274–5
  quantum expression 276
  second-order coherence 274, 336
  shot noise 122, 278
  time series 250
photoelectric effect 118, 173–6
photomultiplier, *see* phototube
photon absorption
  destruction operator 139
  Einstein theory 16–19
  momentum properties 41–2
  quantum calculation 169–70
  transition rate 52–6
photon antibunching
  definition 250
  relation to sub-Poissonian statistics 250,
    396–7
  resonance fluorescence 335–6, 351
  second-harmonic generation 396–7
  two-photon cascade emission 368–9
  see also nonclassical light
photon bunching 249
photon concept 1–2, 9, 224, 227
photon emission
  comparison of theories 172–3
  creation operator 140
  driven atom 331–7
  Einstein theory 16–19
  momentum properties 41–2
  single mode 289–90
  spontaneous
    compared to stimulated 20–2

directional properties 22
  effect on laser light 307–8
  quantum calculation 170–1, 326–7
  related to scattering 22–3
stimulated
  coherence 297
  compared to spontaneous 20–2
  directional properties 22, 170
  quantum calculation 169–70
photon interference 1–2
  Brown–Twiss interferometer 3, 218,
    220–3
  independent sources 227
  Mach–Zehnder interferometer 1, 218–21,
    223–4
  single photon 218–21
  two-photon 260–5
photon momentum 40–2
  see also radiation pressure
photon number
  fluctuations 13–16, 184, 193, 199
  number–phase uncertainty product 196,
    210
  operator 140
  squeezing 230
  see also sub-Poissonian statistics
  state, *see* number state
  thermal equilibrium 11
photon-number distribution
  beam-splitter output 224–8
  chaotic light 14, 199
  coherent state 193–4
  effect of beam splitter 224–6
  factorial moments 15, 194, 200
  interaction with two-level atom
    fixed atomic populations 292–7
    inverted population 296–7
    rate equations 290–2
    second-order coherence 294–7
    steady-state solutions 293
    time dependence 293–7
  laser 305–6
  thermal equilibrium 11, 14–15
  two-photon absorption 420–3
photon pair state
  single-beam 253–6
    creation operator 254
    eigenvalue equation 255
    first-order coherence 255
    photon flux 255
    second-order coherence 256
    two-photon interference 260–2
  two-beam 256–60
    creation operator 250
    eigenvalue equation 257
    interbeam coherence 258
    intrabeam coherence 257–8
    photon flux 257, 259

photon pair state, two-beam *cont'd*
    role as single-photon source 259–60
    two-photon interference 262–5
    wavepacket 258
phototube 117–18, 173–4, 272
    dead time 117, 272
    efficiency 118
    integration time 116, 271, 278, 414
    photocount operator 274
    quantum efficiency 272
picture
    Heisenberg 154–5
    interaction 166–8
    Schrödinger 153–4
Planck
    constant 3
    distribution 11, 14–15
    radiation law 10–13
Poisson
    distribution 121, 193–4, 277
    equation 128
polarization
    linear 61, 63, 156–7, 385, 388
    nonlinear 385, 389, 394, 399–400, 411–
        12
polarization vectors 131, 144
population inversion 31, 296, 298, 319, 321
power broadening 72–6
Power–Zienau–Woolley Hamiltonian 161
Poynting vector 27, 174, 239, 391–3
pressure broadening, *see* collision
        broadening
principal part 327
pumping rate 31, 297
pure state 72, 148

quadrature operators 138, 142
quadrature squeezing
    definition 201
    squeeze operator 201
    see also squeezed vacuum; squeezed
        coherent state
quantized radiation field
    canonical commutation relation 144–5
    creation operator 140
    destruction operator 139
        equation of motion 324
    electric-field operator 141–2, 153, 181,
        238, 240
    energy 7–10
    field commutation properties 145, 147
    Hamiltonian 142–3, 161, 238
    in media 389–93
        electric-field operator 392
        intensity operator 391–3
        magnetic-field operator 392
    intensity operator 173–6, 239–40
    magnetic-field operator 142, 238
    number operator 140, 238

Planck hypothesis 3
Poynting vector 174–5, 239
quadrature operators 142
source-field 328–31
transition from classical theory 141
vacuum state 143, 284–6
vector potential operator 141
zero-point energy 143, 175
quantum efficiency 272
quantum jumps 360–5
    rate equations 361
    second-order coherence 363–5
quantum-mechanical harmonic oscillator,
        see harmonic oscillator
quantum regression theorem 333–4

Rabi
    frequency 71, 171–2, 289, 345
        vacuum 172
    oscillations 71
        radiative damping 74–5
radiation pressure 40–4
    force 42
    optical tweezers 44
    saturation efffects 42–4
radiative
    broadening 60–5
    damping 61–2
    level shift 286, 328
    lifetime
        classical expression 173
        hydrogen atom 57
        quantized field calculation 170–1
        related to *A* coefficient 26, 57, 62,
            171
    linewidth 64
        multilevel atom 376
    transition rate, *see* transition rate
Raman scattering 339–40, 378–81
    see also inelastic light scattering
random-walk theory 84, 105–6, 308–9
rate equation
    atoms in laser 32, 298
    comparison with optical Bloch equations
        79–81
    photon 289–92, 298–9
    three-level atom 32, 361, 366
    two-level atom 18, 23, 80–1
Rayleigh
    radiation law 13
    scattering 339–40, 342–3, 374–8
        see also elastic light scattering
refractive index 28, 388
resonance fluorescence
    cross-section 346
    emitted light
        coherent fraction 346–8, 352
        first-order coherence 331–2, 334, 347–
            9, 352–3

photon antibunching 335–6, 351
second-order coherence 331–2, 334–7, 347, 350–1, 354, 357–60
 spectrum 347–9, 354–7
 sub-Poissonian statistics 335–7, 347, 351, 358
 exactly-resonant incident beam 352–60
 experiments 356, 358–9
 method of calculation 331–5, 344–8
 multiatom effects 358
 rate-equation treatment 335–7
 two-level atom theory 344–8
 weak incident beam 348–52
retarded time 315–16, 330, 393
rotating-wave approximation 53, 69, 165

saturation
 broadening 73
 effect on attenuation 31
 effect on radiation pressure 42–4
 energy density 21
 intensity density 21
 photon number 301
 two-level atom 24, 73
scalar potential 126, 156
 gauge transformation 127
scattering of light, *see* light scattering
Schrödinger picture 153–4
second-harmonic generation 393–8
 harmonic intensity 395–6
 nonlinear susceptibilty 386, 393
 phase mismatch 395
 photon antibunching 396–7
 second-order coherence 396–8
 sub-Poissonian statistics 396–7
second-order
 coherence, *see* degree of second-order coherence
 correlation function, *see* correlation function
second quantization 162
self-focusing 412
self-phase modulation 411–17
 homodyne field 413–16
 noise 414–15
 nonlinear refraction coefficient 412
 nonlinear susceptibility 411
 photon flux conservation 413
 quadrature squeezing 415–16
shelving state 360
shot noise 122, 278
signal, *see* coherent signal
signal-to-noise ratio
 coherent state 195
 definition 183
 degenerate parametric amplifier 409–10
 homodyne detection 281–2
 number state 186

squeezed coherent state 209
travelling-wave amplifier 322–3
travelling-wave attenuator 317–18
slowly-varying amplitude approximation 389
source-field expression 328–31
spatial coherence 101, 117
spontaneous emission, *see* Einstein *A* coefficient; photon emission
squeeze operator 201
squeezed coherent state
 amplitude squeezing 209–11, 213, 230, 416
 number and phase uncertainties 210
 coherent signal 209
 eigenvalue relation 207
 graphical representation 208, 210–12
 noise 209
 parametric amplification 410
 phase squeezing 211–13, 231
 number and phase uncertainties 211
 photon-number mean and variance 207
 quadrature mean and variance 208
 self-phase modulation 416
 signal-to-noise ratio 209
 single mode 206–13
squeezed vacuum
 single continuous-mode
  definition 265
  effect of amplification 323
  effect of attenuation 318
  homodyne detection 282–4
  parametric down-conversion 403–4
  photon flux 267
  wavepacket amplitude 266
 single-mode
  complex squeeze parameter 201
  eigenvalue relation 203
  graphical representation 204–5
  noise 204–6, 213
  number-state expansion 201
  photon-number mean and variance 202
  quadrature uncertainty 203–4
  second-order coherence 203, 229
 two continuous-mode
  definition 268
  first-order coherence 269
  interbeam second-order coherence 270
  intrabeam second-order coherence 269
  parametric down-conversion 403
  photon flux 269
  wavepacket amplitude 268
stationary
 light source 93
 state 47
statistical mixture 76, 148
Stefan–Boltzmann radiation law 13
stimulated emission, *see* Einstein *B* coefficient; photon emission

sub-Poissonian statistics
   definition 199
   direct detection 275
   Mandel Q parameter 230
   relation to photon antibunching 250
   resonance fluorescence 335–7, 347, 351,
      358
   second-harmonic generation 396–7
   self-phase modulation 417
   squeezed coherent state 210–11
   two-photon absorption 419–20, 422, 425
   see also photon-number squeezing
summation related to integration
   one dimension 235, 237, 313
   three dimensions 7, 145, 373
super-Poissonian statistics 199, 203, 230–
   1, 276
susceptibility, *see* linear susceptibility; nonlinear susceptibility

thermal equilibrium
   mode excitation 12–14
      neglect 20
   photon emission 20
   photon-number distribution 11, 14–15,
      19
Thomson scattering 375
three-level atom
   amplifier 31–5
   cascade configuration 365–7
   laser 297–8
   quantum jumps 359–63
   V configuration 359–62
time average, related to ensemble average
   13, 93
time-dependent
   wave equation 46–7
   wavefunction 47
      linear superposition 48, 61
      normalization 48
time-development operator 154
time series 86, 250
transition operator 163–4
   equation of motion 325–8
transition rate
   absorption 52–6
   emission 170–1
   Fermi's golden rule 60
   first-order 60
   light scattering 372–3
   rotating-wave approximation 53
   second-order 371
   two-photon absorption 418–19
transverse
   delta-function 145–6
   vector field 128, 145–7
travelling-wave
   amplification
      quantum theory 319–24

semiclassical theory 31–5
attenuation
   quantum theory 310–18
   semiclassical theory 27–31
two-photon absorption 417–25
   effect on photon-number mean and
      fluctuations 423–5
   nonlinear susceptibility 419
   photon-number rate equation 420, 423
   photon statistics 420
   second-order coherence 177, 419–20,
      422–5
   transition rate 418–19
two-photon cascade emission 365–71
   multiatom effects 368–9
   nonclassical correlation 369–70
   rate equations 367
   second-order coherence 367–9
   source for single-photon state 220, 370
two-photon coherent state, *see* squeezed co-
      herent state
two-photon interference 260–5
   classical input 263
   single-mode input 260–2
   two-mode input 262–5

uncertainty relation
   frequency–time 65
   number–phase product 196, 210
   position–momentum 138
   single-mode field 182

vacuum state
   Casimir force 284–6
   condition 136, 143
   electromagnetic 143–4, 284–6
   quadrature-squeezed 201–6
   see also squeezed vacuum
van der Waals forces 286
vector potential
   classical radiation field 126, 130–2
   gauge transformation 127
   operator 141
virtual intermediate state 371
visibility of interference fringes 100–1
Voigt lineshape 68
von Neumann measurement theory 217

wave fluctuations 122, 199
'which way' experiment 224
Wien
   displacement law 12
   formula 13
Wiener–Khintchine theorem 102–3

Young interferometer 3, 223

zero-point energy 9, 143–4, 284–6

Lightning Source UK Ltd.
Milton Keynes UK
UKHW020203271022
411180UK00001B/1

9 780198 501763